BRUCE ARDEN JOHNSON

GENERAL BIOLOGY
The science of biology

GENERAL BIOLOGY
The science of biology

EIGHTH EDITION

With 699 illustrations

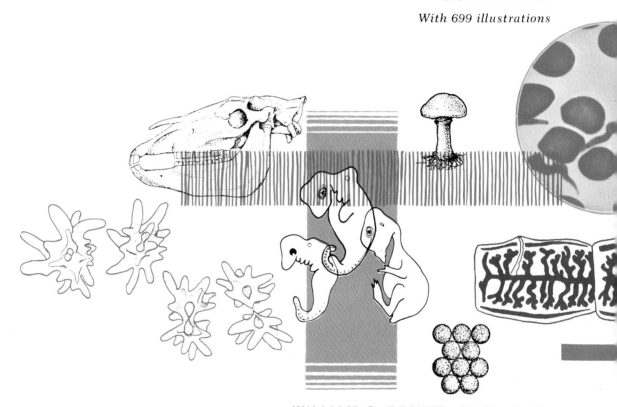

WILLIAM C. BEAVER, **Ph.D., Sc.D.**

Professor Emeritus of Biology, Wittenberg University, Springfield, Ohio

GEORGE B. NOLAND, **Ph.D.**

Professor of Biology, University of Dayton, Dayton, Ohio

Saint Louis

THE C. V. MOSBY COMPANY

1970

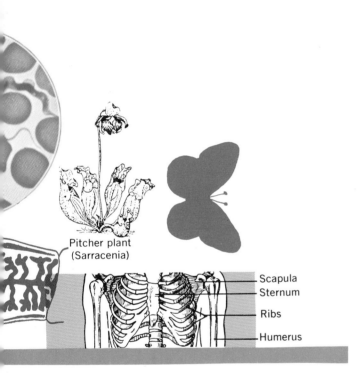

Pitcher plant
(Sarracenia)

Scapula
Sternum
Ribs
Humerus

EIGHTH EDITION

Copyright © 1970 by

THE C. V. MOSBY COMPANY

Previous editions copyrighted 1939, 1940, 1946, 1952, 1958, 1962, 1966

Printed in the United States of America

Standard Book Number 8016-0544-X

Library of Congress Catalog Card Number 74-99910

Distributed in Great Britain by Henry Kimpton, London

Dedicated to our students —
past, present, and future

Preface

Although revisions have been made in most chapters, the general plan of the previous edition of *General Biology, the Science of Biology*, has been retained. The primary effort was directed toward making the text more readable. This was accomplished by rearranging or combining some chapters, rephrasing awkward passages, and eliminating many redundant examples that tended toward the encyclopedic.

Many new illustrations have been added. Most of these are electron micrographs of high quality and sharpness. A number of new tables have been added and several "boxes" (for want of a better name) were used in various chapters.

A total of sixty figures that had been used in previous editions and were somewhat dated were eliminated. One chapter was deleted and two others were combined.

The text now consists of six parts, largely following the previous edition, with emphasis and reorganization as will follow.

Part One, Introductory biology, consists of seven chapters that have been strengthened by rewriting and reorganization. Readers, most of whom are not biology majors, are introduced to biology through science, and learn about the effect of biology on them as citizens and also as students. Then they proceed from a short historical treatment to cells, cell division and organization, and to characteristics of life, some chemical and physical properties, and finally the kinds of life. We hope the instructor will feel free to assign a few readings from *Scientific American* somewhere during this introduction.

In Part Two, Viruses, monera, and plants, several chapters have been extensively revised, but in general the emphasis remains similar to that of the previous edition.

Part Three, Animal biology, remains largely the same, although most chapters now have new material, and approximately 30 illustrations have been dropped. The chapter on the pig has been deleted in this edition.

Part Four, The biology of man, has been changed primarily by addition of new material and deletion of a few illustrations.

Part Five, The continuity of life, has been extensively reworked and a new chapter, Genes and gene action, has been added.

Part Six, Organism and environment, retains its emphasis on ecology in the broad sense. Chapter 33, Interrelationships among organisms, has been expanded and rearranged, with more emphasis on behavior.

Recent references have been added to all chapters. With the appearance of many paperbacks and reprint series, such as those in *Scientific American*, it is possible to add whatever type of emphasis is required by a particular instructor. If desired, any one of the parts may be replaced by pertinent reprints or paperback volumes.

Many of the illustrations are new, and proper credit is given in each instance. We would especially like to acknowledge the fine electron micrographs used by permission of Dr. M. Arif Hayat.

We would like to acknowledge the assistance given by our colleagues and the many helpful suggestions made by readers of previous editions.

William C. Beaver
George B. Noland

Contents

contents

part three

ANIMAL BIOLOGY

contents

contents

part six

ORGANISM AND ENVIRONMENT

INTRODUCTORY BIOLOGY

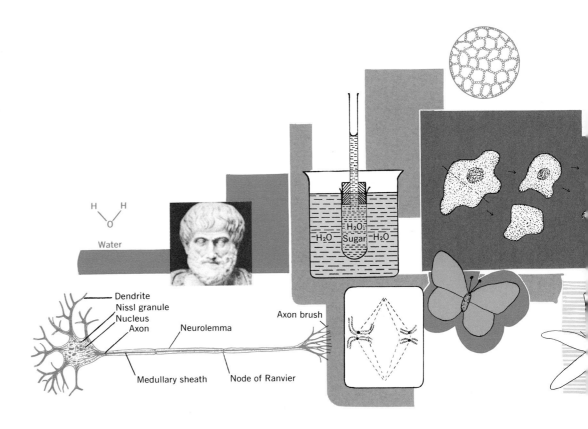

chapter one

The science of biology

No one reading this textbook can possibly be unaware that many of the major problems confronting the world today are biologic in nature. Reports from the newspapers and television stations emphasize that starvation, air and water pollution, expanding population, pesticide poisoning, and destruction of natural resources are now part of our daily lives (see p. 3).

Legal and moral aspects of heart, kidney, and lung transplants continue to plague the average citizen as well as the courts and the churches. The recent replacement of a damaged heart with a mechanical pump to extend the patient's life has added to the controversy. The discovery that extra chromosomes in some people may be associated with criminal behavior has compounded the problems of more than one judge.

WHY STUDY BIOLOGY?

At the risk of being overly dramatic, the answer to this question should be "in order to survive!"

Now that we have your attention it might be well to consider the kinds of students who are reading this text. Two groups can be readily recognized. The first is composed of people who recognize a need to study biology as a preparation for their life work. These are the biology majors and the medical technology, predental, and premedical students.

Most of you, however, are not in this group. Your interest is not in science but in the world of ideas and service. Very likely, the best expression of your attitude is that you do not openly object to biology or, indeed, to science in general.

This attitude is not new or even unusual. The English author C. P. Snow called attention to it in his book, *Two Cultures and the Scientific Revolution.* He felt that there was a growing gap between the scientifically and nonscientifically oriented people and that these two groups were in danger of soon not being able to communicate with each other. Recent events on many campuses indicate that this time is nearer than he thought. Many people believe that the field of biology might be able to bridge this gap.

Everyone, scientists as well as nonscientists, has an interest in the problems mentioned earlier. The survival of our civilization depends on their solution and control.

SHELLFISH-ASSOCIATED GASTROENTERITIS — New Haven, Connecticut*

An outbreak of gastroenteritis occurred on November 16, 1968, following a shellfish sanitation association meeting in New Haven; 17 persons became ill 31 to 53 hours (mean 38.9 hours) after a cocktail party where mixed drinks, raw oysters and hard clams, potato chips, cheese snacks, and a hot sauce were served. The illness was characterized by nausea, vomiting, fever, and diarrhea with 16 of the 17 persons having diarrhea. Of 12 families surveyed with index cases, five reported secondary cases with these same symptoms. One other case occurred in the man who delivered the clams to the party and who consumed a dozen of them on November 13. He became ill 32 hours later on November 14.

Food histories from 23 persons at the party showed that 19 ate clams and 20 ate oysters. The one person who ate only clams became ill while the two persons who ate only oysters did not become ill. Of all food items, only the clams were found to be significantly associated with illness.

The clams were part of 11 1/2 bushels of cherrystones and little necks harvested by a shellfish dealer on November 12 from a bed 3 miles southwest of Norwalk, Connecticut. One bushel was used at the party and the other 10 1/2 were sent to a retail market in Yonkers, New York. On November 22, investigation showed that the clam bed had 1,100 total coliforms per 100 ml of water and, as a result, was closed to further harvesting. From December 2 through 6, samples taken twice a day from the bed also revealed abnormal contamination. In late October, waters near the bed had had acceptable coliform counts. The Yonkers health department was notified, but no increase in gastroenteritis cases was noted in Yonkers in November.

The probable source for the contamination was a sewage treatment plant located in eastern Norwalk that discharges the plant effluent 4.4 miles upstream from the bed. On November 11, the plant had no electricity for 4 1/2 hours. Consequently, because the storage capacity of the system is 2 hours, there was a major overflow of the storage system for at least 2 1/2 hours. During the power failure, attempts were made to chlorinate the major overflow points with hypochlorite powder.

Because of the possibility of hepatitis developing following the ingestion of contaminated raw shellfish, most of the persons at the party received immune serum globulin and were observed for 2 to 6 weeks. Weekly serum specimens for 3 weeks were obtained for SGOT determination from 18 persons at the meeting. Two of these 18 had a previous history of hepatitis infection. None of the 18 had abnormal SGOT's. Frozen stool specimens for virus isolation and rectal swabs for bacterial culture were also solicited, and additional clams were harvested from the contaminated bed for virus studies.

*From Morbidity and mortality weekly report, United States Public Health Service, Feb. 8, 1969.

Not too long ago, the average biology professor felt that it was enough to present a course in biology as simply another scholarly discipline. Many now realize that we must try to meet the humanities and social science students' "demand for relevance" as well as the biology majors' need for depth in his subject.

In your school at the present time many professors in widely varied departments are actively studying and investigating what are essentially biological problems (Table 1-1). Where will you fit in?

No matter what his future profession or work may be, man may be able to live a more complete and happy life if he is familiar with the wonderful phenomena and laws of nature. Biology helps us to appreciate and understand nature and natural laws.

One of the most important and valuable requirements for successful living is an understanding of human beings, both collectively and as individuals. Much of the lack of success in the family, in society, in government, in business, and in the world at large results from a misunderstanding of human beings by other human beings. A study of

Table 1-1
Biology-related problems in other fields

Department	Example of biologically oriented research
History	Effect of epidemics on ancient civilizations
Political science	Regulation of water and air pollution
Psychology	Biochemical control of mental illness
Sociology	Population control and society
Philosophy	Evolution and philosophy of science
Theology	Morality of organ transplants; abortion
Anthropology	Blood groups and genes in native tribes
Engineering	Thermodynamics of smooth muscle contraction
Physics	Physical properties of complex macromolecules
Chemistry	Biochemistry of enzyme-substrate complexes

heredity, endocrine secretions, personal and public health, sanitation, abnormalities and diseases, and human behavior can help materially in our attempt to live happily and successfully. A consideration of the relative effects of environment and heredity on mankind can aid us in our understanding of education, social progress, crime, and human diseases and abnormalities. Biology contributes most significantly to our familiarity with the more important theories and laws that have materially aided in man's progress and thinking. In other words the cultural values of a natural science, such as biology, are immeasurable.

We have recently come to realize the great importance of our natural resources. The study of biology can help us to understand the need for enactment of regulations to conserve our forests, wild animals, fish, and wild plants. Such a study also will help us to learn the importance of animals and plants, particularly as they relate to medicine, industry, agriculture, horticulture, and diseases.

Biology also can serve as a foundation for such professions as medicine, dentistry, pharmacy, nursing, agriculture, forestry, education, entomology, horticulture, landscape gardening, and the "profession of living." In our preparation for such professions we will appreciate the interrelationship of all the sciences, such as chemistry, biology, physics, geology, geography, psychology, paleontology, and many others.

SCIENTIFIC METHOD

A course in biology should give the student an idea of the aim and nature of science, the methods employed, and its value and limitations. Science attempts to observe and describe facts and to relate them to each other. Its conclusions are always subject to revision in the light of newly discovered facts.

There are numerous popular concepts concerning the limitations and advantages of science and what science tries to do, or can do. Some uninformed persons think that science can do anything and can solve all problems. Although this is not completely true, the employment of a scientific method in the solution of most problems will give more logical and accurate answers than if no method is used.

The failure to appreciate and understand the true nature of science has caused much misunderstanding and criticism of the value of its methods. Some persons criticize science because biology cannot explain fully what "life" is. Here, as elsewhere, scientists can use only the tools that are available to them—they can investigate the chemical and physical processes inherent in living things and attempt to explain life in terms of such investigations. This may not give the desired explanation of life, because the investigations are incomplete or because the ultimate problem is not solvable by science.

There may be variations in the steps to be followed in the use of a scientific method, but the following are representative.

Steps in a scientific method

1. Recognition of the problem

Previously unnoticed problems or conditions occur constantly in the laboratory, in our work, or in our daily living. These may be simple, or they may be extremely complex, requiring sophisticated techniques and experiments for their solution. The awareness and clear recognition of a specific problem may be stimulated (1) by mere general curiosity, (2) by an actual need for the solution of the problem, or (3) by reading or thinking about a similar problem.

Indeed, several famous discoveries are reputed to have been stimulated by such things as being hit by a falling apple, by daydreaming while watching bubbles rise in a glass, and by a nightmare after a late evening. Most, however, are the result of hard work.

In the solution of any problem a main objective should be to ascertain the truth. In part, at least, this depends on an attitude in which problems are approached with an open, unprejudiced mind and with as much objectivity and detachment as possible. Misplaced enthusiasms and bias must be held to a minimum, if not completely eliminated.

Scientists are often considered by others to be unemotional, cold, and critical. The scientist even may be considered slow because he sees so many possible angles to a question before committing himself. A scientist is naturally critical of things. This critical attitude is the basis for accurate observations and for the collection of pertinent information.

2. Accurate preliminary observation

Observation includes preliminary studies of available information to determine what is already known about the condition being investigated. These studies might include reading, preliminary investigations, and gathering suggestive information from reliable sources and persons. This step is naturally not exhaustive in its scope but merely lays the foundation for the next step as accurately as possible.

3. Formulation of a hypothesis

A hypothesis might be considered a guess, speculation, or assumption that may be a tentative explanation of the problem. It is sometimes called a working hypothesis because we work forward from it. Sometimes only one hypothesis for a given problem can be suggested, whereas for another many hypotheses may present themselves. No hypothesis, even though it might appear insignificant at the moment, should be ignored. Each is

considered in turn, those that are not proved are eliminated, and possibly new ones are substituted as progress is made.

Another important phase in formulating a hypothesis is the decision on methods of investigation. In some problems devising the proper methods of investigation may require broad practical training, imagination, special techniques, or even elaborate equipment or apparatus. In all cases the data and information must be sufficiently extensive to reduce the chance effects of unusual differences or variations.

4. Testing the hypothesis

A hypothesis may be tested (1) through additional observations and investigations, (2) through controlled scientific experiments, if such can be performed in the particular problem, or (3) by a combination of these.

When using the scientific experimental method of investigation, it is highly desirable to utilize a control group of organisms or data in which a separate group is observed under conditions identical with the experimental group, except that the one condition being examined is not applied to the control group. All factors, except the one we are attempting to discover, are duplicated carefully in the controls. Only one variable should be permitted between the experimental and control groups at any one time.

5. Evaluation of the collected data

As relevant data and information are collected, they must be precisely recorded. All measurements, records, interpretations of data, or "case histories" must be scientifically accurate and sufficiently broad and comprehensive to be reliable. The careful and accurate collection and recording of the facts may make the difference between the problem being solved correctly or incorrectly. The investigator must be honest and faithful, his observations must be accurate, and his records must be complete and contain all data relevant to the problem. Sometimes graphs, tables, and summaries are valuable in this step of the technique (Fig. 1-1).

6. Drawing logical conclusions

After the information and data are recorded in such ways as to give accurate and meaningful revelations, they must be interpreted correctly.

5

This means checking the hypothesis. If the hypothesis is not proved, it may have been wrong. However, it may also have been correct, but the collected data were inaccurate, incomplete, or incorrectly interpreted. Great care must be taken not to draw conclusions that are broader than the collected facts will actually support or warrant.

7. Repeatability and reporting the results

We have all heard the lament of the professor, "publish or perish." The "publish" part of this statement is an absolute insistence that the results of experiments be made available to the scientific community. This is done by presenting papers and discussions before groups of workers in the same

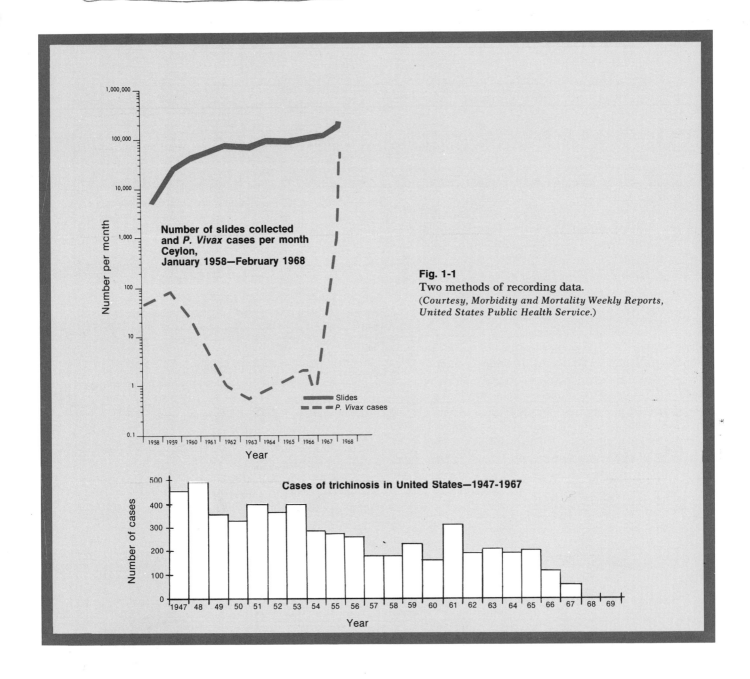

Fig. 1-1
Two methods of recording data.
(Courtesy, Morbidity and Mortality Weekly Reports, United States Public Health Service.)

field. Here, the results may be questioned, observations may be made on the techniques used, and interpretations may be made. The conclusion may then be published in a learned journal, such as *Science, Genetics, American Journal of Botany, Experimental Cell Research,* or any one of several hundred more.

Other workers may attempt to repeat the experiment in their laboratories. If repeatable in these different settings, the results have been verified. The data becomes part of biologic literature and is available to others interested in the same topics.

Much of the material in Chapter 30, Genes and Gene Action, is basically a treatise on the use of the scientific method.

◼ Use of scientific method

In order to illustrate the scientific method we might consider the following example. Alexander Fleming (1929) must have followed steps somewhat similar to the following in the discovery of the antibiotic, penicillin. He observed that a growth (*Penicillium* mold) from a spore accidentally dropped in a plate culture prevented the growth of bacteria in the culture. Fleming might have simply sterilized the culture and stopped, but his curiosity was aroused. He hypothesized that this mold produced something that might cause the death (or at least prevent the rapid growth) of the bacteria and hence be useful in the treatment of diseases. But this one observation would not prove his hypothesis. The lack of bacterial growth around the mold might have some other cause, or be merely coincidental. Fleming tested his hypothesis by repeated experiments with plate cultures of the same bacteria, and he introduced the same mold to observe if the phenomenon occurred repeatedly. After recording his data he organized and interpreted his experimental results and concluded that this mold did produce some antibacterial substance.

Later work by Fleming, Florey, and many other scientists uncovered many details of this phenomenon that led to the production of the antibiotic, penicillin, which was produced in purified form in 1940. This discovery led to searches for other antibiotics, and the results of their use are common knowledge today.

What if Fleming had not carefully observed that original phenomenon and had not been curious about it? Possibly some other investigator, sooner or later, would have given the initial start to this great medical discovery. Possibly the lives of some of you have been saved because of the use of some antibiotic whose discovery was instigated by the curiosity and observation of the Englishman, Alexander Fleming.

SOME SUBDIVISIONS OF BIOLOGY

Biology, which is the "science of living things," is divided into (1) zoology, which deals with the biology of animals, and (2) botany, which deals with the biology of plants. Botany and zoology have grown so extensively that such subdivisions as the following are really sciences in themselves.

anatomy* (Gr. *anatemnein*, to cut up) a study of gross structures, especially by dissection.

biogeography (Gr. *bios*, life; *geo*, earth; *graphein*, to write) the science of geographic distribution of organisms in space or throughout a particular region.

cytology (Gr. *kytos*, cell; *logos*, study) a detailed study of cells and their protoplasm.

ecology (Gr. *oikos*, house or home; *logos*, study) a study of the interrelations of living organisms and their living and non-living environments.

economic biology a study of organisms, which results in the improvement of desirable types or the destruction or control of undesirable ones, including the value of beneficial organisms and the losses because of detrimental ones.

embryology (Gr. *embryon*, embryo; *logos*, study) a study of the formation and development of an embryo.

evolution (L. *e*, out; *volvere*, to unroll or develop) a study of developmental changes undergone by organisms whereby they change throughout time.

heredity or genetics (L. *heres*, heir); (Gr. *genesis*, origin) a study of the inheritance or transmission of characteristics from one generation to another.

histology (Gr. *histos*, tissue; *logos*, study) a microscopic study of tissues.

paleontology (Gr. *palaios*, ancient; *onta*, beings; *logos*, study) the study of the distribution of organisms in time as revealed by their fossil records in the strata of the earth's surface.

pathology (Gr. *pathos*, suffering; *logos*, study) the study of diseases and abnormal structures and functions, including causes, symptoms, and effects.

physiology (Gr. *physis*, function; *logos*, study) a study of the functioning or working of an organism or its parts.

taxonomy (Gr. *taxis*, arrangement; *nomos*, law) the science of systematic classification of organisms.

*Derivations are based on *Webster's New International Dictionary, Henderson's Dictionary of Scientific Terms,* or *Dorland's Illustrated Medical Dictionary.*

Some early contributors to natural philosophy and the beginning of science

Democritus (460-370 B.C.)

A Greek philosopher who, like other early Greeks, enjoyed learning the true nature of things. He developed an atomic theory (atomism) that he applied to natural philosophy. He suggested that all phenomena are to be explained by the incessant movement of atoms, differing only in shape, order, and position. He did this without the benefit of the knowledge of atoms we have today.

Plato (427-347 B.C.)

A Greek philosopher and Aristotle's teacher who interpreted natural phenomena by relying upon intuition or instinct rather than upon reasoning. (*The Bettman Archive, Inc.*)

Plato

Aristotle (384-322 B.C.)

A Greek philosopher who stressed the importance of accurate and direct observations in securing facts and data. He drew his conclusions from facts that he secured by direct observations, and thus he initiated the basis for a scientific method of solving problems. Earlier philosophers had a tendency to reach conclusions and then select data and facts that agreed with their conclusions. (*Historical Pictures Service, Chicago.*)

Johannes Scotus Erigena (815-877 A.D.)

An Irish-Scottish philosopher in France who is thought to have begun a so-called Scholastic Philosophy (scholasticism), the core of which was the doctrine of the continuity and interdependence of the natural with respect to the supernatural order of truth. Scholasticism included the methods and doctrines of the Christian philosophers of the Middle Ages, and its sources were the writings of the church fathers and of Aristotle and his Arabian commentators.

Aristotle

Ibn Roshd Averroes (1126-1198)

An Arabian philosopher and physician who helped to progress the science of his time and has been called "The Aristotle of the Middle Ages." (*Historical Pictures Service, Chicago.*)

Roger Bacon (1214-1294)

An English philosopher who was noted for his scholasticism; yet, his precocious explanations of various phenomena were not appreciated and he was imprisoned.

Francis Bacon (1561-1626)

An English natural philosopher, who reached his conclusions through the process of induction from facts, thus breaking away from contemporary scholasticism. He stated (1620) that experiments are of fundamental importance in acquiring scientific knowledge, since experiments enable us to establish causes that determine an occurrence and enable us to bring about such occurrences when we wish.

René Descartes (1596-1650)

A Frenchman, was a pioneer in systematic philosophy who, because of his views, was forced to flee to Holland. He stated that analysis is the means of establishing the truth of the first principles of all knowledge. He suggested the wisdom of dividing the problem to be solved into as many parts as possible in order better to solve it. He stated "the purpose of analysis is to find out by means of one single truth, or a particular fact, the principles from which it derives."

Giovanni Borelli (1608-1679)

An Italian philosopher, mathematician, and disciple of Galileo who applied the latter's principles of physics to biology, thus suggesting an experimental approach to the science.

Immanuel Kant (1724-1804)

A German philosopher who believed that there was something in nature that united the mechanistic and teleologic (natural design) views in biology, rather than their being opposed to each other.

Averroes

Johann Goethe (1749-1832)

A German philosopher who suggested that life is a constant self-destruction and self-recomposition of living matter and that "life is a flame."

John Stuart Mill (1806-1873)

An English philosopher who elaborated on the philosophy of induction still further, propounding as its basis the law of the Uniformity of Nature.

Review questions and topics

1 List reasons why a study of living organisms should be made.
2 Define a scientific method.
3 List and describe each step to be followed in a scientific method, including enough details to ensure that you know the purpose and correct use of each step.
4 Define biology, zoology, and botany.
5 Define and learn the correct derivation and pronunciation of each subdivision of biology as listed in this chapter. Learn the correct pronunciation and derivation of each new term as you encounter it in your study and include a definition to be sure that you understand the meaning of the term.
6 List each of the subdivisions of biology, including the derivation, the pronunciation of each term, and the special area of biology emphasized in each.

Selected references

Baker, J. J. W., and Allen, G. E.: Hypothesis, prediction, and implication in biology, Reading, Mass., 1968, Addison-Wesley Publishing Co., Inc.

Beveridge, W. I. B.: The art of scientific investigation, New York, 1960, Random House, Inc.

Butterfield, H.: The scientific revolution, Sci. Amer. 203:173-192, 1960.

Cannon, W. B.: The way of an investigator, New York, 1968, Hafner Publishing Co., Inc.

Conant, J. B.: On understanding science, New Haven, 1947, Yale University Press.

Conant, J. B.: Modern science and modern man, New York, 1952 Columbia University Press.

Deevey, E. S.: The human population, Sci. Amer. 203:195-204, 1960.

Glassman, E. (editor): Molecular approaches to psychobiology, Belmont, Calif., 1967, Dickenson Pub. Co., Inc.

Hardin, G. (editor): Population, evolution and birth control, a collage of controversial ideas, ed. 2, San Francisco, 1969, W. H. Freeman and Co. Publishers.

Lerner, I. M.: Herdity, evolution and society, San Francisco, 1968, W. H. Freeman and Co. Publishers.

McDermott, W.: Air pollution and public health, Sci. Amer. 205:49-57, 1961.

9

chapter two

Early history and development of methods in biology

The refinement of methods of observing, collecting, and analyzing data led to the development of biology as a science. Beginning with the unaided eye, the science of biology has progressed to the modern laboratory where such tools as electron microscopes, chemical analyzers, complex recorders, and computers may be used.

EARLY RECORDS (? B.C. -200 A.D.)

The history and development of biology have passed through distinct periods. Since the beginning of history man has "studied" plants and animals, and later he tried to identify, name, classify, and use them. Early man associated closely with plants and animals in his daily living. Because he depended upon them for food, shelter, clothing, and medicines, he had to know something about them.

Systems of medicine were used more than 5,000 years ago. Many names of plants were given to the early Greeks by the guild of root-cutters (*rhizotomoi*) who supplied ingredients for early medicines and many foods.

Many names of animals were supplied by early hunters, fishermen, and priests. Agriculture, hunting, and medicine of one type or another had their origins with early man.

When Greek and Roman civilizations were at their height, the foundations of natural science were laid. The first scientific and systematic work was accomplished in Greece. The early Romans seemed to be interested more in the practical or applied side of science because they probably had to develop agriculture for their food supply.

Long before the Christian era various civilized peoples possessed considerable information about plants and animals that was of value for foods, medicines, and shelter. Some of this information, in the form of pictures and hieroglyphics painted on tombs or carved in stone, is available for study today. These records show that the Egyptians and Assyrians (4000 B.C.) were practical plant scientists of a rather high type, having cultivated medicinal, ornamental, and food plants. Among them were roses, apricots, figs, dates, grapes, olives, wheat, and barley. Early Egyptians had a wealth of knowledge about animals and possessed domesticated cattle, sheep, pigs, ducks, geese, and cats. Some of early man's impressions of animals have survived in the cave paintings of France and Spain. Man decorated pottery, cloth, and tools with animal figures.

The Chinese (2500 B.C.) had acquired a practical knowledge of plant uses, including such cultivated plants as rice, tea, oranges, and a plant that is the source of ephedrine (a medicine).

Ancient races in America also had considerable knowledge about plants and animals. The pre-Inca race of Peru (3000 B.C.) was apparently the first in America to cultivate maize (corn). From Peru its cultivation spread northward and southward, so that by 1492 it was a major crop from the St. Lawrence River to Argentina. The aboriginal Americans domesticated and cultivated such practical plants as potatoes, squashes, cacao trees, and avocados (alligator pear).

THE THIRD TO THE TWELFTH CENTURY (200-1200 A.D.)

After the Greek and Roman periods there was a decline in European science during the Middle Ages, or so-called Dark Ages, which extended from the third century to the twelfth century A.D. For more than 1,000 years after 200 A.D. much of the knowledge of the ancients was lost in Europe, and things were accepted without question, observation, or experimentation. During this time science was opposed by public opinion and authority. In general, science as known by the earlier Greeks did not exist.

The period of the Middle Ages was one of relative inactivity because of political, social, and psychologic adjustments after the fall of the Roman

Empire (476 A.D.) and the domination of much of Europe by barbarian tribes. During this unproductive period the emphasis was placed on "opinions" and "judgments" of a few so-called authorities. Their opinions were accepted, usually without question. Few if any investigators attempted to prove things for themselves. Much of the biologic work of this period involved mythology and superstition, with only a limited knowledge of biologic facts.

There was considerable scientific activity in Arabia (800-1300 A.D.), where much of the biologic interest centered around medicinal plants. Botanical gardens were established, and the medicinal uses of plants and plant products flourished. The Arabian scientific activity was rather extensive, yet it was not original and progressive, since much of the work involved the translation of early Greek and Roman publications into the Arabic language in Baghdad.

THE THIRTEENTH TO THE SIXTEENTH CENTURY (1200-1600 A.D.)

After the many centuries of comparative inactivity during the Dark Ages there was a revival in science and other areas of man's activities. This period is referred to as the Renaissance (Fr. *renaître*, to revive or to be born again), which began in Europe about ten centuries after the fall of the Roman Empire.

There was a gradual revival of the scientific studies of plants and animals that had been started by the Greeks many years before. In Europe, universities were founded in which professors lectured from pulpitlike reading desks, and students in academic robes observed demonstrations by assistants. Students themselves did little or no dissecting or direct observing.

Botanical gardens were popular during the sixteenth century, and most European universities had gardens of medicinal and food plants by the middle of the seventeenth century. During these two centuries many large volumes known as *Herbals* were published. These contained descriptions of food and medicinal plants, illustrations drawn from living plants, and descriptions of many natural phenomena, some of which were purely superstitions and myths. Gaspard Bauhin published excellent descriptions of nearly 6,000 plants in his *Herbals* (1623).

Many fantastic stories and drawings appeared in the *Herbals*, and many descriptions were not very accurate. A so-called *Doctrine of Signatures*, advocated by Paracelsus (1493-1541), a Swiss physician and alchemist, stated that certain plant structures were modeled upon structural principles similar to those of human organs and that these plant structures supposedly constituted remedies for diseases of those human organs that they resembled most closely. For example, the sap of the blood-root plant was used as a blood tonic; walnuts with their numerous ridgelike convolutions were used in the treatment of brain diseases.

Other published volumes known as *Bestiaries* contained descriptions of animals and, although the earliest ones were somewhat allegorical and moralizing in their descriptions of beasts of the Middle Ages, they represented the best efforts of their authors to meet the needs of their times. Scientific societies and the publication of scientific journals also were established in the seventeenth century.

DEVELOPMENT OF THE MICROSCOPE

It is not known who invented the first microscope, but in Nineveh, an ancient city in Assyria, a rock crystal was excavated that may have been used as a "lens" in the eighth century B.C. Euclid (of Megara) (440 B.C.-?), a Greek philosopher, investigated the properties of curved reflecting surfaces. His investigations were used later in connection with studies in magnification. Lucius Seneca (4 B.C.?-65 A.D.), a Roman philosopher, reported that water-filled glass globules would assist one in seeing small objects. Claudius Ptolemy (127-151 A.D.), a Greco-Egyptian astronomer, studied some problems of magnification by using curved surfaces. Even though burning glasses were used, magnifying glasses (lens thicker at the center than at the edge) were probably not used extensively until the invention of eyeglasses in the thirteenth century. Leonardo da Vinci (1452-1519), an Italian painter and architect, stressed the use of lenses in studying small objects. The early microscopes (magnifying glasses) were commonly called "flea microscopes," because a flea was a specimen commonly studied.

11

Fig. 2-1
Exact replica of the compound microscope made by Zaccharias Janssen and his son Hans of Middleburg, Holland, between 1590 and 1610.
(Courtesy The Armed Forces Institute of Pathology, Washington, D. C., No. 53-662.)

Fig. 2-2
Hooke's microscope (1665). The body tube contained a series of lenses that magnified the image in the manner of the compound microscope. Illumination was provided by a lamp and bull's-eye condenser. The instrument was 16 inches high and had a maximum magnification of 42×.
(From The Evolution of the Microscope, American Optical Co., Instrument Division, Buffalo, New York.)

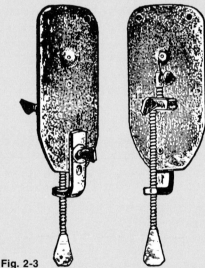

Fig. 2-3
Leeuwenhoek's microscope (1673). This simple microscope consisted of a lens mounted between two flat pieces of metal, with an adjustable point for holding the specimen and for focusing purposes.
(From The Evolution of the Microscope, American Optical Co., Instrument Division, Buffalo, New York.)

Because so much scientific progress of the past has been dependent upon the development and use of microscopes, a brief consideration of early and recent microscopes will be presented.

A compound microscope (Fig. 2-1) is thought to have been invented by Zaccharias Janssen and his son Hans in Middleburg, Holland, between 1590 and 1610. These spectacle makers and lens grinders combined lenses when viewing objects and discovered that a second lens would magnify the enlarged image from a magnifying glass. Their microscope was made of tubes that slid together for focusing, had a size of 2 by 18 inches, and magnified about nine times.

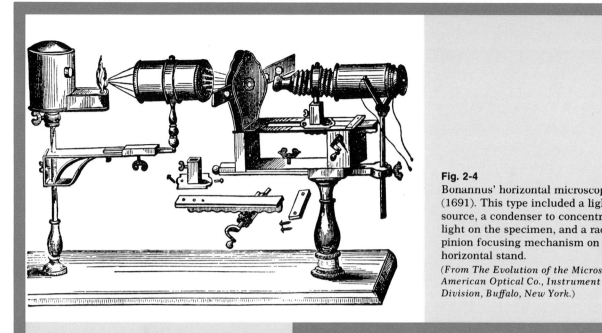

Fig. 2-4
Bonannus' horizontal microscope (1691). This type included a light source, a condenser to concentrate light on the specimen, and a rack-and-pinion focusing mechanism on a horizontal stand.
(*From The Evolution of the Microscope, American Optical Co., Instrument Division, Buffalo, New York.*)

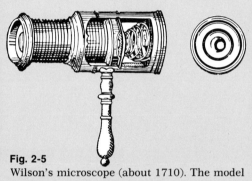

Fig. 2-5
Wilson's microscope (about 1710). The model is made of ivory, and the body is cut open and the ends threaded for the attachment of lenses. The specimen is held by a spring for focusing. The handle is unscrewed when carried in the pocket.
(*From The Evolution of the Microscope, American Optical Co., Instrument Division, Buffalo, New York.*)

Robert Hooke (1635-1703), an English microscopist, constructed a microscope in 1665 that consisted of an objective lens, a field glass, and an eye lens (Fig. 2-2). The latter two magnified the image of the former in the manner of a compound microscope. He provided a lamp for illumination and a bull's-eye condenser for intensifying the light. His microscope had magnifications of 14 to 42 diameters.

Antonj van Leeuwenhoek (1632-1723), a Dutch microscopist, developed a simple microscope (about 1673) by mounting a lens between two flat pieces of metal and adding a pivoted joint for holding the specimen (Fig. 2-3). He ground lenses that had magnifications of up to 300 diameters. With his lenses he studied bacteria, molds, protozoans, red blood corpuscles, plants, animals, and the circulation of blood in the tadpole tail.

Bonannus improved the microscope in 1691 and developed a horizontal type that included a source of light, a condenser to concentrate light, and a rack-and-pinion mechanism for more efficient focusing (Fig. 2-4).

Wilson, about 1710, developed a screwbarrel type of microscope (Fig. 2-5). The body had threads on the observing end into which lenses of different magnifying powers might be screwed. The opposite end had a condensing lens to concentrate light. The specimen was pushed against a spring for focusing. A Wilson type of microscope was received at Harvard College in 1732 and may have been one of the first compound microscopes used in American colleges, although simple microscopes probably were used earlier.

There were only about a dozen microscopes in the United States in 1831. Instructors were using

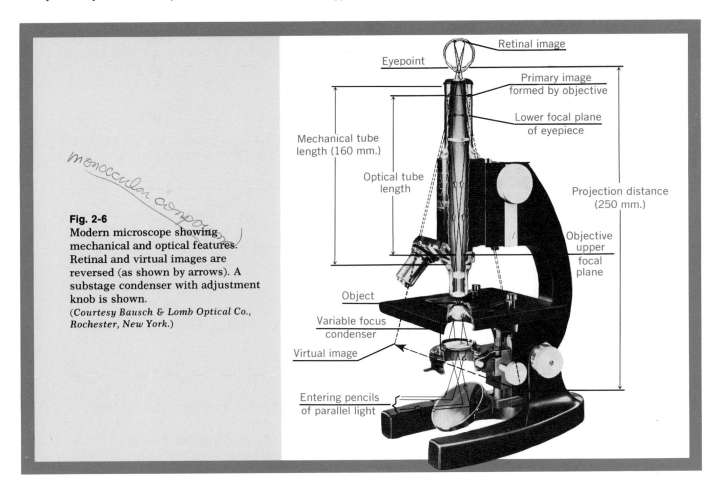

Fig. 2-6
Modern microscope showing
mechanical and optical features.
Retinal and virtual images are
reversed (as shown by arrows). A
substage condenser with adjustment
knob is shown.
(*Courtesy Bausch & Lomb Optical Co.,
Rochester, New York.*)

them by 1850, and students began using them as early as 1875, but they were not in general student use until about 1890.

Charles A. Spencer (1813-1881) built the first American microscope (1847) and several additional models of it. Robert B. Tolles (1824-1883) was another early American microscope builder who started as an apprentice of Spencer but established his own business (1858). He is famous for his improvements of objectives and for inventing the homogenous immersion objectives. In these a drop of the proper type of liquid is placed on the cover slip on the slide and the immersion objective is made to contact the liquid, which acts as a type of lens to assist in higher magnifications. Today, oil immersion objectives are used for high magnifications. Edward Bausch (1854-1944) made his first

microscope in 1872. He was the son of J. J. Bausch (1830-1896), the founder of the Bausch & Lomb Optical Company.

Until the end of the nineteenth century, the making of complete microscopes was largely done by individuals who made one microscope at a time. The metal parts were made by hand and the lenses ground and polished with rather simple equipment. Increasing demands for more microscopes suggested to manufacturers that specialists (scientists, designers, engineers, specialized workers) must be trained, that standards must be set, and that microscopes must be built on an assembly-line basis.

The twentieth century has seen many improvements in the manufacture and usefulness of the various types of microscopes (Figs. 2-6 and 2-7).

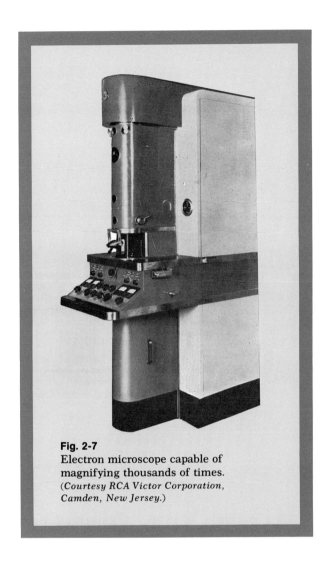

Fig. 2-7
Electron microscope capable of magnifying thousands of times.
(Courtesy RCA Victor Corporation, Camden, New Jersey.)

In an ultraviolet microscope invisible ultraviolet rays (of shorter wavelengths and beyond the visible violet light waves) are used instead of ordinary light. Because of the invisibility of the ultraviolet rays, photographs must be made, since the image cannot be seen. Special quartz lenses must be employed that permit the passage of the ultraviolet rays.

In dark-field microscopy the term "dark field" refers to a method of illuminating a specimen brightly while the surrounding background (field) remains dark. The most practical dark field is obtained by a special dark-field condenser (dark-field illuminator) whereby direct light rays do not enter the specimen. The oblique light rays are focused on the specimen, which thus appears as a luminous body against a dark field. A very small, bright object is more easily seen in a dark background (field) than is a very small, dark object in a bright field. This is similar to the phenomenon of seeing small dust particles in a beam of light when the region at the back of the light beam is dark.

Phase-contrast microscopy, proposed by Zernike of Holland in 1932, permits the study of living organisms and other transparent materials, some of which do not absorb visible light and hence are not visible under an ordinary light microscope.

Electron microscopy employs beams of electrons produced by special apparatus instead of light (Fig. 2-7). Magnetic fields ("electron lenses") are used instead of glass lenses, and photographic films may be used to record the image, which may be magnified over 100,000 times. The specimens being photographed must be very thin and in a vacuum.

The fact that axially symmetrical magnetic and electric fields could be employed as lenses was discovered by H. Busch in 1926. Hence, by the proper use of magnetic fields (acting as lenses), the charged particles (electrons) can be made to do what light waves accomplish in ordinary, optical microscopy. Electron microscopes were made by Knoll and Ruska (Germany) in 1932, by Marton (Belgium) in 1934, and by Prebus and Hillier (Canada) in 1938. The Radio Corporation of America in 1941 manufactured a commercial electron microscope of the magnetic type. Many kinds and models have been made and used in various parts of the world since that time.

Some of the more recent improvements include ultramicroscopes, ultraviolet microscopy, dark-field microscopy, phase microscopy, and electron microscopy.

When particles too small to be seen with a microscope under ordinary conditions are illuminated by a strong beam of light parallel to the surface of the stage (at right angles to the direction of vision through the microscope), they appear as bright specks because of their reflection of light but do not show their outline or shape. The apparatus used for such study is called an ultramicroscope (L. *ultra*, beyond).

Review questions and topics

1 Discuss the conditions under which natural science originated, including factors that may have influenced its development.

2 Discuss the reasons why biology as a science did not originate earlier and why progress was not more rapid and uniform through the years.

3 Why were earlier biologists called natural philosophers? What is meant by the term philosopher? Is there still "philosophy" in present-day biology?

4 Was the lack of progress in biology primarily caused by the lack of scientific equipment? What other factors might have contributed? Why do you say so?

5 Describe the following types of microscopes: simple, compound, monocular, and binocular.

6 List the important stages in the development of microscopes, including the persons and their specific contributions.

7 How may scientific progress depend on the efficient use of microscopes in such fields as medicine, medical technology, nursing, agriculture, industry, and sanitation?

8 What are the chief differences in the following types of microscopes: light, ultraviolet, and electron?

Selected references

Anonymous: The roots of healing; ancient herbals, Natural History, 68:578-591, 1959.

Asimov, I.: A short history of biology, Garden City, New York, 1964, Natural History Press.

Braidwood, R. J.: The agricultural revolution, Sci. Amer. 203:131-48, 1960.

Butterfield, H.: The scientific revolution, Sci. Amer. 203:173-92, 1960.

Clark, G. L.: The encyclopedia of microscopy, New York, 1961, Reinhold Publishing Corp.

Dawes, B.: A hundred years of biology, London, 1952, Gerald Duckworth & Co., Ltd.

Gardner, E. J.: History of biology, Minneapolis, 1965, Burgess Publishing Co.

Lenhoff, E. S.: Tools of biology, New York, 1966, The Macmillan Company.

Moment, G. B. (coordinator): Frontiers of modern biology, Boston, 1962, Houghton Mifflin Company.

Needham, G. H.: The practical use of the microscope, including photomicrography, Springfield, Illinois, 1958, Charles C Thomas, Publisher.

Nordenskiold, E.: The history of biology, New York, 1933, Alfred A. Knopf, Inc.

Ritterbush, P. C.: Overtures to biology, New Haven, 1964, Yale University Press.

Singer, C.: A history of biology, New York, 1959, Abelard-Schuman, Limited.

Sirks, M. J., and Zirkle, C.: The evolution of biology, New York, 1964, The Ronald Press Company.

Washburn, S. L.: Tools and human evolution, Sci. Amer. 203:63-75, 1960.

Waterbalk, H. T.: Food production in prehistoric Europe, Science 162:1093-1102, 1968.

Cells

1. heredity
2. Repair
3. Abnormal (cancer)
4. growth

cells must work as a whole group, to give the organism coordination, even though seperate units

CELL PRINCIPLE

According to the cell principle all living organisms (or those that were once alive) are made of cells, and all phenomena of life are fundamentally cellular in nature. Robert Hooke, an Englishman, studied cells as early as 1665. Matthias Schleiden, the German botanist, and Theodor Schwann, the German zoologist, are commonly given credit for the formulation of the cell principle in 1839, although René Dutrochet, a French physiologist, preceded them with similar views in 1824.

Early investigators used the word *cell* because they emphasized the cell wall and practically ignored the important substance within. To them the cells looked like the "cells" of a honeycomb in which something might be placed. This emphasis on cells led to the founding of the specialized science of cytology (Gr. *kytos*, cell; *logos*, study). The cell principle was verified repeatedly by later investigators.

A study of the principle shows that plants and animals, although different, are really organized and constructed along similar patterns. Fundamentally, the functions of normal plants and animals, as well as those in abnormal, diseased organisms, are but expressions and behaviors of cells. This principle laid a foundation for much of the unified scientific investigations by hundreds of biologists, thereby affecting their research and progress.

The cell is considered the unit of structure of animals and plants, the tissues and organs being made of cells much as the brick is a unit of structure in a brick wall. The cells are also units of function (physiology) because the functions of living organisms are the results of cellular activities. Each cell works somewhat as a unit, but most often groups of cells work together in some common function. There must be a proper interdependence, interfunction, coordination, and subordination if the organism is to function as a whole with efficiency. The cell is also a unit of growth and development as in a complex organism with many cells that has grown and developed through a division of its cells, an increase in their size, and a specialization into tissues. Cells are units of heredity, for it is through them that hereditary materials are received from parents, maintained within the embryo and adult, and passed on to future offspring. During the division of cells each one receives genes that enable the organism to express its specific traits. Cells are also units of regeneration (repair) when tissues or organs are replaced or repaired. It is thought that abnormal cell divisions are responsible for such growths as tumors and cancers.

CELL STRUCTURES AND THEIR FUNCTIONS

All organisms are composed of one cell, or a myriad of cells, so small that an adult, human male, weighing 160 pounds, contains approximately 60,000 billion cells. In order to study microscopic structures in a cell it is necessary to make thin sections of the cell, whether the studies are made with a light microscope or with an electron microscope. Consequently, it must be remembered when studying a thin section that usually only a portion of a complete cell is observed. To ascertain the structure of the complete cell one must put together what is observed in a series of these sections.

There are two main interdependent parts of a cell: the centrally located, spherical or ovoid nucleus and the surrounding cytoplasm. Some of the organelles in cells include the following.

Cell wall

The cell wall is a semirigid, laminated (in layers), nonliving covering that is present in most plant cells, but it is absent in most animal cells. It is

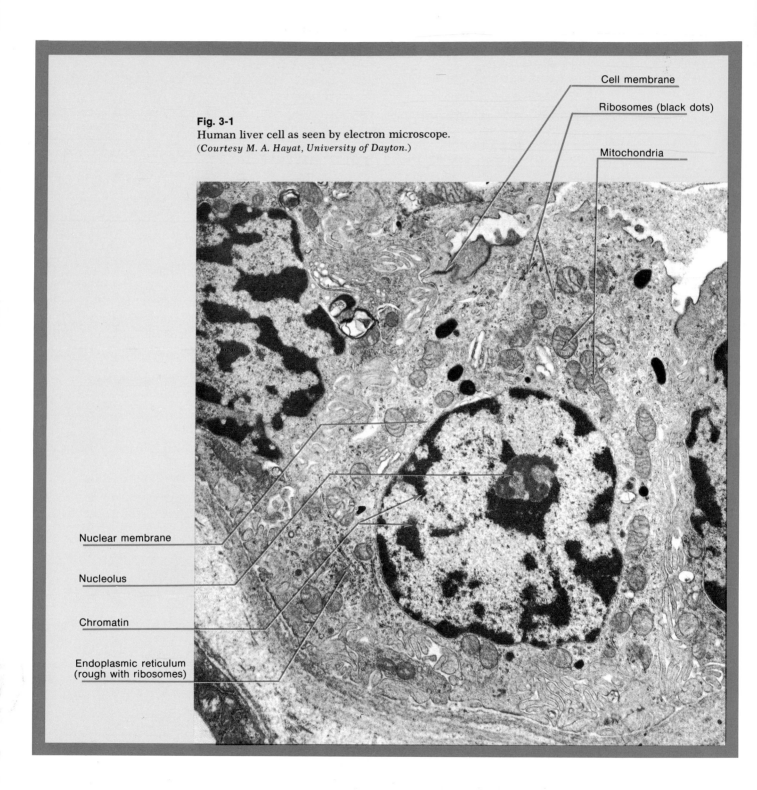

Fig. 3-1
Human liver cell as seen by electron microscope.
(Courtesy M. A. Hayat, University of Dayton.)

Cell membrane

Ribosomes (black dots)

Mitochondria

Nuclear membrane

Nucleolus

Chromatin

Endoplasmic reticulum
(rough with ribosomes)

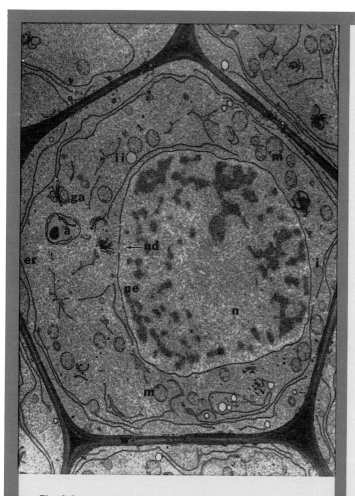

Fig. 3-2

Meristematic cell of the root tip of corn showing the following ultrastructures: n, nucleus showing chromatin material; ne, nuclear envelope; nd, nuclear envelope discontinuity (not a pore); er, endoplasmic reticulum; ga, Golgi apparatus; m, mitochondrion; a, amyloplast (starch forming); i, unidentified cytoplasmic inclusion body; ii, unidentified cytoplasmic inclusion body; w, cell wall. Magnification approximately 8,000×.

(From Whaley, W. G. W., Mollenhauer, H. H. M., and Leech, J. H. L.: The ultrastructure of the meristematic cell, Am. J. Bot. 47:423, 1960.)

Fig. 3-3

Common freshwater plant *Elodea* (*Anacharis*). In these living cells chloroplasts are green and float in the cytoplasm where they photo-synthesize food. The vacuole contains cell sap. The nucleolus may be visible inside the nucleus. The middle lamella (intercellular layer) is the first thin wall layer formed by the protoplasm of the new cell.

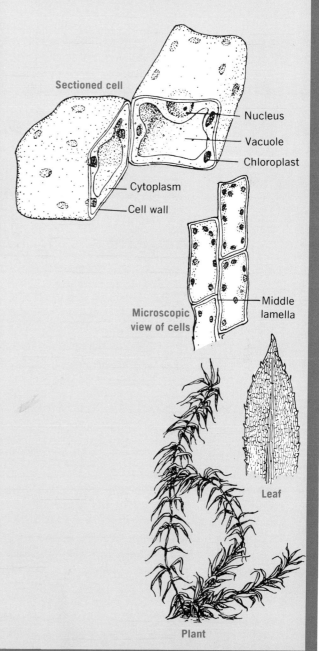

secreted by the cell and gives protection and support, covering the living plasma (cell) membrane beneath. In some plants it contains minute pits through which pass elongate plasmodesmata (plas mo -des' ma ta) (Gr. *plasma*, molded; *desma*, bond) that connect adjacent cells (Fig. 3-4). In plant cells it is composed primarily of cellulose, a complex carbohydrate material that occurs in long threads called fibrils. Occurring between the fibrils are other complex materials, such as lignin and pectin. Enough space remains so that air and fluids pass freely through the cell wall.

Two adjacent plant cells are bound together by a shared cell wall layer, the middle lamella. Cell walls form the chief component of wood and yield fibers used in the manufacture of such things as paper, flax, cotton, and hemp.

Plasma (cell) membrane

The plasma membrane is a living, ultrathin, elastic, porous, semipermeable (permits passage of some molecules but not others) covering that is present in both plant and animal cells, transmitting materials to and from the cell. It is not easily seen with a light microscope, and it is thought to be composed of two protein layers (thin meshwork of fine, elongated protein molecules) with a double layer of lipoid or fatlike molecules between them. The long protein molecules can fold or unfold, thus accounting for the membrane's elasticity. Externally it has an irregular contour with deep craterlike infoldings, whereas the inner surface has numerous saclike inpouchings. These invaginations may form pinocytic vesicles from which fluid may be ingested by the cytoplasm. The plasma membrane grows as the cell enlarges and has a limited ability to repair itself.

Simple sugars, amino acids, potassium ions, and water may pass through the membrane rapidly. Sodium and other substances, even if constructed of small molecules, may not pass readily. It must be stated that the permeability of the membrane is not constant and fixed but is subject to change from one moment to the next. It seems that the cell can determine the structure and behavior of the membrane. This, together with the size and character of the entering molecules, is important in the passage of materials through it.

INTERNAL ENVIRONMENT

Fig. 3-4
Section demonstrating numerous long protoplasmic strands (plasmodesmata) that pass through cell walls and connect adjacent cells, as found in the Philippine persimmon.
(*Courtesy General Biological Supply House, Inc., Chicago, Illinois.*)

Fig. 3-5
Section of chicken heart showing many mitochondria adjacent to the myofibrils of the ventricle.
(*Electron micrograph courtesy M. A. Hayat, University of Dayton.*)

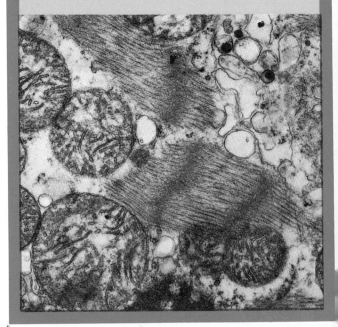

Cytoplasmic matrix (ground substance)

Cytoplasmic matrix fills the space between the plasma membrane and the internally located nucleus and is composed of a highly organized, intricate meshwork of elongated protein molecules responsible for many of the cell functions. Within the matrix are located the endoplasmic reticulum, mitochondria, the Golgi apparatus, the centrosome, and other cell organelles.

Endoplasmic reticulum

The endoplasmic reticulum, discovered and named by Keith Porter in 1956, extends from the nuclear membrane to the peripheral cytoplasmic meshwork and is composed of delicate, slotlike, roughly parallel membranelike structures.

There are two types of endoplasmic reticulum: rough endoplasmic reticulum and smooth endoplasmic reticulum. Rough endoplasmic reticulum is found in all cells except erythrocytes and is especially abundant in glandular cells, such as salivary gland cells. Only rough endoplasmic reticulum has direct connections with the nuclear membrane. It has tiny particles of ribonucleic acid (RNA) called ribosomes clinging to its outer surfaces. It is involved in the synthesis, segregation, and accumulation of secretory proteins and in the biosynthesis of membranes.

Smooth endoplasmic reticulum does not show any association with ribosomes and is less common than rough endoplasmic reticulum. It is abundant in the liver and intestinal epithelium and seems to be involved in detoxification mechanisms, in lipid and cholesterol metabolism, and in the biosynthesis of steroid hormones.

Ribosomes and protein synthesis

Ribosomes are tiny spheres consisting of almost equal amounts of RNA and protein. They usually occur in clumps or clusters that are collectively referred to as a polyribosome or, more simply, as a polysome. They are the site of protein formation.

As will be detailed in Chapter 30, a messenger RNA–ribosome complex is formed. The RNA then moves along the length of the ribosome and attracts transfer RNA (tRNA) to the messenger RNA (mRNA) surface. Amino acids at the end of the tRNA are then brought into contact and peptides are formed. As the ribosome moves along the mRNA, the process is repeated and a polypeptide or protein is formed.

During protein formation, four or five ribosomes are attached to a particular mRNA strand. Apparently, each ribosome is "reading the genetic code" for a particular polypeptide. The attached ribosomes are collectively called the polysome.

Lysosomes

Small membrane-enclosed bodies containing digestive enzymes were first seen by DeDuve in the early 1950s. They were formed on the ribosome-reticulum complex. Because of their digestive function, he referred to them as a sack full of enzymes and as "suicide bags." Lysosomes are rich in lytic enzymes (for example, acid hydrolases) and appear to be involved in the engulfment of foreign materials into the cell, in the destruction of dead cells, and in tissue degeneration. They have been found in many cell types, including spleen, kidney, and liver cells, and in meristematic cells of plants.

In lower animals such as *Hydra* the lysosomes in the cell cytoplasm fuse with ingested food vacuoles. The food is then digested. Lysosomes also appear to function in cell breakdown, such as in resorption of the tail of a tadpole, in aging of cells, and in general cell dissolution.

breakdown fat, proteins nucleic acids into small parts — be taken in by mito.

ADJ (cells growin ¿ ÷ at tips stems plants.

Mitochondria

When a thin section of a plant or animal cell is properly prepared and magnified, numerous round, or rodlike, structures known as mitochondria (mi to -kon' dri a) (Gr. *mitos*, thread; *chondros*, granular) will be seen within the cytoplasm (see Fig. 3-1). Certain cells contain 1,000 or more, and their sizes may vary from 0.2 to 3μ. Mitochondria are surrounded by a thin double membrane of lipoprotein with the inner layer folded to form partition-like ridges called cristae that extend inwardly. Mitochondria may assume different shapes in different organisms, and some of the cristae may run crosswise instead of lengthwise. They may even be tubular as in the mitochondrion of *Paramecium*. The outer membrane is elastic and may swell to increase its size. It is composed of protein mole-

OXIDATIVE ENZYMES

21

top of

Resperation - process by which a living organism or cell takes in oxygen from air or water & utilizes it in oxidation, & gives off products of oxidation, especially carbon dioxide.

$C_6H_{12}O_6$
Monosaccaride

Cellular respiration

Glucose + Oxygen → Carbon dioxide + Water + Energy
(Sugar) O_2 CO_2 H_2O (Released)
       to 200

$C_6H_{12}O_6$

ADP   + Phosphate   + Energy → ATP
(Adenosine   ($-PO_3H_2$)   (Adenosine triphosphate)
diphosphate)   (Energy-rich phosphate)

ATP + Water → ADP + Phosphate + Energy

Fig. 3-6
Release of energy in living organisms.
Adenosine is an organic, phosphate-
containing compound that functions in
energy transfers within cells.

cytochromes - proteins contain iron atoms surrounded by ~~phy~~ porphyrin #similar to ~~clo~~ chlorophyll's porphyrin.

cules that can be extended or folded greatly, thus regulating the sizes of the substances passing in and out. Mitochondria are always in motion and are called the "power houses" of the cell.

Mitochondria are present in large numbers wherever energy is needed. They produce oxidative energy-transfer enzyme systems whereby specific enzymes bring about reactions, so that energy is released in a slow, continuous flow in order that it can be used efficiently for the formation of cell products, the contraction of muscles, and the conduction of nerve impulses. Parts of the mitochondrial enzyme system are formed by such vitamins as riboflavin, nicotinamide, and pantothenic acid. It is known that there are many different enzyme systems in a mitochondrion, each system being composed of many different protein molecules. It is also suggested that as many as 2,000 duplicates of each enzyme may exist in the mitochondrial unit at the same time.

cytochrome part of electron messengers of Photosyn

Production of energy

Much of our present knowledge of cellular energy and cellular respiration began with the discovery by Otto Warburg of the respiratory enzyme called cytochrome oxidase and the rediscovery of a pigment cytochrome in the cells of living organisms. Cytochrome is related to hemo-

globin, the oxygen-carrying substance in red blood corpuscles.

When sugar is broken down, it releases hydrogen that is accepted by the cytochrome. In the presence of free oxygen and cytochrome oxidase the cytochrome gives up hydrogen to the oxygen to form water.

P.49 Cytochrome oxidase + Oxygen + (Cytochrome + Hydrogen) → Water

Every cell contains a group of chemicals that are energy-trapping and known collectively as adenosine phosphates (Fig. 3-6). Adenosine is a nucleotide (see Chapter 6) and is one of the intermediate reaction products in carbohydrate metabolism. Each cell contains at least two types, namely, ADP (adenosine diphosphate) and ATP (adenosine triphosphate). As the name suggests, ATP contains three phosphate groups (H_2PO_3) attached to each adenosine molecule and is the principal molecule in which energy is stored. Under proper conditions ATP will lose a phosphate group to some other substance and release energy, thus becoming ADP with two phosphate groups. A molecule of ADP can be transformed into ATP by adding a third phosphate group, an addition which of course requires energy. Energy-rich ATP is the chief end product of cellular respiration through the action of the mitochondria of cells. ATP diffuses to all parts of a cell where energy is needed.

Carbohydrates, proteins, and fats are broken down outside the mitochondria into their constituents before passing through the outer membrane of the mitochondrion. Inside the mitochondria the oxidation reactions remove carbon atoms from these constituents until carbon dioxide and water are produced. The energy thus released by these successive oxidation steps passes to the ATP by the phosphorylation process.

A sugar, such as glucose, may combine with phosphate, through the process of phosphorylation, to form sugar phosphate (glucose phosphate) and the latter, through a series of steps, eventually ends up as pyruvic acid. These steps are complicated and are considered in more detail elsewhere.

OXIDATION - UNION OF A SUBSTANCE w/air.
OXIDASE - ANY enzyme that acts as an oxidater.

Golgi apparatus

The Golgi apparatus (after Golgi, the Italian histologist) is quite prominent in nerve cells, secretory cells, and germ cells. It has a large surface area, because it consists of a series of membranous structures of variable size and shape. In some cells the apparatus is seen as a pile of joined, crescentic (curved) layers (Fig. 3-1). It is rich in fatty materials and darkens when treated with osmium or silver stains. It may be observed by both light and electron microscopes and is considered a center of cellular synthesis.

Although it was suggested by Ramon y Cajal in 1914 that the Golgi apparatus secreted the mucus that covers the cells lining the intestine, confirmation of this hypothesis was not made until fairly recently. Electron microscope studies by Neutra and Leblond have shown that globules of mucus do, in fact, arise from the Golgi apparatus.

Further studies by several workers indicate that Golgi apparatus function in the formation of very large carbohydrate-protein complexes. Apparently, the Golgi apparatus adds the carbohydrate to the protein that has been formed in ribosomes. It appears that most plant and animal cells have active Golgi complexes responsible for the synthesis and secretion of a wide variety of carbohydrate complexes.

Spherosomes

Spherosomes are ovoid or spherical structures that can be seen both with light and electron microscopes in plant cells. They are bound by a single membrane and contain an osmiophilic material. They appear to be involved in fat production or storage, although many other possible functions have been proposed. It is thought that spherosomes are formed by the endoplasmic reticulum.

Centrosome

The centrosome (Gr. *kentron*, center; *soma*, body) is an area of dense protoplasm usually located near the nucleus of many animal cells. It is not commonly present in cells of higher plants, although it may be present in certain lower plants, such as brown algae. It is usually rather incon-

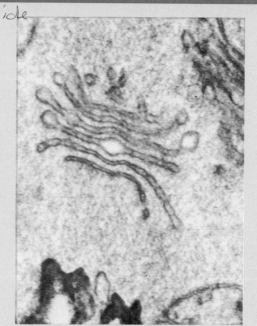

Fig. 3-7
Electron micrograph showing Golgi apparatus.
(*Courtesy, M. A. Hayat, University of Dayton.*)

spicuous in cells except during cell division when it plays an important role.

After the prophase stage of cell division the centrosome contains two rod-shaped, granular centrioles. The electron microscope reveals them to be short, cylindrical structures placed at right angles to each other. One of the earliest signs of cell division (if the centrosome is present) is the movement of the two centrioles away from each other until eventually they take positions at opposite sides (poles) of the nucleus. After the nuclear membrane disappears the centrioles organize cellular proteins into a series of fibers between them that constitute the spindle. More detailed consideration will be given under cell division in Chapter 4.

Fat droplets

Mitochondria are engaged in the breakdown of fats to yield energy and in the build-up of fats when more sugar is available than is needed at that time.

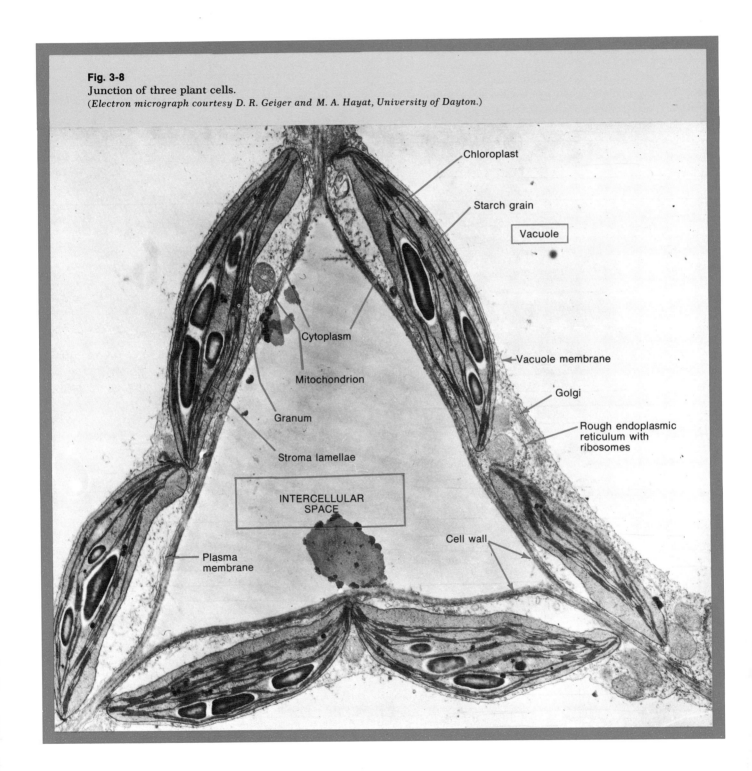

Fig. 3-8
Junction of three plant cells.
(*Electron micrograph courtesy D. R. Geiger and M. A. Hayat, University of Dayton.*)

Chloroplast

Starch grain

Vacuole

Cytoplasm

Mitochondrion

Granum

Stroma lamellae

INTERCELLULAR
SPACE

Plasma
membrane

Vacuole membrane

Golgi

Rough endoplasmic
reticulum with
ribosomes

Cell wall

Spherical fat droplets are found close to mitochondria in cells, which may explain their role in fat metabolism. Fat is a principal reserve supply of energy in the body.

Cytoplasmic vacuoles

Some of the more interesting cytoplasmic vacuoles occur in Protozoa, such as *Amoeba* and *Paramecium*. These contractile vacuoles play an important part in maintaining the water balance of the organism and in expelling waste. Some of them have a series of radiating canals. They can be observed to swell with water, discharge, and fill again.

Most plant cells have a large vacuole bounded with a single membrane and filled with a watery cell sap. This fluid may contain high concentrations of pigments, amino acids, sugars, and other compounds. These plant vacuoles are small in young cells but are large enough in older cells to push the cytoplasm to the periphery of the cell.

chromplast - have
yellow - xanthofil
Ph Red - phycoirriferin

Plastids

Plastids are organized bodies, usually spherical, oval, or ribbon shaped and are present in certain plant cells (Fig. 3-8). Three types found in certain plant cells include: (1) chloroplasts (Gr. *kloros*, green), which contain green chlorophyll and absorb radiant energy to photosynthesize foods; (2) chromoplasts (Gr. *chromo*, color), which are usually yellow, orange, or red and occur in flowers and fruits; (3) leukoplasts (Gr. *leuko*, white), which are colorless and may store energy-rich starch (formed from simple sugars) as in the potato tuber.

Chloroplasts are especially important in the process of photosynthesis. The boundary of a chloroplast is the typical double membrane. Inside is a structure peculiar to the chloroplast called the granum. These structures are platelike lamellae resembling stacks of coins. They are interconnected by structures called the stroma lamellae.

The arrangement of lamellar components is

chloro

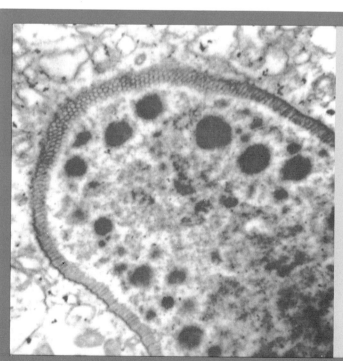

Fig. 3-9
Nuclear membrane of *Amoeba proteus* as shown by the electron micrograph: Note that the specialized membrane is fenestrated (L. *fenestra*, opening) and probably double. This section shows a layer of closely packed, hexagonal prisms ending in precisely centered pores: Note the honeycomb structure. The section is normal to the membrane at lower left and upper right, tangential at upper left.
(From A Scope Monograph on Cytology, The Upjohn Co., Kalamazoo, Michigan; courtesy Dr. G. D. Pappas, College of Physicians and Surgeons, New York, N. Y., and J. Biophys. Biochem. Cytol. 2 (suppl.): 431, 1956.)

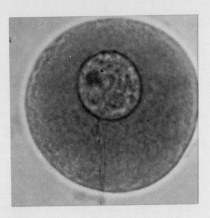

Fig. 3-10
Starfish egg as shown by light microscope. Note the nuclear membrane, nucleolus, and chromatin.

Fig. 3-11
Electron micrograph of plant nucleolus. Numerous black dots are particles of RNA.
(*Courtesy, M. A. Hayat, University of Dayton.*)

very precise in that the protein, lipid, and pigment molecules will not support photosynthesis if disrupted.

Nucleus and nuclear membrane

The nucleus (L. *nucleus*, kernel or nut) contains chromatin materials (genes) that transmit hereditary traits from one generation to another. When chromosomes are formed, they seem to be helically (spirally) coiled internally and often appear as two fine, closely twisted strands. This helical structure of chromosomes resembles the helical structure of the deoxyribonucleic acid (DNA) molecules that compose them.

The nucleus is surrounded by a thin, double-layered nuclear membrane that is composed of proteins and lipids and supplied with minute, pore-like perforations (Figs. 3-2 and 3-9).

The nucleus exerts regulatory effects on the cell processes by moving ribonucleic acid (RNA) from the nucleus to the surrounding cytoplasm. RNA acquires part of the hereditary pattern from the DNA of the chromosome, migrates through a pore of the nuclear membrane, and then acts as a template (pattern) for protein synthesis in the ribosomes.

Additional material on nuclear activity will be found in the section on mitosis and meiosis in Chapter 4.

Nucleolus

A conspicuous nucleolus (L. *nucleolus*, "little nucleus") is present within the nucleus (when the cell is not dividing) (Fig. 3-10). It is composed of RNA and a high concentration of protein. Sometimes several nucleoli are present in a cell. A nucleolus is formed by a specific region of a particular chromosome, which is called the "nucleolar organizer."

Some contributors to the knowledge of the biology of cells

Marcello Malpighi (1628-1694)

An Italian scientist and physician, studied the structure of plants and animals and may have referred to cells when he spoke of "globules" and "saccules" (1661). He is considered the "Father of Microscopic Anatomy." He discovered the existence of blood capillaries, whose existence had been predicted by William Harvey about thirty years earlier.

Antonj van Leeuwenhoek (1632-1723)

A Dutch microscopist, studied the cellular structure of plants and animals, including bacteria, protozoans, and sperm.

Robert Hooke (1635-1703)

An Englishman, microscopically studied and described many types of natural objects, and from his investigations of cork tissues of plants, he observed minute, box-like structures that he called "cells" (1665). He gave great impetus to microscopic biology.

Nehemiah Grew (1641-1712)

An English physician, microscopically studied and described (1672) the cells and tissues of plants. He and Malpighi described the microscopic structure of plants so well that few additional contributions were made for over a century.

Robert Brown (1773-1858)

A Scotchman, discovered the general occurrence of the nucleus in plant cells (1831).

Matthias Schleiden (1804-1881)

Schleiden, a German botanist, and Theodor Schwann (1810-1882), a German zoologist, through microscopic studies of many plants and animals, promulgated the "Cell Principle" (1839) in which they stated that all living organisms are composed of cells, or of cells and their products. (*Historical Pictures Service, Chicago.*)

Schleiden

René Dutrochet (1776-1847)

A French physiologist, who preceded Schleiden and Schwann with similar views on cells (1824). He stated that "growth results from both the increase in the volume of cells and from the addition of new little cells."

Edmund B. Wilson (1856-1939)

He was an American cytologist whose text, *The Cell in Development and Heredity* (1924), was an outstanding work in this field. He directed attention to the cellular study and explanation of biologic phenomena.

Review questions and topics

1 State the cell principle, including when and by whom it was formulated.
2 Discuss the various ways in which cells may be considered as units, with examples of each.
3 Discuss the detailed structures and functions of the following: cell wall, plasma membrane, cytoplasmic matrix, endoplasmic reticulum, lysosomea, mitochondria, Golgi apparatus, spherosomes, centrosome, fat droplets, cytoplasmic vacuoles, plastids, nucleus, nuclear membrane, nucleolus.
4 Contrast and give examples of chloroplasts, chromoplasts, and leukoplasts.
5 Discuss such structures and materials as plasmodesmata, ribonucleic acid (RNA), deoxyribonucleic acid (DNA), adenosine triphosphate (ATP), adenosine diphosphate (ADP), phosphorylation, and centriole.

Selected references

Brachet, J.: The living cell, Sci. Amer. 205:50-61, 1961.

De Robertis, E. D. P., Nowinski, W. W., and Saez, F. A.: Cell biology, Philadelphia, 1965, W. B. Saunders Company.

Giese, A. C.: Cell physiology, Philadelphia, 1968, W. B. Saunders Company.

Hokin, L., and Hokin, M.: The chemistry of cell membranes, Sci. Amer. 213:78-86, 1965.

Hurry, S. W.: The microstructure of cells, Boston, 1965, Houghton Mifflin Company.

Kennedy D. (editor): The living cell. Readings from Scientific American, San Francisco, 1965, W. H. Freeman and Co. Publishers.

Loewy, A. G., and Siekenitz, P.: Cell structure and function, ed. 2, New York, 1969, Holt, Rinehart & Winston, Inc.

McElroy, W. D.: Cell physiology and biochemistry, ed. 2, Englewood Cliffs, New Jersey, 1964, Prentice-Hall, Inc.

Porter, K., and Bonneville, M.: An introduction to the fine structure of cells and tissues, ed. 2, Philadelphia, 1964, Lea & Febiger.

Stern, H., and Nanney, D. L.: The biology of cells, New York, 1965, John Wiley & Sons, Inc.

Swanson, C. P.: The cell, ed. 2, Englewood Cliffs, New Jersey, 1964, Prentice-Hall, Inc.

chapter four

Cell division and cellular organizations

An increase in the number of cells of a tissue or in the number of individuals of a single-celled organism is accomplished through a process called mitosis (Gr. *mitos*, thread). This process involves the replication of the chromosomes and is usually followed by a division of the cytoplasm, called cytokinesis (Gr. *kytos*, cell; *kinesis*, movement). Mitosis is most easily studied in rapidly growing tissues, such as those of an animal embryo or the root tip of a plant. Special chemical and physical techniques are necessary to investigate the details of the process.

MITOSIS

Although mitosis is a continuous process, it is usually considered during intervals called interphase, prophase, metaphase, anaphase, and telo-phase (Fig. 4-1). These phases are followed by periods of growth.

The total length of time required for mitosis and cytokinesis varies in different cells and in different species. The rate is influenced by the type and age of the tissue involved and the temperature of the surroundings. However, the complex processes of chromosome condensation, spindle formation, chromosome splitting, and cell wall formation (in plants) occur rather quickly when the details are taken into consideration. (See Table 4-1 for examples.)

The body cells of each plant and animal species have a constant chromosome number and chromosome structure. Chromosomes occur in pairs, both members of which are identical in form. One of each pair comes from one original parent and the other from the other parent. The two chromosomes of a pair are called homologous (Gr. *homo*, same; *logos*, discourse) chromosomes. (For specific numbers of chromosomes in various species refer to Table 4-2.)

In the following discussion animal mitosis will be used to show the general phases of mitosis.

Interphase

Interphase (L. *inter*, between; Gr. *phasis*, appearance) is the period when the cell is not obviously dividing and is often referred to as a metabolic phase to emphasize the normal activities of living. In the animal cell an obvious and distinct nucleus complete with nucleoli and surrounded by a nuclear membrane is present. Chro-

Table 4-1
Approximate time requirements for phases of mitosis

	Prophase (min.)	Metaphase (min.)	Anaphase (min.)	Telophase (min.)	Total (min.)
Fruit fly (embryo)	4	1/2	1-2	1	6½-7½
Chick (embryo)	30-60	2-10	2-3	3-12	37-85
Human being	30-60	2-5	3-15	30-60	65-140

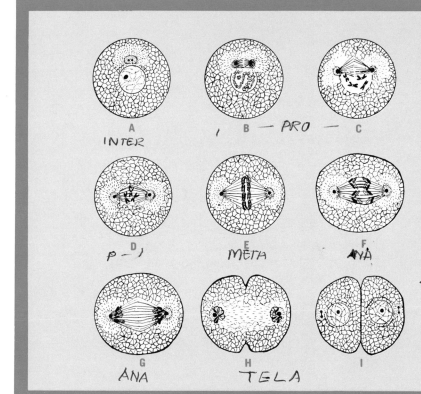

Fig. 4-1
Mitosis in animal cell. A, *Interphase.* Organized nucleus with membrane, nucleolus, and diffuse chromatin is present. B, C, and D, *Prophase stages.* B, Centrioles move apart, asters form, and chromatin threads coil; C, spindle forms between centrioles, nuclear membrane and nucleolus are gone, definite chromosomes are present; D, spindle is complete between polar centrioles, chromatids and centromeres are present in chromosomes that move toward equatorial plane. E, *Metaphase.* Chromosomes lined up by centromeres and equatorial plane. F and G, *Anaphase.* Chromatids separate and move toward opposite sides at poles. H and I, *Telophase.* Cytokinesis begins, chromosomes reach poles, nuclear reorganization commences, chromosomes become threadlike and not so obvious, cell division is complete.

mosomes as such are not visible, but a diffuse, deeply staining chromatin does exist. It can be shown by chemical analysis that the DNA content doubles just prior to prophase. A centrosome with two centrioles can be seen just outside the nucleus.

Prophase

In prophase (Gr. *pro,* before) the first indication of mitotic activity is usually the movement of the centrioles away from each other and toward opposite sides of the cell. During the movement to these poles distinct, raylike fibers called asters surround the centrioles.

Fine, threadlike fibers connect the two migrating asters and form the mitotic spindle, which progressively increases in size, being broadest at the center of the cell—the equatorial plane—and tapering toward either end. The thin nuclear membrane begins to disappear, and the spindle occupies the position of the original nucleus. The nucleoli also are no longer visible.

The nuclear materials condense into a darkly staining, coiled network of chromatin threads. The chromatin threads, also called chromonemata,

thicken and shorten into distinct chromosomes. At a definite point in each chromosome is a clear area called the centromere (Gr. *kentron,* center) or kinetochore (Gr. *kinein,* to move; *choros,* place) that seems to assist in the orientation and division of the chromosomes.

Metaphase

Metaphase (Gr. *meta,* after) begins when the chromosomes reach the equator of the spindle. Each chromosome is divided lengthwise into two chromatids and is attached to a spindle fiber by means of its centromere. The centromeres now divide and separation of the chromatids begins.

Anaphase

During anaphase (Gr. *ana,* up) the chromatids are completely separated, and the newly formed daughter chromosomes move to opposite poles of the spindle. At this time the spindle seems to elongate, and often the cell itself becomes longer, so that the two groups of daughter chromosomes eventually lie far apart at opposite ends of the cell (Fig. 4-2).

29

Table 4-2

Number of chromosomes in various organisms

Animal	Diploid number of chromosomes per body (somatic) cell
Horse roundworm (Ascaris)	1 pair
Snail (Heiix)	24 pairs
Crayfish (Cambarus virulus)	100 pairs
Fruit fly (Drosophila melanogaster)	4 pairs
Frog (Rana)	13 pairs
Pigeon (Columba)	8 pairs
Rhesus monkey (Macaca mulatto)	21 pairs
Man (Homo sapiens)	23 pairs
Plant	
Green alga (Spirogyra)	12 pairs
Pear moss (Sphagnum)	20 pairs
Fern (Dryopteris cristata)	164 pairs
Pine tree (Pinus)	12 pairs
Pea (Pisum)	7 pairs
Wheat (Triticum vulgare)	21 pairs

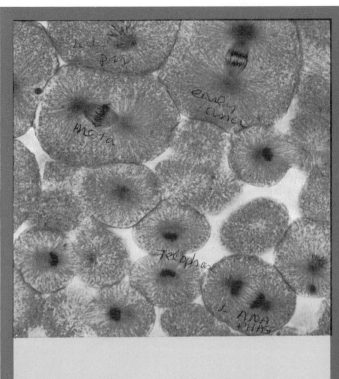

Fig. 4-2
Photograph of a section of the embryo of whitefish showing cells in various stages of mitosis. Note particularly the chromosomes, spindle, and asters.
(*Courtesy General Biological Supply House, Inc., Chicago, Illinois.*)

Telophase

In telophase (Gr. *telos,* end) the chromosomes elongate, and the coils of the chromonemata loosen and gradually assume a threadlike nature. The spindle and asters disappear, but the centrosomes persist, one with each daughter nucleus. There may be one centriole in each centrosome, or the centriole may have divided to form two per centrosome. A nuclear membrane re-forms around the nucleus. Nucleoli reappear, and each daughter nucleus is again in the interphase period.

In animal cells cytokinesis usually begins as a furrow, or indentation, in the plasma membrane that deepens gradually to divide the cell. Cytokinesis usually begins during telophase, but in some cases, it may take place much later.

Plant mitosis

Any comparison of plant and animal mitosis must first consider the differences in the cells. Plant cells do not have centrioles and therefore do not form asters. They do form spindles, however. Other details of nuclear division are essentially the same as in the animal form already discussed (Fig. 4-3).

As the nuclei reorganize, the parts of the spindle near the new nuclei tend to disappear, and the portion of the spindle at the equatorial plane widens until it extends nearly across the cytoplasm. As the equatorial part of the spindle widens, thickenings appear on the spindle fibers of this region. These thickenings enlarge and finally fuse to form a continuous cell plate across the cell, thus sepa-

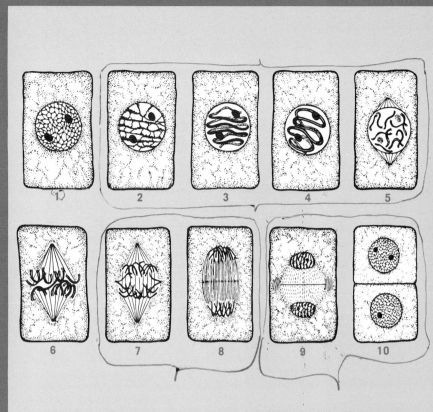

Fig. 4-3
Mitosis in plants: 1, Interphase, before actual division begins, with a chromatin network and nucleoli. Prophase stages: 2 to 5, network disappears and chromatic strands thicken and shorten, forming chromosomes consisting of two identical chromatids. In stage 5 the mitotic spindle is forming and the nucleus is disappearing. Metaphase stage: 6, chromosomes line up on the equatorial plane of the spindle. Anaphase stage: 7, each chromosome has divided lengthwise (chromatids separate) into two identical halves. New (daughter) chromosomes move along the spindle to opposite poles; 8, chromosomes at the poles; the cell wall begins to form at minute swellings on the spindle fibers at the equatorial plane. Telophase stages: 9, chromosomes become threadlike; the nuclear membrane and nucleoli reappear; the cell wall continues to develop; the spindle disappears; 10, the division of the cell is complete; two cells similar to the original will grow to normal size.
(Courtesy General Biological Supply House, Inc., Chicago, Illinois.)

rating the cytoplasm into two parts. Upon either side of the cell plate are deposited the new cell walls that separate the newly formed cells. The formation of new cell walls completes cytokinesis, and thus two new cells are formed from the original one.

In some plant cells cytokinesis is accomplished by a furrowing of the parent cell wall as in animal cells.

MEIOTIC CELL DIVISION

A special type of cell division occurs in the life cycle of organisms that reproduce by the fusion of egg and sperm. This process is called meiosis (Gr. *meioun*, to reduce) and is accompanied by a reduction in the number of chromosomes in the daughter cells.

The time at which meiosis occurs in the life cycle varies with different organisms, but is constant for each particular species. There are three types of meiosis, depending on the place in the life cycle when it occurs. It may occur (1) during the formation of spores, as is common in plants; (2) during the formation of gametes, which is common in animals (and in a few lower plants); or (3) immediately after the fertilization of an egg by a sperm (at the beginning of the cleavage of the egg), which occurs in certain lower plants.

A cell whose nucleus contains the number of pairs of chromosomes typical of a particular species is referred to as a 2N or diploid cell. After the reduction division the resulting cells are called N or haploid (monoploid) cells. 2N and diploid refer to cells with pairs of chromosomes. N and haploid refer to cells with single members of a pair of chromosomes.

Except for the kind of cell resulting from the process of meiosis, the phenomenon is similar in both sexes, following the same phases or stages in each. These phases of meiosis follow a sequence similar to those in mitosis.

Reduction in chromosome number is really two

31

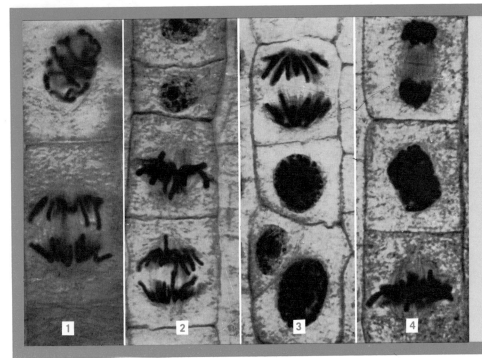

Fig. 4-4
Cell division in the root of an onion. 1, Prophase (top) and anaphase; 2, metaphase (middle) and anaphase (bottom); 3, late anaphase (top); 4, telophase (top) and metaphase (bottom).
(*Courtesy Carolina Biological Supply Co., Elon College, North Carolina.*)

nuclear divisions accompanied by only one division of chromosomes, which results eventually in the formation of four cells whose nuclei contain the haploid number of chromosomes. The phases (prophase, metaphase, anaphase, and telophase) occur twice in meiotic division. The two nuclear divisions may be designated as meiotic divisions I and II and the phases as Prophase I, Metaphase I, Anaphase I, Telophase I; Prophase II, Metaphase II, Anaphase II, and Telophase II. These may be summarized briefly as follows.

Prophase I

The threads of chromatin in the nucleus are present in diploid number and form chromosomes, each with two chromatids. These chromosomes are relatively long and thin. They may appear to be single rather than double (Fig. 4-5).

During this stage the chromosomes move, bringing together the pairs of homologous chromosomes (homologues). The active pairing called synapsis (Gr. *synapsis,* union) begins and eventually unites, but does not fuse, them along their entire length. They may even twist around each other. After synapsis is complete it appears as if the chromosomes are duplicated. This duplication is caused by longitudinal splitting of each chromosome as the chromatids become distinct. The shape of the chromosome is determined by a constriction at a point where the two "arms" of a chromosome meet. Within the constriction is a centromere that seems to be related functionally to chromosome movements.

The paired chromosomes tend to separate from each other as the attraction forces of synapsis decrease. However, complete separation of homologues does not occur at this time. At one or several points along their length they may remain in contact by means of chiasmata (Gr. *chiasma,* cross). Each chiasma results from an exchange (crossing over) of chromatids between two homologues, which may be of great significance in heredity. Each chromosome consists of two chromatids, and thus the chromosome pairs consist of four chromatids, which might be called, collectively, a tetrad (Gr. *tetra-,* four).

When only one chiasma is formed, the paired chromosomes may appear as a cross; when two

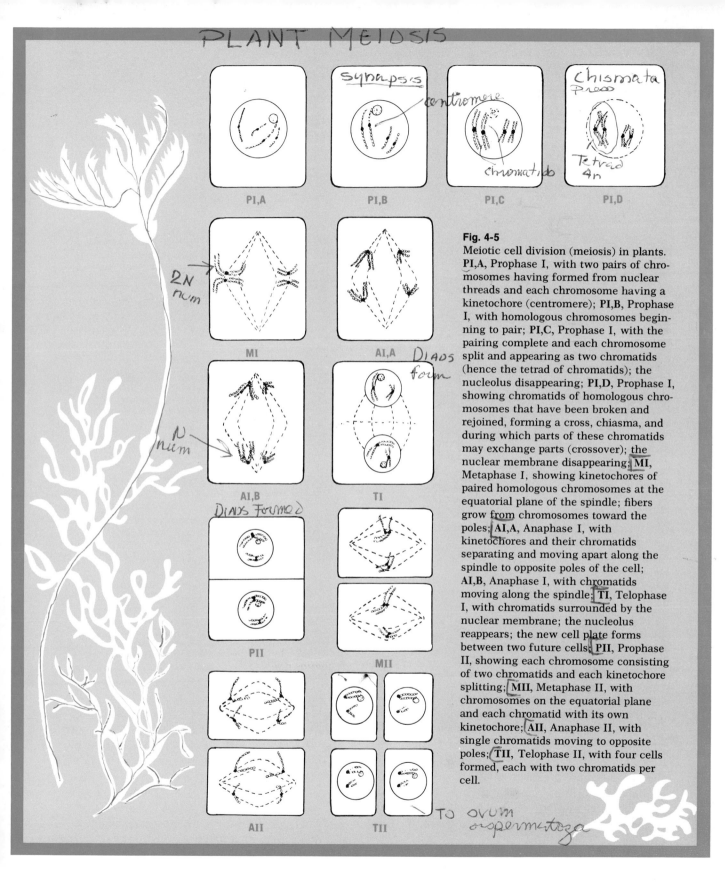

PLANT MEIOSIS

PI,A   PI,B synapsis centromere   PI,C chromatids   PI,D Chismata Press Tetrad 4n

MI   2N num

AI,A Diaos form

AI,B   N num   TI

Diaos Formed

PII   MII

AII   TII   To ovum or spermatozoa

Fig. 4-5
Meiotic cell division (meiosis) in plants.
PI,A, Prophase I, with two pairs of chromosomes having formed from nuclear threads and each chromosome having a kinetochore (centromere); **PI,B**, Prophase I, with homologous chromosomes beginning to pair; **PI,C**, Prophase I, with the pairing complete and each chromosome split and appearing as two chromatids (hence the tetrad of chromatids); the nucleolus disappearing; **PI,D**, Prophase I, showing chromatids of homologous chromosomes that have been broken and rejoined, forming a cross, chiasma, and during which parts of these chromatids may exchange parts (crossover); the nuclear membrane disappearing; **MI**, Metaphase I, showing kinetochores of paired homologous chromosomes at the equatorial plane of the spindle; fibers grow from chromosomes toward the poles; **AI,A**, Anaphase I, with kinetochores and their chromatids separating and moving apart along the spindle to opposite poles of the cell; **AI,B**, Anaphase I, with chromatids moving along the spindle; **TI**, Telophase I, with chromatids surrounded by the nuclear membrane; the nucleolus reappears; the new cell plate forms between two future cells; **PII**, Prophase II, showing each chromosome consisting of two chromatids and each kinetochore splitting; **MII**, Metaphase II, with chromosomes on the equatorial plane and each chromatid with its own kinetochore; **AII**, Anaphase II, with single chromatids moving to opposite poles; **TII**, Telophase II, with four cells formed, each with two chromatids per cell.

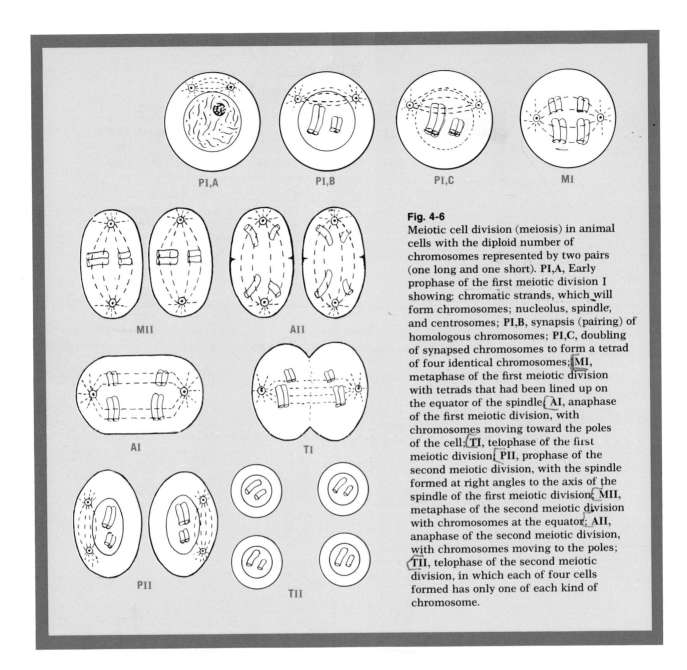

Fig. 4-6
Meiotic cell division (meiosis) in animal cells with the diploid number of chromosomes represented by two pairs (one long and one short). PI,A, Early prophase of the first meiotic division I showing: chromatic strands, which will form chromosomes; nucleolus, spindle, and centrosomes; PI,B, synapsis (pairing) of homologous chromosomes; PI,C, doubling of synapsed chromosomes to form a tetrad of four identical chromosomes; MI, metaphase of the first meiotic division with tetrads that had been lined up on the equator of the spindle; AI, anaphase of the first meiotic division, with chromosomes moving toward the poles of the cell; TI, telophase of the first meiotic division; PII, prophase of the second meiotic division, with the spindle formed at right angles to the axis of the spindle of the first meiotic division; MII, metaphase of the second meiotic division with chromosomes at the equator; AII, anaphase of the second meiotic division, with chromosomes moving to the poles; TII, telophase of the second meiotic division, in which each of four cells formed has only one of each kind of chromosome.

chiasmata are formed, the chromosome is usually ring shaped; when three or more are formed, the homologues appear as loops. In general, the longer the chromosome, the more chiasmata.

The chromosomes tend to shorten during Prophase I. The loss of the nuclear membrane and the formation of the spindle terminate Prophase I of meiosis.

Metaphase I

The tetrads arrange themselves on the spindle, each member located so that the centromere of each lies on either side of and equidistant from the equatorial plane.

Anaphase I

The paired double chromosomes separate from one another and move along the spindle toward

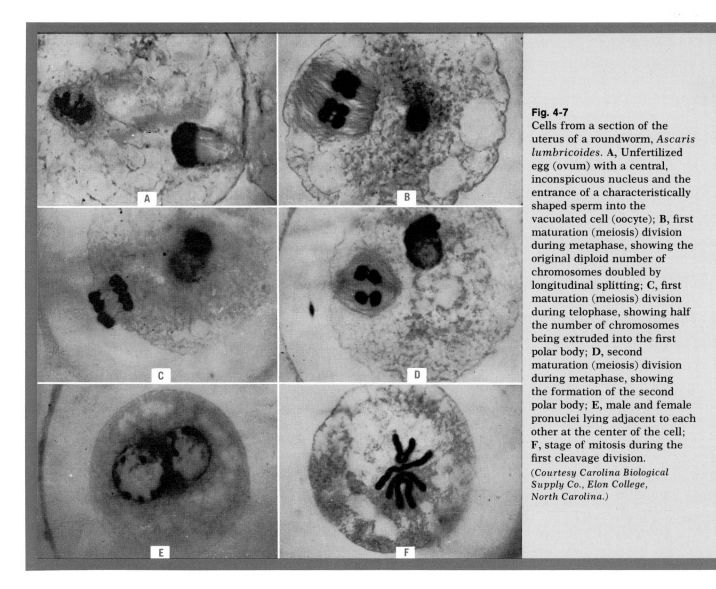

Fig. 4-7
Cells from a section of the uterus of a roundworm, *Ascaris lumbricoides.* **A,** Unfertilized egg (ovum) with a central, inconspicuous nucleus and the entrance of a characteristically shaped sperm into the vacuolated cell (oocyte); **B,** first maturation (meiosis) division during metaphase, showing the original diploid number of chromosomes doubled by longitudinal splitting; **C,** first maturation (meiosis) division during telophase, showing half the number of chromosomes being extruded into the first polar body; **D,** second maturation (meiosis) division during metaphase, showing the formation of the second polar body; **E,** male and female pronuclei lying adjacent to each other at the center of the cell; **F,** stage of mitosis during the first cleavage division.
(*Courtesy Carolina Biological Supply Co., Elon College, North Carolina.*)

opposite poles. The centromeres are undivided as they and their chromosomes move along the spindle. Hence, a haploid number of chromosomes will be located at each pole. Each chromosome consists of two chromatids, united only at their centromeres. The separation of homologues is called reduction division (meiosis).

Telophase I

The chromosomes uncoil, the nucleus forms, and the meiotic cell is divided by a newly formed cell membrane or wall.

Prophase II

After Telophase I there may be a long or short interphase, or none at all, depending on the species. The chromosomes in each of the two haploid cells enter the second meiotic division (equational division). Regardless of the presence or absence of an interphase, DNA content does not double, the chromosomes are unchanged, no reproduction of chromosomes occurs during interphase, and each centromere now divides. A new spindle forms in each of the two cells at right angles to the axis of the spindle of meiosis I. The rest of the process is essentially the same as mitosis.

Metaphase II

The chromosomes line up on the equatorial plane of the second spindle.

Anaphase II

The chromosomes divide, and the resulting haploid set of single chromosomes (chromatids) moves along the spindle toward opposite poles.

Telophase II

The chromosomes become longer and thinner; the nuclear membrane forms and the cytoplasm divides, forming two cells (with their cell membranes or walls). Hence, the original cell with its diploid number of chromosomes results, after two meiotic cell divisions, in the formation of four cells, each with only one of each kind of chromosome. Thus, the resulting cells are produced with the reduced number of chromosomes. Since the chromatids in each of the four cells formed may have been modified as a result of crossing over during synapsis, each of the four cells may be genetically different concerning certain traits.

Meiotic division occurs in plants during the formation of special reproductive cells, such as pollen grains of higher plants or spores in mosses and ferns. In general the process of meiosis is similar in fungi, flowering plants, worms, insects, and man (Figs. 4-6 and 4-7).

Gametogenesis

The stages of meiosis are similar in both male and female sexes, but there are differences, particularly in the unequal division of the cytoplasm when eggs are formed and in the equal distribution of cytoplasm when sperms are produced. This phenomenon can be observed in animals by studying prepared slides of the female reproductive system of a roundworm, *Ascaris,* in which gametes are produced by gametogenesis (Gr. *gamos,* marriage; *genesis,* origin). Specifically, the formation of sperms is called spermatogenesis, and the formation of eggs (ova) is called oogenesis.

The similar stages of spermatogenesis and oogenesis may be summarized briefly as follows.

In spermatogenesis there are numbers of undifferentiated germ cells (primordial cells) present in the male reproductive organs that produce numbers of cells called spermatogonia through the process of mitosis. A similar phenomenon occurs in the female reproductive organs except that the undifferentiated germ cells form oogonia (Gr. *oon,* egg) by the process of mitosis.

Each spermatogonium produces a primary spermatocyte in which the pairs of homologous chromosomes occur. Each oogonium produces a primary oocyte.

Each primary spermatocyte produces two secondary spermatocytes through the first meiotic division. In a similar manner a primary oocyte gives rise to a secondary oocyte and a smaller first polar body that receives less cytoplasm than the secondary oocyte, but does receive one member of each pair of chromosomes of the primary oocyte. The first polar body may divide to form two additional and equal polar bodies.

Each secondary spermatocyte produces two spermatids through the second meiotic division, in which there is a separation of the chromatids of each chromosome. The secondary oocyte produces a large ootid and a small second polar body. Each spermatid forms a mature sperm, whereas the ootid forms a mature egg (ovum).

Thus there are formed four sperms from each primary spermatocyte, but only one egg (and three polar bodies) from the primary oocyte. It will be noted that there has been a reduction of the chromosomes during the first and second meiotic divisions, thus resulting in haploid (N) gametes being formed from diploid (2N) spermatogonia and oogonia cells. Also, there is little if any food placed in the nonfertilizable polar bodies, which are eventually reabsorbed and hence their genic materials are lost. Another important fact is that a crossing over of genic materials between homologous chromosomes may occur during synapsis, with the possible change in the combination of traits that a particular offspring may inherit.

CELLULAR ORGANIZATIONS

An increase in the number of cells of an organism is usually accompanied by the organization of different types of tissues, that is, groups of similar cells with a common function. These tissues in turn may become further specialized as compon-

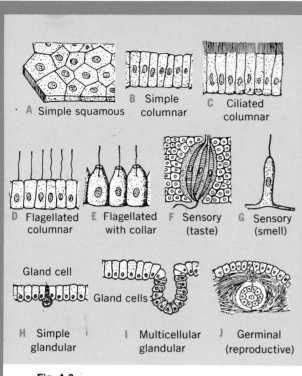

Fig. 4-8
Epithelial tissues. **A,** Simple squamous; **B,** simple columnar; **C,** ciliated columnar; **D,** flagellated columnar; **E,** flagellated with collar; **F,** sensory (taste), in which cells are connected with the nerve below and have sensitive hairs above; **G,** sensory (smell) with a sensitive hair; **H,** simple glandular (secretory); **I,** multicellular glandular (secretory); **J,** germinal (reproductive), in which sex cells originate from epithelial cells.

ents of an organ, which is then part of an organ system. Thus a simple cell may be part of an epithelial tissue lining the heart, an organ composed of muscular, epithelial, connective, nervous, and vascular tissues. This heart, together with the veins and arteries forms the circulatory system.

The preparation of sections of tissues and organs for microscopic study is known as histologic technique and requires much time and skill. In brief, this technique follows such steps as (1) rapid removal, killing, and fixation of the specimen; (2) removal of water by the use of alcohol; (3) treatment with some type of a clearing agent; (4) infiltration of the specimen with liquid paraffin (or collodion) and then embedding it in a block of paraffin; (5) cutting of sections of the hardened paraffin block that contains the embedded tissue, usually by a precision cutting machine called a microtome (Gr. *mikros*, small; *tome,* cut); (6) affixing the sections to a slide and staining. (There are many variations to the procedure as outlined.)

Animal tissues

There are five types of animal tissues—epithelial, connective, blood, vascular, and nervous tissues.

Epithelial tissue

Epithelial (Gr. *epi,* upon; *thele,* nipple) (Fig. 4-8) consists of a layer of flat, cuboidal, or column-shaped cells, depending upon the type of epithelium. There is a minimum of intercellular space between cells, but epithelial cells maintain a continuity between adjacent ones through delicate strands called protoplasmic bridges. Epithelial tissue forms the outer protective surface of the body, and it lines many internal cavities, such as the digestive and urinary systems. Epithelial tissue absorbs substances from the exterior and eliminates substances to the outside. It may produce special structures (glands) for secreting purposes. It may also become modified to produce sex cells within the testes and ovaries; it also forms parts of the sense organs.

Squamous (pavement) cells

Squamous (L. *squama,* scale) cells are thin and flat and are arranged like stones in pavement. Squamous tissue consists of three types of cells. (1) Simple squamous tissue is made up of single layers of cells that line internal cavities. Endothelium (Gr. *endo,* within), which lines the heart, blood vessels, and lymph vessels and mesothelium (Gr. *mesos,* middle), which lines the abdominal, lung, and heart cavities are made up of simple squamous cells. (2) Stratified squamous tissue is composed of more than one layer of cells. Stratified tissue lines the mouth and the esophagus, and it forms the outer layers of skin of vertebrates.

Columnar cells

Columnar (L. *columna,* pillar) cells are column-shaped, and often have tapering ends next to the

37

underlying tissues. In simple columnar tissue there is one layer of cells. This tissue lines the stomach and intestine of most higher animals, where it has a secretory function.

Ciliated, flagellated, and collared cells

When the free surface of the epithelium bears hairlike cilia, whiplike flagella, or "collars" it is named accordingly. Ciliated epithelium is present in gills of clams, the roof of the frog's mouth, and the lining of the air passages of vertebrates, where the cilia move materials from the surface. Flagellated epithelium is found in the inner layer of *Hydra*. Collared epithelium is found in certain canals of sponges.

Sensory epithelium (neuroepithelium)

Certain specialized column-shaped cells help to form sense organs (receptors) that are affected by different stimuli. Sensory epithelium is found in the retina of the eye, the lining of the nose, the taste buds of the tongue, and the auditory cells of the ear.

Secretory (glandular) cells

Certain glands are unicellular cells, while others are complex and multicellular. Usually, modified columnar epithelial cells produce specific secretions, such as the unicellular goblet cells of the intestine and the multicelluar salivary glands.

Germinal (reproductive) cells

Germinal epithelial cells are modified for the formation of sex cells in the reproductive organs.

Connective (supportive) tissue

Connective (L. *com*, together; *nectere*, to bind) tissue (Figs. 4-9 and 4-10) is common in most parts of the body. Fibers are usually present. There are large intercellular spaces that may contain non-living intercellular substances, such as fibers, cartilage, and bone (depending upon the type of connective tissue). Connective tissue binds the body parts together. Some types form semirigid or rigid structures for the protection and attachment of other tissues and organs.

White fibrous (collagenous) tissue

The white collagenous (Gr. *kolla*, glue; *gen*, to produce) tissue is composed of long, white or color-

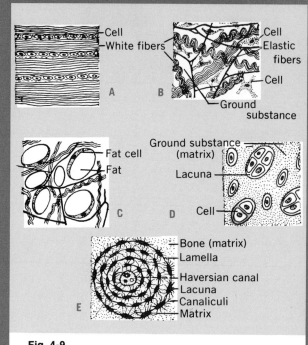

Fig. 4-9
Connective tissues. **A,** White fibrous (from a longitudinal section of a tendon); **B,** areolar (from beneath the skin); **C,** adipose (fat); **D,** hyalin cartilage (from the end of a bone); **E,** bone (osseous) (cross section of a haversian system).

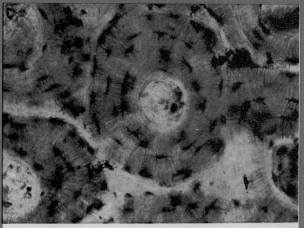

Fig. 4-10
Human bone (cross section): Note the arrangement of lacunae in concentric lamellae around the haversian canals. Observe the threadlike canaliculi associated with the lacunae. Bone cells are not clearly visible in the lacunae of such a ground section of bone.
(*Courtesy General Biological Supply House, Inc., Chicago, Illinois.*)

less, wavy, rather inelastic fibers. This tissue is present in both ligaments, which attach bones to bones, and in tendons, which attach muscles to bones. In tendons there are many white fibers that parallel each other with rows of nuclei between the bundles of fibers. White fibers are called collagenous because they produce gelatinous material called collagen when boiled.

Areolar tissue

Areolar (L. *areola*, dim. of *area*, space) tissue is composed of numerous white collagenous fibers and a few yellow elastic fibers, which branch and form loose networks. This tissue is one of the most widely distributed types of connective tissue. It is found in the fascia (L. *fascia*, band) that surrounds and connects muscles.

Adipose tissue

Adipose (L. *adipis*, fat) tissue is a special type of areolar tissue in which the intercellular spaces are filled with fat contained in hollow "signet-ring" cells (the nucleus is pushed to one side). This tissue is common beneath the skin and around certain organs.

Hyalin cartilage

Hyalin (Gr. *hyalos*, glassy or clear) tissue consists of a clear, flexible ground substance, call matrix (L. *mater*, mother). There are scattered lacunae (L. *lacuna*, cavity) in which there are one or more rounded cartilage cells that secrete matrix. This tissue is found in the ends of long bones, the ribs, and nose.

Bone

Bone consists of matrix hardened with calcium salts. Units of bone construction known as the haversian systems (after Havers, and English anatomist) consist of (1) a central canal with an artery, vein, and nerve; (2) lamellae (L. *lamella*, small plate), made of layers of bony plates that form rough-walled canals that are arranged concentrically in circles or ovals around the central canal; (3) lacunae or enlarged spaces associated with lamellae, which contain the irregularly shaped bone cells; (4) tiny, wavy, canal-like canaliculai (L. *canaliculus*, small channel) that radiate from lacunae and connect them with each other and with the central canal; and (5) hard matrix (bone) that occupies spaces not previously described and which is secreted by the bone cells. Bones support, protect, assist in locomotion, serve for the attachment of muscles and other tissues, and assist in hearing (ear bones).

Blood (vascular) tissue

Blood or vascular (L. *vasculum*, small vessel) tissue (Fig. 4-11) is sometimes considered as a type of connective tissue. It consists of free cells suspended in an intercellular fluid (plasma). The colorless plasma contains three general types of blood corpuscles. (1) Erythrocytes (Gr. *erythros*, red) are the red blood corpuscles. When mature they are biconcave disks. They are nonnucleated in mammals but nucleated in frogs. The erythrocytes are about 7.6μ in diameter, and there are approximately 5,000,000 per cubic millimeter. Erythrocytes contain an iron-containing protein, hemoglobin, (Gr. *haima*, blood; *globos*, sphere) for carrying oxygen. (2) Leukocytes (Gr. *leukos*, white) are the white corpuscles. The leukocytes vary in size, shape, size of the nucleus, and the kinds of

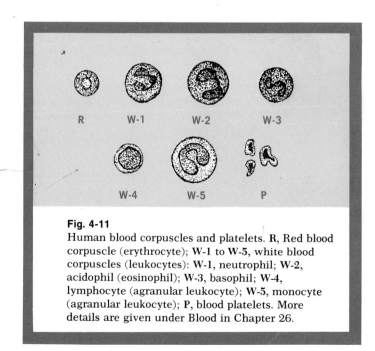

Fig. 4-11
Human blood corpuscles and platelets. **R**, Red blood corpuscle (erythrocyte); **W-1** to **W-5**, white blood corpuscles (leukocytes): **W-1**, neutrophil; **W-2**, acidophil (eosinophil); **W-3**, basophil; **W-4**, lymphocyte (agranular leukocyte); **W-5**, monocyte (agranular leukocyte); **P**, blood platelets. More details are given under Blood in Chapter 26.

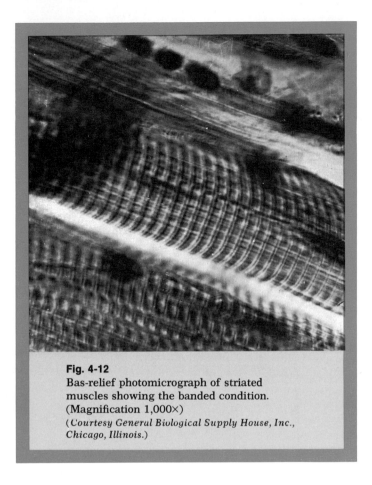

Fig. 4-12
Bas-relief photomicrograph of striated muscles showing the banded condition. (Magnification 1,000×)
(*Courtesy General Biological Supply House, Inc., Chicago, Illinois.*)

Muscular (contractile) tissue

Muscle (L. *musculum*, muscle) fibers or cells (Fig. 4-12) are usually elongated and have one or several nuclei per cell, depending upon the type of muscle. There are special internal contractile fibers (myofibrils) within the cells that cause the muscle to contract when stimulated. Muscle cells move various body parts or the body as a whole.

Skeletal (striated) tissue

Skeletal tissue is associated with the skeleton and is composed of cylinder-shaped cells, each with several nuclei near the surface of the cell. This tissue makes up the voluntary muscles (under the control of the will). The thin internal myofibrils (Gr. *myos*, muscle) have alternate dark and light bands, thus giving the muscle cell its striated nature. This tissue is found in the muscles of arms, legs, and body walls.

Visceral (unstriated or smooth) tissue

Visceral muscle occurs in the walls of the viscera (internal organs). The cells are long and spindle shaped, and each has one central nucleus. These muscles are involuntary (not under the control of the will). The myofibrils are not striated.

Cardiac tissue

Cardiac muscle occurs in the walls of the heart. It is indistinctly striated and has several interior nuclei per cell. Cardiac muscle is involuntary. The internal contractile fibers have alternate dark and light bands that are not as distinct as those in the skeletal muscles. The cells of cardiac tissue are often branched.

Nervous

Nervous (L. *nervus*, fiber) tissue (Figs. 4-13 and 4-14) is highly specialized animal tissue consisting of neurons (nerve cells) that may be held in position by a special type of tissue called neuroglia (Gr. *neuron*, nerve; *glia*, glue). Each typical neuron has a nucleus and processes arising from the cell body. These processes are dendrites (Gr. *dendron*, tree or branched) and axons (L. *axon*, axis). The dendrites are branched and carry impulses to the neurons. There are different types, which vary in length from 1 mm. to 1 meter. The axons carry impulses away from the neuron, and

granules in the cytoplasm. There are from 7 to 14μ in diameter, and there are approximately 8,500 per cubic millimeter. Some types destroy foreign materials by phagocytosis. (3) Blood platelets are small, irregular, nonnucleated masses of protoplasms in mammals. They are comparable to the nucleated spindle cells of the frog. They are about 3μ in diameter and there are approximately 300,000 per cubic millimeter. The platelets assist in the formation of blood clots.

The vascular tissue carries foods, wastes, oxygen, carbon dioxide, and hormones to all parts of the body. The blood also equalizes temperature and maintains the acid-alkaline balance between the various parts of the body. It carries antibodies (chemical substances), which assist in defending against certain diseases. The blood destroys bacteria and other foreign particles by phagocytosis. Modified plasma outside the circulatory system is called lymph (L. *lympha*, liquid).

40

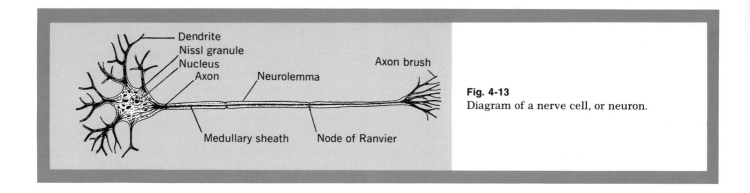

Fig. 4-13
Diagram of a nerve cell, or neuron.

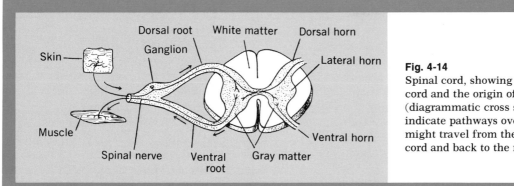

Fig. 4-14
Spinal cord, showing pathways through the cord and the origin of spinal nerves (diagrammatic cross section). Arrows indicate pathways over which impulses might travel from the skin through the cord and back to the muscle.

are unbranched. Nissl's granules in the cytoplasm are probably nutritive, as they tend to disappear after prolonged neuron activity. Nervous tissues receive, interpret, and redirect nerve impulses.

The junction of one neuron with the next, a simple contact between two plasma membranes, is called a synapse (Gr. *synapsis*, union). Where nerve impulses are temporarily delayed in their transmission (no longer than 0.006 sec.). Synapse acts as a polarizing zone, since the impulse will travel across it in only one direction; a chemical activating substance, acetylcholine, has been discovered in synapses.

Within neurons are threadlike neurofibrils that extend into dendrons and axons, apparently for impulse conduction. Neurons are highly specialized, long lived, and cannot divide to replace themselves. The may die but their place is taken by some of enormous excesses that are present in certain parts of the nervous system. They use only a small fraction of those present, so do not miss the loss of an occasional one. Neurons cannot tolerate the absence of oxygen, even for a short time.

Neurons may be classed according to the number of processes as follows: (1) unipolar, which have one process that may be divided into two as found in the dorsal ganglia of the spinal cord (Fig. 4-14); (2) bipolar, which have two processes, one a dendron and the other an axon, as found in the ganglia of the inner ear; (3) multipolar, which have several dendrons and an axon, as found in motor cells of the ventral horn of the spinal cord.

The brain is composed of the cerebrum (L. *cerebrum*, brain) and the cerebellum. The cerebrum has an outer cortex region (gray matter) that contains various sizes of pyramid-shaped neurons. The cerebellum (L. diminutive of *cerebrum*, brain), consists of an outer cortex of gray matter with an outer layer of scattered stellate (L. *stella*, star) cells and fibers; a middle layer of treelike (Purkinje) neurons; and an inner granular layer of small neurons.

The spinal cord (Fig. 4-14) is composed of a central, H-shaped column of gray matter (seen best in cross section) surrounded by white matter. On each side of "H" are two so-called horns: the ven-

41

tral horn, from which originates the ventral roots of the spinal nerves, and the dorsal horn, from which originates the dorsal roots of the spinal nerves. Motor neurons are found in the larger ventral (anterior) horns. White matter consists of nerve fibers (sensory and motor) that run lengthwise of the spinal cord to transmit nerve impulses up and down.

Organs and organ systems

Because the bodies of single-celled animals are not organized into tissues and organs, all functions necessary for life are carried on by a single cell. As animals become more complex, both structurally and functionally, tissues eventually become present and finally organ systems develop, with division of labor for the principal life functions (Fig. 4-15). The organ systems and their general functions in man are briefly given as follows.

1. The integumentary system, which covers and protects the body.

2. The skeletal system, which supports, protects, and assists in movement and locomotion from place to place.

3. The muscular system, which together with the skeletal system produces movement and locomotion.

4. The digestive system, which takes in foods, changes them chemically into smaller molecules, and absorbs them into the circulatory system.

5. The circulatory system, which transports foods, wastes, oxygen, carbon dioxide, endocrine secretions, and other materials throughout the body.

6. The respiratory system, which permits the exchange of oxygen and carbon dioxide.

7. The excretory system, which excretes the waste products of metabolism.

8. The nervous system, which receives, conducts and redirects nerve impulses throughout the body, thus assisting in the coordination of the activities of the other systems.

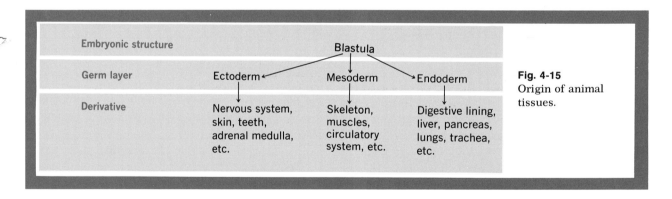

Fig. 4-15
Origin of animal tissues.

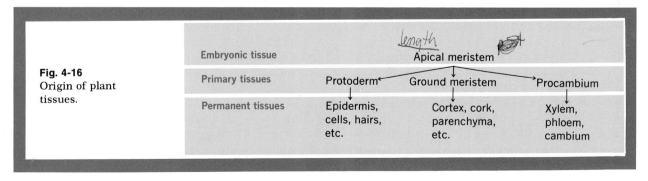

Fig. 4-16
Origin of plant tissues.

9. The endocrine system, which through its ductless gland secretions also assists in coordinating the activities of other systems.
10. The sensory system, which receives stimuli in various parts of the body.
11. The reproductive system, which provides for the continuation of the individual, and hence the race.

■ Plant tissues

The formation of tissues in a higher plant begins in growing regions called meristems (Gr. *merizein*, to divide), where active mitosis takes place. The resulting cells at first form temporary tissues, which may then become modified into the characteristic permanent tissues (Fig. 4-16).

Temporary tissues

Meristematic

Meristematic tissue cells are small, thin walled, usually cubical, and may divide frequently. They are the basis of the growing regions of the plant and lead to the formation of the other temporary tissues.

grows 2 ways
length- Apical (Apex)
Dia (lateral)

Protoderm

The protoderm (Gr. *protos*, before; *derma*, skin) makes up the outermost layer of cells. It is one cell thick and develops into the epidermis.

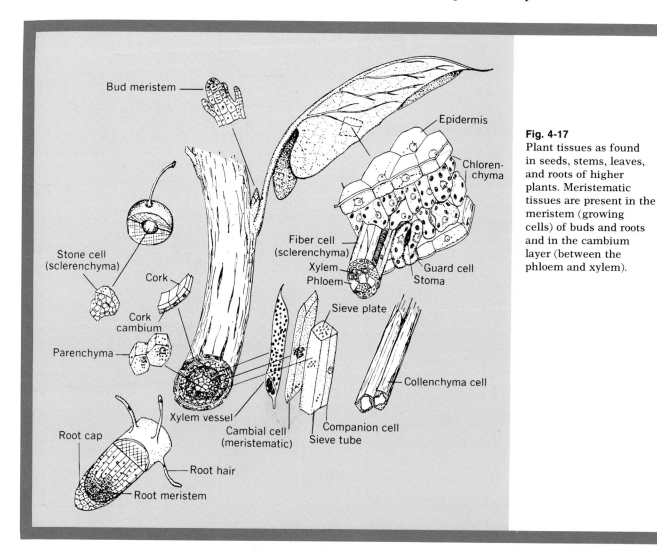

Fig. 4-17
Plant tissues as found in seeds, stems, leaves, and roots of higher plants. Meristematic tissues are present in the meristem (growing cells) of buds and roots and in the cambium layer (between the phloem and xylem).

Bud meristem
Epidermis
Chloren-chyma
Stone cell (sclerenchyma)
Fiber cell (sclerenchyma)
Cork
Xylem
Phloem
Guard cell
Stoma
Cork cambium
Sieve plate
Parenchyma
Collenchyma cell
Xylem vessel
Cambial cell (meristematic)
Companion cell
Sieve tube
Root cap
Root hair
Root meristem

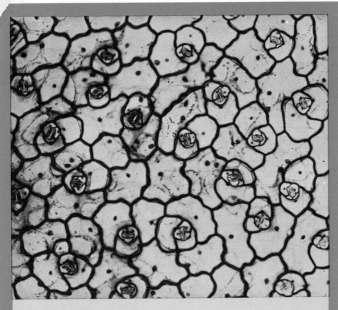

Fig. 4-18
Surface view of the epidermis of a plant, *Sedum*,
showing stomata with surrounding guard cells.
*(Courtesy General Biological Supply House, Inc.,
Chicago, Illinois.)*

Ground meristem

Ground meristem forms the greater portion of
the developing meristems and may persist for some
time as parenchyma or pith cells.

Procambium

The procambium (L. *cambiare*, to exchange)
cells appear as small patches in the developing
stem or root and retain the capacity for cell division. They give rise to the permanent primary vascular tissues. ~~meristem tissue~~

Permanent tissues

Permanent tissues usually do not change into
other kinds of tissues, but in most cases they retain
their structural and functional characteristics
throughout life.

Simple tissues

Simple tissues are composed mostly of one kind
of cell and all are constructed similarly.

Epidermis. Epidermis (Gr. *epi*, upon; *derma*,

skin) (Figs. 4-17 and 4-18) is usually one cell thick
and may contain such pigments as blue, red, or
purple. The outer cell walls are coated by a waterproof, waxy cutin; however, guard cells that control gas movements through epidermal pores (Fig.
4-18) called stomata (Gr. *stoma*, opening) possess
chlorophyll in bodies called chloroplasts.

Epidermal tissues are found on the surface of
leaves, flower parts, and younger stems and roots.
They function as a protection against mechanical
injuries, the effects of parasites, and assist in conserving internal moisture.

Parenchyma. Parenchymal (Gr. *para*, beside;
en, in; *chein*, to pour) cells are usually ovoid or
spherical, but sometimes cylindroid. There is a
large central vacuole, and the cell walls are thin.
There are usually numerous intercellular (L. *inter*,
between) spaces. The protoplasm may remain alive
for a long time. Parenchyma cells may form tissues
by themselves, but they may be mixed with other
types of cells in complex tissues.

Parenchymal cells are very abundant in higher
plants. It is found in the pith of roots, stems, and
fruits. It stores water and foods in potato tubers,
tomato pulp, and watermelon. Green, chlorophyll-
bearing parenchyma cells of leaves are called
chlorenchyma (Gr. *kloros*, green) tissues.

Sclerenchyma. Sclerenchymal (Gr. *skler-*, hard;
en, in; *chein*, to pour) cell walls are greatly thickened with cellulose and lignin to give strength
and rigidity (corn stem). The protoplasm is not long
lived. There are two types of sclerenchyma: (1)
fibers that are tough, pliable, strong, elongated cells
with tapering ends, and (2) sclereids (Gr. *skleros*,
hard) or stone cells, in which cells are not elongated but give strength and support (the hull of
walnuts).

These tissues give mechanical support and
strength; because of the cohesive powers of the
fibers, they can be used to make threads, ropes,
and textiles; fibers are common in plant stems;
sclereids are common in shells of nuts and as gritty
masses in pears.

Collenchyma. Collenchyma (Gr. *kolla*, glue; *en*,
in; *chein*, to pour) cells frequently are somewhat elongated. The cell walls are thickened at
corners or elsewhere and protoplasm remains alive
for a long time (celery "stalk"). These tissues

strengthen both younger and older parts of plants.

Cork. Cork (Span. *alcorque,* cork) cells frequently are rectangular and regularly arranged. The cell wall contains waterproof, waxy material called suberin (L. *suber,* cork). Protoplasm dies rather soon.

Cork is present in the outer bark of stems and roots of woody plants where it serves as a protection against mechanical injuries and excessive loss of internal moisture (cork oak tree).

Complex tissues

Complex tissues consist of several kinds of cells that usually engage in a group of closely related activities (Fig. 4-17).

Xylem. Xylem (Gr. *xylon,* wood) is composed of tracheids and vessels. Tracheids are elongated, tapering cells with short-lived protoplasm. The cell walls usually are strengthened and thickened by spirals or rings of lignocellulose, and often with thin areas (pits). The vessels are long, rather large tubular structures made of series of cells whose walls may be thickened and with pits. Other tissues in xylem may include fibers, xylem parenchyma, and ray cells.

Xylem tissues give strength and conduct water, dissolved mineral salts, and sometimes foods upward through stems (also laterally). Xylem tracheids, the chief conducting tissues, are in gymnosperms (as in pine trees). Xylem vessels are the principal conducting structures in angiosperms (as in oak trees).

Phloem. Phloem (Gr. *phloios,* bark) always consists of two types of cells: sieve tubes and phloem parenchyma. The sieve tubes, like xylem vessels, are elongated rows of cylindroid cells whose end walls contain sieve plates containing numerous pores like those in a sieve. The protoplasm remains alive. Phloem parenchyma have the same characteristics described under parenchyma. Other tissues in phloem include fibers and ray cells. In flowering plants there are elongated, living, nucleated cells, known as companion cells, which border sieve tubes and possibly assist in conduction or in food storage.

Phloem tissues give strength and mainly conduct foods downward from leaves through stems and roots. Phloem parenchyma stores food.

Some contributors to the knowledge of tissues, organs, and cell division

Marcello Malpighi (1628-1694)
An Italian scientist and physician who studied tissues and organs microscopically. He is considered the founder of microanatomy. He related anatomy and physiology to medicine, and his studies included the detailed structure of lungs, kidneys, spleen, and other organs, and the capillary circulation in frogs (1660).

Antonj van Leeuwenhoek (1632-1723)
A Dutch microscopist who first published an account of sperms.

Jan Swammerdam (1637-1680)
A Dutch microscopic anatomist who studied the anatomy and life histories of insects and also injected blood vessels.

Nehemiah Grew (1641-1712)
An English scientist and physician who microscopically studied cells, tissues, and organs of plants. His book, *Anatomy of Vegetables* (1672), started the work in plant histology.

Marie F. X. Bichat (1771-1802)
A French surgeon and anatomist who recognized that the organs of animals were composed of masses of substance to which the term tissue was applied. He studied without a microscope, and his work is remarkable considering that he died at an early age. He is the "Founder of Histology."

Hugo von Mohl (1805-1872)
He observed cell division in plant root tips and buds (1835), recording the presence of a cell plate between daughter cells.

Rudolph Albert von Kölliker (1817-1905)
A German biologist who demonstrated that sperms and eggs are cellular products of organisms (1845) from which a new organism is derived by cell division. He placed histology on a cellular basis and applied it to embryology.

Rudolph L. K. Virchow (1821-1902)
A German cytologist who established the law that "all cells arise from pre-existing cells" (*omnia cellula e cellula*). He also stated that pathology was a cellular science (1858), which gave a new emphasis to the explanation of disease conditions in animals. (*The Bettman Archive, Inc.*)

Virchow

Some contributors to the knowledge of tissues, organs, and cell division—cont'd

Walther Flemming (1843-1905)
He described cell division in living and fixed cells of salamanders and developed improved methods of fixing and staining. He discovered the longitudinal reduplication of chromosomes (1879) and coined the term "mitosis" (1882).

Eduard A. Strausburger (1844-1912)
He coined such terms as prophase and metaphase in connection with cell division (1884). He stated that the essential factor in both plant and animal fertilization was the fusion of the gametic nuclei of paternal and maternal origin.

Eduard van Beneden (1845-1910)
A Belgian cytologist, demonstrated the constancy of chromosome numbers for the species, the reduction of their number during maturation of germ cells, and the restoration of the double number at fertilization.

August Weismann (1834-1914)
A German biologist who postulated that a reduction in chromosome number occurred in the germ cells of plants and animals in such a manner as to separate the diploid chromosomes into two haploid groups without a longitudinal division of each chromosome taking place. His prediction has been verified, and reduction division (meiosis) is found to be coexistent with bisexuality. (*Historical Pictures Service, Chicago.*)

Weismann

M. M. Rhoades (1903-   )
He has contributed much information concerning chromosomes, heredity, meiosis, and crossing over in numerous publications since 1931.

C. D. Darlington (1903-   )
He demonstrated that synapsis may begin at any of several places along the length of a chromosome in the lily (*Fritillaria*).

C. P. Swanson (1911-   )
He has contributed much information concerning meiosis, chromosomes, heredity, and cellular structure in numerous publication since 1940.

Review questions and topics

1 Give in detail the distinguishing characteristics and an example of each of the different types of plant tissues.
2 Give the functions of each type of plant tissue.
3 Contrast permanent and temporary plant tissues, with examples of each.
4 Contrast simple and complex plant tissues, with examples of each.
5 Describe in detail each phase in the process of mitosis, explaining the significant events in each.
6 List some factors that might influence the rate of mitosis.
7 Distinguish between mitosis (karyokinesis) and cell division (cytokinesis).
8 Describe in detail each phase in meiosis, explaining the significant events in each.
9 List the general characteristics of the following groups of animal tissues, including examples and functions of each: epithelial, connective, blood, muscular, and nervous.
10 What is the origin and significance of the paired condition in chromosomes?
11 Do higher animals necessarily have the greater number of chromosomes?
12 When does meiosis occur in the life cycle of animals that reproduce sexually? Why?
13 Explain the significance of synapsis and doubling of chromosomes. What significant hereditary phenomenon may occur during synapsis?
14 Contrast haploid (monoploid) and diploid numbers of chromosomes, with examples of each.

Selected references

Bloom, W., and Fawcett, D. W.: A textbook of histology, Philadelphia, 1962, W. B. Saunders Company.

Esau, K.: Plant anatomy, New York, 1953, John Wiley & Sons, Inc.

Levene, L. (editor): The cell in mitosis, New York, 1963, Academic Press Inc.

Mazia, D.: How cells divide, Sci. Amer. **205**: 100-120, 1961.

Mazia, D.: Mitosis and the physiology of cell division. In Brachet, J., and Mirsky, A. E. (editors): The cell, vol. 3, New York, 1961, Academic Press Inc.

Mazia, D., and Tyler, A.: General physiology of cell specialization, New York, 1963, McGraw-Hill Book Company.

Rhoades, M. M.: Meiosis. In Brachet J., and Mirsky, A. E. (editors): The cell, vol. 3, New York, 1961, Academic Press Inc.

Sager, R., and Ryan, F. J.: Cell heredity, New York, 1961, John Wiley & Sons, Inc.

Windle, W. F.: Textbook of histology, New York, 1969, McGraw-Hill Book Company.

Zeuthen, E. (editor): Synchrony in cell division and growth, New York, 1964, Interscience Publishers, Inc.

chapter five

Characteristics of life

With so many different characteristics it would seem that living organisms could easily be distinguished from nonliving objects. However, this is not always the case, for in some instances the borderline between living and nonliving is quite narrow. For example, living seeds, spores, and eggs at times may not seem to manifest many of the characteristics of life. Likewise, certain properties of nonliving substances at times seem to approximate similar characteristics in living organism.

The studies of viruses in recent years have demonstrated the difficulty in distinguishing between the nonliving and the living. Viruses have the ability to reproduce, to grow, and to undergo abrupt hereditary changes called mutations. They are smaller than bacteria and exist only in living tissues. Possibly viruses are an intermediate connecting link between ordinary molecules and living protoplasm.

It is impossible to establish absolute rules for distinguishing living from nonliving in all cases. The following brief consideration consists of generalizations that show some areas in which distinguishing may seem easy, whereas in other respects the differentiation is much more difficult.

ORGANIZATION AND INDIVIDUALITY

Organization may be defined as a systematic arrangement of parts in support of the structure and activities of the whole. This involves organization within cells, in tissues, organs, and systems whereby efficiency is increased. In some cases this may involve differentiation, specialization, and division of labor.

All plants and animals consist of either one cell or of groups of cells, which may number many millions in the higher organisms.

Thus higher organisms are composed of organ systems that are coordinated to perform certain functions. Each system is composed of a number of organs, each of which is constructed of various tissues. Each tissue is made of cells similar in structure and specialized to perform a particular function. Each cell is composed of parts that are organized to perform the cell's many activities.

For example, a human body is composed of eleven organ systems, each consisting of several organs.

The human digestive system consists of mouth, teeth, tongue, esophagus, stomach, small and large intestines, and digestive glands, such as the liver and the pancreas. An organ such as the stomach is made of such tissues as epithelial, connective, muscular, and nervous. Each tissue is composed of numerous similar cells that are constructed of different chemical compounds. The chemicals are constructed of molecules and atoms that are organized along certain patterns and controlled by natural laws.

In a higher plant such as a tree there are organs associated together for performing certain functions. For example, there are roots, stems, leaves, and flowers, each performing particular functions and all contributing to the structure and activities of the entire plant.

Each organism has an individuality that characterizes it. Each animal and plant possesses an individuality concerning its construction and activities. A living organism of whatever type may fluctuate within rather narrow limits as to its form and size, there being upper and lower limits; nonliving things show greater variations in their size and form. The protoplasm of living organisms is far more complex in structure than most nonliving substances. There are no chemical elements unique to protoplasm alone. Some are found also in nonliving substances, but the organization and

47

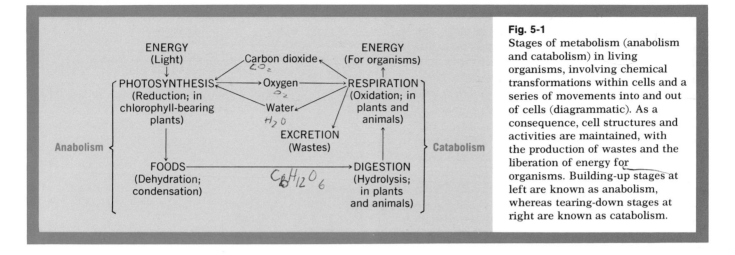

Fig. 5-1
Stages of metabolism (anabolism and catabolism) in living organisms, involving chemical transformations within cells and a series of movements into and out of cells (diagrammatic). As a consequence, cell structures and activities are maintained, with the production of wastes and the liberation of energy for organisms. Building-up stages at left are known as anabolism, whereas tearing-down stages at right are known as catabolism.

patterns of construction are unique for protoplasm.

All parts of living organisms are so integrated that the entire organism is a unit or an individual. In each living organism, there are (1) interdependence and systematic correlation of parts, (2) a variable susceptibility to environmental influences, (3) inherent self-regulatory tendencies, and (4) a centralized control.

METABOLISM

Metabolism (Gr. *metabole,* change) (Fig. 5-1) refers to all the chemical changes and functional activities that occur in a living organism that result in growth, repair, respiration, use of food, and secretion. That part of metabolism that results in building up of rather complex materials from simpler ones is called anabolism (Gr. *anabole,* a throwing up), whereas the tearing down and changing of more complex substances into simpler ones is called catabolism (Gr. *katabole,* a throwing down). In anabolism there is a storage of energy and an increase of protoplasm, but in catabolism there is a release of energy.

Life processes constantly involve both anabolism and catabolism, but in younger organisms anabolism predominates, whereas in older forms catabolism may be in excess. Nothing comparable to metabolism is known in the nonliving world. Certain nonliving materials may form more complex substances at times and may be oxidized or broken down, yet these simple processes are not organized as they are in the metabolism in plants and animals.

Metabolism is especially significant in such activities as nutrition, respiration, excretion, coordination, and reproduction. In respiration, excretion, and coordination, energy is expended primarily. However, even when new protoplasm is synthesized or new individuals are reproduced, there is a limited expenditure of energy. When energy expenditures exceed the building up of energy, disintegration and death of the organism or its parts may follow.

There is no time during life when chemical changes do not occur incessantly. In general, plants are primarily anabolic organisms, whereas animals are primarily catabolic, when the total metabolic activities of each are considered.

The actual process of converting nonliving materials into living protoplasm is called assimilation (L. *assimilatio,* conversion). During this process rather simple materials are built into more complex compounds through the action of enzymes (Gr. *en,* in; *zyme,* leaven) produced by the protoplasm. Assimilation may replace protoplasm or add to the original, thus resulting in growth.

The characteristic ability of living protoplasm to duplicate itself by synthesizing complex mole-

OXIDASE- any enzyme that acts as an oxidator.

cules from simpler ones is called autosynthesis. The duplication of genes, chromosomes, and similar phenomena illustrates autosynthesis. As stated previously, many of the activities of protoplasm result from the presence of enzymes, or organic catalysts, as they are called. The ability of such catalysts to create more molecules of their own kind, thus gradually increasing the speed of the action, may be considered autocatalysis. The disintegration of cells or tissues by the actions of autogenous enzymes (Gr. *autos*, self; *genesis*, origin) that they produce may be considered autolysis. Life seems to be the result of many interacting forces and factors, only a few of which have been considered even briefly.

RESPIRATION AND RESPIRATORY PIGMENTS (CYTOCHROMES)

Respiratory pigments, called cytochromes, have been found in muscles and other tissues of almost all groups of animals. Cytochromes are composed of large molecules with high molecular weights. MacMunn suggested (1886) that these pigments were allied to hemochromogens (Gr. *haima*, blood; *chroma*, color; *gen*, form) and were associated with respiration because of their being capable of oxidation and reduction. Keilin, of Cambridge University, stated (1925) that spectroscopic evidence indicated that cytochrome was a mixture of iron-porphyrin compounds (Gr. *porphyra*, purple) that form the basis for respiratory pigments in living organisms. Recent researchers suggest that there are at least seven such iron-porphyrin proteins.

Cytochromes function in cellular metabolism as terminal intermediates for the transfer of hydrogen or electrons to the enzyme cytochrome oxidase (Warburg, 1930). Hence, they are regarded as reversibly oxidizable or reducible substances, or as hydrogen or electron carriers, in the long series of enzymatically catalyzed reactions leading from the substrate (the substance acted upon) to molecular oxygen.

For aerobic cells that utilize oxygen or that are not destroyed by its presence, cytochrome oxidase is the agent that binds oxygen in the process of cytoplasmic oxidation. The oxidase does not act on the initial metabolic substrates (such as carbohydrates, fats, or proteins), but the oxidizable substrates yield electrons to a long series of interlocking intermediate enzymes, much like a conveyor belt system, of which cytochrome oxidase is the terminal enzyme. The free protons, the transmitted electrons, and the activated molecular oxygen are united chemically on the enzyme surface to form water. The enzyme is bound to particles of the cell cytoplasm that have molecular weights approximating several millions. The problems of energy release through enzymatic actions are considered in more detail elsewhere (see Chapter 6).

GROWTH

Growth in living organisms consists of an increase in volume by the formation of new protoplasm, with or without the formation of new cells. In this complex phenomenon the synthesized living protoplasm is peculiar to each organism and is made from relatively simple chemical materials. For example, green, chlorophyll-bearing plants photosynthesize foods, using light as a source of energy and giving off oxygen in the process. These foods may be combined in the protoplasm into more complex compounds. For example, starches and other carbohydrates, oils and fats, and protein may be formed by the addition of nitrogen and amino acids.

During the process of growth, as the carbohydrates, fats, and proteins are increased, there is an increase in the amounts of water and salts. Animals can secure water and salts from outside sources, but they cannot synthesize carbohydrates, fats, and proteins and are dependent upon green plants for these body-building and energy-supplying materials. Green plants are thus responsible for their own growth and life needs and must supply these major requirements for animals.

DIFFERENTIATION

The phenomenon of differentiation stamps each living cell with its own uniqueness of structure and function. A generalized cell is transformed or differentiated into a specialized cell by a series of progressive changes. For example, growing human cells are differentiated into the many types of cells that compose the various organs and systems of the body.

49

aerobic - able to live only where free oxygen is.
pigment - any coloring matter in the cells or tissues of plant or animals.

Whereas growth is a rather endless process of multiplication of similar units, differentiation is the specialization of some of these units for different purposes. Differentiation begins before any visible changes may be apparent in cells and is preceded by chemical alterations. Specializations are influenced and limited by the genetic capabilities of the particular cell in question. Examples of differentiations in living organisms include muscle cells, nerve cells, sex cells, pollen grains, and conducting cells of higher plants.

IRRITABILITY

Irritability is a fundamentally inherent property of all protoplasm. It consists of a sensitivity to stimulation, an ability to transmit the excitation, and an ability to react to the various stimuli that the organism encounters. A stimulus (L. *stimulare*, to incite) may be considered as any factor, or change in some feature of the environment, that causes an organism to respond in some way. Changes in such variables as the direction or intensity of light, in the temperature, and in the chemical composition of the surroundings are common stimuli.

Stimuli may be internal or external, such as light, contact, gravity, and temperature. A stimulus does not supply energy for a reaction but merely initiates it. The particular response of an organism depends on the quantity and quality of the stimulus and the condition of that organism at the time of stimulation.

A movement in an organism may be initiated by different kinds of stimuli, but the extent and nature of the resulting response depends on the specific stimulus. When an organism, or some part of it, moves toward the source of the stimulus, a positive reaction results, whereas a movement away from the source constitutes a negative reaction. Organisms have a tendency to restore themselves to their original conditions after responding to a stimulus.

Even though all cells are irritable, in many cases certain cells or parts of cells are particularly sensitive to special kinds of stimuli. In many lower animals the entire body surface is sensitive to stimuli. In most higher animals, groups of cells are specialized to be affected by particular stimuli; for example, certain retinal cells of an eye are light sensitive, certain cells of the inner ear are sound

sensitive, whereas certain epithelial cells of the nose and mouth are sensitive to certain chemicals.

In most animal reactions the stimuli are received in one part of an organism or organ, but the resulting reaction occurs in some other part. This necessitates a transmission of excitation (conduction) from the place of its origin to the region of reaction. This can be accomplished by means of nerve cells (if they are present), or by means of chemical substances, or by a combination of the two. If a tentacle of a *Hydra* is touched with a glass rod, other tentacles, or even the entire *Hydra*, may contract, showing that there has been conduction, possibly by means of its nerve net. Likewise, if a leaf of a sensitive plant, *Mimosa* (Fig. 5-2), is touched, the result of the stimulation may be conducted to adjacent parts of the plant.

Living animals have developed two systems of coordination, a chemical coordination and a nervous coordination. In simpler types only the former method is utilized, but in higher forms both are employed. In plants the chemical method is used. In chemical coordination simple molecules may be used in the process, and in higher animals and plants substances called hormones (Gr. *hormaein*, to excite) are used. In chemical coordination the action is usually long lasting, relatively slow, and may be quite extensive. In nervous coordination the actions are usually rapid, of short duration, and somewhat localized. Thus in higher animals the two systems supplement each other in their attempts to adapt to fluctuations in the external and internal environments.

Sometimes when certain structures have been injured or destroyed, other parts may take on compensatory activities. If a human kidney is removed, the other adjusts to compensate for the loss. If more oxygen is needed, the circulatory and respiratory systems attempt to supply the extra need. Such adaptations are common in living organisms. Through some internal recording and regulations, the reactions and efforts of living organisms sometimes serve as a kind of "experience" by which they attempt to avoid similar actions in the future.

MOVEMENT

Probably the most easily demonstrated activities of a living organism are the movements of parts or

Normal

Fig. 5-2
Irritability and conduction as shown by the
sensitive plant *Mimosa*. When stimulated
properly, leaves or even stems may respond
(collapse), showing effects of the stimulus
and even conduction to other parts of the
plant from the point of stimulation. After
a period of recovery the living plant may
resume its normal condition.

After stimulation

the organism's locomotion from place to place.
Living protoplasm is rarely, if ever, motionless. The
quality and quantity of movement depends on
many influences, such as age, water content,
hereditary factors, and composition. Sometimes
organisms may appear to be inactive, but careful
investigation may reveal streaming movements
within the protoplasm, slight movements of the
circulating fluids, and muscle contractions.

REPRODUCTION AND HEREDITY

For any species of plant or animal to successfully
maintain its existence it must possess some inherent means of reproduction or duplication of its
kind. There are two general methods of reproduction in the living world: asexual, without the formation of specialized sex cells (gametes), and sexual,
in which specialized sex cells are produced. In
asexual reproduction a new offspring arises from
a part of an older individual. This part may be a
single-celled spore as in ferns, a many-celled bud
as in certain sponges, or an organism may divide
into two parts as in the case of such single-celled
animals as *Amoeba* and *Paramecium*. In asexual
methods the offspring are very much like the
parent.

In sexual reproduction two specialized sex cells
unite to form a single fertilized cell from which a
new individual develops. In sexual reproduction
something more than mere propagation occurs,
because the new individual has arisen from the
union of sex cells from two different parents, and
each parent has contributed some of its traits.
Hence, the new individual is usually not like either
parent but possesses traits from both. When this
occurs generation after generation, endless varieties of offspring are formed, thus making it
possible for them to fit into a great variety of environmental conditions.

In certain organisms a new individual may
develop from an egg that has not united with a
sperm, and hence this is a special type of sexual
reproduction called parthenogenesis (par the no
-jen′ e sis) (Gr. *parthenos*, virgin; *genesis*, birth).

REGENERATION

A phenomenon called regeneration concerns the
replacement of certain parts that have been lost,

worn out, or used up by the activities of the organism. Information has been secured concerning the role of the nucleus in the control of cell activities by experiments on the single-celled, marine alga, *Acetabularia*. This plant (between 5 and 10 cm. long) has basal rootlike rhizoids and a stalk with a cup-shaped "umbrella" at the tip (Fig. 5-3). A single nucleus is present near the base of the stalk. When reproduction occurs, the umbrella divides into reproductive cells. Two species of *Acetabularia* differ in the shapes of their umbrellas.

Hämmerling, a German biologist, experimentally cut across the stalk and found that the lower, nucleated part would regenerate the distinctive umbrella, whereas the upper part eventually died without regenerating stalk and rhizoids. In other experiments Hämmerling cut the stalk just above the nucleus and made a second cut just below the umbrella. The middle stalk section, thus isolated, grew a partial or complete umbrella when placed in sea water.

This might suggest that a nucleus is not necessary for regeneration. However, when this second umbrella was cut off, the resulting stalk did not form an umbrella. It was concluded that the nucleus supplies a substance that passes up the stalk and instigates umbrella formation. Some of this substance seems to remain in the stalk after the cuts are made and in sufficient quantity to produce a new umbrella, but when the substance is used, no second regeneration can occur in the absence of a nucleus. The nucleus apparently produces a substance that regulates cell growth.

From what we know today we might infer that mRNA had been formed and had led to umbrella formation. However, once it was all used, a nucleus was necessary for further regenerative activity.

It is possible to graft the stalk of one species of *Acetabularia* onto the nucleus-containing rhizoid of another species. When the umbrella regenerates, it is always characteristic of the species that contributes the nucleus, not of the species that contributes the stalk. When the nuclei of two different species are present in a graft, the umbrella formed is intermediate in shape. The role of the nucleus is shown in these phenomena.

52 The role of the nucleus for cytoplasmic survival

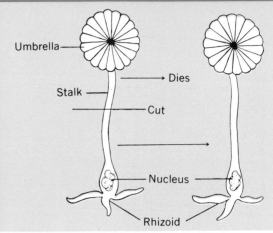

Fig. 5-3
Role of the nucleus in regeneration and other life processes as shown by a single-celled, marine alga, *Acetabularia*. When cut as shown, the upper portion dies and the lower part with its nucleus regenerates a new individual.

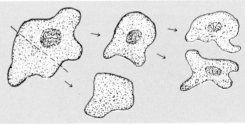

Fig. 5-4
Role of the nucleus in regeneration and cell activities as shown when a single-celled protozoan, *Amoeba*, is bisected experimentally. The part without the nucleus dies eventually.

can be shown experimentally by cutting a single-celled protozoan, such as *Amoeba*, into two halves so that one half has a nucleus and the other does not (Fig. 5-4). The nucleated half carries on like a normal ameba while the enucleated half eventually dies. The ameba without a nucleus may move and feed for a time, but it does not grow or reproduce. Eventually it dies, because the long-range effects of the original nucleus are gone. Here again, we infer that mRNA has been depleted. Just as the cytoplasm depends on the nucleus, the nucleus depends on the cytoplasm, because it is the site of respiration, food management, and synthesis. A naked, isolated nucleus free from cytoplasm will eventually die, because of a lack of food materials

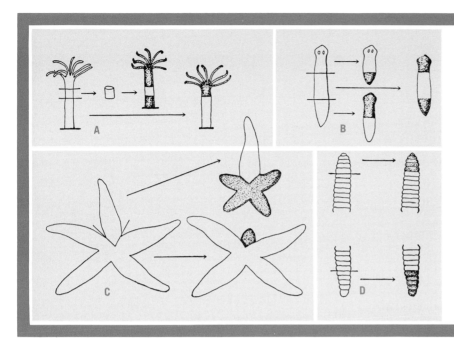

Fig. 5-5
Regeneration of animals as shown by **A**, *Hydra*; **B**, *Planaria*; **C**, starfish; **D**, earthworm. Lines show the position where parts were removed, and darker (stippled) areas are those that have been regenerated: Note that in the starfish the missing ray is regenerated and the missing ray also regenerates the four missing rays. The kind of regeneration that develops depends largely upon where the cut is made.

and energy. This shows the nucleus-cytoplasm interactions that are necessary for normal cellular activities.

When the arm of a starfish is removed, it may be regenerated (Fig. 5-5); the removed arm may also regenerate an entire animal, thus producing two individuals. Certain other structures in living organisms may be replaced if lost, thus maintaining the organism near normality. In general, regeneration in higher and more complex organisms is somewhat limited and is more of a repair or replacement than a regeneration. Human skin from a finger may be replaced, but the tip of an entire finger may not. Some salamanders actually regenerate a new tail. Other examples include the healing of wounds in plants and animals, the repair of broken bones, the replacement of red blood corpuscles, the replacement of lost plant sap, and the renewal of bark in trees. In all regenerative phenomena many factors are influential and must be considered in explaining the differences in the various living organisms, or even in different parts of the same organism.

EVOLUTION

One of the basic characteristics of living plants and animals is their ability to evolve or change. These changes, evolution, are controlled heredity mechanism. Throughout the distant past living organisms have changed, to a greater or lesser degree, just as they do now and as they probably will in the future. Evolution is the theme that is followed by all living organisms.

Review questions and topics

1 Discuss each of the activities of living protoplasm, including (1) organization and individuality, (2) metabolism, (3) respiration, (4) growth, (5) movement, (6) irritability, conduction, coordination, and adaptation, (7) reproduction and heredity.
2 Define life in your own words.
3 Contrast, with examples of each: metabolism, anabolism, catabolism, and assimilation.
4 Explain autosynthesis and autocatalysis, with examples of each.
5 Discuss the different types of stimuli, with examples of each.
6 Explain the importance of (1) chemical coordination and (2) nervous coordination; give the attributes of each.
7 Discuss various ways in which organisms may attempt to adapt themselves to meet changing environmental conditions.
8 Explain the methods of reproduction, with the significance of each type.
9 Explain the phenomenon of regeneration and its importance.
10 Discuss viruses and their significance on the borderline between living and nonliving.

53

Some contributors to the knowledge of certain activities of protoplasm

Spallanzani

Loeb

Aureolus B. von Hohenheim (Paracelsus) (1493-1541)
A Swiss physician who seemed to gain occasional glimpses of fundamental truth but usually through a haze of astrologic and alchemical absurdities. He may have had some ideas on metabolism and the cycle of elements.

Charles Bonnet (1720-1793)
A Swiss naturalist who discovered parthenogenesis (development of an egg without a male sperm) in certain insects and inferred incorrectly that the females carried many miniature young for succeeding generations in their bodies.

Lazaro Spallanzani (1729-1799)
An Italian naturalist, philosopher, and priest who described the regeneration of organs in such animals as worms, snails, and frogs. (*Historical Pictures Service, Chicago.*)

Joseph Priestley (1733-1804)
An English naturalist and divine who studied the respiratory interdependence of plants and animals. (*The Bettman Archive, Inc.*)

Robert Brown (1773-1858)
A Scotchman who discovered the so-called "Brownian movement" of minute particles (1828), which phenomenon provides a basis for demonstrating molecular motion.

Priestley

Jacques Loeb (1859-1927)
A German-American biologist who experimentally studied sea urchin eggs in which he induced parthenogenesis by artificial means (1910). (*Historical Pictures Service, Chicago.*)

Charles M. Child (1869-1954)
An American biologist who proposed the "Axial Gradient Theory" to explain the differences in rates of metabolism along the axes of living organisms (1921).

C. A. MacMunn
By studying various tissues from echinoderms to man (1886), through the use of microspectrography, he found pigments whose spectra showed a series of absorption bands that varied little from tissue to tissue or from organism to organism.

David Keilin (1887-1963)
A Cambridge University biologist who rediscovered (1925) the same pigments, now called cytochromes ("cellular pigments"), that had been found by MacMunn, thus verifying the latter's contentions.

Otto H. Warburg (1883-   )
A German biochemist who discovered an iron-porphyrin protein known as "cytochrome oxidase" (1930). He was awarded a Nobel Prize (1931) for his contributions to cellular metabolism.

Selected references

Chambers, R., and Chambers, E. L.: Exploration into the nature of the living cell, Cambridge, 1961, Harvard University Press.

Comfort, A.: Ageing; the biology of senescence, New York, 1964, Holt, Rinehart & Winston, Inc.

Hardin, G. (editor): 39 steps to biology. Readings from Scientific American, San Francisco, 1968, W. H. Freeman and Co., Publishers

Haynes, R. H., and Hanawalt, P. C. (editors): The molecular basis of life. Readings from Scientific American, San Francisco, 1968, W. H. Freeman and Co. Publishers.

Johnson, W. H., and Steere, W. C. (editors): This is life, New York, 1962, Holt, Rinehart & Winston, Inc.

Spratt, N. T.: Introduction to cell differentiation, New York, 1964, Reinhold Publishing Corp.

Stanley, W. M., and Valens, E. G.: Viruses and the nature of life, New York, 1961, E. P. Dutton & Co., Inc.

Walker, B. S., Boyd, W. C., and Asimov, I.: Biochemistry and human metabolism, ed. 3, Baltimore, 1957, The Williams & Wilkins Co.

chapter six

Chemical and physical properties of life

Many scientific explanations for biologic phenomena come from a knowledge of chemistry and physics. In fact, it is difficult to mention a process in any organism in which chemical and physical factors are not involved. Progress in these fields and their utilization in biology has become so extensive that the sciences of biochemistry and biophysics have caused a revolution in biologic thinking.

PHYSICAL PROPERTIES OF MATTER

Matter is anything that occupies space (has size) and is affected by gravity (has weight). Matter is the substance of which all things are made. All matter exists in three states (phases): solid, liquid, or gas. In the solid state matter has a definite shape; in the liquid state it takes the shape of the container in which it is placed; and in the gaseous state it has no definite shape (the individual particles are free to move about in space).

When masses of two or more states of matter are placed together it is called a mixture, in which the masses are not combined firmly and need not be in any definite proportion. When something is mixed with a liquid, one of three kinds of mixtures is formed as show in Table 6-1.

Atoms and ions

An atom (Gr. *atomos*, indivisible) (Fig. 6-1) is the smallest particle of an element capable of taking part in a chemical reaction. There are as many different kinds of atoms as there are elements, and vice versa. Atoms are so small they are invisible and consist of still smaller units that are arranged somewhat like a miniature solar system, in which much of the atom is supposedly "empty

Table 6-1
Mixtures of solids and liquids

True (molecular) solution	Colloidal solution	Suspension
Molecules or ions added to water (solvent) are uniformly dispersed among molecules of water (dissolved molecules or ions called solute)	Solid particles intermediate between molecular state and suspension; particles will not settle out	Solids, when added to water remain in masses larger than molecules; upon standing tend to settle in bottom of container
Transparent in appearance	Either transparent or somewhat cloudy in appearance	Turbid in appearance
Boiling point higher and freezing point lower than pure water	Boiling point and freezing point same as pure water	Boiling point and freezing point same as pure water
Particles smaller than 0.001μ	Particles range from 0.001 to 0.1μ	Particles are larger than 0.1μ and can be seen with light microscope

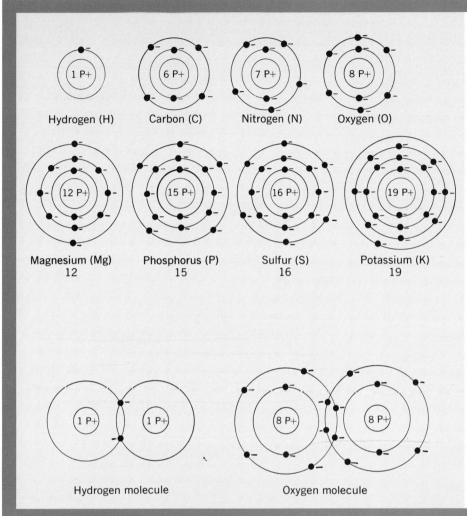

Fig. 6-1
Atomic structure of certain elements that may be present in protoplasm (diagrammatic). The atomic number, which is the same as the number of nuclear protons, is given below the name; symbols in parentheses follow the name. P, Proton (positive electrical charge); black circle with dash, electron (negative electrical charge). The inner circle represents the nucleus of the atom; outer rings represent one or more orbits ("shells") of electrons.

Fig. 6-2
Molecular structure of the gases, hydrogen and oxygen (diagrammatic). In hydrogen mutual utilization of one electron from each atom (total, two electrons) is involved in the combination. In oxygen two electrons from each atom (total, four electrons) are involved in the combination. P, Proton (positive electrical charge); black circle with dash, electron (negative electrical charge). The inner circle represents the nucleus of the atom; outer rings represent one or more orbits ("shells") of electrons.

ELECTRONS = PROTONS

space." Atoms are made up of protons, electrons, and neutrons. The units that compose an atom are as follows.

A central atomic nucleus consists of particles that are smaller but heavier than the electrons that revolve about the atomic nucleus. These particles of the nucleus are the (1) protons that are charged with positive electricity, and (2) neutrons that are uncharged electrically (neutral). Each proton has the power to hold one of the whirling, negatively charged electrons in its orbit. Thus, the number of electrons is determined by the number of positively charged protons in the atomic nucleus. The atom as a whole is electrically neutral (neither positive nor negative). For example, the nucleus of

the oxygen atom contains eight positively charged protons that hold the eight negatively charged electrons in the two orbits (Figs. 6-1 and 6-2). The nuclear protons and neutrons are held together by intra-atomic forces.

A series of negatively charged electrons revolve at great speed in one or more concentric orbits ("shells") about the atomic nucleus. The electrons can be considered to constitute "whirlpools of energy." Depending on the kind of atom, there may be from one to seven concentrically arranged orbits, and for each orbit there is a maximum of electrons that it can accommodate, although an orbit may not be completely filled. In general, the inner orbit must be filled to capacity before a second appears. The maximum numbers of electrons in various orbits are suggested in Table 6–2.

Each element is given an <u>atomic number equal</u> to the <u>number of protons in the atomic nucleus.</u> The atomic weight of an element is the weight of a typical atom of that element relative to the weight of carbon 12. The atomic weights of isotopes are slightly higher than of the typical atom of the element.

In some cases some atoms of an element may have additional neutrons in the nucleus. These are isotopes (Gr. *isos*, equal; *topos*, place) of the element. Examples are hydrogen and its isotopes deuterium (^2H) and tritium (^3H), carbon 12 and its isotope carbon 14, and oxygen 16 and its isotopes ^{11}O and ^{12}O.

Artificially produced radioactive isotopes are extremely valuable in the study of certain biologic problems. The use of such radioactive isotopes as "tracers" is possible, because they emit certain radiations whose presence can be detected by sensitive counters. Hence, the rate of absorption of iodine by the thyroid gland can be determined by the use of radioactive iodine, and this has assisted in the treatment of goiter. Radioactive phosphorus has been traced to the stems and certain parts of the leaves of tomato plants, whereas radioactive zinc concentrates in tomato seeds. The use of radioactive isotopes may be of great value in the study of animal and plant metabolism and in the diagnosis and treatment of certain diseases (Table 6-3).

The chemical and physical behaviors of atoms are determined largely by the number and arrangements of the orbital electrons. Proper bombardment of certain atoms (by neutrons and protons) results in the release of tremendous amounts of atomic energy by the process of nuclear fission. For example, the energy released by the fission of one pound of ^{235}U (a fissionable isotope of uranium) is roughly equivalent to that secured from burning 10,000 tons of coal.

Atoms with less than one half the maximum number of electrons in the outer orbit may, under certain conditions, even lose those that they have, whereas atoms with more than one half the maximum number of electrons in the outer orbit may add electrons until the outer orbit is filled to its

Table 6-2

Maximum number of electrons in atomic orbits

Orbit	Number of electrons (maximum)
First	2
Second	8
Third	18
Fourth	32
Fifth	18
Sixth	12
Seventh	2

Table 6-3

Some uses of radioactive isotopes

Element	Isotope	Use
Carbon	^{14}C	Fate of labeled nutrients
Carbon	^{14}C	Fate of injected drugs
Sulfur	^{35}S	Use of amino acids
Iron	^{59}Fe	Iron metabolism
Sodium	^{24}Na	Rate of sodium transfer
Phosphorus	^{32}P	Blood cell production
Potassium	^{42}K	Permeability of cell membranes
Cobalt	^{60}Co	Gamma ray treatment
Strontium	^{90}Sr	Radiation accumulation
Iodine	^{131}I	Thyroid metabolism
Iodine	^{131}I	Location of brain tumors
Gold	^{198}Au	Cancer treatments

maximum. The additions or losses of negatively charged electrons in the orbits does not affect the structure of the atomic nucleus, but it can no longer be electrically neutral after such changes. Normally, the positively charged protons and the negatively charged electrons are balanced. Hence, the loss of electrons makes the atom positively charged and the addition of electrons makes the atom negatively charged. <u>Such a charged atom is called an ion (Gr. *ion*, going).</u> Atoms that have <u>gained electrons have a negative electrical charge and are called anions;</u> atoms that have lost electrons have a positive electrical charge and are

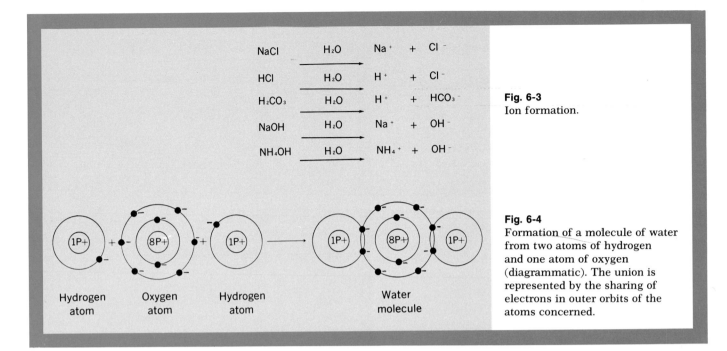

NaCl $\xrightarrow{H_2O}$ Na$^+$ + Cl$^-$

HCl $\xrightarrow{H_2O}$ H$^+$ + Cl$^-$

H$_2$CO$_3$ $\xrightarrow{H_2O}$ H$^+$ + HCO$_3$$^-$

NaOH $\xrightarrow{H_2O}$ Na$^+$ + OH$^-$

NH$_4$OH $\xrightarrow{H_2O}$ NH$_4$$^+$ + OH$^-$

Fig. 6-3
Ion formation.

Hydrogen atom — Oxygen atom — Hydrogen atom → Water molecule

Fig. 6-4
Formation of a molecule of water from two atoms of hydrogen and one atom of oxygen (diagrammatic). The union is represented by the sharing of electrons in outer orbits of the atoms concerned.

called cations. Since unlike charges attract each other, anions and cations combine. The amount of combining that atoms can undergo is determined by the number of electrons in the outer orbit and the number that can be gained or lost. Hence, hydrogen with one electron tends to lose it, becoming a cation (positive). Oxygen, with six electrons in its outer orbit, tends to gain two, becoming an anion (negative). Thus two atoms of hydrogen and one atom of oxygen combine to form a molecule of water, since hydrogen loses only one electron and oxygen gains two electrons (Fig. 6-4).

Many substances in living organisms are soluble in the large quantity of water present in the protoplasm, and in solution many of them dissociate to form ions. A molecule of common salt, sodium chloride (NaCl), dissociates into (1) a positive sodium ion (Na$^+$) while losing its single outer electron and (2) a negative chlorine ion (Cl$^-$), gaining one extra outer electron (Fig. 6-3).

A substance that dissociates to form ions is called an electrolyte (Gr. *elektron*, amber; *lysis*, a loosening), because of its ability to conduct electric currents.

A substance that dissociates to form hydroxyl ions (OH$^-$) is called a base or alkali. A substance releasing hydrogen ions (H$^+$) is called an acid. A compound resulting from the reaction of an acid and a base is a salt. Compounds differ in the extent to which they form ions. As a result, some acids in solution form more H$^+$ ions than others and are said to be strong acids. Likewise, a strong base forms more OH$^-$ ions in solution than a weak base.

The relative strength of an acid or base is expressed in terms of pH, which is an expression of the number of free H$^+$ in a solution as compared to pure water. On a scale of 0 to 14, then, we find the following:

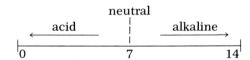

As the strength of H$^+$ ions increases, the pH gets lower; as the strength of OH$^-$ ions increases, the pH gets higher. As the concentration of H$^+$ and OH$^-$ equalize, the pH approaches neutrality.

Elements and compounds

An element is a simple form of matter that has not been decomposed to form other substances by ordinary chemical means. All of the thousands of different substances that are known are composed of about 100 fundamental substances called elements. The chemical elements are designated by symbols that usually are the first or first few letters of the name of the element. For example, the symbol for hydrogen is H; for oxygen, O; for carbon, C; for magnesium, Mg. In some cases the symbol is derived from the Latin name of the element. For example, the Latin name for iron is *ferrum*, and the symbol is Fe; the Latin name for potassium is *kalium*, and the symbol is K; the name for sodium is *natrium*, and the symbol is Na. All of the elements are grouped in a Periodic Table and arranged according to their atomic structure. Each element is given a specific number, known as its atomic number. The number of protons (or electrons) in the atom of an element is called the atomic number of that element. Hydrogen has the atomic number 1; helium, 2; carbon, 6; nitrogen, 7; oxygen, 8; magnesium, 12; mercury, 80; and uranium, 92. Some of the more important elements found in protoplasm are considered in Table 6-6.

When two or more elements are firmly united in definite proportions, they are said to form a compound. With so many elements and with their various combinations so large, it seems probable that the number of compounds to be formed would be enormous. Some of the more common compounds in protoplasm, such as carbohydrates, fats, proteins, inorganic salts, water, and enzymes are considered later.

Molecules

All matter is made of extremely small, invisible molecules (L. *molecula*, little mass). A molecule is composed of the union of two or more atoms and is the smallest unit of matter capable of a separate, distinct physical existence. A molecule of free oxygen consists of two atoms. A molecule of water consists of two atoms of hydrogen and one atom of oxygen, hence its molecular formula is H_2O.

Molecules are too small to be visible with the highest magnification of an ordinary microscope that uses light rays, because the smallest particle visible with such a light (optical) microscope must have a diameter of 150 mμ (1 millimicron is one-millionth of a millimeter and is abbreviated mμ). The largest molecules probably have a diameter of only 1 mμ. However, an electron microscope (see Fig. 2-7) that uses beams of electrons instead of beams of light can be used to photograph the larger molecules.

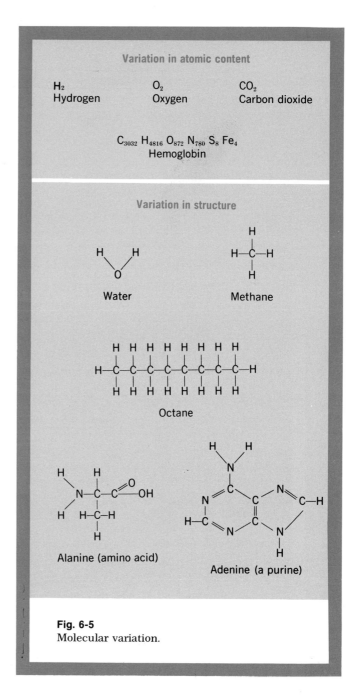

Fig. 6-5
Molecular variation.

Molecules may vary in complexity from the simple water molecule to the extremely complex long carbon "chains" and "rings" present in many living materials. Molecules may contain only one kind of atom (such as O_2), or they may contain atoms of two or more different kinds (such as CO_2); in the latter case they form a compound (Fig. 6-5).

59

Atoms are held together either by the attraction of opposite ions or by the sharing of an electron by two different atoms (Figs. 6-2 and 6-4).

Molecules are constantly in rapid motion, moving about in their intermolecular spaces. Their speed depends upon certain conditions and varies over wide ranges, but an average speed is thought to be approximately 2 million·million times their own diameter per second (about 20 miles per minute) in such a substance as a gas. No matter how sparsely distributed, molecules cannot travel without colliding with other molecules also in motion. This energy of molecular movement is an example of kinetic energy. The intermolecular spaces between molecules in a gas are greater than the spaces between molecules in a liquid or solid. In the atmosphere the several kinds of molecules must move about one thousand times their own diameters before colliding with other molecules. The motion of molecules in a gas is greater than in a solid, since in the latter the molecules merely vibrate back and forth because of the mutual attraction between adjacent molecules and, probably, the closer association of the molecules.

If a single molecule could be completely isolated and remain so, its kinetic energy would remain constant. However, the kinetic energy of molecules is influenced by the kinetic energy of surrounding molecules. When molecules increase their speed, they exert greater pressure on other molecules, so that the average distance between them is increased. Consequently, when heat is applied to certain substances, the molecules increase their speeds and the substance expands. Kinetic energy is measured in terms of temperature. Likewise, contraction usually is the consequence of reduced molecular speed.

If molecules of two kinds are placed together, they tend to mix with each other through a process called diffusion (L. *diffusia*, spread). If a drop of perfume volatilizes (becomes a gas) in a room, its molecules will move and mix with the various molecules of the atmosphere, and the odor will diffuse so as to be detectable some distance away. Our noses are affected by the molecules of the perfume so we detect the odor. It is not detected at a distance immediately, because it takes some time for the perfume molecules to move toward us, and the various molecules of the atmosphere also offer resistance (because of collisions). The continual bombardment of an enclosing membrane or wall by molecules exerts a molecular pressure, which varies with the number of molecules, their speeds or movements, and the temperature. A chemical substance will also diffuse through water in which it is placed, as is shown by placing a crystal of copper sulfate in water and observing the spread of the blue color. These phenomena of diffusion through gases, liquids, and solids are common in the living and nonliving worlds and probably play important roles there.

Membrane permeability and osmosis

Permeability is the property of a membrane or partition that determines its penetrability. The permeability of a membrane depends on (1) the size of the pores of the membrane, (2) the size of the particles of the substance attempting to pass through that membrane, and (3) the solubility of the substance in the membrane. A membrane may be permeable to small molecules but impermeable to large molecules. Another membrane may be permeable to ions but impermeable to even the smallest molecules.

Protoplasm

The boundary of cells consists of fatty substances and other materials that influence its solubility properties, which, in turn, at least partially determine its permeability. Living membranes, such as the plasma membrane of cells, have a selective permeability. Living membranes usually permit the passage of small molecules and certain ions, whereas larger molecules, such as protein molecules, and colloidal particles are restrained. Different cells vary in the permeability of their boundaries. Each has its specific type of permeability, and the plasma membrane of each individual cell plays an important role in regulating the activities of the protoplasm within that cell.

Osmosis

The force exerted by the pressure of moving molecules in a solution against a membrane is known as osmotic pressure. The passage of water through a semipermeable membrane is known as

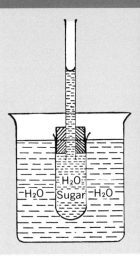

Fig. 6-6
Demonstration of osmosis, in which a semipermeable membrane separates the sugar solution and water. Pores of the membrane are of a size that permits the passage of water molecules but not sugar molecules. Hence, the passage of water is sufficient to cause it to rise in the upright tube. Water molecules pass in either direction, but they pass faster into the tube than out of it because of a greater concentration of water molecules on the outside. The liquid will rise in the upright tube until it reaches a level at which its hydrostatic pressure, because of its weight, is equal to the osmotic pressure produced by the sugar solution.

(From Roe, J. H.: Principles of chemistry, ed. 9, St. Louis, 1963, The C. V. Mosby Co.)

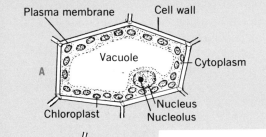

Plasma membrane   Cell wall
Vacuole
A
Cytoplasm
Chloroplast   Nucleus   Nucleolus

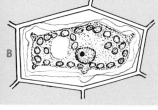

B

Fig. 6-7
Turgor and plasmolysis as shown by a cell of the plant *Elodea*. **A,** Turgid cell when surrounded by ordinary pond water, with an osmotic equilibrium between the internal protoplasm and surrounding pond water; **B,** cell when plasmolyzed by the addition of a little salt to the pond water. This disturbs the osmotic equilibrium, and water osmoses out of the protoplasm through the plasma membrane into the surrounding salt solution; this reduces the volume of protoplasm, and the plasma membrane shrinks away from the cell wall.

osmosis (Gr. *osmos*, push) (Fig. 6-6). The measurable force within living cells is considerable and usually keeps the cell membrane distended.

A solution with greater concentration (less water) than the protoplasm, which draws water from the protoplasm of the cell, is known as a hypertonic solution. In this case water will pass out of the cell in an attempt to equalize the pressure. Under such circumstances (loss of water) animal cells will tend to shrink as a whole because of their delicate cell membrane, whereas the protoplasm of plant cells shrinks away from the rather rigid, resistant cell wall. Such shrinking of protoplasm from the cell wall or membrane during the loss of water is called plasmolysis (Gr. *plasma*, molded; *lysis*, loosening) (Fig. 6-7). A solution with less concentration (more water) than the protoplasm, which places water into the protoplasm of the cell, is known as a hypotonic solution. In this case the addition of water to the protoplasm causes a condition known as turgor (L. *turgere*, to swell). If carried to extreme, the cell may be destroyed. A solution that has the same concentration as the protoplasm and in which there is no net movement of water through the cell membrane is known as an isotonic solution. In this case pressures are equal on both sides of the cell membrane, and there is no shrinking or swelling. There can be no passage of materials to or from a cell under such conditions. It is quite clear that hypertonic and hypotonic solutions around a cell determine the passage of materials out of the cell and into the cell. The obtaining of foods and the elimination of wastes probably are accomplished in this way.

Colloids

Colloids in protoplasm are composed of relatively large particles in the form of large molecules, or clumps of molecules, that are suspended in a liquid medium and often present a sticky gluelike consistency, hence the name colloid (Gr. *kolla*, glue). Changes in protoplasm from the fluid sol state to the semisolid gel state and back again may be explained on the basis of the distribution of the colloidal particles. If the particles are somewhat evenly distributed in the liquid, the mixture is in the sol state and flows easily. If the particles are arranged in a meshwork with the liquid enclosed

61

Table 6-4

Common varieties of colloidal systems

Dispersed phase (discontinuous or internal phase)	Dispersion medium (continuous or external phase)	Examples
Gas	Liquid	Foams; carbonated water
Liquid	Gas	Fog or mist (water droplets in air)
Liquid	Liquid	Emulsions (oil in water; cream; homogenized milk)
Solid	Gas	Smoke (carbon particles in air)
Solid	Liquid	Ink; colloidal gold in water
Liquid	Solid	Gel
Solid	Solid	Stained glass

within the meshes, the mixture is in a gel state and does not flow.

Colloidal mixtures are not unique to protoplasm because they are often found in nonliving substances. A colloidal mixture of gelatin in water is in a sol state when warm but becomes a gel when cooled. This condition can be changed back and forth (reverse of phases) just as it can be in protoplasm. However, not all colloids are able to reverse their phases.

Molecules in the interior of a colloidal particle are attracted equally in all directions by other surrounding molecules, whereas those on the surface of a colloidal particle are subject to unequal forces of attraction (similar to the unequal attraction of molecules on the surface of a liquid). Because colloidal particles are so small and numerous, they possess a great total surface area, so that there are great numbers of molecules on the surface of each particle. As a result, these surface molecules are able to attract and hold other molecules, atoms, or ions through the process of adsorption (L. *ad*, to; *sorbere*, to draw in). This property plays important roles in the living and nonliving worlds. Certain colloids are less selective regarding their

adsorption of ions and tend to adsorb molecules of the medium in which they are dispersed (dispersion medium). As these colloidal particles adsorb the molecules of the dispersion medium, they may swell until the entire colloidal system becomes more viscous, or even semisolid, as in jellies, and gelatin desserts. The swelling of dried fruits also illustrates this phenomenon.

In a colloidal system the particles that are dispersed are larger than ordinary molecules but are not as large as those of a suspension, in which the particles may be large enough to be visible with an ordinary microscope (see Table 6-1). The sizes of colloidal particles vary between 0.000001 and 0.0001 mm. Colloidal particles may be demonstrated with an ultramicroscope in which a strong beam of light is directed parallel to the stage, thus illuminating the dispersing liquid. Particles too small to be seen under ordinary conditions appear as bright specks, because they reflect the strong light that is brought in from the side.

Colloidal systems may be classified according to the physical state of each of the so-called phases, whether solid, liquid, or gas. In protoplasm a liquid may be dispersed in a liquid, or solid particles may be dispersed in a liquid. Many of the proteins are examples of solid particles dispersed in a liquid. In protoplasm the dispersing phase (external or continuous phase) is usually water, and the dispersed phase (internal or discontinuous phase) may be carbohydrates, fatty materials, or proteins. Some of the more common varieties of colloidal systems are given in Table 6-4.

Colloids may also be classified according to the affinity that exists between the dispersion medium and the dispersed phase. When two substances of the two phases do not mix readily, as in the case of water and oil, the bright specks appear distinct and bright under the ultramicroscope. When the substances of the two phases tend to mix, as in the case of starch and most proteins and water, the specks may be somewhat indistinct under the ultramicroscope. The latter phenomenon is thought to result from the indistinct outlines of the particles because of the adsorption of water molecules on their surface.

When a strong beam of light is passed through a colloid, the small colloidal particles suspended in the liquid reflect the light, and the path of light appears as a visible cone known as Tyndall's cone, named for John Tyndall, the British physicist (1820-1893) who discovered it. In this same manner the effect is observed when a beam of light passes through smoke or fog. However, if the beam of light is passed through a true solution of a substance, no such cone is visible.

When colloidal particles are placed between two electrodes of a battery with relatively high voltage, the particles migrate toward either the positive electrode or the negative electrode, depending upon the specific colloid. The electrical charge borne by colloidal particles is caused by the somewhat selective adsorption of positive or negative ions from the surrounding medium, the specific type of ion adsorbed depending upon the particular colloid.

If the proper colloidal particles suspended in a liquid are viewed through an ordinary microscope, the light reflected from the particles reveals them to be moving in an unordered manner, back and forth, in all directions. This is called Brownian movement because it was first discovered in 1827 by the Scottish botanist, Robert Brown. If the particle is fairly large, the movement may appear to be little more than an irregular vibration, but if it is sufficiently small (0.002 mm. or less) it will go in one direction, then in another, in a zigzag course, yet stay near the same spot. The movement of the particles is not caused by the currents in the liquid but by the bombardment of the particles by the moving molecules of the water. Although water molecules are too small to be seen with highly magnifying microscopes, they move with great force and speed, and when they hit the suspended particles, the latter are moved.

Some of the properties and reactions of matter that are in a colloidal state depend upon the great surface displayed by the enormous numbers of minute colloidal particles that constitute that particular matter. The great amount of surface exposed by small particles is illustrated as follows: A cube of matter having edges 1 cm. long has an exposed surface of 6 sq. cm. (six surfaces, each 1 sq. cm. in area). If this cube were divided into similar, smaller cubes, each having edges only 0.01 cm. long, the total number of small cubes

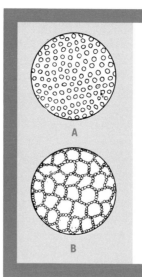

Fig. 6-8
Colloidal states. A, Fluid sol state in which particles move freely; B, gel state in which particles may form a continuous network, thus rendering the substance jellylike (semisolid).

would be 1 million. Each small cube has a surface area of 0.0006 sq. cm., and the total surface area of all the small cubes is 600 sq. cm., or an area one hundred times greater than the original larger cube. However, if the original large cube were divided into extremely minute cubes, each with the size of an average colloidal particle (0.000001 cm. diameter), there would result 1 million billion cubes (each having edges 0.000001 cm. long), and the total surface areas of all the colloidal-sized cubes would be 6,000,000 sq. cm., or one million times as great as the original cube. These 6,000,000 sq. cm. are the equivalent of over 6,500 sq. ft., or a city lot 65 by 100 ft. It must be recalled that the original cube was only 1 cm. square; however, there is an enormous surface exposure when even a small block of matter is properly divided into particles of colloidal size.

In general, colloids are of great importance, because all vital processes of plants and animals are associated with colloidal materials. The living protoplasm of plants and animals is colloidal in nature. As a result, many vital phenomena are based upon the activities of their colloids. The growth of plants, the germination of seeds, and many similar phenomena are associated with colloids and their properties.

Among some of the important roles of colloids in living protoplasm are the following: (1) their enormous surface exposures that permit a great number of chemical reactions; (2) their inability to pass through membranes, which influences the stability of the protoplasm in cells; (3) their ability to reverse phases so that protoplasm can carry on diverse functions and change its physical properties during metabolic activities; (4) their ability to reverse the phase of the colloidal structure of the plasma membrane, thereby affecting its selective permeability and consequently influencing the passage of substances into and out of cells; and (5) their ability to undergo solation or gelation that permits protoplasm to expand and contract, such as is shown in the movements in *Amoeba*.

Surface tension

In surface tension the tension or attraction between molecules on the surface of a liquid is greater than between those beneath. All molecules of a substance exert enormous attraction for each other — a property called cohesion. In the deeper portions of a volume of liquid each molecule is attracted by adjacent molecules with equal force in all directions. However, on the surface of the liquid, the liquid molecules are attracted downward by the lower molecules of the liquid and attracted upward by the molecules of the gases of the air. The attraction of the liquid molecules for each other is greater than the attraction of the gas molecules for the liquid molecules. Hence, the attraction forces on the surface molecules of the liquid are unequal. Equilibrium is attained only when the surface is made as small as possible by reducing the number of liquid molecules on the surface. This produces a tendency for the surface to occupy the least amount of space. When a droplet of oil is immersed in water, the former will assume a spherical shape, and the boundary, known as the interface, between the oil and water is in a state of tension and therefore represents an equilibrium between forces. This tendency for surfaces to contract because of tension is known as surface tension. Naturally, surface tension differs widely among various materials.

Any substance that reduces surface tension is likely to accumulate at the surface. When ether is added to water, the ether molecules accumulate in greater numbers at the surface of the water than elsewhere in the water. The amount of potential energy at the surface of an ether-water mixture is much less than at the surface of pure water. If the area of the surface of a substance is reduced, there is a release of energy. Surface tension in living protoplasm is constantly being reduced by the presence of fats. In protoplasm the energy relation of the interfaces (boundaries) between the colloidal particles and their suspending medium are constantly changing. In the living process new compounds are constantly formed, and different sorts of molecules appear and disappear so that the interfaces are changing almost constantly. These changes in surface tension at the interfaces are associated with many living phenomena.

Energy and matter

The universe is composed of matter and energy. Energy is the ability to produce change or motion in matter, that is, the ability to perform work (Table 6-5). The abilities to produce changes and do work are attributes of living protoplasm. Energy in protoplasm is measured ordinarily by the amount of change or work performed.

Energies are divided into potential and kinetic. Potential energy (stored energy) is the ability to perform work because of the position or condition of atoms, molecules, or larger bodies. Examples of stored potential energies are coal and wood before they are burned and carbohydrates before they are digested. Chemical digestion of foods results in changing the foods' potential energy into heat, light, electricity, or energy of movement. A stationary ball at the top of an inclined plane has potential energy, but it displays kinetic energy as it rolls (motion) down the incline. Kinetic energy (Gr. *Kinein*, to move) is the energy possessed by virtue of motion (active energy). Kinetic energy may become potential, and potential energy may become kinetic. Energy required to form a molecule of substance becomes inactive potential energy when stored in that molecule, but it is converted into active kinetic energy when the molecule is broken down.

All chemical reactions involve changes in energy distribution. Certain chemical reactions require some form of energy, usually heat, whereas others release energy in some form. When a sugar is built,

Table 6-5

Transduction (L. *transducere*, to convey over) of energy by living organisms*

Energy transduction	Energy transduced by
Radiant energy to chemical energy	Chlorophyll
Radiant energy to electrical energy	Eyes
Chemical energy to radiant energy	Luminescent substance (fireflies, certain bacteria)
Chemical energy to electrical energy	Nerves; electric organs (electric fish)
Chemical energy to mechanical energy	Muscles
Chemical energy to osmotic energy	Cell membranes
Oxidative energy to usable chemical energy	Mitochondria of cells
Sonic energy to electrical energy	Ears

*Only a few of the several kinds are shown.

energy is required; when it is broken down, energy is released. The construction and destruction of other foods reveal similar phenomena. The ultimate natural source of energy of foods produced by green chlorophyll-bearing plants is the sun. The energy value of a food is measured by a unit called a Calorie, which is the amount of heat required to raise the temperature of 1 kg. (1,000 gm.) of water 1° C. One gram of fat produces about 9 Calories of heat; 1 gm. of carbohydrate, about 4 Calories; 1 gm. of protein, about 4 Calories.

Use of energy by organisms
Production and use of heat

When energy is released, there is an accompanying production of more or less heat. In some instances this heat is used to regulate chemical activities or control body temperature, whereas in others the heat is a waste product that is no longer of use to the organism. In the formation of certain chemical compounds there is often some heat produced. In some instances, as in the spontaneous combustion of hay, the amount of heat produced is sufficiently great to start a fire. In the destruction of chemical compounds, usually by oxidation, a certain amount of heat is liberated. For example, the oxidation of such foods as carbohydrates, fats, and proteins releases heat for use by the living organism.

A living organism that generates large amounts of heat through its activities is frequently not very efficient in this respect. In such cases much more heat is liberated than is required by that organism. In general, plants are more efficient in this respect than animals. Much of the heat acquired by plants is absorbed from the surroundings. So-called cold-blooded animals attempt to maintain a body temperature somewhat similar to that of their environment, whereas warm-blooded animals generate and conserve heat so as to maintain a rather constant temperature, regardless of environmental factors. Animals lose heat (1) by conducting it to other objects, (2) by radiating it, (3) by losing it through feces and urine, and (4) by evaporation from the lungs and skin.

Bioelectric phenomena

Protoplasm contains numerous electrolytes. The ions into which acids, bases, and salts dissociate confer charges on surfaces on which they may accumulate. Hence, colloidal particles, each bearing a minute charge, may be changed as chemical changes occur in the protoplasm or as ionizing substances are introduced from the outside. The effects on colloidal particles of protoplasm by inorganic and organic substances brought to it may assist in explaining the many variations in living phenomena.

In certain species of fishes there are modified muscle cells that are arranged in series to serve as

65

electric organs. In such organs the electricity is produced, stored, and discharged into the surrounding water for offensive and defensive purposes. In these electric organs the positive pole of one cell is arranged against the negative pole of the next, so that the voltage produced is determined by the number of cells arranged in the series.

Bioluminescence and light

Bioluminescene (Gr. *bios,* life; L. *luminescere,* to produce light) is a phenomenon of light production, displayed by certain organisms. In some bacteria and fungi part of the energy released during respiration is used in the production of so-called "cold light" with little heat (1%) being formed. Bioluminescent bacteria are found in decaying wood and leaves, in fish, and in salt water. The production of this type of visible light is controlled by the action of the enzyme, luciferase, acting on a substrate called luciferin (L. *lux,* light; *ferre,* to carry), and the production may be continuous if sufficient oxygen and proper foods are available. Luminous bacteria in the eyes of certain fishes in East India emit light constantly. The light emitted by plants is the result of oxidative metabolism, and its usefulness to such plants is an unsolved problem. No green plants have the ability to produce such light.

Luminescence is displayed by such animals as the firefly (beetle), the glowworm, certain squids and fishes, certain jellyfishes and shrimp, and certain species of protozoans. In the firefly, photogenic organs containing localized masses of fatty substances produce light by oxidizing the fatty substances. The photogenic organs are well supplied with oxygen by a copious quantity of tracheal tubes. The greenish yellow light has few nonluminous rays. Its emission is controlled by regulating the oxygen supply. The light seems to be associated with sexual attraction, the female generally producing flashes of a longer duration. In luminous squids and fishes there are organs, lenses, and reflectors to reflect the glow. In the jellyfish (*Pelagia noctiluca*), the surface of the umbrella is covered with glowing granules. In the protozoan, *Noctiluca,* the luminous granules remain inside the cell.

Light affects animals in several ways. The earthworm has no eyes, yet it moves away from light because of light-sensitive cells near the surface. Certain protozoans, planarians, clams, snails, and certain crustaceans are also affected by light. The simple eyes of insects and spiders are influenced by light intensities. The compound eye of arthropods is constructed like a bundle of hollow tubes arranged in the form of a cone. The tubes are isolated from each other by black pigment, and together they produce a reduced image of the object being viewed. The outer end of each tube contains a lens and a facet that are seen on the surface of the compound eye. The inner ends of these tubes possess light-sensitive materials connected with nerves. These eyes also give the organism an interpretation of movement of objects.

The eyes of vertebrate animals act somewhat like a camera. The lens focuses and forms an image on the black, light-sensitive retina on the inside of the eyeball. The retina consists of enormous numbers of nerve cells with chemicals that are changed temporarily by light. Each temporary image produces chemical changes in the nerve cells that vary with the quantity of light on each cell. These chemical changes stimulate other nerve cells that send impulses over the optic nerve to the brain.

Production and reception of sound

The vibration of some sounding body produces sound waves that are borne to and interpreted by a specialized organ, such as the ear of higher animals. Plants and lower animals do not produce sounds in the accepted sense, although some may be affected by sound waves.

In several higher animals sounds are produced and interpreted in some manner. Almost every insect that has sound-receiving mechanisms also has sound-producing (stridulating) organs. In the common locust there are two types of stridulation. When at rest, certain species draw the femoral joint of the hind leg across a specialized vein of the wing cover to produce sound. When flying, they produce a crackling sound by rubbing wings and wing covers together. Tympanic membranes connected by nerves to the nervous system are assumed to be auditory organs.

The female mosquito produces a characteristic

sound by vibrating her wings 512 times per second. In male mosquitoes the hairs on the antennae are auditory. The hairs are adjusted during flight so that the two plumelike antennae are stimulated equally by the wing sounds produced by the female thus directing the male toward the female.

In the cicada the male has a pair of large, ridged, parchmentlike drumheads on the first abdominal segment beneath the wings. The drumheads are vibrated by a pair of muscles. A pair of cavities within the body act as resonators for the sounds produced. The female cicada has no sound-producing or sound-receiving apparatus. In the katydid the stridulating organs consist of a rough file and a scraper on the wing covers. The sound-receiving apparatus consists of a series of tympanic chambers with membranous tympana that pick up the sound vibrations (chirp) and transmit them to the nervous system. The chambers intensify the sound.

The honeybee produces its humming sound by moving its wings 190 times per second. The housefly produces its buzzing sound by completing 330 wing strokes per second. Many insects, especially those with heavy, sclerotized exoskeletons, possess spines and hairs attached to nerves by means of which they recognize or "feel" sound vibrations.

Lower fishes are frequently affected by stimuli produced by changes in the position of the fish. Hence, such fishes maintain a typical position with respect to their surroundings. Well-developed vocal cords and organs of hearing appear only in terrestrial vertebrates. Eardrums and vocal cords are absent in fishes.

Amphibia (frogs and toads) have the simplest of vocal cords. It seems that the developments of sound-production and sound-reception mechanisms go together and that they are rather closely correlated. Male and female frogs (*Rana pipiens*) produce different kinds of croaking sounds by forcing air back and forth from the mouth cavity and lungs across their vocal cords. Frogs produce a "pain scream" when caught, a "grunting" sound when satisfied, and an "alarm cry" when startled.

In the ears of higher vertebrates the semicircular canals function as an organ of equilibrium, and the cochlea ("snail shell") receives sound waves and sends them over the auditory nerve to the brain. The human vocal apparatus and ears are described elsewhere.

CHEMICAL COMPOSITION OF ORGANISMS

An analysis of various organisms reveals that protoplasm is composed of about thirty-six elements, with perhaps twelve being predominant.

From Table 6-6 it will be noted that the first four common elements make up about 99% of protoplasm, whereas eight other common elements make up approximately 1%. It has been estimated that an average human body contains about 95 pounds of oxygen, 30 pounds of carbon, 15 pounds of hydrogen, 10 pounds of calcium phosphate, 7 ounces of sodium chloride, 4 ounces of sulfur, 3 ounces of potassium, $\frac{1}{4}$ ounce of iron, and tiny amounts of iodine, copper, and a few other elements.

These elements usually occur in the form of inorganic salts, water, and organic compounds as in Table 6-7.

Table 6-6

Elements present in protoplasm

Element	Percent (by weight)
Oxygen (O)	76.0
Carbon (C)	10.5
Hydrogen (H)	10.0
Nitrogen (N)	2.5
Phosphorus (P)	0.3
Potassium (K)	0.3
Sulfur (S)	0.2
Chlorine (Cl)	0.1
Sodium (Na)	0.04
Calcium (Ca)	0.02
Magnesium (Mg)	0.02
Iron (Fe)	0.01

In addition there may be traces of copper, cobalt, manganese, zinc, and others, depending upon the protoplasm

Table 6-7

Percentages of compounds in protoplasm

Organic compounds (percent by weight)		Inorganic compounds (percent by weight)	
Proteins	15	Water	80
Fats	3	Inorganic salts	1
Carbohydrates	1		

Table 6-8

Common ions in protoplasm

Positive	(cations)	Negative	(anions)
Sodium	Na^+	Chloride	Cl^-
Potassium	K^+	Bicarbonate	HCO_3^-
Calcium	Ca^{++}	Carbonate	CO_3^{--}
Magnesium	Mg^{++}	Hydrogen phosphate	HPO_4^{--}
Iron	Fe^{+++}	Phosphate	PO_4^{---}

Table 6-9

Some common carbohydrates

Type	Sugar	Molecular formula
Monosaccharides		
Pentose		
(5 carbon)	Ribose	$C_5H_{10}O_5$
	Deoxyribose	$C_5H_{10}O_4$
Hexose		
(6 carbon)	Glucose	$C_6H_{12}O_6$
	Fructose	$C_6H_{12}O_6$
	Galactose	$C_6H_{12}O_6$
Disaccharides	Lactose	$C_{12}H_{22}O_{11}$
	Maltose	$C_{12}H_{22}O_{11}$
	Sucrose	$C_{12}H_{22}O_{11}$
Polysaccharides	Starch	60 to 80 glucose units
	Glycogen	12 to 16 glucose units
	Cellulose	100's of glucose units

Inorganic salts and water

Most inorganic constituents of protoplasm occur in the form of ions. Most of the inorganic compounds in protoplasm occur in the form of electolytes, and many of the organic compounds occur in the form of nonelectrolytes. When electrolytes dissolve in water, they break up into ions by the process of ionization (Table 6-8). Nonelectrolytes, when in solution, occur as molecules dispersed among molecules of the solvent. Many of the vital processes are influenced by the presence of inorganic substances in the protoplasm and body fluids, even though they may be present in very small quantities.

Among the functions of common inorganic salts are the following: Iodine helps to form thyroxin in the thyroid gland; iron helps to form hemoglobin in red blood corpuscles; calcium helps to construct bones and teeth. If calcium is deficient in the blood of mammals, convulsions and death may occur; if calcium, sodium, and potassium are not balanced properly, heartbeats may be abnormal.

Water constitutes about 80% of average protoplasm and is a very important compound because it is the solvent in which most other compounds are dissolved or suspended, in which they interact, and in which they move. In a sense water is the arena in which life processes perform. Water also has a capacity to absorb and conduct heat and thus protects protoplasm against sudden temperature changes.

Organic content

Most compounds containing carbon are referred to as organic compounds. Carbohydrates, lipids (including fats), proteins, nucleotides, nucleic acids, and enzymes are among the more important types.

Carbohydrates

Carbohydrates (L. *carbo*, carbon or coal; Gr. *hydro-*, water) contain only three elements—carbon, hydrogen, and oxygen. There are many different kinds of carbohydrates, differing only in the number and arrangement of the three kinds of atoms in their molecules (Fig. 6-9). In carbohydrates the ratio of hydrogen to oxygen atoms is two of hydro-

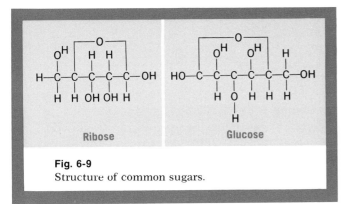

Fig. 6-9
Structure of common sugars.

gen to one of oxygen, the same as in water. Certain carbohydrates may even have the same chemical formula yet possess different properties, as shown in Table 6-9. In fact, in all organic substances the arrangement of the atoms is of great significance. Carbohydrates supply energy for protoplasmic activities by being oxidized, that is, when their atoms are combined with oxygen or when hydrogen is taken away from them.

The significant occurrence is that food in the form of glucose can be oxidized to provide energy for life in the cell. The liberated energy is used for vital activities. Carbohydrates are usually stored as glycogen in animals and as starch in plants (Fig. 6-10).

Simple sugars, such as glucose, may combine to form disaccharides (double sugars) by losing a molecule of water. For example, sucrose is the result of the enzymatic union of a molecule of glucose with a molecule of fructose. In the process a molecule of water is formed.

$$\underset{\text{Glucose}}{C_6H_{12}O_6} + \underset{\text{Fructose}}{C_6H_{12}O_6} \xrightarrow{\text{Enzyme}} \underset{\text{Sucrose}}{C_{12}H_{22}O_{11}} + \underset{\text{Water}}{H_2O}$$

The foregoing is also the general equation for the formation of maltose from two molecules of glucose, and of lactose from glucose and galactose. Note that these simple sugars differ in structure, although their molecular formula is identical.

Fats

Fats are compounds formed by the enzymatic reaction of glycerin ($C_3H_8O_3$) with three molecules of fatty acids ($C_nH_{2n}O_2$). In the process, three molecules of water are split out (Fig. 6-11). Fats are energy sources and play an important part in membrane formation in cells.

The length of the fatty acid chains determines the physical state of the fat, that is, whether it is an oil, a semisolid, or is tallowlike and hard. (Fig. 6-12) Since there are many different fatty acids, and three combine with glycerin, there are many different types of fats (Table 6-10).

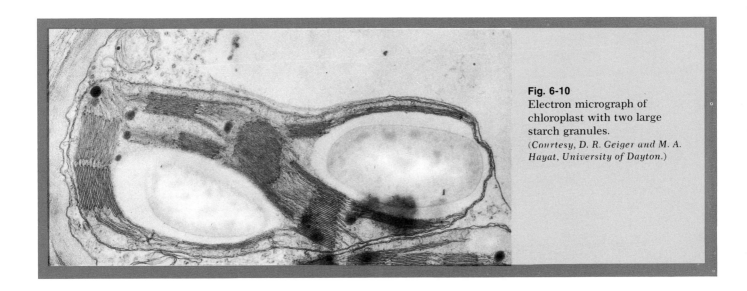

Fig. 6-10
Electron micrograph of chloroplast with two large starch granules.
(*Courtesy, D. R. Geiger and M. A. Hayat, University of Dayton.*)

Glycerin + 3 Fatty acids ⟶ 1 Molecule fat + $3H_2O$

Glycerin + 3 Stearic acid ⟶ 1 Molecule tristearin + $3H_2O$

Fig. 6-11
Formation of a fat molecule.

The oxidation of a typical fat takes place as follows:

$$2\ C_{51}H_{98}O_6 + 145\ O_2 \xrightarrow{\text{Enzyme}} 102\ CO_2 + 98\ H_2O + \text{Energy}$$

Tripalmitin

Proteins

Basic combinations of carbon (C), hydrogen (H), oxygen (O), nitrogen (N), sulfur (S), and sometimes other elements are found in proteins as in Table 6-11. These are united in an amino acid as follows:

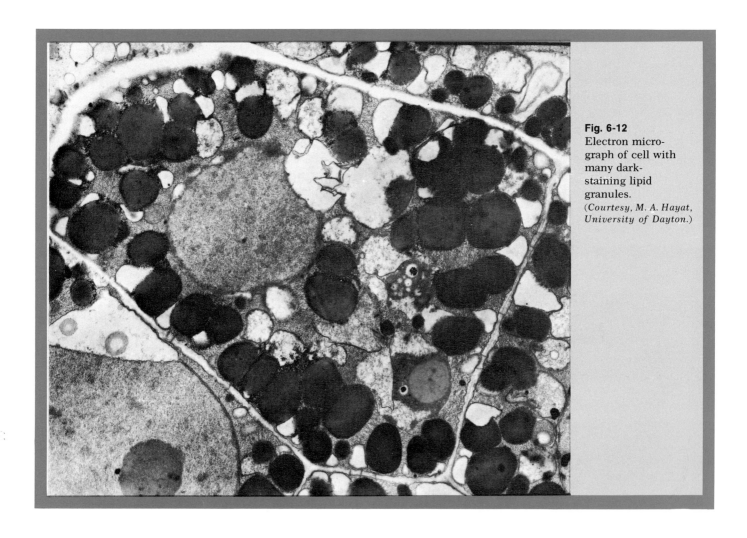

Fig. 6-12
Electron micrograph of cell with many dark-staining lipid granules.
(Courtesy, M. A. Hayat, University of Dayton.)

In this the NH_2 is the amino group, the COOH, usually called a carboxyl group, represents an organic acid, and the R represents the variety of chemical combinations that are found in the twenty or so naturally occurring amino acids.

Some molecules contain as many as nineteen different amino acids. In large protein molecules the number of combinations of amino acids may be enormous. Some of the common proteins are given in Table 6-11.

The molecules of amino acids are both acidic (because of the carboxyl group) and basic (because of the amino group) and thus can unite with each other to form larger and more complex molecules. When an amino acid is placed in water, its molecules may rearrange, and a hydrogen ion becomes detached from the carboxyl group, leaving it nega-tive, and it becomes attached to the amino group, making the latter positive. This gives the amino acid molecule an electric polarity whereby the positive pole of one amino acid may attract the negative pole of another amino acid, and their combining may form one molecule with the loss of a molecule of water where they join. The newly formed molecule is also polarized, and the union or linkage is called peptide linkage. A compound formed by the union of any two amino acids is called a dipeptide and the dipeptide molecule has a COOH group and an NH_2 group, similar to the original amino acid. Consequently, a dipeptide can unite with a third amino acid molecule to form a tripeptide. When still larger numbers of amino acids unite with each other, polypeptides originate. A number of polypeptides uniting give rise pro-gressively to peptones, proteoses, and proteins. In this the molecules become increasingly more complex and the molecular weights larger.

Proteins may be of two kinds: (1) a fibrous form, in which the so-called foundation is extended and forms a threadlike axis of the molecule, and (2) a globular form, in which the coiled threads are bent upon themselves to form a mass and held in posi-tion by weak bonds. Examples of fibrous proteins are cell walls, muscles, connective tissues, and fibers of silk and wool; examples of globular pro-teins include the albumin of egg and many living materials.

Proteins play important roles in the construction and activities of many parts of cells. Next to water, proteins are the most common constituents of protoplasm. Proteins may release energy when used as foods and may assist in forming enzymes and hormones. It is believed that the cell membrane is an infinitesimally thin meshwork of fine, elon-gated protein molecules enmeshing fatty materials. The cytoplasm of cells is considered to be a highly organized, intricate meshwork of protein mole-cules that is responsible for many of the functions of cells. Proteins are common in such animal materials as meat, fish, milk, eggs, and in such plant materials as beans, peas, and nuts.

Nucleic acids

Nucleic acids are formed by the union of phos-phoric acid (H_3PO_4), a 5-carbon (pentose) sugar,

Table 6-10
Some common fats

Fat	Molecular formula
Triacetin (simple fat, in cod-liver oil)	$C_9H_{14}O_6$
Tripalmitin (common in fats and oils)	$C_{51}H_{98}O_6$
Tristearin (common in mammals)	$C_{57}H_{110}O_6$

Table 6-11
Some common proteins

Protein	Molecular formula
Albumin (white of egg)	$C_{239}H_{389}O_{78}N_{58}S_2$
Zein (corn)	$C_{736}H_{1161}O_{208}N_{184}S_3$
Hemoglobin (oxygen-carrying blood pigment)	$C_{3032}H_{4816}O_{872}N_{780}S_8Fe_4$

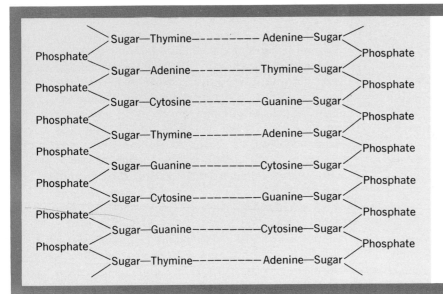

Fig. 6-13

Structure of DNA (deoxyribonucleic acid), showing the complementary structure of two chains. Sugar is deoxyribose. Thymine ($C_5H_7N_2O_2$) and cytosine ($C_4H_5N_3O$) are known as pyrimidine bases. Adenine ($C_5H_5N_5$) and guanine ($C_5H_5N_5O$) are known as purine bases. Two complementary chains twisted around one another are connected by hydrogen bonding (– – – –). (For the structure of bases see Fig. 6-14; for the structure of sugar see Fig. 6-15.)

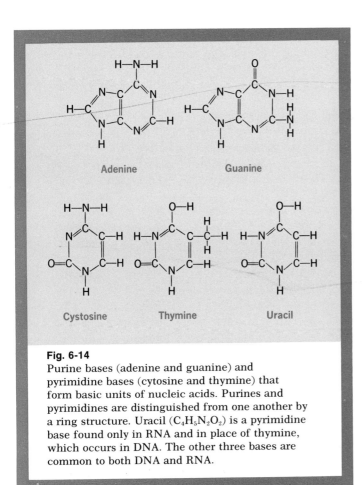

Fig. 6-14

Purine bases (adenine and guanine) and pyrimidine bases (cytosine and thymine) that form basic units of nucleic acids. Purines and pyrimidines are distinguished from one another by a ring structure. Uracil ($C_4H_5N_2O_2$) is a pyrimidine base found only in RNA and in place of thymine, which occurs in DNA. The other three bases are common to both DNA and RNA.

and bases such as adenine, guanine, cytosine, and thymine or uracil (Figs. 6-13 and 6-14). Two forms of nucleic acids are known as deoxyribonucleic acid (DNA) and ribonucleic acid (RNA), which are similar chemically (Table 6-12). Each consists of a long chain of molecules of phosphate and pentose sugar with small side groups called bases attached to the sugars. DNA and RNA both contain the bases (adenine, guanine, and cytosine), but DNA contains a fourth base (thymine) and RNA contains a fourth base (uracil). The bases along the backbone of the nucleic acid follow a sequence (Fig. 6-13) that in each case has a particular meaning and determines the function of the molecule, just as the sequence of letters in words and sentences convey a certain meaning. This sequence makes a gene specific for a particular polypeptide formation (see Chapter 30 for details). The two pentose sugars vary slightly in DNA and RNA (Fig. 6-15). Nucleic acids occur in each living cell and seem to direct the manufacture of proteins and control the hereditary constitution of all living organisms.

Chemically, nucleic acids are made of units called nucleotides (nu′ kle o tides). A nucleotide is a molecule composed of joined phosphate, a pentose sugar (either deoxyribose or ribose) and a so-called purine base (adenine or guanine), or a pyrimidine base (cytosine, or thymine or uracil). Hundreds of nucleotides may be joined together by means of the phosphoric acid to form a mole-

Table 6-12

Relationships between DNA and RNA

DNA (Deoxyribonucleic acid)	RNA (Ribonucleic acid)
Contains pentose (5-carbon) sugar called deoxyribose	Contains pentose (5-carbon) sugar called ribose
Contains bases adenine, guanine, cytosine, and thymine	Contains bases adenine, guanine, cytosine, and uracil
Contains phosphoric acid (phosphate) that connects various sugars with one another	Contains phosphoric acid (phosphate) that connects various sugars with one another
DNA is genetic material of life	RNA present in large amounts in nucleoli produced by certain chromosomes; at certain times stored RNA of nucleoli may be discharged into cytoplasm
DNA always associated with chromosomes (genes); each set of chromosomes seems to have fixes amount of DNA	RNA is found mainly in combination with proteins in ribosomes in the cytoplasm as messenger RNA and as transfer RNA

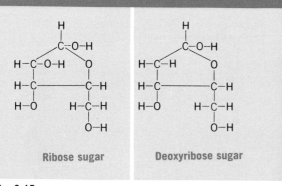

Fig. 6-15
Sugars (pentose or 5-carbon types) that form basic units of nucleic acids. Ribose sugar is found in ribonucleic acid (RNA) and deoxyribose sugar is found in deoxyribonucleic acid (DNA).

cule of nucleic acid. For example, DNA is formed by the repetition of many units held together like a "chain" and thus is said to be polymerized. Polymerization is a joining together of several small molecules to form a larger unit known as a polymer (Gr. *polys*, many; *meros*, part). In fact, DNA is called a polynucleotide. The distinctive part of each nucleotide is its purine or pyrimidine base, the sugar acting as a backbone to which the bases are attached.

By x-ray diffraction studies and by special electron microscope techniques, DNA is found to be a double molecule, with one chain twisted around the other in a helical manner. The bases of one chain fit onto the bases of the other chain, and a given base on one chain is opposite a specific base on the other. For example, thymine pairs only with adenine, and cytosine only with guanine. The sequence of bases on one chain determines the sequence of bases on the opposite chain. The two chains seem to be held together by hydrogen bonds between the bases.

Enzymes

The presence of enzymes in living matter ensures that the rate of chemical reaction is sufficient to maintain life. Enzymes then may be considered as organic catalysts. Some of them are entirely protein, while others consist of a protein fraction,

73

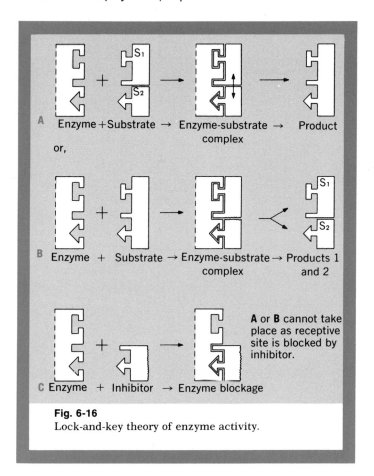

A Enzyme+Substrate → Enzyme-substrate → Product
complex

or,

B Enzyme + Substrate → Enzyme-substrate → Products 1
complex   and 2

C Enzyme + Inhibitor → Enzyme blockage

A or **B** cannot take place as receptive site is blocked by inhibitor.

Fig. 6-16
Lock-and-key theory of enzyme activity.

may prevent the formation of an enzyme-substrate complex by blocking an active site on the molecule (Fig. 6-16).

A particular enzyme is able to accelerate a reaction in either direction, since it does not determine the direction of the reaction. The enzyme converting glucose into starch is also active in the breakdown of starch into glucose.

Generally, two systems of naming enzymes are in use. In one system the name of the substrate with an -ase ending is used to denote the enzyme, as in lactase and lactose and in lipase and lipids.

In the second system the type of activity involved is used as the name, again with the -ase ending. Thus, one group involved in the splitting of a molecule with the aid of water is called hydrolases. For example, the digestive process is often referred to as one of enzymatic hydrolysis.

Since the sum of metabolic activity is enormous and the rate of reaction is very fast, we may infer the presence of hundreds of kinds of enzymes, differing in the amino acid sequence of their proteins, in their coenzymes, and in the conditions in which they will react.

the apoenzyme, and an additional part, the coenzyme.

Since enzymes are primarily protein, they show the same properties, being denatured by such things as heat, change of pH, and ion changes. They also are involved in very exact reactions, a property known as specificity. As is true with catalysts in general, enzymes are effective in extremely small quantities, being capable of influencing many times their weight in reactants. Also, they are not used up in the reaction. These reactants are referred to as the substrate, and while reacting a temporary enzyme-substrate complex is formed. It is thought that the interaction is possible because of the molecular shape of the enzyme, the protein structure forming a type of "lock" into which a substrate "key" will fit. Enzyme inhibitors

Review questions and topics

1 Define matter, atom, ion, element, compound, molecule, and colloid, with examples of each.
2 Contrast molecular solutions, colloidal solutions, and suspensions, with examples of each.
3 Discuss the roles of colloids in living protoplasm.
4 Discuss each of the physical properties of protoplasm, including its significance.
5 Discuss the chemical composition of protoplasm, including the common elements present.
6 Distinguish between organic and inorganic compounds, with examples.
7 Discuss the important characteristics of carbohydrates, fats, and proteins, and the roles of each in the metabolic activities of protoplasm.
8 Define a Calorie, including its significance from a dietary standpoint.
9 Discuss the functions of inorganic salts in the living processes of organisms.
10 Discuss the importance of water as a constituent of living protoplasm.
11 Discuss the general characteristics and properties of enzymes, including their importance in living processes.

Some contributors to the knowledge of protoplasm

Purkinje

Johannes Purkinje (1787-1869)
A Bohemian physiologist, applied the term protoplasm (Gr. *protos*, first; *plasma*, liquid) to the jellylike substance (1840), although the meaning of the term was probably somewhat different from the meaning in later usage. (*The Bettman Archive, Inc.*)

Félix Dujardin (1801-1860)
A French zoologist, observed the jellylike material in animal cells (1835) and applied the term sarcode (Gr. *sarx*, flesh) to it. This substance was later found in living plant cells.

von Mohl

Hugo von Mohl (1805-1872)
A German botanist, found that plant cells were composed of the living substance (1846) and used the term protoplasm much as we do today. (*Historical Pictures Service, Chicago.*)

Max Schultze (1825-1874)
A German cytologist, enunciated the Protoplasmic Theory and stated that the jellylike protoplasm was similar in plant and animal cells (1861). He concluded that "the cell is an accumulation of living substance or protoplasm definitely delimited in space and possessing a cell membrane and nucleus."

Huxley

Thomas H. Huxley (1825-1895)
An English biologist, referred to protoplasm as "the physical basis of life" (1868). (*Historical Pictures Service, Chicago.*)

Otto Bütchli (1848-1920)
A German protozoologist and physiologist, made experimental observations on nonliving substances (mixtures of olive oil, potassium carbonate, and glycerin) that simulated some of the movements and appearance of living protoplasm.

Max Verworn (1862-1921)
A German physiologist, suggested that chemical particles with special chemical actions are what we really mean by life.

Selected references

Baker, J. J. W., and Allen, G. E.: Matter, energy and life, Palo Alto, Calif., 1965, Addison-Wesley Publishing Co., Inc.

Barrington, E. J. W.: The chemical basis of physiological regulation, Glenview, Ill., 1968, Scott, Foresman and Company.

Calvin, M., and Jorgenson, M. J. (editors): Bio-organic chemistry. Readings from Scientific American, San Francisco, 1968, W. H. Freeman and Co. Publishers.

Christensen, H.: pH and dissociation, Philadelphia, 1964, W. B. Saunders Company.

Crick, F. H. C.: Nucleic acids, Sci. Amer. 197:188-200, 1957.

Epstein, H. T.: Elementary biophysics, Reading, Mass., 1963, Addison-Wesley Publishing Co., Inc.

Glassman, E. (editor): Molecular approaches to psychobiology, Belmont, Calif., 1968, Dickenson Pub. Co., Inc.

Haynes, R. H., and Hanawalt, P. C. (editors): The molecular basis of life. Readings from Scientific American, San Francisco, 1968, W. H. Freeman and Co., Publishers.

Ingram, V. M.: The biosynthesis of macromolecules, New York, 1965, W. A. Benjamin, Inc.

Jellinck, P. H.: The cellular role of macromolecules, Glenview, Ill., 1967, Scott, Foresman and Company.

McElroy, W. D.: Cell physiology and biochemistry, Englewood Cliffs, New Jersey, 1964, Prentice-Hall, Inc.

White, E. H.: Chemical background for the biological sciences, Englewood Cliffs, New Jersey, 1964, Prentice-Hall, Inc.

Whittingham, C. P.: The chemistry of plant processes, New York, 1964, Philosophical Library, Inc.

Chapter seven

The kinds of living things

Late in February, 1969, a very large "thing" washed ashore on a Mexican beach. Its sudden appearance and large size caused considerable commotion among the local populace. Several reactions were immediately evident. The first of these was "What is it?" As no answer was immediately forthcoming, more practical questions were asked, such as, "Can we eat it?" Several intrepid souls cut off chunks and proceeded to find out. No report is available as to their findings. Another group proceeded to cut off desirable features as souvenirs, such as pieces of a protruding tusk or tooth and parts of flippers.

News of the find was slow in reaching scientific circles. The creature had begun to rot and to suffer from sun and water when the first scientists arrived on the scene. To date it has not been identified.

The same type of report was given out not too long ago concerning the discovery of a strange fish in the Indian Ocean. After some study it proved to be a coelacanth, a lobe-finned fish, formerly believed to be extinct. The species of fish is still occasionally being found since the discovery of the first one in 1939. There is some thought that the famous "Loch Ness Monster" will turn out to be one of these "extinct" forms.

Similar discoveries are made frequently by specialists who study various kinds of plants or animals.

The reason for citing these examples is to point out that the question "What is it?" is a very human

inquiry. Beyond the layman's curiosity lies the domain of the taxonomist who is concerned with the various kinds of living things, their similarities and differences, and their place in nature. The taxonomist's scientific interest can be traced to the first time early man noticed that some animals have horns and others do not, that some plants made him sick and others did not. The earliest taxonomists, therefore, were very practical men.

The world of life is a vast assemblage of living things, differing in many ways and similar in many others. The number of different organisms is well over one million, with some estimates being over two million. In order to understand them we must know something about classification.

Any system of classification, regardless of the subject, may have several objectives. Perhaps the simplest of these is just keeping track of things. Another aspect is more descriptive, that is, the recognition of the fact that various groups have something in common. Still another objective might be shown by the way the groups are considered.

An example of this is the grouping of students. A class of English majors may be distinct from a class of history or biology majors. All students in one of the majors have something in common with each other, and each group is different in some ways from the other groups. However, when it is also specified that these are all sophomore or junior classes, we recognize that all of them have something in common. This same situation exists with systems of biologic nomenclature and classification.

CLASSIFICATION OF ORGANISMS

The basic group of organisms in biologic classification is the species. The species may be considered a group or population of organisms having many characteristics in common and differing from all other groups in some significant ways. A species is considered to be capable of interbreeding to produce fertile offspring. Normally members of two different species do not interbreed.

Two or more species having some characteristics in common form a genus (pl. genera). Similarily, two or more genera with common characteristics form a family. Families are grouped into orders, orders into classes and classes into phyla. In dealing with plants, the term *phylum* is replaced by division. All of the phyla (or divisions) constitute a kingdom.

Binomial nomenclature

Because language barriers present problems to scientists of different nationalities who are studying the same species, the scientific name of the species is used. The scientific name of a particular organism consists of the genus and species and, hence, is called binomial (two names) nomenclature. Examples of bionomial nomenclature are: man, *Homo sapiens;* housefly, *Musca domestica;* corn, *Zea mays;* and pink bread mold, *Neurospora crassa.* In strict practice the binomial is followed by the name of the person who first described the species. Thus the full scientific name for the sparrow is *Passer domesticus* Linnaeus (often abbreviated as Linn. or L.)

The first real basis of modern nomenclature is due to the work of Carolus Linnaeus. He developed the binomial system through ten editions of his *Systema Naturae.* This tenth edition, published in 1758, is generally accepted as the starting point of modern biologic nomenclature. Many others have contributed to the modern concept of taxonomy and nomenclature, and much discussion is still going on.

THE KINGDOMS OF LIFE

Even though they are both very large, no one would mistake a pine tree for an elephant. However, when there are no obvious differences between two organisms, problems in taxonomy and nomenclature arise. Everyone is agreed on the designations plant and animal, and these designations are easy to accept in the case of the pine tree and the elephant.

Many microscopic forms, however, possess characteristics of both plants and animals and, hence, are classified as both. An alternative treatment places all of the problem species in a third kingdom, the Protista. This classification really just avoids the problem, since the Protista kingdom becomes an obvious catch-all. Still another kingdom, the Monera, is reserved for the bacteria and blue-green algae. Yet, these classifications manage to avoid the question, "What is a virus?"

Undoubtedly each instructor has his own preference in classifying problem organisms. In this text we will consider all organisms as being in three kingdoms: Monera, the bacteria and blue-green algae; Plantae, the plants, obvious and not so obvious; and Animalia, the animals.

Monera

The bacteria and blue-green algae are a strange group of organisms, having in common the lack of many "typical" cell structures. They lack nuclear membranes, mitochondria, plastids, lysosomes, Golgi apparatus, and endoplasmic reticula. Since they are considered in detail in the next chapter, further discussion of them is deferred.

The plant kingdom

Over 300,000 species of plants have been studied. It would be impossible to consider large numbers of them in a beginning course. However, the best way to learn something about plants is to study a few representative types from the major groups into which the plant kingdom is divided. Naturally, the greater the number of representative plants studied, the more complete the knowledge of the plant world would be.

One system of classification divides the plant kingdom into (1) subkingdom Thallophyta (tha-lof' i ta) (Gr. *thallos,* sheetlike; *phyta,* plants) and (2) subkingdom Embryophyta (em bri -of' i ta) (Gr. *embryon,* embryo; *phyta,* plants) (Table 7-1). It will be noted that the Thallophyta contain mostly simple plants, such as algae and fungi, that do not produce many-celled embryos, whereas the Embryophyta contain more complex plants that do produce multicellular embryos. In this classification the subkingdoms are divided into groups that will be given scientific names in our future studies.

Subkingdom Thallophyta
General characteristics

The thallophytes are simply constructed, some species being unicellular (one celled), some consisting of a linear series of cells or in some cases "balls" of cells, whereas others are composed of sheetlike masses of cells. They usually live in water or moist places and are among the oldest of plants. They are without true leaves, stems, or roots, al-

Table 7-1
Some distinguishing characteristics of plants

Plant	Chloro-phyll	Multi-cellular embryos	True leaves, stems, and roots	Conducting tissues (phloem and xylem)	Flowers	Seeds exposed (naked)	Seeds enclosed
Subkingdom Thallophyta							
Fungi	Absent	Absent	Absent	Absent	Absent	Absent	Absent
Algae	Present	Absent	Absent	Absent	Absent	Absent	Absent
Subkingdom Embryophyta							
Liverworts and true mosses	Present	Present	Absent	Absent	Absent	Absent	Absent
Club "mosses," horsetails, and ferns	Present	Present	Present	Present	Absent	Absent	Absent
Gymnosperms (conifers and allies)	Present	Present	Present	Present	Absent	Present	Absent
Angiosperms (flowering plants)	Present	Present	Present	Present	Present	Absent	Present

though in certain species there may be structures that resemble them in a general way. Certain higher types of thallophytes are multicellular, some being more than 200 feet long. In general, they do not possess rigid tissues so they can not grow upright to any great extent. In contrast to higher plants thallophytes do not possess true vascular (conducting) tissues, phloem and xylem, for conducting water and foods. Since many of them live in water, they do not need xylem for water conduction, nor do they need supporting tissues. Food is passed from cell to cell.

Reproductive spores are formed in sporangia (spor -an′ ji a) (Gr. *spora*, spore; *angios*, case), which are usually unicellular structures. When sex cells (gametes) are formed, they are produced in gametangia (gam e -tan′ ji a) (Gr. *gamos*, marriage), which are usually unicellular structures. When an egg is fertilized by a sperm, a single-celled zygote is produced. This zygote does not develop into a multicellular embryo, as does the zygote of the Embryophyta.

Many thallophytes are of economic importance, either beneficially or detrimentally. This will be discussed in Chapter 9.

The thallophytes may be divided into two groups: (1) the algae (al′ ji) (L. *alga*, seaweed) and (2) the fungi (fung′ ji) (L. *fungus*, mold). Algae possess chlorophyll that can combine carbon dioxide and water in the presence of energy-supplying light to produce carbohydrates through the process of photosynthesis (fo to -sin′ the sis) (Gr. *phos*, light;

syntithenai, to build). Fungi lack chlorophyll, so must depend on outside sources for their foods. Other characteristics of algae and fungi will be discussed in Chapters 9 and 10.

Subkingdom Embryophyta

General characteristics

Embryophyta produce a multicellular embryo (Table 7-1) that is parasitic for some time within a multicellular female sex organ called an archegonium (Gr. *archegonos*, first of a race). A multicellular male sex organ is called an antheridium (Gr. *anthos*, flower). The sex organs are surrounded by a sterile, protective jacket layer of tissues.

The multicellular spore cases called sporangia also have a sterile, jacket layer. Embryophytes reproduce by oogamy in which unlike sex cells unite, and in which the female sex cell is nonmotile. In the reproductive cycle the embryophytes possess an alternation of generations, in which a spore-producing multicellular plant called a sporophyte alternates with a gamete-producing multicellular plant called a gametophyte.

Embryophytes are essentially terrestrial (land plants), although some types may live in water. The chlorophyll is present in bodies called plastids (Gr. *plassein*, to make). The aerial parts of the plants may be protected by a waxlike cutin (L. *cutis*, skin). Embryophyta include such plants as true mosses, ferns, cone-bearing plants, and flowering plants. These will be considered in Chapter 11.

As we study various groups of plants throughout the kingdom Plantae, it is evident that many have certain characteristics in common, whereas other characteristics may be used to distinguish between them. Among such distinguishing characteristics are the presence or absence of chlorophyll; the production or nonproduction of multicellular embryos; the absence or presence of true leaves, stems, and roots; the absence or presence of conducting (vascular) tissues such as phloem and xylem; the absence or presence of flowers; the failure or ability to produce seeds; and the presence of exposed (naked) or enclosed seeds (protected by an ovary, or fruit). Table 7-1 shows how these characteristics are useful in distinguishing different groups of plants.

Table 7-2
Approximate number of species

Plant	Approximate number	
Subkingdom Thallophyta (simple plants that do not form a multicellular embryo)		
Euglenoids	500	
Green algae	6,500	
Yellow-green algae, golden brown algae, and diatoms	10,700	
Brown algae	1,500	
Red algae	3,500	
		22,700*
Slime molds	500	
True fungi	31,300	
		31,800*
Subkingdom Embryophyta (more complex plants that do form a multicellular embryo)		
Liverworts	9,000	
True mosses	14,400	
		23,400*
Club "mosses"	1,200	
Horsetails (scouring rushes)	25	
True ferns	10,000	
Gymnosperms (naked seed plants)	500	
Angiosperms (flowering plants)	200,000	
		211,725*
		290,000*

*Does not include certain groups.

When the total number of plant species is considered, it is a striking fact that some groups contain a large number of species, whereas other groups contain a small number (Table 7-2). Evidently, some groups have changed more in time than others.

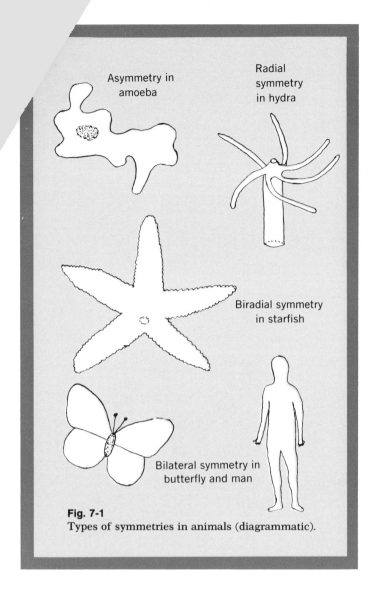

Asymmetry in amoeba

Radial symmetry in hydra

Biradial symmetry in starfish

Bilateral symmetry in butterfly and man

Fig. 7-1
Types of symmetries in animals (diagrammatic).

The animal kingdom

There are over 1,000,000 species of animals, and it is impossible to study more than a few typical representatives of the major groups.

There are different ways in which biologists may classify the various members of the animal kingdom. As studies are made, the following traits will be considered: (1) Protozoa (unicellular or acellular animals, composed of one cell or associations of cells to form a colony) and Metazoa (composed of many cells, with the formation of tissues [lower types of animals] or with tissues and organs [higher types]); (2) diploblastic (having two cellular layers) and triploblastic (having three embryonic cellular layers); (3) radial symmetry (several parts arranged around a central point like spokes on a wheel), biradial symmetry (several parts arranged opposite each other), and bilateral symmetry (two parts arranged opposite each other) (Fig. 7-1); (4) acoelomate (without a true body cavity, or coelom) and coelomate (with a true coelom, lined with mesoderm tissue); (5) unsegmented (without a linear series of segments) and segmented (with a linear series of segments); and (6) invertebrate (without a vertebral column) and vertebrate (with a vertebral column). Some of the characteristics of animals are listed in Table 7-3.

Protozoans

Protozoans are unicellular (acellular) animals. Although the unicellular condition is very common, several protozoan genera have species in which colonies are formed by aggregates of cells. Being without chlorophyll, protozoan nutrition is accomplished by (1) ingesting solid foods by a process called holozoic nutrition (Gr. *holos*, whole; *zoon*, animal), (2) saprozoism (Gr. *sapros*, dead), or (3) symbiosis (Gr. *symbioun*, to live together).

The protozoan cell is either naked or surrounded by a nonrigid cuticle, composed of chitinlike materials. In some instances shells of various inorganic materials are secreted as external skeletons. Many protozoans are free living, while several in each class are symbiotic and, particularly, parasitic. Members of the class Sporozoa are exclusively parasitic.

There are probably more than 100,000 species of protozoans. Most adult protozoans are motile. Three types of locomotion in adults are common: by (1) fingerlike pseudopodia, or "false feet," (2) whiplike flagella, and (3) short, hairlike cilia. Because the sporozoan types are parasitic, their active locomotion is greatly restricted. All four classes of protozoans may undergo encapsulated states as temporary phases in their life cycles. In free-living flagellated and ciliated types the digestive gullets are usually well developed. In rhizopod types the fingerlike pseudopodia are used as feeding (ingesting) structures. Contractile vacuoles for waste elimination and water balance are present in most

Table 7-3

Some distinguishing characteristics of animals (kingdom Animalia)

Phylum	Germ layers	Symmetry	Segmented body (metameres)	True body cavity (coelom)	Skeleton	Vertebral column
Protozoa	Absent	Absent or radial	Absent	Absent	Absent*	Absent
Porifera	2	Absent or radial*	Absent	Absent	Spicules, fibers of spongin*	Absent
Coelenterata	2	Radial	Absent	Absent	Perisarc, mesoglea, limy*	Absent
Ctenophora	3	Biradial*	Absent	Absent	Absent	Absent
Platyhelminthes	3	Bilateral	Absent	Absent	Absent	Absent
Nematoda (Aschelminthes)	3	Bilateral	Absent	Absent	Absent	Absent
Annelida	3	Bilateral	Present	Present	Absent	Absent
Mollusca	3	Absent or bilateral*	Absent	Present (small)	Calcareous shells*	Absent
Arthropoda	3	Bilateral	Present	Present	Chitin	Absent
Echinodermata	3	Radial, biradial*	Absent	Present	Calcareous plates*	Absent
Chordata	3	Bilateral	Present	Present	Cartilage, bone*	Present*

*In general, the characteristics are typical for a majority within the phyla.

instances. Most protozoans contain only one nucleus, but some types are multinucleated.

Metazoans

Metazoan animals are multicellular with bodies usually composed of distinct tissues, organs, and often with several organ systems. Metazoan cells contain centrioles, and the cells are naked and without cell walls or cuticles. Metazoans possess multicellular reproductive organs, and development passes through distinct embryonic and, typically, larval stages.

It is thought that metazoans evolved from some ancestral, unicellular type of life and then became multicellular, retained and improved their ability to locomote, but lost any photosynthetic abilities, if any were present originally. In general, metazoan animal bodies possess a structural architecture that reflects a way of life based on alimentation and locomotion, while plant body architecture reflects a way of life based on photosynthesis and attachment, or sessilism. There are exceptions to this general principle in both plants and animals. Metazoans may be classified in several different ways, as will be noted in successive chapters. For general orientation purposes the following may be

Table 7-4
Approximate number of species in certain phyla of the animal kingdom

Phylum	Approximate number of species
Protozoa (pro to -zo′ a) (Gr. *protos,* first; *zoon,* animal)	100,000
Porifera (po -rif′ er a) (Gr. *poros,* pore; *ferre,* to bear)	15,000
Coelenterata (*Cnidaria*) (se len ter -a′ ta) (Gr. *koilos,* hollow; *enteron,* digestive cavity) (ni -da′ ri a) (Gr. *knide,* nettle)	10,000
Platyhelminthes (plat i hel -min′ thez) (Gr. *platys,* flat; *helmins,* worm)	10,000
Nematoda* (Aschelminthes) (Gr. *askos,* cavity; *helmins,* worm)	10,000
Annelida (a -nel′ i da) (L. *annellus,* little ring; *eidos,* like)	10,000
Mollusca (mo -lus′ ka) (L. *mollis,* soft)	100,000
Arthropoda (ar -throp′ o da) (Gr. *arthron,* joint; *pous,* appendage or foot)	1,000,000
Echinodermata (e ki no -dur′ ma ta) (Gr. *echinos,* spiny; *dermos,* covering or skin)	6,000
Chordata (kor -da′ ta) (L. *chorde,* chord or string)	50,000
	1,311,000†

*Classified as Nematoda in this textbook. Phylum (superphylum) Aschelminthes includes many groups; only roundworm will be considered.

†Does not include all phyla, groups, or species.

helpful—refer to them in the future as you study the various types of animals.

Classification of metazoa
Superphylum Radiata (radiates) (L. *radius,* ray) (sponges and coelenterates)
  Characteristics:
    Middle germ layer not well developed; absent in some cases
    If middle layer is present, it is formed from the ectoderm (outer germ layer); consists largely of a mass of jellylike material within which a few scattered cells are embedded
    Radial symmetry (parts arranged in a raylike manner around an imaginary line)
    Saclike alimentary system; single opening to the exterior
  Representative phyla:
    Porifera (sponges)
    Coelenterata (Cnidaria) (coelenterates)
Superphylum Acoelomata (flatworms) (Gr. *a,* without; *koilos,* hollow or cavity)
  Characteristics:
    Mesoderm develops from the ectoderm; this middle tissue remains as a solid layer
    No true body cavity or coelm
    Bilateral symmetry
  Representative phylum
    Platyhelminthes (flatworms)
Superphylum Pseudocoelomata (Gr. *pseudo,* false)
  Characteristics:
    Mesoderm arises largely from the ectoderm; does not form a solid layer
    Tissues collect in limited areas between the ectoderm and endoderm
    False body cavity (coelm) (enclosed by ectoderm on the outside and by endoderm on the inside, not by mesoderm on both sides as in true coelm)
    Some cells of adult tissue have lost their boundary membrane and hence are syncytial
  Representative phylum:
    Nematoda (Aschelminthes) (roundworms)
Superphylum Schizocoelomata (Gr. *schizein,* split)
  Characteristics:
    Embryonic mesoderm has two sources: the early mesoderm arising from ectoderm (characterizes larval stages); in the adult the early

Table 7-5
Summary of the origins of mesoderm

Superphylum	Origin of mesoderm	Characteristics
Radiata	Ectoderm	Mesoderm absent in some species; when present, it is not well developed but consists largely of a mass of jellylike materials with few cells embedded in it; often called mesoglea (Gr. *mesos*, middle; *gloia*, jelly)
Acoelomata	Ectoderm	Mesoderm remains as solid layer
Pseudocoelomata	Ectoderm	Mesoderm does not form solid layer, but tissues collect in limited areas between ectoderm and endoderm
Schizocoelomata	Early mesoderm from ectoderm; later mesoderm from endoderm	Mesoderm eventually splits into two layers, and space between them forms true coelom
Enterocoelomata	Early mesoderm from endoderm	Mesoderm forms pouches that fill space between ectoderm and endoderm; pouches separate off and form true coelom

mesoderm degenerates and new mesoderm develops from endoderm

Adult mesoderm eventually splits into two layers; the outer layer lies against the inner surface of the body wall and the inner layer surrounds the alimentary tract.

Possess a true coelom

Larvae (called trochophores) are ciliated and free-swimming

Representative phyla:

Annelida (segmented worms

Mollusca (mollusks)

Arthropoda (arthropods)

Superphylum Enterocoelomata (Gr. *enteron*, gut)

Characteristics:

Embryonic mesoderm arises from endoderm; forms pouches which eventually fill space between ectoderm and endoderm and separate from the endoderm that formed them

True coelm (pouches of mesoderm enclose cavities)

Representative phyla

Echinodermata (echinoderms)

Hemichordata (lower chordates) (marine, wormlike animals)

Chordata (chordates, vertebrates)

Some contributors to the knowledge of classification

Theophrastus (370-285 B.C.)
A Greek who was a student of Aristotle and is called the "Father of Botany." He described about 500 plants that he classed as herbs, undershrubs, shrubs, and trees. He wrote a *History of Plants*.

Pliny the Elder (23-79 A.D.)
A Roman literary man and general who described nearly a thousand plants, many of which were useful for medicinal purposes. His *Natural History* of thirty-seven volumes contained half-true data on natural history collected from his predecessors and is not considered too important in the history of botany.

Konrad von Gesner (1516-1565)
A Swiss naturalist and zoologist who published *Historia animalium* and founded the first zoologic museum.

Andrea Cesalpino (1519-1603)
An Italian physician and herbalist who classified plants into fifteen classes, largely on the basis of fruits and flowers (1583).

Linnaeus

Cuvier

Gaspard Bauhin (1560-1624)
A Swiss naturalist who published excellent descriptions of nearly 6,000 species of plants (1623). He used a system of binominal nomenclature (a genus name and a species name) and, by means of his accurate descriptions, attempted to clear up the confusion of plant classification that had existed for years.

Mathias de Lobel (Lobelius) (1538-1616)
He pointed out that leaves are valuable in plant classification and divided plants into two groups on this basis. His great work *Kruydtboeck* (1581) contained many excellent wood engravings.

John Ray (1628-1705)
An English biologist who wrote the *Historia Planatarum* (1686-1704) and used the terms dicotyledons and monocotyledons (embryonic seed leaves). He grouped animals on the basis of opposite traits. His catalog of plants was standard reference and laid the foundation for Linnaeus.

Joseph P. de Tournefort (1656-1708)
A French botanist who was the first to provide the genera of classification systems with careful descriptions, thus setting genera apart from species quite definitely.

Carolus Linnaeus (1707-1778)
A Swedish biologist who is considered the "Father of Modern Classification" because of his revisions and reorganizations of biologic nomenclature. He used the binary (two name) system in the descriptions of hundreds of species much better than previous workers. He wrote *Systema Naturae* (1735) and *Genera Planatarum* (1737). (*Historical Pictures Service, Chicago.*)

Johann F. Blumenbach (1752-1840)
A German anthropologist and anatomist who classified animals into five classic varieties (1775) and called attention to the taxonomic problem of man himself.

Georges L. Cuvier (1769-1832)
A French naturalist who attempted to show the relationship between structure and function, not only in living animals but also from fossil remains. (*The Bettman Archive, Inc.*)

Augustin P. de Candolle (1778-1841)
A Frenchman who published his *Théorie élémentaire de la botanique* (1813) in which he laid down the laws of plant classification so definitely that the natural system was permanently established.

Constantine S. Rafinesque (1784-1840)
A Frenchman who made a classification of medical plants. He came to the United States in 1802 and in 1815 was Professor of Botany at Transylvania College, Kentucky.

Louis Agassiz (1807-1873)
A Swiss naturalist and teacher at Harvard University who did much to focus attention on living organisms.

Adolphe T. Brongniart (1801-1876)
He proposed (1843) a system of classifying plants, dividing them into *Cryptogamae* (without flowers) and *Phanerogamae* (with flowers). The latter contained monocotyledonous and dicotyledonous types.

Gray

Asa Gray (1810-1888)
An American who, using European systems of classification, discovered and described the plants of the United States in the nineteenth century. He was the first widely known botanist in the United States. He improved the system and wrote *Gray's Manual of Botany* (1848). The Gray Herbarium of Harvard University is named in his honor. (*Historical Pictures Service, Chicago.*)

Adolph Engler (1844-1930)
A German who proposed a system of classifying plants that formed a basis for modern systems of taxonomy. He was assisted by K. Prantl at the University of Berlin.

Libbie H. Hyman (1888–1969)
An American zoologist who spent years in research and in writing her extensive treatise of several volumes, *The Invertebrates*. Special reference is given to the anatomy, embryology, physiology, ecology, and taxonomy of invertebrate animals.

Review questions and topics

1 Contrast the subkingdoms Thallophyta and Embryophyta in as many ways as possible.

2 Define (a) sporangia, (b) gametangia, (c) archegonium, (d) antheridum, (e) gametophyte, (f) sporophyte, (g) egg, (h) sperm, and (i) zygote.

3 Give reasons for using scientific names and classification rather than common names and personal systems of classification.

4 What is meant by binominal nomenclature, and what is its importance?

5 In the construction of a scientific classification, explain each of the following: kingdom, subkingdom, phylum, class, order, family, genus, and species.

6 What is meant by "true" leaves, stems, and roots? Why are certain structures that resemble the foregoing not considered to be true stems, leaves, and roots?

7 What are the chief characteristics that distinguish each of the following groups from each other: (a) algae and fungi, (b) liverworts and true mosses, (c) club "mosses," horsetails, and ferns, and (d) gymnosperms and angiosperms?

8 How many species are there in each phylum? What is the total number of species of animals? Explain how this number may vary.

9 Can you give some explanations why there are so few species in one phylum and so many thousands in another?

10 Why are the phyla of animals listed in a particular sequence? Could this arrangement be changed? Explain why this might be possible.

11 Contrast acoelomate and coelomate; unsegmented and segmented; invertebrate and vertebrate; diploblastic and triploblastic; radial, biradial, and bilateral symmetry.

12 What do you think the status of biology might be today if we did not have a scientific method of naming, classifying, and identifying animals?

Selected references

Barnes, R. D.: Invertebrate zoology, ed. 2, Philadelphia, 1969, W. B. Saunders Company.

Barrington, E. J. W.: Invertebrate structure and function, Boston, 1967, Houghton Mifflin Company.

Blackwelder, R. E.: Classification of the animal kingdom, Carbondale, Illinois, 1963, Southern Illinois University Press.

Blair, W. F., and others: Vertebrates of the United States, New York, 1957, McGraw-Hill Book Company.

Bold, H. C.: The plant kingdom, Englewood Cliffs, New Jersey, 1964, Prentice-Hall, Inc.

Boughey, A. S. (editor): Population and environmental biology, Belmont, Calif., 1967, Dickenson Pub. Co., Inc.

Cronquist, A.: The divisions and classes of plants, Bot. Rev. 26:425-482, 1960.

Dittmer, H. C.: Phylogeny and form in the plant kingdom, Princeton, New Jersey, 1964, D. Van Nostrand Co., Inc.

Fingerman, M.: Animal diversity, New York, 1969, Holt Rinehart and Winston, Inc.

Gottlieb, J. E.: Plants: adaptation through evolution, New York, 1968, Reinhold Publishing Corp.

Hyman, L. H.: The invertebrates; vol. I, Protozoa through Ctenophora, 1940; vol. II, Platyhelminthes and Rhynchocoela, 1951; vol. III, Acanthocephala, Aschelminthes, and Entoprocta, 1951; vol. IV, Echinodermata, 1955; vol. V, Smaller coelomate groups, 1959, New York, McGraw-Hill Book Company.

Jensen, W. A., and Kavaljian, L. G. (editors): Plant biology today, Belmont, Calif., 1963, Wadsworth Publishing Co., Inc.

Meglitsch, P. A.: Invertebrate zoology, New York, 1967, Oxford University Press, Inc.

Moore, J. A. (editor): Ideas in modern biology, Garden City, New York, 1965, Natural History Press, Division Doubleday & Company, Inc.

Pennak, R. W.: Freshwater invertebrates of the United States, New York, 1953, The Ronald Press Company.

Rounds, H. D.: Invertebrates, New York, 1968, Reinhold Publishing Corp.

Scagel, R. F., and others: An evolutionary survey of the plant kingdom, Belmont, Calif., 1965, Wadsworth Publishing Co., Inc.

Simpson, G. G.: Principles of animal taxonomy, New York, 1961, Columbia University Press.

Tippo, O.: A modern classification of the plant kingdom, Chronica Botanica 7:203-206, 1942.

part two

VIRUSES, MONERA, AND PLANTS

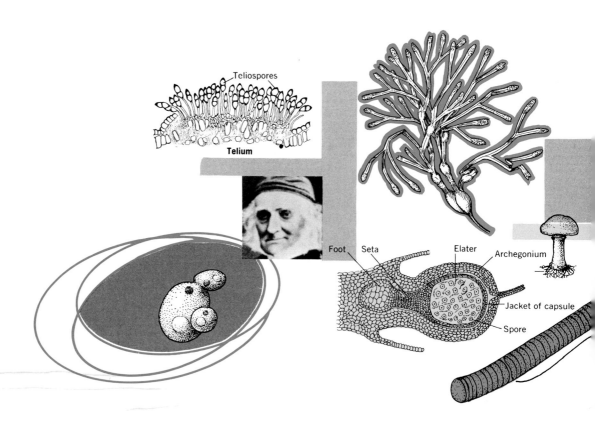

Teliospores

Telium

Foot Seta Elater Archegonium

Jacket of capsule

Spore

chapter eight

Viruses, blue-green algae, and bacteria

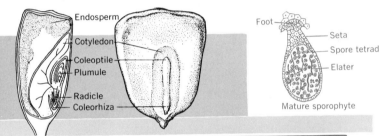

Endosperm
Cotyledon
Coleoptile
Plumule
Radicle
Coleorhiza

Foot
Seta
Spore tetrad
Elater

Mature sporophyte

Cell

Hormogonium

The topics considered in this chapter comprise a rather strange collection about which there is much doubt and discussion.

The viruses are macromolecular complexes that appear to have only the process of reproduction in common with truly living things. Bacteria and blue-green algae are simple living organisms that lack many of the characteristics of more complex cells.

VIRUSES

A virus is a submicroscopic structure consisting of a nucleic acid core surrounded by a protein coat. Either DNA or RNA may be found in the viral core, although DNA is more common. The coat is composed of one or more kinds of proteins, usually forming a symmetrical pattern of various types. Some viruses appear rather simple (Fig. 8-1), while others have a more complex structure with several types of projections. Since viruses range in size from less than 10 to more than 200 mμ, they cannot be seen with an ordinary microscope. Information about their structure has been obtained through the use of the electron microscope and through such techniques as x-ray diffraction.

The question of whether viruses are living or nonliving is still subject to discussion, although the controversy has not stopped virology from becoming one of the fastest growing areas of biology.

About all that can really be said about "living" viruses is that they are the smallest macromolecules capable of reproducing themselves. While viruses possess DNA or RNA, any reproduction must take place in the presence of components of a living cell. Viruses, then, may be called obligate parasites; that is, they exist only in the presence of a living cell.

Viruses may be crystallized, yet they are capable of assuming typical activity later. If they are disrupted chemically or physically, they can reassemble various parts into an intact virus.

Viruses produce many kinds of diseases in plants, animals, and human beings. Many viruses that cause animal diseases show an affinity for specific tissues and organs, such as nerves, skin, respiratory organs, and internal organs.

Among the human viral diseases are measles, mumps, smallpox, chickenpox, influenza, poliomyelitis, yellow fever, warts, cold sores, shingles, virus pneumonia, infectious hepatitis, and infectious mononucleosis. Effective vaccines have been developed against many human virus diseases, such as smallpox, yellow fever, and poliomyelitis.

Among viral diseases of animals are foot-and-mouth disease of cattle, rabies, hog cholera, fowl pox, parrot fever (psittacosis) of birds (and man), cattle plague, and dog and cat distemper. Effective vaccinations have been developed for many of these diseases.

Among the viral diseases of plants are tobacco mosaic disease, tomato mosaic, cucumber and melon mosaic, the yellows disease of peach, aster yellows, curly top of beets, and many others.

Bacteriophage

Bacteriophage (bacteria; Gr. *phagein*, to destroy) is a type of virus that may rapidly destroy certain bacterial cells (Figs. 8-2 and 8-3). Some bacteriophages seem to attack only one species or strain of bacteria, whereas others attack several. The phage particles enter the bacterial cell, multiply, and eventually destroy the cell. Even bacteria have their destructive, disease-producing pathogens!

A virus attaches itself to the bacterium, enters quickly, and minutes later the bacterium ruptures and out come many new viruses, each an exact replica of the original invading virus. By the use of radioactive tracers it is possible to find out how the virus makes these living replicas of itself from the materials available. By labeling with radioactive atoms either the substances of the virus or the medium in which it grows, experimenters can trace the materials and events resulting in the construction of new virus offspring.

The virus of the T_2 strain that infects the intestinal bacterium *Escherichia coli* has a hexagonal head and a tail and is approximately 0.0007 of an inch long. The outer layer consists of protein, which has the ability to attach itself to the surface of the bacterium of the proper species. Inside the head the core contains DNA.

Some of the most fascinating experiments with viruses are studies of their heredity by means of tracer techniques. Our knowledge of how bacterial viruses reproduce themselves may be summarized

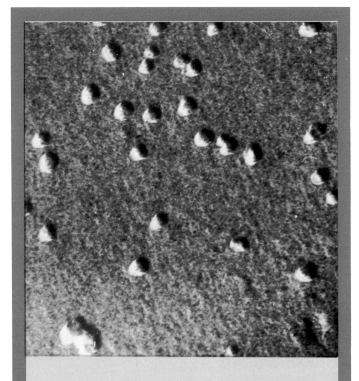

Fig. 8-1
Internal cork virus (ICV) of sweet potato.
(Electron micrograph courtesy M. A. Hayat, University of Dayton.)

briefly as follows: (1) the virus can attach itself to the surface of a bacterial cell by means of its protein coat; (2) the virus then pours its DNA into the cell, the emptied protein coat remaining outside and serving no further purpose; (3) the DNA within the cell makes replicas of itself, using as raw materials the nucleic acids of the bacterium and the fresh substances absorbed by the bacterium from its surroundings; and (4) the DNA induces the synthesis of new protein in the cell, and the units of protein combine with DNA replicas to form about 200 exact copies of the original (parent) virus. About 40% of the parent virus DNA is saved and appears in the descendants.

These facts give only a brief outline of the process, but they assist in solving the problem of how organisms build structural copies of themselves and pass on their heredity from generation to generation.

89

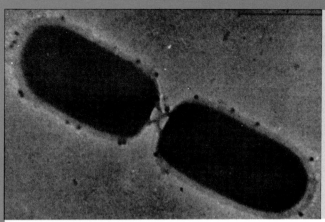

Fig. 8-2
Bacteriophage attached to bacterial cells
(*Escherichia coli*) as shown by an electron
micrograph. Bacterial cells are shown dividing by
fission, and nineteen phage particles are shown
adsorbed to the cell walls.
(*Courtesy Society of American Bacteriologists;
by permission of Dr. S. E. Luria*)

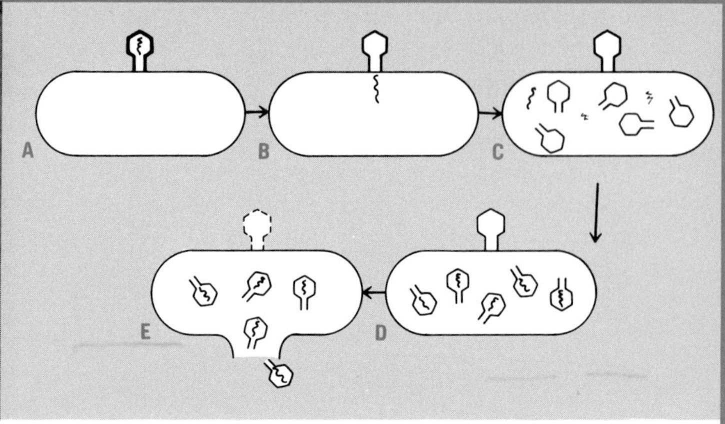

Fig. 8-3
Bacteriophage life cycle. A, Virus particle attached to the
bacterium; B, nucleic acid core migrates into the cell; C,
DNA replicates and induces the formation of new protein
coats from the cell contents; D, new virus particles are
assembled; E, cell bursts, releasing a new generation of
the virus.

MONERA

Many of the characteristics of cells described in
earlier chapters should have been prefaced by the
phrase "except in blue-green algae and bacteria."
These two groups of organisms are procaryotes
(Gr. *pro*, before; *karyon*, kernal); that is, they lack
an organized nucleus, since they do not have a
nuclear membrane or complex chromosomes (see
Table 8-1).

They also lack other membrane-bounded structures, such as mitochondria, lysosomes, plastids,
Golgi apparatus, and endoplasmic reticula. However, it is probable that the enzymatic functions
of these membranes may take place along indenta-
tions of the cell membrane.

If flagella are present in bacteria, they do not
have fibrils arranged in a pattern of 2 in the center,
surrounded by a circle of 9 more. This 9-2 pattern
is present in flagella of most other organisms.

Because of these missing structures, the blue-
green algae and the bacteria are often placed in the
kingdom Monera. In older systems of classification,
they are called plants or placed in the catch-all
category Protista.

Phylum Cyanophycophyta (blue-green algae)

The blue-green algae are placed in the phylum
Cyanophycophyta (Gr. *kyanos*, blue; *phyta*, plant).
In addition to the missing structures indicated
above, they have certain special characteristics.
They are simple, unicellular forms, often forming
a colony of similar cells with little differentiation
among the cells. The cells do not have an organized
nucleus, although the DNA content may be concentrated in one area. The cells are often surrounded
by a slimy, gelatinous sheath. No flagellated cells
are present. Furthermore, the cells do not contain
the large watery vacuoles found in higher plant
cells.

characteristic

Table 8-1

Comparison of procaryote and eucaryote* cells

Cell component	Procaryotic cells	Eucaryotic cells
Cell wall	Contains amino sugars and muramic acid	Amino sugars and muramic acid not present
Nuclear membrane	Absent	Present
Chromosomes	Contains DNA only	Contain DNA and protein
Mitochondria	Absent	Present
Golgi apparatus	Absent	Present
Lysosomes	Absent	Present
Endoplasmic recticulum	Absent	Present
Photosynthesis	Contains chlorophyll and other pigments, but not in chloroplast	Chlorophyll, if present, is in chloroplasts
Flagella	Lacks 9-2 fibrillar structure	Have characteristic 9-2 fibrillar structure

*With a complex nucleus.

Like the photosynthetic plants, blue-green algae cells contain **chlorophyll a** and **beta carotene**. Two other carotenoid pigments, myxoxanthine and myxoxanthophyll, are often present. Two special phycobilin pigments are also present in the cells of the blue-green algae—phycocyanins and phycoerythrins. These pigments appear to function in the energy transfers that take place during photosynthesis. While plastids do not appear to be present within the cell, organized photosynthetic lamellae, similar to the grana of green plants, can be seen with the electron microscope.

Most blue-green algae live in fresh water, although many are marine and others are found on wet rocks and in other moist places. At times, they become so abundant that they produce "water blooms," resulting in foul odors, in the murky appearance and the strange taste of the water.

Some species grow in hot springs with temperatures of over 80° C., where they precipitate calcium and magnesium salts to produce travertine, a chalky material with brilliant colors. Calcium carbonate in lakes may be precipitated by the algae to form marl, an earthy material formerly used on lime deficient soils.

One colorless species of blue-green algae, *Beggiatoa*, which is sometimes called a bacteria, is a sulfur oxidizing form. The chemical process relative to this oxidation may be illustrated as follows:

$$2\ H_2S + O_2 \rightarrow 2\ S + H_2O$$

Minute granules of elemental sulfur accumulate

91

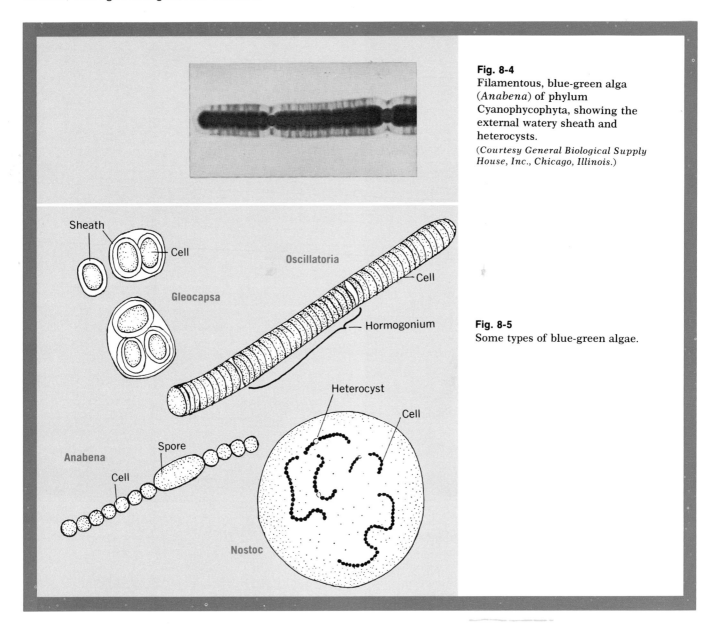

Fig. 8-4
Filamentous, blue-green alga
(*Anabena*) of phylum
Cyanophycophyta, showing the
external watery sheath and
heterocysts.
(*Courtesy General Biological Supply
House, Inc., Chicago, Illinois.*)

Fig. 8-5
Some types of blue-green algae.

in the filaments of *Beggiatoa* as a result of this process.

Several species of blue-green algae, including *Nostoc* and *Anabena* (Figs. 8-4 and 8-5) are able to fix atmospheric nitrogen as do certain species of bacteria. Possibly these species also play a role in maintaining the fertility of certain soils.

●Asexual reproduction occurs by (1) fragmentation of filaments or colonies, (2) spores, or (3) fission. Filamentous types may have occasional thick-walled cells, known as heterocysts (Gr. *heteros*, different; *kystis*, sac), in which the protoplasm becomes colorless. Fragmentation of the filament usually occurs in the heterocysts. In other filamentous types fragmentation occurs at points where two vegetative cells are separated by separa-

Table 8-2
Some characteristics of blue-green algae and bacteria

Characteristic	Blue-green algae	Bacteria
Photosynthesis and pigments	By chlorophyll a that, like chlorophyll b of certain other plants, absorbs light from the red portion of the spectrum; oxygen given off	Some species, for example, photosynthetic purple-sulfur bacteria and green-sulfur bacteria, contain chlorophyll-like pigments that differ only in minor structural detail from chlorophyll a and b
	Other pigments include carotenoids and phycobilins	Obligate anaerobic organisms that require reduced sulfur (usually H_2S) for photosynthetic metabolism
		Bacterial chlorophylls principally absorb longer wavelengths (outside visible light spectrum in near infrared region); no oxygen given off
Nucleus	Not organized (amorphous)	Not organized (amorphous)
Cell wall	Distinct	Distinct
Number of cells	Unicellular or in colonies	Unicellular or in colonies
Slimy, sheath-like covering	Often present	Present in some species
Flagella	Absent	Present in certain species
Fixation of atmospheric nitrogen	By some species	By some species
Average size	Usually larger than bacteria	Coccus types (about 1μ diameter) Rod types (about $0.5 \times 2\mu$) Spiral types ($2 \times 25\mu$)
Reproduction	Asexual: by (1) fission; (2) spores; (3) fragmentation (of filaments or colonies)	Asexual: by (1) fission; (2) endospores
	Sexual: not known	Sexual: by conjugation

tion disks of gelatinous material. The parts of the filament that are separated (by disks or heterocysts) are called hormogonia (singular, hormogonium) (Gr. *hormos*, chain; *gonos*, offspring). Colonies may also fragment because of the failure of the gelatinous sheath to hold them together.

Spores may be of two kinds. Numerous endospores (Gr. *endon*, within; *spora*, spore) may be formed by the repeated division of the cell contents. These endospores are released by a rupture of the cell wall. An akinete (Gr. *a-*, not; *kinein*, to move) is a resting cell in which the cell wall is enlarged and thickened before germination occurs under favorable conditions. Certain species reproduce asexually by fission that results from simple cell division.

Little is known about sexual reproduction in blue-green algae, if it occurs at all.

In addition to the general characteristics given, the following traits apply specifically to these typical species.

Gleocapsa

Gleocapsa (Gr. *gloios*, jelly; L. *kapsa*, box) (Fig. 8-5) has a bluish green region, the color resulting from the diffused pigments (chlorophyll and phycocyanin). Several individual cells may be associated together and surrounded by a jellylike sheath. *Gleocapsa* is common on wet rocks and in other moist places. Reproduction is by cell division and colony fragmentation.

Oscillatoria

Oscillatoria (os i la -to′ ria) (L. *oscillare*, to swing) (Fig. 8-5) consists of a series of independent plants associated in an unbranched, filamentous colony. The chlorophyll and phycocyanin are located in the outer region of the cell. Living filaments may swing back and forth, or oscillate, hence the name *Oscillatoria*. Soft gelatinous areas, called separation disks, may develop between cells and break the filament into hormogonia. A hormogonium may form a new colony. *Oscillatoria* is common in damp earth, on moist stones, and on flowerpots.

Nostoc

Nostoc (nos′ tok) (Fig. 8-5) is composed of a chainlike colony of individual cells arranged like a twisted string of beads. At intervals there are thickwalled, transparent heterocysts that serve to break the filament into hormogonia, as in *Oscillatoria*. A heterocyst may germinate to form a new filament. Numerous filaments may aggregate into a spherical colony surrounded by a sheath. *Nostoc* is common in fresh water and in soils.

Anabena

Anabena (an a -be′ na) (Gr. *anabainein*, to go up) (Figs. 8-4 and 8-5) resembles filaments of *Nostoc* but differs in that certain thick-walled, elongated spores (akinetes) may separate to form a threadlike colony. Heterocysts may be formed as in *Nostoc*. *Anabena* occurs chiefly as free-floating colonies in ponds and lakes. The gelatinous sheath is quite watery.

Phylum Schizomycophyta (bacteria)

Bacteria are placed in the phylum Schizomycophyta because they reproduce by splitting cells. Their numbers increase very rapidly as a result of this process, doubling as often as every 20 minutes during active growth.

Most people are aware of bacteria primarily in relation to disease. While it is true that there are many pathogenic (disease-producing) species, most are indifferent to man, and others are even beneficial.

Studies of bacteria began with the work of Leeuwenhoek, an early Dutch inventor of microscopes. With his primitive microscope (see Fig. 2-3), which was really not much more than a hand lens, he observed many types of bacteria and reported their activities to the scientific community of his day.

Today bacteria are studied with more complex light and electron microscopes, by biochemical tests, and by various physiologic and growth methods.

Morphology

Bacteria are quite small, ranging in size from about 5μ (1 micron is 1/250,000 of an inch) to about 0.1μ. Most bacteria have one of three basic shapes: spherical, cylindrical or rod-shaped, or helical (spiral). Cells with these shapes are called **coccus, bacillus,** and **spirillum,** respectively. Bacterial

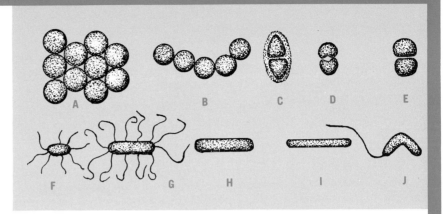

Fig. 8-6
Various types of bacteria (coccus or spherical, rod shaped and spiral) of phylum Schizomycophyta. **A,** *Micrococcus (Staphylococcus) pyogenes* (0.9μ), boils, abscesses, pus; **B,** *Streptococcus pyogenes* (0.8μ), infections; **C,** *Diplococcus pneumoniae* (in capsule) (0.7μ), pneumonia; **D,** *Neisseria gonorrhoeae* (gonococcus) (0.7μ), gonorrhea; **E,** *Neisseria meningitidis* (meningococcus) (0.9μ), meningitis; **F,** *Escherichia coli* (colon organism) (0.5 × 1μ), intestinal organism; **G,** *Salmonella typhosa* (0.6 × 2μ), typhoid fever; **H,** *Corynebacterium diphtheriae* (0.5 × 3μ), diphtheria; **I,** *Mycobacterium tuberculosis* (0.2 × 4μ), tuberculosis; **J,** *Vibrio comma* (0.5 × 2μ), Asiatic cholera. Organisms drawn somewhat on a proportionate scale. Average dimensions are given in microns (μ). 1 micron = 1/25,000 inch. Flagella are shown in F, G, and J.

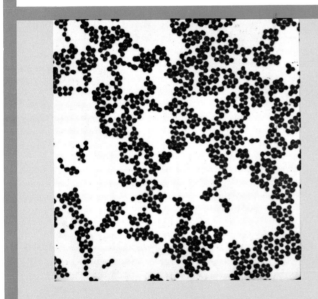

Fig. 8-7
Staphylococcus: Note the characteristic grapelike clusters.

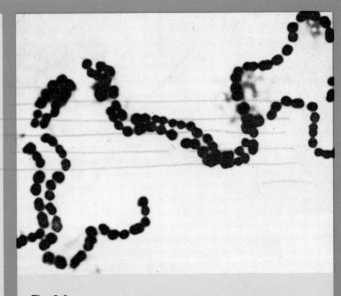

Fig. 8-8
Streptococcus: Note the characteristic chain formation.

cells may also occur in colonies, such as in clusters (as in *Staphylococcus*), in chains (*Streptococcus*), or in pairs, fours, and other groupings (Figs. 8-6 to 8-9).

As indicated earlier, bacteria do not have a very complex internal organization. They do, however, contain DNA, RNA, various proteins, enzymes, sugars, lipids, and amino acids as well as more complex molecules.

The bacterial cell is surrounded by a wall composed of various kinds of mucopeptides, amino acids, and carbohydrate derivitives. These sub-

95

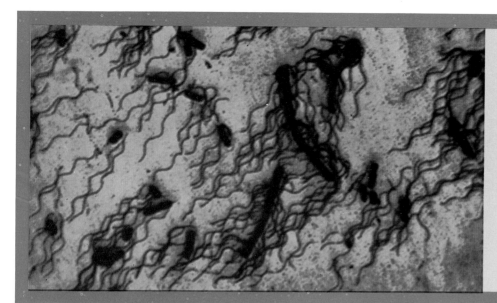

Fig. 8-9
Typhoid organism (*Salmonella typhosa*) stained to show flagella extending from the surface of rod-shaped bacteria. (*Courtesy General Biological Supply House, Inc., Chicago, Illinois.*)

stances add to the strength of the wall and assist in other functions. Because of their presence, bacterial cells react differently to various stains. The well-known Gram stain, named after its developer, takes advantage of this selective staining property of bacteria. Many activities of a particular bacterium can be inferred by its reaction to this stain.

Some bacteria secrete a complex polysaccharide or peptide that forms a gelatinous capsule around the cell wall. Many of the more virulent types of bacteria form this capsule. Studies of one of these types, a *Pneumococcus* that causes pneumonia, helped determine that DNA was the genetic substance of a cell (see Chapter 30).

Motility

Movement in bacteria is by means of slender flagella. The number and position of these flagella vary with the species. Although the flagella are extensions of the cytoplasm, they differ structurally from the more complex organelles in other organisms, for bacteria lack the internal 9-2 arrangement of fibrils.

Metabolism

Most bacteria, like higher plants, require free oxygen for their normal activities and are known as aerobes (a′ er ob) (Gr. *aer*, air; *bios*, life). Other types can thrive only in the absence of free oxygen, hence are known as anaerobes (an -a′ er ob) (Gr. *an*, without). The anaerobes secure their oxygen from oxygen-containing foods, such as carbohydrates. Respiration in cells is similar to that considered in Chapter 26.

Various types of bacteria grow at different temperatures. Those growing best at temperatures of of about 20° C. or below are known as psychrophiles (Gr. *psychros*, cold; *philein*, to love); those having an optimum temperature between 20° and 40° C. are mesophiles (Gr. *mesos*, middle); while those that grow best above 45° C. are called thermophiles (Gr. *thermo*, heat). In spite of the high temperatures, numbers of thermophilic bacteria are found in a great variety of places, including hot springs. Their cytoplasms seem to be especially constructed to live under such temperatures. Psychrophilic organisms are common in cold, deep waters, where they exist as saprophytes. Psychrophiles may decompose foods in cold storage. Most bacteria are mesophiles. Saprophytic mesophiles are common in soils, water, and decomposing materials.

A few species of bacteria (Fig. 8-11) are able to emit light, which may continue for a time after they die. They may be present in decaying leaves, wood, meat, fish, and in salt water. They are commonly called luminescent or photogenic bacteria. Light production is thought to be caused by an oxidizing enzyme called luciferase (L. *lux*, light; *ferre*, to bear), which acts on its substrate material known as luciferin. The reaction is thought to be one in which the reduced luciferase combines with oxygen to emit light. When certain species are grown in a large volume, they produce light in such quantities that they may be photographed by their own light. One may read in a dark room with light produced under such conditions.

🐦 Bacteria vary widely in their methods of securing foods and energy. Most bacteria lack the chlorophyll that characterizes higher plants and, in general, cannot carry on photosynthesis. However, some species contain chlorophyll-like pigments by means of which they manufacture their foods.

From food and energy standpoints bacteria may be classified briefly as follows.

Autotrophic bacteria

Autotrophic (Gr. *autos*, self; *trophe*, nourishment) bacteria require inorganic substances only, from which they manufacture their foods. They are usually classified as chemosynthetic or photosynthetic bacteria.

Chemosynthetic organisms are without photosynthetic pigments and obtain energy from chemical reactions (oxidations).

Among the chemosynthetic organisms are certain iron bacteria that utilize carbon dioxide of the atmosphere as a source of carbon and derive their energy from the oxidation of ferrous iron to basic ferric sulfate or insoluble ferric hydroxide. Some of the chemical processes relative to such iron bacteria may be illustrated as follows:

$$4\ FeCO_3 + O_2 + 6\ H_2O$$
Ferrous
carbonate

$$\rightarrow 4\ Fe(OH)_3 + 4\ CO_2 + Energy$$
Ferric
hydroxide

Another group of chemosynthetic bacteria oxidize certain sulfur compounds and utilize

Fig. 8-10
Colonies of bacteria growing on an agar plate. Each group started with one cell.

Fig. 8-11
Luminescent bacteria (*Photobacterium fischeri*) photographed in total darkness using only the light produced by a growing culture of rod-shaped organisms. (*Courtesy Carolina Biological Supply Co., Elon College, North Carolina.*)

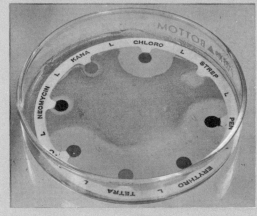

Fig. 8-12
Test for the sensitivity of bacteria to various antibiotics. The clear zones indicate an inhibition of bacterial growth as the antibiotic diffuses from the disk; darker areas are colonies of bacteria.

some of the released energy in the manufacture of their foods. Two of the chemical processes relative to such sulfur bacteria may be illustrated as follows:

$$2 H_2S + O_2 \rightarrow 2 S + 2 H_2O + \text{Energy}$$

Hydrogen sulfide — Elemental sulfur

$$2 S + 2 H_2O + 3 O_2 \rightarrow 2 H_2SO_4 + \text{Energy}$$

Sulfuric acid

Photosynthetic organisms contain chlorophyll-like pigments and obtain energy from sunlight.

Among the photosynthetic organisms are certain sulfur bacteria that contain bacteriochlorophyll pigments and grow in environments containing sulfur compounds. Yellow and red carotenoid pigments may also be present in some species. In the presence of light energy, they photosynthesize foods, but do not liberate oxygen as is true of higher chlorophyll-bearing plants. Some of the chemical processes relative to such sulfur bacteria may be illustrated as follows:

$$2 H_2S + CO_2 + \text{Light energy}$$

Hydrogen sulfide

$$\rightarrow \text{Carbohydrate} + 2 S + H_2O$$

Elemental sulfur

$$2 S + 3 CO_2 + 5 H_2O + \text{Light energy}$$

$$\rightarrow 2 \text{ Carbohydrate} + 2 H_2SO_4$$

Sulfuric acid

Heterotrophic bacteria

Heterotrophic (Gr. *heteros*, different) bacteria require an organic source of carbon as food. They usually exist as saprophytes or parasites.

Saprophytes (Gr. *sapros*, dead or rotten; *phyta*, plants) generally live entirely on nonliving organic matter and are not ordinarily involved in disease production.

Parasites or pathogens (Gr. *para*, beside; *sitos*, food) (Gr. *pathos*, suffering; *gen*, origin) can and do live in or upon other living organisms and may cause disease, but may or may not be able to live as saprophytes.

Reproduction

The primary method of reproduction in bacteria is asexual fission (cell division), in which the parent cell divides crosswise into two unicellular organisms. In coccus types fission may occur in various planes, depending upon the species. Often after fission many newly formed cells remain together to form a colony. Different species produce colonies of different sizes, shapes, and even colors. Each species forms a rather definite type of colony (Fig. 8-10) whose specific characteristics are used for identification purposes. Under favorable growth conditions, fission may occur in approximately 20 to 30 minutes. Theoretically, a bacterial cell dividing at its maximum rate would, with continuous fissions, produce 4,700,000 quadrillion offspring in 24 hours. This mass would weigh approximately 2,000 tons. This never happens because sufficient food is not available in any one place, and the accumulation of waste products of metabolism prevents such reproduction.

Some bacteria under such environmental conditions as extremes of temperature and lack of moisture may form internal endospores. Bacterial spores may occur at the end of a rod (polar), at the center of a rod (central), or between these two points (subpolar or subterminal). Spores may be smaller than the diameter of the rod or they may be larger than the diameter of the rod at the time spores are formed, in which case the rods may appear spindle shaped, or club shaped. The size and location of a spore within a cell are definite for a particular species.

Spores are formed by condensing the cytoplasm into a spherical or ovoid mass that is surrounded by a resistant wall. A loss of water accompanies spore formation. When a spore encounters favorable environmental conditions, it absorbs water and germinates to form an active cell.

Sexual reproduction

Three types of genetic recombination are known in bacteria—conjugation, transformation, and transduction. Conjugation is related to the sexual process in other organisms in that the parental cells actually come together.

Conjugation has been studied most in *Escherichia coli*, a common intestinal bacterium. This bacterium has a single chromosome with genes arranged in a definite order.

In 1946, Joshua Lederberg and Edward Tatum studied mutants of *E. coli* that were unable to synthesize certain amino acids. When two different mutants were grown on the same medium, some new combinations with characteristics of both "parent" strains were seen. Since this only happened when two bacteria were in contact, some sort of a sexual process was thought to have occurred.

Studies with the electron microscope in the 1950's showed that when bacteria from two different sexes (or mating types) come together, a cytoplasmic bridge is formed between them, the chromosome of the male is transferred to the female, and, presumably, the process of genetic recombination takes place. The male cell dies after the transfer. In many cases only part of the chromosome is transferred, and the new traits are governed by whatever genes are on that part.

In transformation a bacterium picks up strands of DNA that have come from dead cells. In transduction the new DNA is carried from one cell to another by viruses.

Harmful activities of bacteria

Bacterial diseases of man

Louis Pasteur is generally credited with advancing the idea that bacteria cause disease. The work of Robert Koch, who studied the causes of tuberculosis and anthrax, helped this concept to gain acceptance.

Koch developed a procedure to illustrate that a particular organism was related to a particular disease. These methods have survived to this day in the form of Koch's postulates. In general, they are as follows:

1. The bacterium in question must always be found in cases of the particular disease.
2. It must be isolated from the diseased organism and grown in culture.
3. When injected into a healthy organism, it must produce the disease.
4. It must be isolated from the second organism and compared with the bacteria from the first organism.

The use of this procedure has helped in establishing the causes of the diseases shown in Table 8-3.

Bacterial diseases of other animals

Many bacterial diseases of man seem to have their counterparts in lower animals, although certain animal diseases do not affect human beings. Among the bacterial diseases of animals might be included such forms as tuberculosis of cattle and hogs; anthrax of sheep; chicken cholera; glanders of horses, sheep, and goats; pneumonia; septicemia in cattle; botulism in chickens and other animals; tularemia in rabbits; rat plague; and Bang's disease (brucellosis or undulant fever) in cattle.

Bacterial diseases of plants

It may seem surprising that there are many plant diseases caused by bacteria, but the list is growing constantly as further research is done in this highly important field of bacteriology. By his research on the fire blight of pears in 1879, Burrill of the University of Illinois proved that bacteria could cause plant diseases. Among the symptoms by which some of the bacterial plant diseases are recognized are blights, in which there is a rapid death of blossoms, young leaves, and stems, usually when they are not fully developed; rots, which are soft discolorations in which cell walls of the affected tissues are destroyed; wilts, in which the aerial parts of plants wilt and dry out, sometimes caused by the interference with the conduction of water in the xylem tissues; and cankers, which are dead depressions in the surface tissues caused by bacterial activities. Examples of plant diseases caused by specific bacteria are soft rot of carrots, celery, cabbage, cucumbers, and eggplant; cotton root rot; pineapple rot; wilt diseases of tomato, potato, cantaloupe, squash, cucumber, and corn; citrus canker; fire blight of pears and apples; crown galls of apples, grapes, and raspberries; bacterial blight of beans; and bacterial blight of walnut trees. Annual losses of crops because of bacterial diseases reach millions of dollars.

Spoilage of foods

Bacteria cause spoilage of foods as illustrated by rotting of meats, spoilage of butter, souring of milk and other dairy products, and spoilage of fresh

99

BAJ
viruses, monera, and plants
viruses, blue-green algae, and bacteria

Table 8-3
Bacterial diseases of man

Bacterial disease	Causal agent
Boils, abscesses, pus	*Micrococcus (Staphylococcus) pyogenes*
General infections	*Streptococcus pyogenes*
"Sore throat" (certain types)	*Streptococcus pyogenes (hemolyticus)*
Meningitis	*Neisseria meningitidis*
Gonorrhea	*Neisseria gonorrhoeae*
Pneumonia	*Diplococcus pneumoniae*
Anthrax	*Bacillus anthracis*
Diphtheria	*Corynebacterium diphtheriae*
Typhoid fever	*Salmonella typhosa*
Paratyphoid	*Salmonella paratyphi*
Tuberculosis	*Mycobacterium tuberculosis*
Leprosy	*Mycobacterium leprae*
Undulant fever	*Brucella abortus*
Plague ("black death")	*Pasteurella pestis*
Tularemia ("rabbit fever")	*Pasteurella tularensis*
Whooping cough	*Hemophilus pertussis*
Tetanus ("lock jaw")	*Clostridium tetani*
Gaseous gangrene	*Clostridium perfringens (welchii)*
Botulism (toxic food poisoning)	*Clostridium botulinum*

and canned wines, fruits, and vegetables. These damages can result in rendering the foods unfit for use or in the production of detrimental substances in the foods as a result of bacterial activities. As an example of this, certain anaerobic bacteria (*Clostridium botulinum*) may attack some types of foods (for example, meat, canned corn, and beans) and produce a very potent toxin which may cause death if even a small quantity is ingested. This disease is known as botulism (bot′ u lism) (L. *botulus*, sausage) because it was originally discovered in sausage.

Beneficial activities of bacteria

Probably a preponderance of the common knowledge of the roles of bacteria pertains to their harm-ful effects. However, their activities can be utilized quite beneficially, as will be shown.

Manufacture of foods

Through their products and metabolic activities bacteria may play important roles in the production of sauerkraut and pickles, the manufacture of butter and certain kinds of cheese, the production of vinegar by the oxidation of alcohol to acetic acid, and the curing of coffee, cocoa beans, and black tea.

Industrial and agricultural activities

Through increased bacteriologic research the roles of bacteria in a great variety of industrial and agricultural activities are being realized. Among

100

such applications are the "tanning" of skins to produce leather; the removal of flax fibers from the stems of flax plants; the fermentation of shredded green plants for the production of ensilage for animal feed; bacterial actions on sugars, proteins, and other organic materials that form such commercially important products as acetone (used in the manufacture of photographic films), and explosives; the production of butyl alcohol used in the manufacture of lacquers; the production of lactic acid used in the "tanning" of skins; the production of citric acid used in lemon flavoring; and the production of vitamins used in food and medicinal preparations.

Increase of soil fertility

Bacteria may decompose proteins, fats, carbohydrates, and other organic compounds from the bodies of animals and plants, or from their waste products, thereby removing much of the organic debris from the earth and returning the simple products thus formed to the soil to help maintain its fertility. The anaerobic bacterial decomposition of nitrogenous organic compounds usually results in the formation of malodorous materials, chiefly sulfur compounds. This process is known as putrefaction (L. *putrefacere*, to make rotten). The aerobic bacterial decomposition of organic compounds in the presence of oxygen and without the development of malodorous materials is called decay (L. *de*, down; *cadere*, to fall). In the decomposition of organic substances several species of bacteria, some aerobic and some anaerobic, may carry the process through numerous successive stages until there is a complete breakdown into simpler materials that will be usable by future plants.

The various bacteria involved in the important nitrogen transformations to maintain soil fertility include ammonifying bacteria, nitrifying bacteria (both nitrite and nitrate bacteria), and nitrogen-fixing bacteria. These important phenomena will be considered in greater detail elsewhere in the text.

The sulfur bacteria in soils may convert hydrogen sulfide, a product of protein destruction, through a series of stages to eventually form sulfuric acid. This is transformed into sulfates, which are an essential, major source of sulfur for the metabolism of green plants. Thus the supply of sulfates is maintained.

Digestive and other physiologic activities in animals

Some species of bacteria perform certain beneficial physiologic activities in the intestines of animals. Cellulose-digesting enzymes secreted by bacteria assist in the digestion of cellulose in herbivorous animals, such as cattle and horses. Bacteria in the large intestine and in the lower part of the small intestine of man may synthesize vitamins. About one-third of the dry weight of human feces consists of the bodies of microorganisms, primarily bacteria.

Review questions and topics

1 What are viruses? Why are they considered to be on the borderline between living and nonliving?
2 List several diseases of plants and animals caused by viruses.
3 What are bacteriophages and what are some of their more important characteristics?
4 Define bacteria in your own words.
5 Why should bacteria be classified in the phylum Schizomycophyta?
6 Why are bacteria placed in the kingdom Monera?
7 Describe all the ways in which bacteria may be (1) beneficial and (2) harmful.
8 Contrast and give examples of heterotrophic and autotrophic nutrition.
9 Do any bacteria photosynthesize food even though they do not possess chlorophyll? Explain in detail.
10 Do bacterial cells possess nuclear materials? Do any of them possess an organized nucleus?
11 Explain the methods of reproduction of bacteria.
12 Describe the common types of true bacteria, giving examples of each type.
13 Explain the following terms: (1) bacterial locomotion, (2) flagella, (3) chromogenic bacteria, (4) cytochromes, (5) luminescent or photogenic bacteria, and (6) luciferase.
14 Contrast and give examples of parasitic and saprophytic bacteria.
15 Explain the physiologic differences between aerobic and anaerobic bacteria.
16 Explain and give examples of the two groups of autotrophic bacteria.
17 Explain these terms: psychophiles, mesophiles, and thermophiles.
18 Discuss the characteristics of the blue-green algae.
19 In what ways are bacteria and blue-green algae alike? Different?

Some contributors to the knowledge of viruses*

Louis Pasteur (1822-1895)
He artificially attenuated (lowered the pathogenicity) rabies virus (1885) but preserved its immunizing property. (*Historical Pictures Service, Chicago.*)

Pasteur

Martinus Beijerinck (1851-1931)
A Dutch microbiologist, substantiated the evidence of filtrable viruses in living organisms (1899). He discovered the root-nodule bacteria that fix free atmospheric nitrogen in certain plants (1888). He also studied the fixation of free atmospheric nitrogen by bacteria (*Azotobacter*) in pure culture (1908). He suggested a "*contagium vivum fluidum*" as the cause of the mosaic disease of tobacco plants (1899).

Walter Reed (1851-1902)
Reed and members of the American Army Commission in Havana, Cuba, described the viral cause of yellow fever (1900).

* See Chapter 9 for contributors and references to blue-green algae.

Wendell M. Stanley (1904-)
An American biochemist who isolated from the juice of tobacco plants infected with mosaic disease, a protein of great molecular size. It displayed the properties of a virus and from it he isolated a crystalline form by sedimentation in ultracentrifuges (1935).

John E. Enders (1897-)
An American microbiologist and virologist, has given us a powerful tool to produce immunization against certain diseases. He first demonstrated how to grow the dangerous polio virus in a place other than nerve tissue, for which he received a Nobel Prize. Jonas Salk has been credited for developing a polio vaccine; this was made possible because of the studies of Enders and his associates.

Joshua Lederberg (1925-)
An American biologist and winner of a Nobel Prize (1958), showed that viruses can alter the heredity of bacteria. With his teacher, Edward Tatum of Yale University, he demonstrated that bacteria may have sex.

Selected references

Brieger, E. M.: Structure and ultrastructure of microorganisms, New York, 1963, Academic Press Inc.

Echlin, P.: The blue-green algae, Sci. Amer. 214:75-81, 1966.

Edgar, R. S., and Epstein, R. H.: The genetics of a bacterial virus, Sci. Amer. 212:70-8, 1965.

Fritsch, F F..: The structure and reproduction of the algae, New York, 1945, Cambridge University Press.

Horne, R. W.: The structure of viruses, Sci. Amer. 208:48-56, 1963.

Pelczar, M. J., and Reid, R. D.: Microbiology, New York, 1965, McGraw-Hill Book Company.

Sistrom, W. R.: Microbial life, New York, 1962, Holt, Rinehart & Winston, Inc.

Stanier, R. Y., Doudoroff, M., and Adelberg, E. A.: The microbial world, Englewood Cliffs, New Jersey, 1962, Prentice-Hall, Inc.

Stent, G. S.: The multiplication of bacterial viruses, Sci. Amer. 188:36-39, 1953.

Thimann, K. V.: The life of bacteria, New York, 1963, The Macmillan Company.

Umbreit, W. W.: Modern microbiology, San Francisco, 1962, W. H. Freeman and Co., Publishers.

Some contributors to the knowledge of bacteria

Leeuwenhoek

Antonj van Leeuwenhoek (1632-1723)
A Dutch microscopist, was probably the first to see bacteria (1676), and his descriptions were quite good considering the low magnifications (300 diameters) of the 247 microscopes that he constructed. (*Historical Pictures Service, Chicago.*)

Louis Pasteur (1822-1895)
A French bacteriologist and chemist, proved that fermentations and decompositions of substances resulted from the activities of microbes. He proved that certain diseases were caused by bacteria, and he is often called the "Father of Bacteriology."

Koch

Robert Koch (1843-1910)
A German bacteriologist and physician, devised a plate method for growing bacteria on solid media (1881). He proved that specific bacteria cause the diseases of anthrax (1877) and tuberculosis (1882). He formulated Koch's Postulates for proving that a particular microbe causes a specific disease. (*Historical Pictures Service, Chicago.*)

William H. Welch (1850-1934)
An American and a student of Koch, discovered the "gas bacillus" in infected wounds after his return to America. Welch is called the "Dean of American Medicine." Based on Koch's methods he gave the first course in bacteriology in the United States at Johns Hopkins University, Baltimore, Md.

Sergius Winogradsky (1856-1913)
A Russian microbiologist, studied the physiology of sulfur bacteria (1899). He identified the bacteria of the soil that oxidize ammonia to nitrites and nitrates (1890).

Alexander Fleming (1881-1955)
An English bacteriologist, discovered the antibiotic penicillin (1929). This led other investigators to discover many additional antibiotics with which to treat numerous diseases today.

Selman A. Waksman (1888-   )
An American, discovered the antibiotic streptomycin (1944) and received a Nobel Prize (1952).

chapter nine

Simple plants— the algae

Plants without true leaves, stems, or roots; not forming multicellular embryos; without true vascular (conducting) tissues; subkingdom Thallophyta*

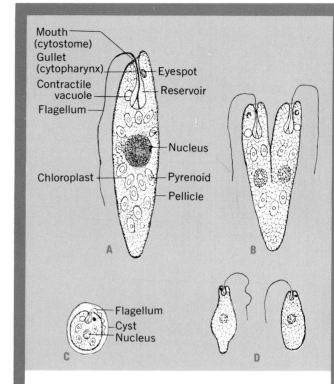

Fig. 9-1
Euglena viridis, a flagellated euglenoid of phylum Euglenophycophyta. **A,** Free-swimming adult (highly magnified and somewhat diagrammatic); **B,** reproduction by longitudinal fission; **C,** *Euglena* rounded and protected by a cyst; **D,** shapes (in outline) assumed during euglenoid movements.

Algae is a common term that may be applied to those thallophytes that carry on photosynthesis in the presence of chlorophyll. The algae vary among themselves (Tables 9-1 and 9-2) but have several characteristics in common. Algae may live in fresh water (aquatic) or in salt water (marine). Some may be free-swimming or free-floating, and others may live on the bottom of a body of water. Certain species of algae live in moist soils, on moist rocks and trees, in snow and ice, or in hot springs. Some species are parasites on other living plants or animals. Certain algae known as epiphytes (Gr. *epi,* upon; *phyta,* plants) gain physical support from other objects, but manufacture their own food. Algae may serve as food for aquatic animals. In a special group of plants known as lichens certain algae and fungi are intimately associated for mutual benefit, the algae manufacturing food and the fungi furnishing water and protection.

PHYLUM† EUGLENOPHYCOPHYTA (EUGLENOIDS)

Euglenoids are placed in the phylum Euglenophycophyta (Gr. *eu,* good; *glene,* eyeball; *phyta,* plants), and almost all are found in fresh water. They resemble the green algae in that the chloro-

phyll is grass green and may be combined with other pigments, such as the carotenoids and several xanthophylls. Euglenoids have an organized nucleus, and the chlorophyll is localized in definite plastids known as chloroplasts (Gr. *chloros,* green). They differ from green algae in that the stored food is paramylum, a starchlike carbohydrate.

Most species are naked, single-celled organisms, and have one or two whiplike flagella. Reproduction occurs by longitudinal cell division. Some species may form cysts (thick-walled resting cells) that protect them from unfavorable conditions, such as drought. Encysted cells germinate into swimming cells. Euglenoids form an important part of plankton (Gr. *planktos,* wandering), which consists of free-swimming or floating aquatic

---

*Reference to "without true leaves, stems, or roots" is in comparison to the more complex plants, such as pine, elm, and tulip.
†Botanical usage replaces phylum with division.

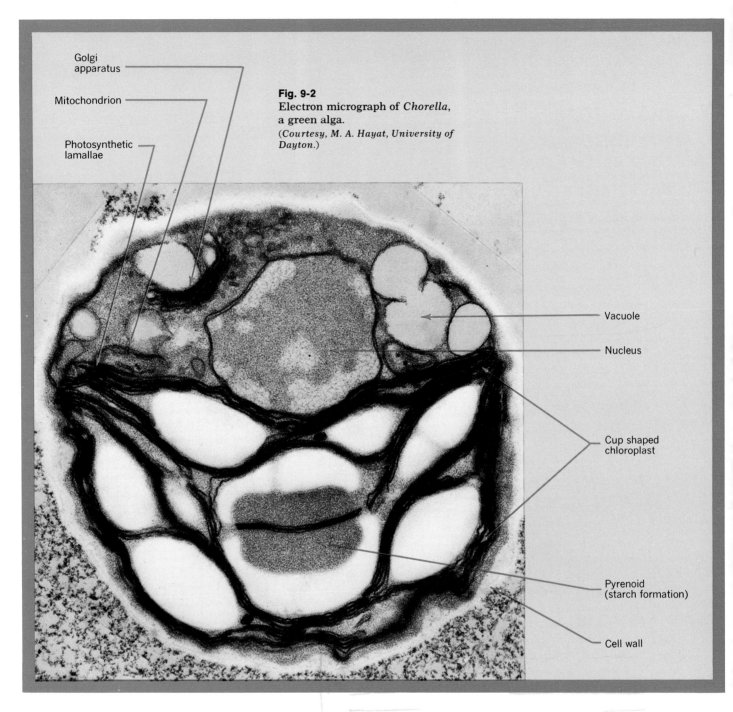

Golgi
apparatus

Mitochondrion

Photosynthetic
lamallae

Fig. 9-2
Electron micrograph of *Chorella*,
a green alga.
*(Courtesy, M. A. Hayat, University of
Dayton.)*

Vacuole

Nucleus

Cup shaped
chloroplast

Pyrenoid
(starch formation)

Cell wall

organisms, an important part of the diet of fishes and other aquatic organisms. *Euglena* (Fig. 9-1) is a typical example.

Euglena

Euglena and its related forms are classified by some botanists as algae, since most of them pos-sess chlorophyll with which to photosynthesize foods, as do higher green plants. However, *Euglena* and related forms are classified by zoologists as single-celled protozoa because they have tubular gullets. Furthermore, some of them may ingest solid·foods, as do most animals. Euglenoids might be considered to be plants that possess animal-

Table 9-1

Some characteristics of algae (see also Table 9-2)

Plants	Organized nucleus	Cell wall	Cytoplasmic strands between cells	Motile cells
Euglenoids (Euglenophycophyta)	Present	Absent; flexible pellicle; cell-ulose in some species	Absent	1 to 3 anterior flagella (unequal or equal length depending on species
Green algae (Chlorophycophyta)	Present	Cellulose*	Absent (cytoplasmic strands in such types as *Volvox*)	When present, 2 to 8 anterior flagella usually of equal length
Yellow-green algae including diatoms (Chrsophycophyta)	Present	Silica†; pectin*	Absent	When present, 1 to 2 anterior flagella (unequal or equal length depending on species)
Brown algae (Phaeophycophyta)	Present	Cellulose*; algin‡	Absent	When present, 2 lateral, unequal flagella
Red algae (Rhodophycophyta)	Present	Cellulose*; pectin*	Present	Absent

*Complex carbohydrates $(C_6H_{10}O_5)_n$.

†Silicon dioxide (SiO_2), stony.

‡Pectic, water-holding, colloidlike substance.

like methods of securing foods or to be animals that show some plant characteristics. This is one reason for using the catch-all kingdom Protista.

Euglena is spindle shaped and has a large, centrally located nucleus. An anterior flagellum propels it through the water. Numerous small, green chloroplasts carry on photosynthesis. Paramylum granules are scattered throughout the cytoplasm. Pyrenoids (colorless plastids) may be present in some instances, assisting in food production.

At the anterior end of the *Euglena* is a tubular gullet (cytopharnyx) that probably does not ingest solid foods. Near the gullet is a light-sensitive,

red eyespot and a contractile vacuole. During reproduction the nucleus undergoes mitosis, after which the cell divides into two new individuals by longitudinal fission. A thick-walled, resistant cyst may be formed around the cell during adverse conditions. A few species lack chloroplasts and are probably saprophytes, absorbing materials by diffusion through the cell surface.

Euglena viridis (L. *viridis*, green) is about 100μ (1/250 of an inch) in length. *Euglena* may move in a spiral rotary manner by the vibratile movements of the flagellum. It may also move by a twisting of the body ("euglenoid movement").

Table 9-2
Some characteristics of algae (see also Table 9-1)

Plants	Plastids	Pigments	Stored food
Euglenoids	Present or absent; pyrenoids present or absent (depending on species)	Chlorophyll a, b; carotenoids (carotene)*; xanthophylls†	Paramylum (starchlike)
Green algae	Present or absent (depending on species)	Chlorophyll a, b; carotenoids (carotene)*; (depending on species)	Starch $(C_6H_{10}O_5)_n$; oils (sometimes)
Yellow-green algae, including diatoms	Present or absent (depending on species)	Chlorophyll a; chlorophyll c (some species); carotenoids (carotenes)*; certain xanthins (diatoms only)	Leucosin (insoluble carbohydrate); and/or oils
Brown algae	Present	Chlorophyll a; chlorophyll c; carotenoids (carotene)*; fucoxanthin (yellow brown); xanthophylls†	Laminarin (soluble carbohydrate); sometimes mannitol (6-atom alcohol) or fats
Red algae	Present	Chlorophyll a, d; carotenoids (carotenes)*; phycoerythrin (reddish)‡ phycocyanin (bluish)‡	"Floridean starch" (starchlike); floridoside (soluble sugar)(depending on species)

*Carotene $(C_{40}H_{56})$; carotenes vary from creamy white to orange red.
†Xanthophylls $(C_{40}H_{56}O_2)$, producing yellowish colors.
‡Water-soluble, protein pigments, insoluble in fat solvents.

PHYLUM CHLOROPHYCOPHYTA (GREEN ALGAE)

Green algae are placed in the phylum Chlorophycophyta. The chlorophyll is localized in definite bodies called plastids or, more specifically, in chloroplasts (Fig. 9-2). Additional orange-reddish pigments known as carotenoids (L. *carota,* carrot; *en,* in; *eidos,* like) may also be present. The cell wall contains cellulose, and starch is the stored food. Starch is formed by special structures known as pyrenoids (Gr. *pyren,* fruit stone; *eidos,* like) that are associated with the chloroplasts. The nucleus is organized, as in all the algae except the blue-green group. Green algae vary in structure; they may be unicellular, colonial, or multicellular, depending upon the species. When the vegetative (body) cells or the reproductive cells are motile, each bears two to eight anterior flagella, usually of equal length. Many of the characteristics are summarized in Tables 9-1 and 9-2.

Reproduction occurs (1) asexually by fission, by fragmentation, by motile spores, or by nonmotile spores; or (2) sexually by isogamy (Gr. *isos,* equal; *gamos,* marriage) with the union of gametes of equal size; by heterogamy (Gr. *heteros,* different)

107

with the union of gametes of unequal size; or by
oogamy (o -og' a my) (Gr. *oon*, egg), which is a
special type of heterogamy in which the female
gamete is a nonmotile egg. The particular methods
of reproduction vary with the species. The gametangium that produces the sex cells is unicellular,
hence it cannot be called a sex "organ" in the true
sense, and it does not have sterile jacket cells surrounding it.

Green algae as well as other types supply foods
for freshwater and marine animals. Most green
algae live in fresh water, but a few species live in
the ocean. Others live in soils, on rocks and trees,
and in ice and snow. Certain species live in salt
waters in which the concentration of salt is much
greater than that in ocean water. Some species
live in other plants or animals. Some marine green
algae in association with red algae may form lime
salts that contribute to the formation of reefs in
the ocean.

Chlamydomonas

In addition to the general characteristics of green
algae the following details apply to *Chlamydomonas* (klam i -dom' o nas) (Gr. *chlamys*, cloak; *mo-
nas*, one) (Fig. 9-4). Each ovoid plant contains a
single, cup-shaped chloroplast with a pyrenoid; a
pigmented stigma (eyespot); and two anterior
flagella of equal length. There are two contractile
vacuoles near the anterior end for excretion. Some
investigators classify this organism with the single-celled animals.

At times there are formed within the cell two,
four, or eight flagellated zoospores (motile spore,
or swarm spores) that look like miniature parent
cells and that swim out to form a new generation.
In some instances the parent cell divides into
eight, sixteen, or thirty-two isogametes (sex cells
of equal size). In the water two isogametes unite
to form a single-celled, tetraflagellated zygote.
When this is surrounded by a thick, resistant wall,
it is called a zygospore. The single nucleus of the
zygospore produces four nuclei, one for each of
the four zoospores produced. As these are formed,
the number of chromosomes is reduced by the
process of meiosis from the double (2N) number
to the haploid (N) number. The zoospores and
gametes look somewhat alike except that the

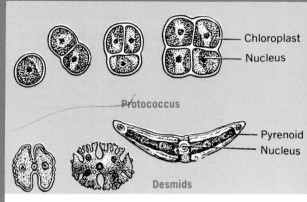

Fig. 9-3
Green algae of phylum Chlorophycophyta.

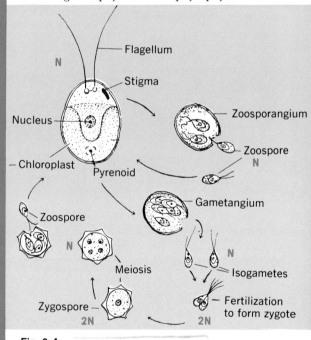

Fig. 9-4
Green alga (*Chlamydomonas*) of phylum
Chlorophycophyta, showing methods of
reproduction. An adult is shown at top. Numbers
of chromosomes are represented by diploid (2N)
and haploid (N). By a process of meiosis there
is a reduction division of chromosomes from
2N to N.

Fig. 9-5
Life cycles of certain green algae. Numbers of
chromosomes are represented by (N) and (2N).

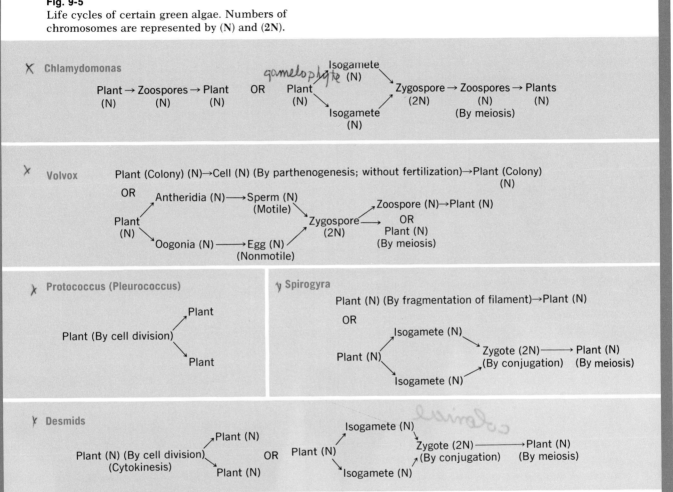

X Chlamydomonas

gameloplyte

Plant → Zoospores → Plant   OR   Plant   Isogamete (N) → Zygospore → Zoospores → Plants
(N)   (N)   (N)   (N)   Isogamete (N)   (2N)   (N)   (N)
                                                                              (By meiosis)

X Volvox

Plant (Colony) (N)→Cell (N) (By parthenogenesis; without fertilization)→Plant (Colony)
                                                                                      (N)
OR   Antheridia (N)——→Sperm (N)
                                (Motile)   Zoospore (N)→Plant (N)
Plant   Zygospore——   OR
(N)   (2N)   Plant (N)
             Oogonia (N)——→Egg (N)   (By meiosis)
                             (Nonmotile)

X Protococcus (Pleurococcus)

                         Plant
Plant (By cell division)
                         Plant

γ Spirogyra

Plant (N) (By fragmentation of filament)→Plant (N)
OR
            Isogamete (N)
Plant (N)   Zygote (2N)———→Plant (N)
            Isogamete (N)   (By conjugation) (By meiosis)

γ Desmids

                              Plant (N)   Isogamete (N)
Plant (N) (By cell division)   OR Plant (N)   Zygote (2N)———→Plant (N)
          (Cytokinesis)   Plant (N)   (By conjugation) (By meiosis)
                                                    Isogamete (N)

γ Ulothrix

Plant (N) (By fragmentation of filament)→Plant (N)
OR
Plant (N)→Zoosporangia (N)→Zoospores (N)→Plant (N)   OR   Plant (N)→Sporangia (N)→
                            (Motile)   Aplanospores→Plant
OR   (N)   (N)
Plant (N)→Gametangium (N)→Isogamete (N)   (Nonmotile)
                           (Motile)   Zygote (2N) ——— Zoospores (N)→Plant (N)
Plant (N)→Gametangium (N)→Isogamete (N)   (By meiosis) (Heterothallic,
                           (Motile)   or + and −)

Fig. 9-6
Volvox globator.
**Daughter colonies
of several ages.**
(*Courtesy, Carolina
Biological Supply
Company.*)

former are larger. *Chlamydomonas* may occur in fresh water, in damp soil, and even in snow.

■ Volvox   *colonial*

Volvox (L. *volvere*, to roll) (Figs. 9-6 and 9-7) is a colonial green alga in which hundreds to thousands of individual cells are arranged in a single layer. These cells are held together by a gelatinous secretion and joined by protoplasmic strands that establish physiologic (functional) continuity between them. The individuals form a hollow sphere that is 1 to 2 mm. in diameter and is composed of 500 to 50,000 cells. The individual cells have much the same structure as *Chlamydomonas*. Division of labor is accomplished by two kinds of cells, the biflagellated somatic (body or vegetative) cells and the reproductive (germ) cells. Most of the cells are vegetative. Each body cell has two flagella, chlorophyll in chloroplasts, a contractile vacuole,

and a light-sensitive eyespot (stigma). The latter may assist in orientation for photosynthetic purposes.

A young colony consists primarily of somatic cells, but a mature colony contains, in addition, a smaller number of cells capable of forming new individual colonies, either asexually or sexually. In asexual reproduction certain body cells divide and redivide until a mass of cells is formed. These cells arrange themselves into a hollow sphere that escapes from the parent to form a new colony.

Colonies are usually of different sexes, but in some species (for example, *Volvox globator*) both types of gametes (sperms and eggs) are formed by different cells of one colony. Numerous biflagellated sperms are produced by certain cells (Fig. 9-7), which are the male sex structures and called antheridia or male gametangia. Other cells, called oogonia or female gametangia, produce single eggs,

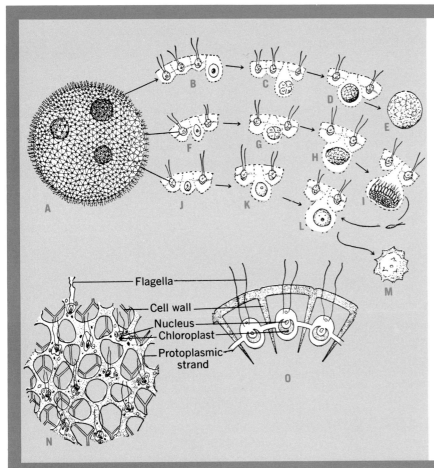

Fig. 9-7

Volvox. Methods of reproduction and certain structures. **A,** Colony with hundreds of vegetative (somatic) cells, each with flagella and showing three immature daughter colonies within; **B to E,** formation of a new colony by parthenogenesis (without fertilization), in which the cell loses its flagella and divides repeatedly to form a new colony; **F to I,** formation of a bundle of flagellated sperms (male gametes); **J to L,** formation of an egg (female gamete) and its fertilization by a sperm to produce a resistant zygospore, **M,** from which a new colony develops; **N,** surface view of the colony; **O,** side view of the colony: Note the protoplasmic strands that connect adjacent cells. Most somatic cells contain an eyespot and contractile vacuoles. In *Volvox* there are numerous somatic (body) cells, which die a natural death, and cells for purposes of reproduction, which continue the species by producing new individuals.

which are large, are without flagella, and contain a large amount of stored food.

A sperm swims to the egg and fertilizes it. This fusion, involving a smaller, motile sperm and a larger, nonmotile egg, is called oogamy. The resulting zygote develops a thick wall, thus forming a resistant zygospore. The zygospore undergoes meiosis or reduction division, after which the protoplast forms a new colony, or else a single zoospore (motile spore) is formed that makes a new colony. *Volvox* occurs in freshwater pools or lakes.

Volvocine series of green algae

The volvocine line (Table 9-3) of certain green algae shows a progressive development in cell number increase, sex differentiation, and distinction between somatic cells and reproductive cells. It is thought that the ancestral motile green cells from which many of the more advanced types may have arisen were much like *Chlamydomonas*. Study of the series will demonstrate that motile, unicellular types progress to smaller colonies and finally to larger colonies in which some of the cells are vegetative and some are reproductive. Also to be noted is the origin and progressive development of sexual reproduction from similar reproductive cells (isogamy) to unlike reproductive cells (heterogamy). *heterothallus*

Protococcus (pleurococcus)

Protococcus (pro to -kok′ us) (see Fig. 9-3) is a unicellular, thick-walled, round green alga common on trees and in moist places. Each cell has a nucleus and a large, thin chloroplast. Reproduction occurs by fission (cell division), and occasionally several individuals may remain together to form a colony.

111

Table 9-3
Volvocine series of green algae

Alga	Number of cells	Form of colony	Somatic cells (nonreproductive)	Gametes (sex cells)	Miscellaneous
Chlamydomonas	1	None	Absent	Isogametes (alike)	Cell wall of cellulose; single nucleus; two flagella of equal length; chloroplast; pyrenoid to make starch
Gonium sociale	4	Platelike; gelatinous	Absent	Isogametes (alike)	As above; cytoplasmic connective between cells
Gonium pectorale	16	Platelike; gelatinous	Absent	Isogametes (alike)	As above
Pandorina morum	16	Hollow sphere encased in gelatinous matrix	Absent	Heterogametes (unlike)	As above
Eudorina elegans	32	Hollow sphere with two kinds of cells	Some?	Heterogametes (unlike)	As above
Pleodorina illinoisensis	32	Hollow sphere with two kinds of cells	Four (in anterior end of colony)	Heterogametes (unlike)	As above
Pleodorina californica	64 or 128	Hollow sphere with two kinds of cells	Half of cells (in anterior end of colony)	Heterogametes (unlike)	As above
Volvox perglobator	500 to thousands	Sphere preceded by a platelike colony of 16 cells; somatic (body) and reproductive cells	All but about 100 cells	Heterogametes (unlike)	As above

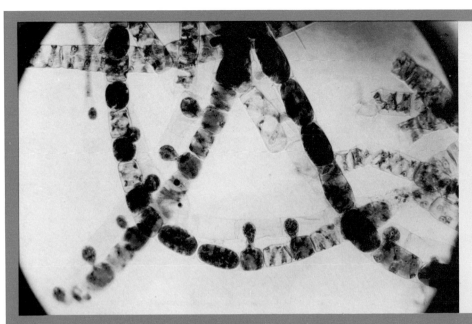

Fig. 9-8
Spirogyra, a green alga of phylum Chlorophycophyta. Transfer of protoplast (protoplasm of cell) through the conjugation tube, from one filament to the other, during the process of conjugation.
(Courtesy General Biological Supply House, Inc., Chicago, Illinois.)

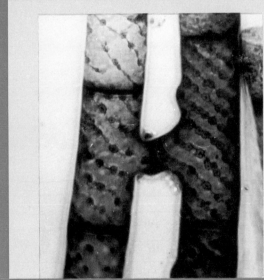

Fig. 9-9
Spirogyra. Conjugation tube formation.

Fig. 9-10
Spirogyra. Zygote formation.

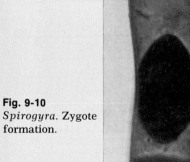

Spirogyra

Spirogyra (Gr. *speira*, spiral; *gyros*, curved) (Figs. 9-8 to 9-12) is common in fresh water, and may be called "pond scum" or "water silk." Each unbranched filament is composed of a linear series of cells and is covered with a mucilaginous sheath.

One or more spiral, ribbon-shaped chloroplasts with several pyrenoids may be present in each cell, depending upon the species. An organized nucleus near the center of the cell is surrounded by a layer of cytoplasm. Strands of cytoplasm also extend to the pyrenoids that are located on the chloroplasts.

113

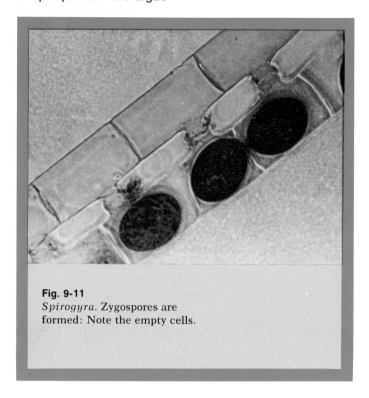

Fig. 9-11
Spirogyra. Zygospores are
formed: Note the empty cells.

Reproduction is accomplished asexually by fragmentation and by cell division; sexually by conjugation. In conjugation the cells known as gametangia in two adjacent filaments form a conjugation tube between them, and the contents of one cell pass through the tube in an ameboid manner to the cell of the other filament. This union, or fertilization, forms a resistant zygospore (zygote) from which a new filament can develop. Even though the nonmotile gametes are of equal size (isogamous), the one that migrates might be considered as male and the other as female. Sometimes the cells in a single filament unite.

Desmids

Desmids (des' mid) (Gr. *desmos*, chain) (Fig. 9-13) are frequently found floating in somewhat acid fresh water and may be unicellular, in filaments, or in irregular colonies. In most species each cell is divided into two halves that are joined by a connecting isthmus containing a nucleus. Each half cell contains a chloroplast. Reproduction occurs by cell division and in some species by con-

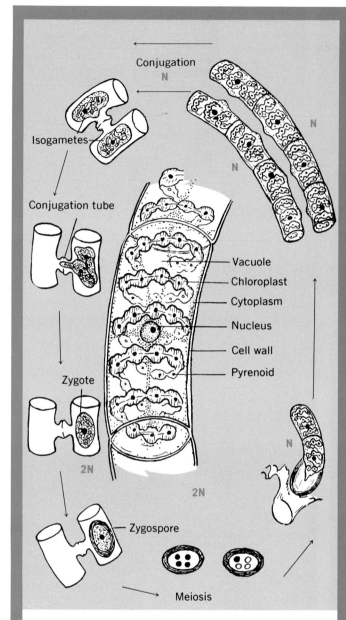

Fig. 9-12
Spirogyra. Reproduction by sexual conjugation. A vegetative cell is shown in the middle. Two conjugating filaments are shown in upper right. The numbers of chromosomes are shown by diploid (2N) and haploid (N). A germinating zygospore is shown in lower right.

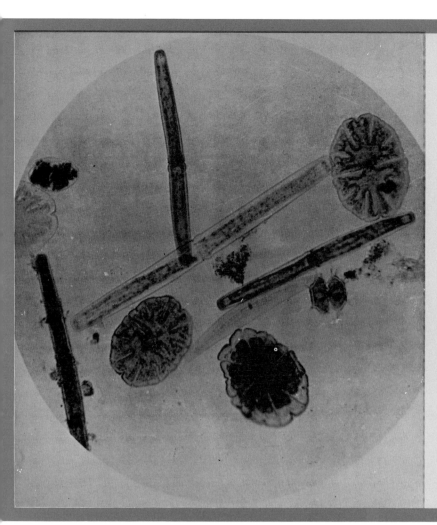

Fig. 9-13
Desmids, green algae of phylum
Chlorophycophyta.
(*Courtesy General Biological Supply House, Inc.,
Chicago, Illinois.*)

jugation, similar to *Spirogyra*. Some species of desmids contain vacuoles at either end with granules of gypsum (plaster). It is theorized that these granules function as statoliths (Gr. *statos*, stationary; *lithos*, stone) or organelles of equilibrium for keeping balance.

■ Ulothrix

Ulothrix (Gr. *oulos*, woolly; *thrix*, hair) (Fig. 9-14) is a freshwater, filamentous, unbranched, multicellular green alga that has a basal holdfast cell for attachment. Each vegetative cell has a nucleus and a chloroplast, which resembles an open band or ring and which contains pyrenoids. Each cell has a large central vacuole.

Certain unicellular reproductive structures, known as zoosporangia, produce from two to thirty-two large, motile zoospores, each with four flagella. Each zoospore forms a new filament by cell division. Other unicellular reproductive structures called gametangia produce from eight to sixty-four gametes, each of which bears only two flagella and is smaller than the four-flagellated zoospores. The gametes arise from different filaments. The union *heterothallus* of isogametes by the process of isogamy produces a single-celled zygote. Each zygote divides by meiosis to produce four zoospores, each of which will form a new filament through repeated cell divisions. Asexual reproduction may occur by fragmentation or by the formation of nonmotile

115

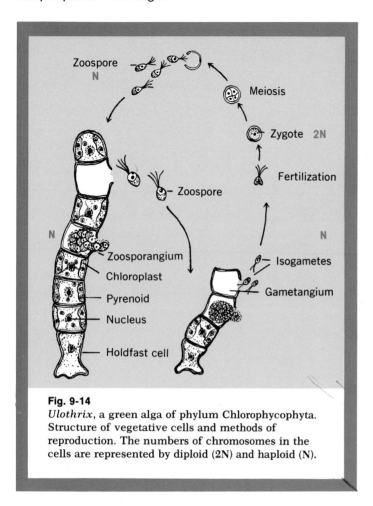

Fig. 9-14
Ulothrix, a green alga of phylum Chlorophycophyta. Structure of vegetative cells and methods of reproduction. The numbers of chromosomes in the cells are represented by diploid (2N) and haploid (N).

aplanospores (Gr. *aplanes*, not moving) in which the cell contents round up and secrete a new wall. The vegetative cells of *Ulothrix* are differentiated, interdependent, and may be considered to be above a true colony type of green alga in the evolutionary scale.

PHYLUM CHRYSOPHYCOPHYTA (YELLOW-GREEN ALGAE, GOLDEN BROWN ALGAE, AND DIATOMS)

The phylum Chrysophycophyta (Gr. *chrysos*, gold) contains algae in which there is a greater proportion of yellow or brown carotenoid pigments than there is chlorophyll, hence their color. The pigments and chlorophyll are contained in plastids. Stored foods are oils or leucosin, a carbohydrate. Flagella

may be present or absent, depending upon the species. The nucleus is organized. These algae may be unicellular, colonial, or even multicellular. Asexual reproduction occurs by cell division, motile zoospores, or by nonmotile spores (aplanospores). Sexual reproduction may occur by oogamy.

Yellow-green algae are primarily freshwater forms. However, some live in soils, on trees, or in damp places. Golden brown algae are predominately freshwater types and may contaminate water supplies. Diatoms occur in fresh and salt waters, in soils, on other algae and plants, and even in hot springs. Since diatoms are so widely distributed, they will be considered in more detail here.

■ Diatoms

Diatoms (Gr. *dia*, across or two; *tome*, to cut) are unicellular, but some may form filaments or other types of colonies (Figs. 9-15 and 9-16). The cells of various species may be shaped like rods, disks, triangles, or boats. Their innumerable shapes and delicate beauty label them as the "jewels of the plant world." The cell walls are usually composed of two overlapping valves (halves) and are often impregnated with silica, making them glasslike, fragile, and transparent. They do not decay after the plant dies. The plastids, in a layer of cytoplasm just within the cell wall, impart different colors to different species. One organized nucleus is present. Stored foods are oils and/or leucosin.

The siliceous cell walls are beautifully patterned with tiny dots or perforations. These are so symmetrically arranged that the efficiency of lenses may be tested by viewing them. The markings on the cell walls of the round types are usually radially arranged, whereas on the elongated types the markings assume a bilateral pattern. In the cell wall a longitudinal slit, called the raphe (Gr. *rhaphe*, seam), may be observed. Vegetative cells do not possess flagella.

Reproduction occurs by cell division, each newly formed cell remaining in the two original valves (Fig. 9-16). Each new cell then secretes a new valve inside each of the old valves. This results in two new cells. One is about the size of the original and the other is slightly smaller than the parent cell. Hence, some of these cells in succeeding divisions

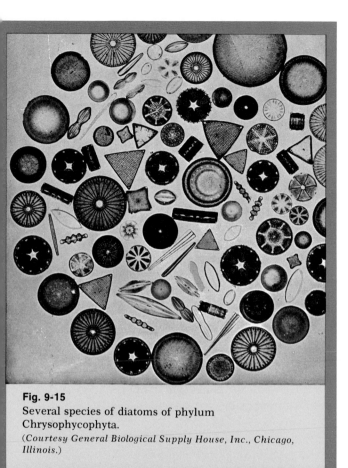

Fig. 9-15

Several species of diatoms of phylum Chrysophycophyta.

(*Courtesy General Biological Supply House, Inc., Chicago, Illinois.*)

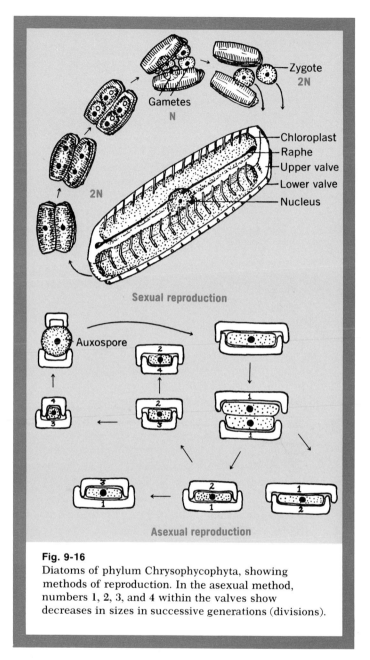

Fig. 9-16

Diatoms of phylum Chrysophycophyta, showing methods of reproduction. In the asexual method, numbers 1, 2, 3, and 4 within the valves show decreases in sizes in successive generations (divisions).

will become smaller and smaller. Finally, these smaller cells regain their original size by the production of rejuvenescent cells called auxospores. In the sexual method diatom cells form gametes, and the fusion of two gametes forms a zygote that acts as an auxospore. This zygote enlarges to maximum cell size for that species, and two new valves are then formed.

Diatoms are important food supplies for aquatic animals. Large accumulations of their siliceous valves are known as diatomaceous earth, which is frequently used in manufacturing polishes and tooth powders and pastes, for insulation, and for filtration purposes. Diatomaceous earth is used in dynamite as an absorbent for the liquid nitroglycerin. When it is added to concrete, strength and workability are improved. Bricks may also be sawed from the earth deposits that in certain parts of the world are hundreds of feet thick. Geologists suggest that diatoms may have assisted in the formation of petroleum. Diatomaceous earth can be molded into hollow cylinders to be used in making bacteriologic filters.

117

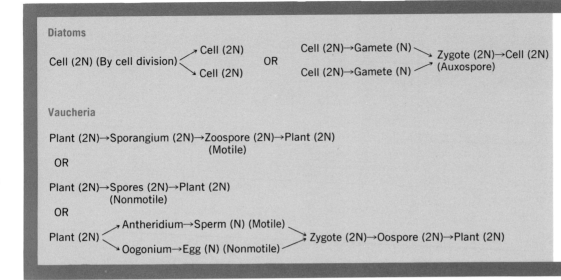

Diatoms

Cell (2N) (By cell division) → Cell (2N)
                              → Cell (2N)   OR   Cell (2N)→Gamete (N)
                                                   Cell (2N)→Gamete (N) → Zygote (2N)→Cell (2N)
                                                                            (Auxospore)

Vaucheria

Plant (2N)→Sporangium (2N)→Zoospore (2N)→Plant (2N)
                            (Motile)

OR

Plant (2N)→Spores (2N)→Plant (2N)
            (Nonmotile)

OR

Plant (2N) → Antheridium→Sperm (N) (Motile)
           → Oogonium→Egg (N) (Nonmotile) → Zygote (2N)→Oospore (2N)→Plant (2N)

Fig. 9-17
Life cycles of certain yellow-green algae and diatoms. Numbers of chromosomes are represented by (N) and (2N).

Vaucheria

Vaucheria (after Vaucher, a Swiss botanist), a golden brown filamentous alga (Fig. 9-18) is present in fresh water, but some are often found as a feltlike mass on moist soils. The plant (thallus) is long, tubelike, and normally not divided by cross cell walls. Numerous chloroplasts and nuclei are present in the cytoplasm that surrounds a central vacuole. Reserve food is stored as oil. The multinucleated filament is known as a coenocyte (se' no site) (Gr. *koinos*, shared; *kytos*, cell) and develops by the elongation of the cell in which the nucleus divides repeatedly without forming cross cell walls.

Asexual reproduction may occur in any one of three ways, depending on the habitat. In aquatic species a sporangium is produced by the formation of a cross wall near the tip of a branch. The protoplast of this cell becomes a motile, multiflagellated, multinucleated zoospore that develops into a new filament. In species on moist soil asexual nonmotile spores may be formed. In other terrestrial species the coenocytic protoplasts may divide into numerous segments, around which heavy walls are secreted to form thin-walled, asexual resting cells.

Sexual reproduction is accomplished by oogamy, in which sex cells of unequal size unite, the egg being nonmotile. Both sexes are present on the same filament. The antheridium is a hooklike branch in which numerous biflagellated sperms, each with a single nucleus, are produced. The oogonium is an enlarged structure containing an egg supplied with food and a single nucleus. The fertilized egg develops into a thick-walled, resistant oospore that later germinates to form a new filament.

PHYLUM PHAEOPHYCOPHYTA (BROWN ALGAE)

The brown algae belong to the phylum Phaeophycophyta (Gr. *phaeo-*, brown) (Figs. 9-19 to 9-22). They are multicellular, some types of kelps being more than 100 feet long. The plant body may show considerable differentiation, with large rootlike holdfasts for attachment, a stemlike stipe, and one or more leaflike blades. Gas-filled bladders (floats) are often present. Brown algae are marine and are usually present along the shores in colder waters. The golden brownish pigment called fucoxanthin (fu ko -zan' thin) masks the chlorophyll. Usually there are several plastids per cell, but no pyrenoids are present. Each cell has an organized nucleus and may contain vacuoles. Some cells are so highly organized that they resemble cells of higher plants.

Brown algae show alternation of generations, in which a gamete-producing gametophyte alternates with a multicellular, spore-producing sporophyte. The brown alga, *Sargassum*, reproduces by asexual fragmentation, and its sexual reproduction is much like the life cycle of the rockweed (*Fucus*). The so-called Sargasso Sea of the Atlantic Ocean, between the West Indies and Africa, has great masses of these plants.

Depending upon the species of brown algae, sexual reproduction may occur by isogamy, heterogamy, or oogamy. When present, the pear-shaped reproductive cells bear two lateral flagella of unequal length.

Certain brown algae are sources of iodine, potassium, fertilizers, and foods for animals and man.

Kelp (Laminaria)

The kelps (Fig. 9-19), which belong to the genus *Laminaria* (L. *lamina*, thin plate or blade), are called "devil's aprons" and occur on both coasts of the United States. In addition to the general characteristics of brown algae already given the following details apply more specifically to *Laminaria*. The large sporophyte plants bear patches of zoosporangia on the blade surfaces. The biflagellated zoospores formed in the zoosporangia produce two types of microscopic gametophytes. The microscopic male gametophyte is a simple, branched filament, bearing one-celled antheridia that produce biflagellated *sperms* with flagella of unequal length. The small, less-branched female gametophyte also is a short filament, bearing oogonia that produce nonmotile eggs. Fertilization by oogamy produces a zygote that eventually develops into a mature sporophyte. Hence, there is alternation of generations between the more conspicious sporophyte and the microscopic gametophyte.

Rockweed (Fucus)

The brown alga called the rockweed or *Fucus* (fu′ kus) (Figs. 9-21 and 9-22) is commonly attached to rocks along sea coasts. The sporophyte plant is leathery, frequently branched, and attached by a disk-shaped holdfast. Some branches bear gas-filled bladders (floats) for buoyancy in the water. The enlarged tips of the branches are called receptacles and are covered with small openings (ostioles) leading into internal cavities called conceptacles, within which are the sex organs. In some species the antheridia are borne on one plant and

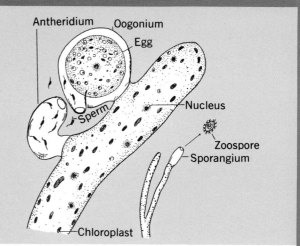

Fig. 9-18
Tubular type of golden brown alga (*Vaucheria*) of phylum Chrysophycophyta, showing methods of reproduction. Observe the numerous nuclei and chloroplasts in the tubular filament, which is usually not divided by cross cell walls. Sexual reproduction by oogamy is shown above, and asexual reproduction by zoospores is shown in lower right.

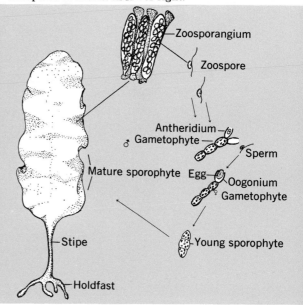

Fig. 9-19
Multicellular marine brown alga (*Laminaria*) of phylum Phaeophycophyta. Bladelike thallus with rootlike holdfasts. Alternation of generations exists between the large bladelike sporophyte (left) and gametophyte stages. The mature sporophyte bears patches of zoosporangia on its surface. Motile zoospores produce two types of small gametophytes (♂ gametophyte and ♀ gametophyte).

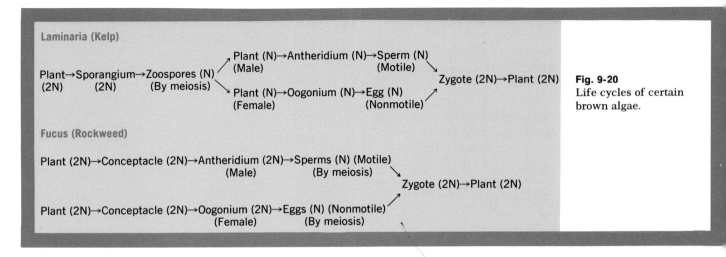

Laminaria (Kelp)

Plant→Sporangium→Zoospores (N) ⟶ Plant (N)→Antheridium (N)→Sperm (N)
(2N)   (2N)   (By meiosis)   (Male)   (Motile)
    ⟶ Plant (N)→Oogonium (N)→Egg (N)   Zygote (2N)→Plant (2N)
    (Female)   (Nonmotile)

Fucus (Rockweed)

Plant (2N)→Conceptacle (2N)→Antheridium (2N)→Sperms (N) (Motile)
    (Male)   (By meiosis)
        Zygote (2N)→Plant (2N)
Plant (2N)→Conceptacle (2N)→Oogonium (2N)→Eggs (N) (Nonmotile)
    (Female)   (By meiosis)

Fig. 9-20
Life cycles of certain brown algae.

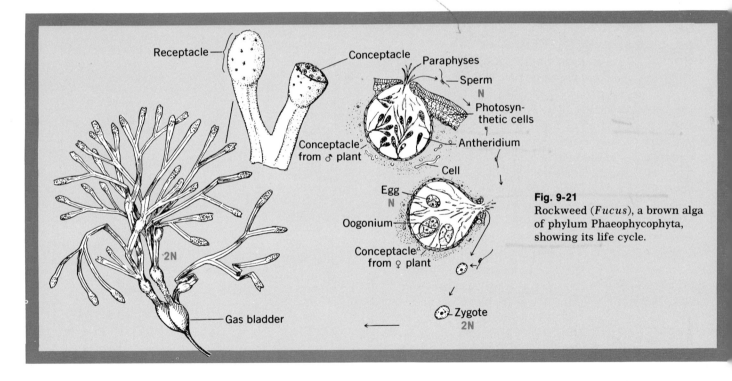

Fig. 9-21
Rockweed (*Fucus*), a brown alga of phylum Phaeophycophyta, showing its life cycle.

the oogonia on another plant; whereas in other species the male and female organs are located on the same plant, even within the same conceptacle. When mature, each oogonium contains eight eggs and is surrounded by a series of multicellular, branched, hairlike paraphyses (Gr. *para*, beside; *physis,* growth). The enlarged, terminal *antheridia* that are borne on the paraphyses produce numer-ous pear-shaped sperms, each with two unequal, lateral flagella. Within the water the larger egg is fertilized by the motile sperm to form a zygote through the process of heterogamy. A zygote eventually produces a new *Fucus* plant by numer-ous cell divisions. Unfertilized eggs may be in-duced to develop artificially, without fertilization by a sperm, through treatment with solutions of

120

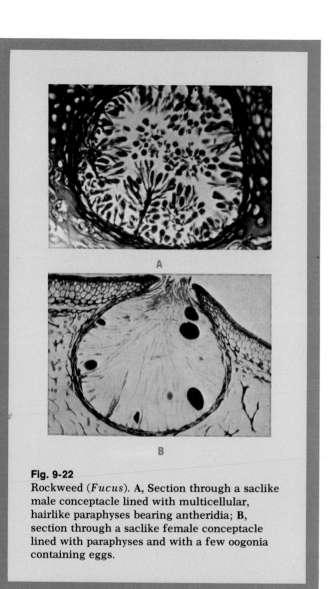

Fig. 9-22
Rockweed (*Fucus*). **A**, Section through a saclike male conceptacle lined with multicellular, hairlike paraphyses bearing antheridia; **B**, section through a saclike female conceptacle lined with paraphyses and with a few oogonia containing eggs.

acetic or butyric acids. This phenomenon is known as artificial parthenogenesis (Gr. *parthenos*, virgin; *genesis*, birth). Asexual reproduction may occur by fragmentation.

PHYLUM RHODOPHYCOPHYTA (RED ALGAE)
Red algae belong to the phylum Rhodophycophyta (Gr. *rhodon*, red) and sometimes are erroneously called sea "mosses" because they seem to resemble certain true mosses. Red algae are primarily marine, especially in warmer waters, although a few spe-

cies live in fresh waters. The chlorophyll is associated with a red pigment called phycoerythrin (Gr. *phykos*, seaweed; *erythros*, red) and sometimes with a blue pigment called phycocyanin. Most species are sessile (attached), multicellular, and may be relatively simple or branched in the form of a ribbon or a sheet. Sizes vary from a few inches to several feet in length. Each cell contains a nucleus, central vacuoles, and one or more plastids, some of which possess pyrenoids. Broad cytoplasmic strands that connect adjacent cells are features of red algae. Stored foods are "starches," which are carbohydrate intermediates between dextrose (a sugar) and true starch.

Most red algae reproduce sexually. None of the sex cells or asexual reproductive cells bear flagella. Many red algae show an alternation of generations between a free-living sporophyte and a free-living gametophyte. The specific methods of reproduction of such typical forms as *Nemalion* and *Polysiphonia* are considered later in this chapter.

Some of the red algae grow on other algae as epiphytes, within the bodies of plants, or even as parasites. One species is associated with a green alga among the hair of the three-toed sloth. Some species grow in the sea to a depth of 600 feet, probably because the phycoerythrin acts as a photosynthetic pigment in the blue light of deep sea water. Certain red algae are sources of agar, which is used as a solid medium for the cultivation of bacteria as well as for a medicine. A jellylike food may be obtained from the Irish "moss" (*Chondrus*) (Fig. 9-25). Some types become encrusted with lime and thus help to form reefs, atolls, and even islands. Other red algae are sources of jellylike substances used in shoe polishes, hair dressings, shaving creams, cosmetics, and various lubricating jellies. Some species serve as foods for fish, cattle, sheep, and even man.

Nemalion
Nemalion (Gr. *nema*, thread) is a cylindrical, forked, marine red alga found attached to rocks along the sea coast (Fig. 9-23). The body (thallus) is composed of a mass of interwoven, branched, threadlike structures surrounded by a gelatinous material. Some branches bear brushlike filaments whose tips are divided into short antheridia (sperm-

121

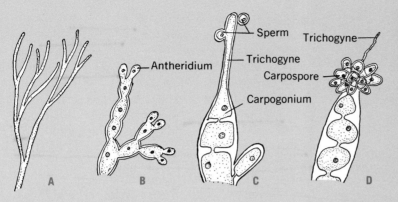

Fig. 9-23
Red alga (*Nemalion*) of phylum Rhodophycophyta. **A,** Portion of forked, cylindrical body (thallus) that grows attached to rocks along the seacoast; **B,** portion of a branch bearing brushlike filaments whose tips divide into antheridia, each of which contains a nonmotile sperm (spermatium); **C,** portion of a branch bearing a basal carpogonium (with an egg) and an elongated trichogyne; a nonmotile sperm, carried by water, descends the trichogyne to unite with the egg to form a zygote; **D,** portion of a carpogonium showing asexual, nonmotile carpospores produced by a cluster of filaments; released carpospores develop into new *Nemalion* plants.

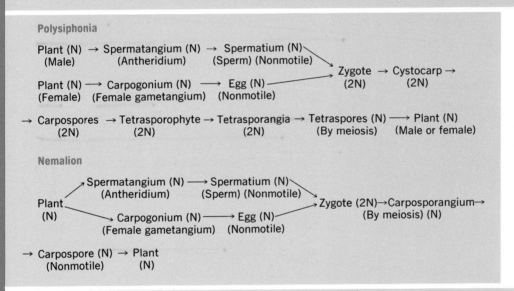

Fig. 9-24
Life cycles of certain red algae.

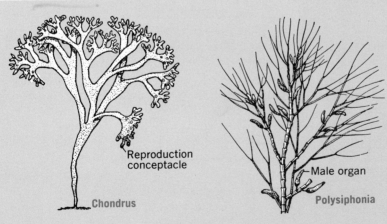

Fig. 9-25
Red algae of phylum Rhodophycophyta. *Chondrus* is sometimes erroneously called Irish "moss." *Polysiphonia* is also shown in Fig. 9-26.

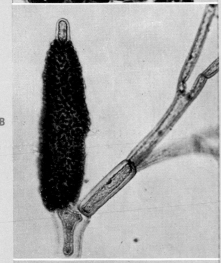

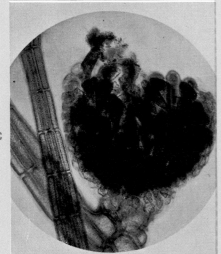

atangia), each containing one nonmotile sperm (spermatium). The tips of other branches bear female structures, each with an enlarged, basal carpogonium (Gr. *karpos*, fruit; *gone*, offspring) that bears an egg and an elongated, hairlike, tubular trichogyne (trik′ o jin) (Gr. *thrix*, hair; *gyne*, female) to receive the sperm that is brought by water. The sperm nucleus descends within the trichogyne to the carpogonium where it fuses with the egg nucleus to form a zygote. From the carpogonium are formed a cluster of short filaments whose tips produce the asexual, nonmotile carpospores. These carpospores germinate to form a new *Nemalion* plant.

There are a few freshwater red algae that are sometimes found in clear, colder waters, which are similar to marine types in many ways but are usually smaller.

Polysiphonia

Polysiphonia (Gr. *poly-*, many; *siphon*, tube) is a marine, multi-branched red alga that grows on rocks of the sea coast (Figs. 9-25 and 9-26). The main axis and larger branches consist of a central core made of a single row of elongated core cells and surrounded by a layer of jacket cells. The elon-

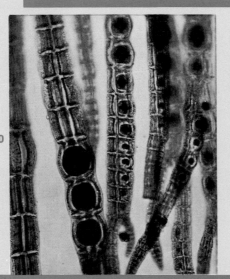

Fig. 9-26
Marine red alga (*Polysiphonia*) of phylum Rhodophycophyta. A, Antheridial (spermatangial) clusters; B, antheridial branch; C, cystocarp (carpogonium) with an enclosed gonimoblast; D, tetraspores.
(*Courtesy General Biological Supply House, Inc., Chicago, Illinois.*)

gated core cells are connected with each other by strands of cytoplasmic connectives that form tube-like siphons, hence the name *Polysiphonia*. Each cell has a nucleus and numerous dislike, red plastids that contain phycoerythrin and mask the chlorophyll.

This red alga is diecious (di -e′ shus) (Gr. *di-*, two; *oikos*, house), the male gametes being produced by one plant and the female gametes by another plant. The lateral branches of the male plants bear clusters of antheridia (spermatangia) that produce numerous nonmotile sperm (spermatia). The side branches of other plants bear female carpogonia. Each carpogonium (female gametangium) bears an elongated trichogyne to receive the sperm that is brought by water. The nucleus of the sperm travels down the trichogyne to the carpogonium where it fuses with the egg nucleus to form a zygote. After numerous cell divisions, many filaments are eventually produced whose tips form many carpospores. Other filaments of the plant form an urn-shaped enclosure (cystocarp) around the carpospores. When the carpospores are released through an opening in the enclosure, they produce new plants that form tetrasporangia, each with four nonmotile tetraspores. The tetrasporangia are borne on the central core and lie just beneath the jacket cells. When liberated, each tetraspore produces a *Polysiphonia* plant that bears either antheridia or carpogonia. This complex life cycle consists of (1) male and female plants, (2) zygote and carpospores, and (3) tetrasporophyte plants with tetrasporangia that produce the tetraspores. The number of chromosomes is reduced by one half (reduction division) when the tetraspores are formed.

Some contributors to the knowledge of algae

A. L. de Jussieu (1748-1836)
A French botanist who proposed a natural system of plants (1789) from which the term algae dates in a somewhat modern sense.

Friedrich Kützing (1807-1893)
A German botanist who first distinguished between diatoms and desmids (1833).

W. H. Harvey (1811-1866)
An Irish botanist who divided the algae into three groups that correspond quite well with present-day groups of blue-green, green, brown, and red algae (1836). (*The Bettmann Archive, Inc.*)

Karl W. von Naegeli (1817-1891)
A Swiss botanist who distinguished the blue-green from the green algae (1853).

Harvey

Gottlob L. Rabenhorst (1806-1881)
A German botanist who recognized the green algae group (Chlorophyta) in approximately its present-day sense (1863).

C. J. Friedrich Schmitz
He made important contributions to our knowledge of the structure and classification of red algae (1883-1889).

W. G. Farlow
He made extensive studies and published a volume on the marine algae of the south coast of New England (1873).

Georg Klebs (1857-1918)
A German botanist who suggested that several groups of flagellated organisms are related to different groups of algae and are not all to be included with the unicellular animals as they had been. This proposal caused **Adolph Pasher (1887-1945)**, a German botanist, to work out a system classifying the major groups of algae along present-day lines (1910-1931). He proposed the term Chrysophyta (1914).

Harald Kylin (1879-1949)
A Swedish botanist who first clearly recognized the major groups of brown algae, based primarily on life cycles (1933). He also modified Schmitz' classification of red algae (1923-1932).

1 Describe in your own words that group of plants commonly referred to as algae.

2 Why are the algae placed in the subkingdom Thallophyta?

3 List the distinguishing characteristics by which the following phyla may be differentiated: Cyanophycophyta, Euglenophycophyta, Chlorophycophyta, Chrysophycophyta, Phaeophycophyta, Rhodophycophyta. Give examples of each phylum.

4 What are the relationships among algae, chlorophyll, and photosynthesis?

5 Define: plankton, epiphyte, and parasite.

6 Describe the special kind of plants known as lichens.

7 Discuss the economic importance of each phylum of algae, including beneficial and harmful conditions.

8 List the asexual methods of reproduction found in algae, describing each.

9 List the sexual methods of reproduction found in algae, describing each.

10 What is meant by a life cycle? Make a diagram of a typical life cycle of an alga of each phylum.

11 Describe the increase in complexity of structures, methods of reproduction, and functions from the simpler to the higher types of algae.

12 Define, giving an example of each: isogamy, heterogamy, and oogamy.

13 List the various types of pigments found in the various algae, giving the specific alga in which each is found.

14 List the various types of stored foods found in various algae, giving the specific alga in which each is found.

Selected references

Alexopoulos, C. J., and Bold, H. C.: Algae and fungi, New York, 1967, The Macmillan Company.

Chapman, V. J.: The algae, London, 1962, Macmillan & Co., Ltd.

Dawson, E. Y.: How to know the seaweeds, Dubuque, Iowa, 1956, William C. Brown Company, Publishers.

Fritsch, F. E.: The structure and reproduction of the algae, vol. 1 and 2, Cambridge, 1935 and 1945, Cambridge University Press.

Gibor, A.: Acetabularia: a useful giant cell, Sci. Amer. 215:118-24, 1966.

Guberlet, M. L.: Seaweeds at ebb tide, Seattle, 1956, University of Washington Press.

Jackson, D. F. (editor): Algae and man, New York, 1964, Plenum Press.

Lamb I. M.: Lichens, Sci. Amer. Oct., 1959.

Lewin, R. A. (editor): Physiology and biochemistry of algae, New York, 1962, Academic Press Inc.

Milner, H. W.: Algae as food, Sci. Amer. 189:31-35, 1953.

Prescott, G. W.: How to know the freshwater algae, Dubuque, Iowa, 1954, William C. Brown Company, Publishers.

Prescott G. W.: The algae: a review, Boston, 1968, Houghton Mifflin Company.

Smith, G. M.: Cryptogamic botany, vol. 1, New York, 1955, McGraw-Hill Book Company.

Tiffany, L. H.: Algae, the grass of many waters, Springfield, Illinois, 1958, Charles C Thomas, Publisher.

Weiss, F. J.: The useful algae, Sci. Amer. 187:15-17, 1952.

chapter ten

Phylum Mycota— nonphotosynthetic plants (fungi)

Nonphotosynthetic plants without true leaves, stems, or roots; do not form multicellular embryos; without true vascular (conducting) tissues; subkingdom Thallophyta

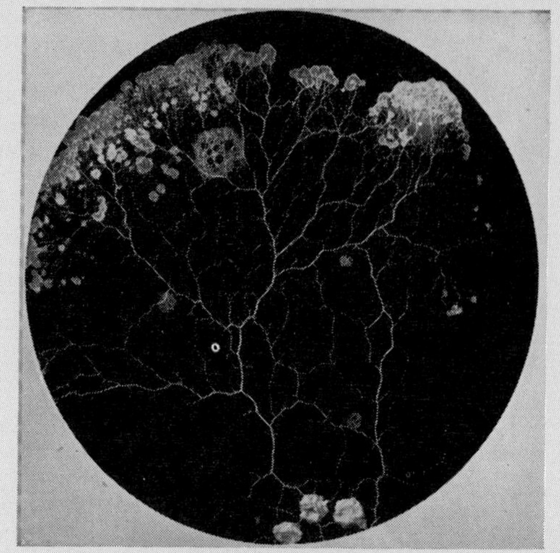

Fig. 10-1
Plasmodium of slime mold (*Physarum*) of class Myxomycetes.
(*Courtesy General Biological Supply House, Inc., Chicago, Illinois.*)

SUBPHYLUM MYXOMYCOTINA (SLIME MOLDS)

Slime molds are placed in the class Myxomyetes (Gr. *myxos*, slime; *mykes*, fungus) and are so named because of their slimy appearance. They are primarily saprophytes and derive foods from nonliving organic materials in soils and from decaying plant materials. A few species are parasitic, some even pathogenic (disease-producing). They resemble certain true fungi in their methods of spore formation. In addition they resemble certain lower animals because of their slimy, amebalike bodies, their ameboid methods of locomotion by forming pseudopodia (Gr. *pseudo*, false; *pous*, foot), and their ingestion of solid foods. The plant body is a thin mass of naked, slimy protoplasm called a plasmodium (Gr. *plasma*, liquid; *eidos*, form) (Fig. 10-1), which contains numerous nuclei and moves by a flowing of the protoplasm that forms the pseudopodia. The plasmodium may be colorless, red, yellowish, violet, or some other color. The plasmodium produces a number of spore cases, called sporangia, which may vary in size and form in different species. The sporangium may be colorless, orange, brown, or purple, depending upon the species. In some species the sporangia are borne on stalks; in others they are stalkless. They may be spherical, ovoid, or some other shape. As a sporangium matures, the internal protoplasm forms a network of fine fibers called a capillitium (L. *capillus*, hair) in the meshes of which are formed numerous unicellular, nonmotile spores. Germinated spores form motile, flagellated cells known as isogametes. Two isogametes unite to form a motile, ameboid zygote, which later loses the flagella and becomes ameboid. The zygote moves over the substratum, engulfing bacteria and organic particles that are digested in vacuoles. The multinucleated plasmodium does not contain a cellulose wall.

Among the common slime molds are *Stemonitis*, *Lycogala*, and *Physarum*. A summary of the characteristics of slime molds and true fungi is given in Table 10-1.

SUBPHYLUM EUMYCOTINA (TRUE FUNGI)

True fungi are placed in the subphylum Eumycotina (Gr. *eu*, true; *mykes*, fungus). They are very diverse in size, form, physiology, and methods of

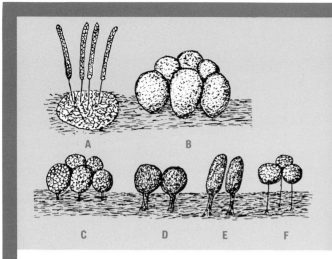

Fig. 10-2
Fruiting bodies (sporangia and stalks) of slime molds. A, *Stemonitis;* B, *Lycogala;* C, *Physarum;* D, *Hemitrichia;* E, *Diachea;* F, *Lamproderma.* (Not drawn to scale and somewhat diagrammatic.)

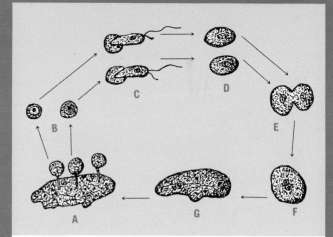

Fig. 10-3
Life cycle of a representative slime mold. A, Plasmodium with spore cases (the latter together with stalks are called fruiting bodies) and with a diploid number of chromosomes (2N); B, spores with a haploid number of chromosomes (N) produced by meiosis; C, swarm spores with two unequal flagella and with a haploid number of chromosomes (N); D, myxamoebae with a haploid number of chromosomes (N); E, fusion of two myxamoebae; F, zygote with a diploid number of chromosomes (2N); G, plasmodium. (Not drawn to scale and somewhat diagrammatic.)

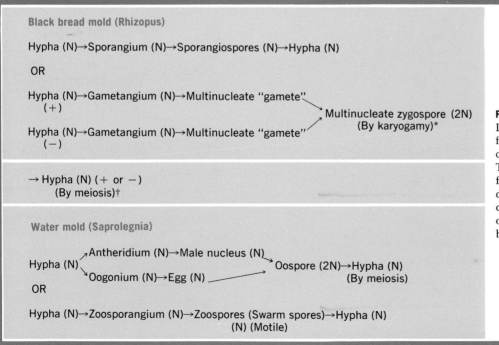

Black bread mold (Rhizopus)

Hypha (N)→Sporangium (N)→Sporangiospores (N)→Hypha (N)

OR

Hypha (N)→Gametangium (N)→Multinucleate "gamete"
  (+)
                                                Multinucleate zygospore (2N)
Hypha (N)→Gametangium (N)→Multinucleate "gamete"   (By karyogamy)*
  (−)

→ Hypha (N) (+ or −)
  (By meiosis)†

Water mold (Saprolegnia)

Hypha (N) ⟋Antheridium (N)→Male nucleus (N) ⟍
                                           Oospore (2N)→Hypha (N)
       ⟍Oogonium (N)→Egg (N) ⟋   (By meiosis)
OR

Hypha (N)→Zoosporangium (N)→Zoospores (Swarm spores)→Hypha (N)
                                   (N) (Motile)

Fig. 10-4
Life cycles of certain true fungi, showing numbers of chromosomes (N) and (2N). The asterisk (*) refers to a fusion of two nuclei. The dagger (†) refers to the degeneration of all but one of the fusion nuclei and all but one of the meiotic nuclei.

Table 10-1

Summary of distinguishing characteristics of slime molds and true fungi

Characteristic	Slime molds (subphylum Myxomycotina)	True fungi (subphylum Eumycotina)		
		Algalike fungi (Phycomycetes)	Ascus fungi (Ascomycetes)	Club fungi (Basidiomycetes)
Multicellular embryos	Absent	Absent	Absent	Absent
Plastids and chlorophyll	Absent	Absent	Absent	Absent
Organized nucleus	Present	Present	Present	Present
Filamentous hyphae	Absent	Present	Present*	Present
Septate hyphae	Absent	Absent†	Present	Present
Ameboid plasmodium	Present	Absent	Absent	Absent
Locomotion (adult stage)	Pseudopodia	Nonmotile	Nonmotile	Nonmotile
Asexual reproduction	Sporangiospores Motile swarm spores (myxamoebae)	Sporangiospores Motile swarm spores (zoospores) (aquatic species)	Ascospores Conidiospores* Budding*	Conidiospores* Chlamydospores* Uredospores* Telispores* Pycnicspores* Aeciospores*
Sexual reproduction	Isogamy	Isogamy* Heterogamy*	Ascospores* (by fusion)	Basidiospores (by fusion)

*Certain species.
†Except certain older hyphae.

reproduction. All true fungi possess heterotrophic nutrition, since they are either saprophytes or parasites. The bodies of most higher fungi consist of filaments called hyphae (Gr. *hyphe,* web). In some species the hypha is unicellular; in others it is multicellular. In some species certain hyphae contain yellow, orange, red, or other types of pigments. A mass of hyphae is called a mycelium (my -se′ li um) (Gr. *mykes,* fungus). All true fungi reproduce by means of some type of microscopic spores, and other methods of reproduction may also

occur in various species. Spores are usually dispersed by air currents, water, animals, or in other ways. Eumycotina require abundant moisture, a favorable temperature, and a sufficient supply of proper organic foods. Most of them are aerobic, but a few are anaerobic.

Class Phycomycetes

Class Phycomycetes (Gr. *phykos,* alga; *mykes,* fungus) is so named because many of these true fungi are threadlike or filamentous, like many of the algae, except that the filaments are without chlorophyll in fungi. Although modern practice is to divide this class into six new classes, the name Phycomycetes will be retained here for simplicity. The characteristics of this class will be shown in the following typical representatives.

Black bread mold

Rhizopus nigricans (Figs. 10-5 to 10-8) is a common saprophytic mold that lives on moist, organic materials, such as bread, fruits, animals, and animal dung. The hyphae are whitish or grayish and form a weblike mycelium. Young hyphae are branched, are without cross walls (nonseptate), and contain numerous nuclei. Older hyphae, especially when reproducing sexually, may have cross walls (septate). The rootlike hyphae, known as rhizoids (Gr. *rhiza*, root; *eidos*, like), absorb nourishment from the substratum and anchor the plant. Other hyphae grow over the surface and are called stolons from which arise spore-forming sporangiophores (Gr. *spora*, spore; *angios*, case; *phorein*, to bear). Each sporangiophore bears a globular spore case (sporangium) at its tip that becomes darker as it matures (Fig. 10-6). An enlarged, basal, internal structure within the sporangium is called the columella (L. *columella*, small column), which may be visible when the sporangium ruptures. The airborne, asexual, nonmotile sporangiospores germinate to form branching hyphae and eventually form a new mycelium.

Sexual reproduction might be considered isogamous. Small branches, known as gametangia, form between two adjacent hyphae, which fuse at their tips. *Rhizopus nigricans* is said to be heterothallic (Gr. *heteros*, different; *thallos*, shoot); that is, two types of hyphae, called "plus" and "minus," are necessary for this sexual fertilization. Each gametangium forms a multinucleated "gamete." The two "gametes" unite to form a thick-walled, multinucleated zygospore (Figs. 10-7 and 10-8), which under proper conditions will germinate to form a new hypha. According to modern classification, the production of a zygospore places a fungus in the class Zygomycetes. Thus *Rhizopus* is placed in this class. Some forms of bread mold are homothallic (Gr. *homos*, same; *thallos*, shoot), both types of gametes being produced on the same plant.

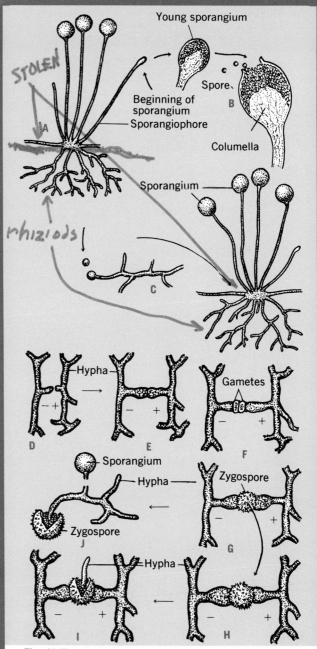

Fig. 10-5
Black bread mold (*Rhizopus nigricans*). **A**, Portion of the mycelium; **B**, sporangium enlarged and with escaping spores; **C**, germination of the spore to form the mycelium; **D to H**, conjugation (union) of hyphae (+ and −) to form the zygospore; **I and J**, germination of the zygospore to form hyphae and eventually sporangia. Stages **A to C** show asexual spore formation; stages **D to H** show sexual spore (zygospore) formation.

Table 10-2
True fungi

Characteristic	Phycomycetes (algalike fungi)	Ascomycetes (sac or ascus fungi)	Basidiomycetes (club or basidium fungi)
Spore formation	Sporangium with indefinite number of spores (sporangiospores)	Saclike ascus, with 8 ascospores (usually)	Club-shaped basidium, with 4 basidiospores
Other methods of reproduction	Isogamy; heterogamy	Conidiospores (in certain species); budding in yeasts	Chlamydospores (in certain species); uredospores, teliospores, pycniospores, aeciospores (in certain species)
Organization of hyphae	When present, typically nonseptate (coenocytic) (may be septate when mature and reproducing)	When present, typically septate	Septate
Mycelium	Loose, weblike mass of hyphae	When present, loose hyphae (certain species); compact hyphae (in some species)	Loose hyphae (in certain species); compact hyphae (in some species)
Motile cells produced	In many aquatic species (not in others)	Absent	Absent
Sexual fruiting body	Unicellular and without sterile envelope	Multicellular and larger; with sterile envelope	Similar to Ascomycetes in general way
Size and form	Range in size from microscopic, single-celled forms to those with hyphae, which are most common	Great diversity of size and construction; yeasts unicellular and without hyphae (usually)	Great diversity of size and construction; some hyphae small, whereas others such as mushrooms, and shelf fungi, complex and larger

Fig. 10-6
Mass of sporangia of black bread mold (*Rhizopus nigricans*). Each tiny dot is a sporangium containing hundreds of spores. The culture was started by placing a few spores on a nutrient agar medium.

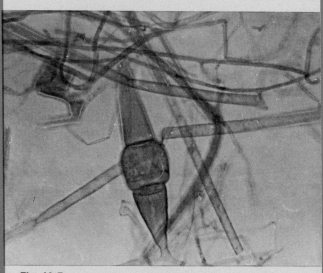

Fig. 10-7
Young zygospore of *Rhizopus nigricans*: Note the strands of mycelium in the background.

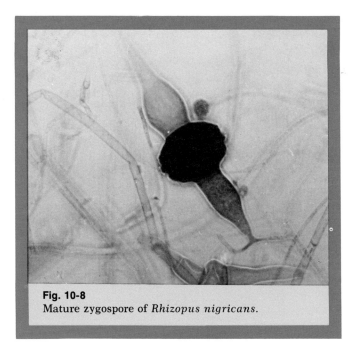

Fig. 10-8
Mature zygospore of *Rhizopus nigricans*.

Water mold

The fungi of the water mold group are primarily aquatic saprophytes (Fig. 10-9), securing foods from organic matters, especially from dead plants and animals, thus having a heterotrophic nutrition. A few species cause serious damage by parasitizing fishes, amphibia, and turtles. Fish in aquaria may become affected by the whitish mycelia of water molds.

Saprolegnia (Gr. *sapros*, rotten; *legnon*, edge) has slender hyphae with tapering ends that penetrate the substratum and also branched hyphae whose tips bear enlarged zoosporangia. The zoospores (swarm spores) with two terminal cilia are liberated by the zoosporangia and swim ("swarm") in the water, eventually losing their terminal cilia and surrounding themselves with a wall. Later, each escapes from this wall, but now has two lateral cilia. Each of these zoospores forms a new hypha.

Sexual reproduction is heterogamous whereby unicellular, enlarged oogonia and clublike male antheridia are developed. Each oogonium produces several eggs. The antheridia develop antheridial tubes that penetrate the oogonia. Male nuclei (unorganized sperms) are discharged through the antheridial tubes into the oogonia. An egg fertilized

131

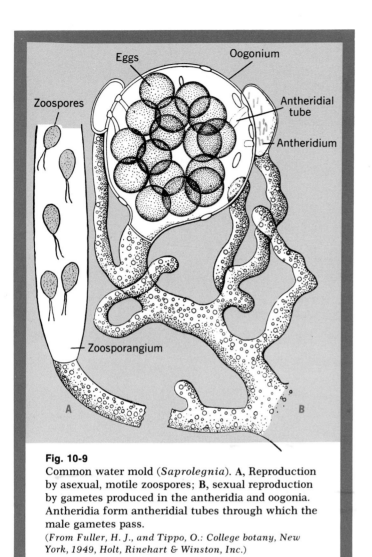

Fig. 10-9
Common water mold (*Saprolegnia*). **A**, Reproduction by asexual, motile zoospores; **B**, sexual reproduction by gametes produced in the antheridia and oogonia. Antheridia form antheridial tubes through which the male gametes pass.
(From Fuller, H. J., and Tippo, O.: College botany, New York, 1949, Holt, Rinehart & Winston, Inc.)

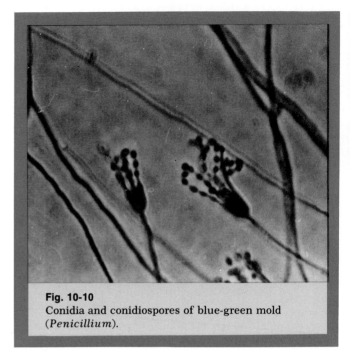

Fig. 10-10
Conidia and conidiospores of blue-green mold (*Penicillium*).

by a male nucleus forms a thick-walled oospore that develops into new hyphae. In some species eggs develop without fertilization. *Saprolegnia* produces both antheridia and oogonia on the same plant, hence is said to be homothallic. In new systems of taxonomy *Saprolegnia* is placed in the class Oomycetes.

■ Class Ascomycetes

Class Ascomycetes (Gr. *askos*, sac; *mykes*, fungus) comprises true fungi that at some time in their life cycle produce ascospores in a saclike ascus. Other methods of reproduction may be present, depending upon the species. The Ascomycetes are filamentous except for certain types of yeasts. When filaments are present, they have cross walls (septate), and the cells possess organized nuclei. Additional characteristics of this class are shown in the following representatives.

Penicillium

Penicillium (L. *penicillus,* painter's brush) is a bluish green mold whose loosely arranged hyphae grow in or on such materials as damp leather, foods, and fruits (Fig. 10-10). *Penicillium* produces ascospores in saclike asci. In addition the spore-bearing hyphae, called conidiophores, bear chains of small, colored spores (conidia) at their tips. Masses of these conidiophores resemble tiny brushes, hence the name *Penicillium*.

Various species of *Penicillium* cause the spoilage of such foods as bread, apples, pears, grapes, and citrus fruits, and cause the destruction of paper, leather, and lumber. A few members of this group of fungi may cause skin infections in man. Certain types are used in the manufacture of Camembert cheese (*Penicillium camemberti*) and of Roquefort cheese (*Penicillium roqueforti*). The characteristic flavors and odors of these cheeses

132

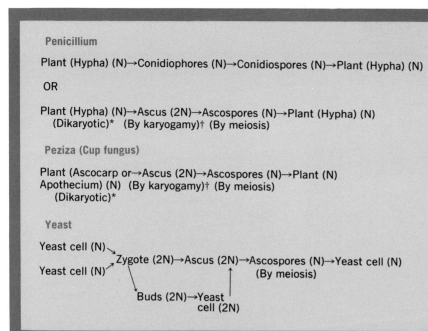

Penicillium

Plant (Hypha) (N)→Conidiophores (N)→Conidiospores (N)→Plant (Hypha) (N)

OR

Plant (Hypha) (N)→Ascus (2N)→Ascospores (N)→Plant (Hypha) (N)
(Dikaryotic)* (By karyogamy)† (By meiosis)

Peziza (Cup fungus)

Plant (Ascocarp or→Ascus (2N)→Ascospores (N)→Plant (N)
Apothecium) (N) (By karyogamy)† (By meiosis)
(Dikaryotic)*

Yeast

Yeast cell (N)
Yeast cell (N) →Zygote (2N)→Ascus (2N)→Ascospores (N)→Yeast cell (N)
(By meiosis)
Buds (2N)→Yeast cell (2N)

Fig. 10-11
Life cycles of certain true fungi, showing numbers of chromosomes (N) or (2N). The asterisk (*) refers to two separate nuclei, each (N). The dagger (†) refers to the fusion of two nuclei.

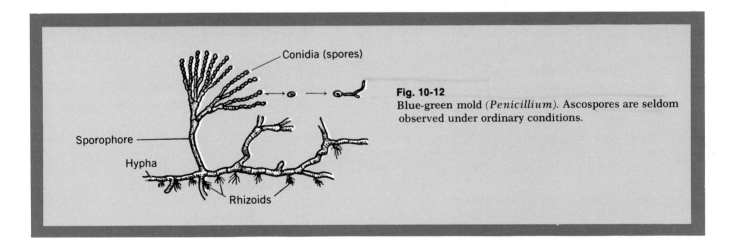

Conidia (spores)

Sporophore

Hypha

Rhizoids

Fig. 10-12
Blue-green mold (*Penicillium*). Ascospores are seldom observed under ordinary conditions.

result from the activities of these molds. The bluish green areas in the cheese are masses of conidia. Certain species of *Penicillium* are used in the manufacture of the antibiotic penicillin.

Antibiotics (an ti bi -ot′ iks) (Gr. *anti*, against; *bios*, life) are organic substances that are synthesized by one type of organism and inhibit or destroy another type of organism. This antagonistic inhibition between two species is called antibiosis. In 1940 the medicinal value of penicillin and other antibiotics was established.

Yeasts

Yeasts are typically unicellular fungi, usually without hyphae, although a few species may develop short ones. Each cell is usually ovoid in shape and contains a nucleus (Fig. 10-13). Asexual reproduction is often accomplished by budding, during which small protuberances (buds) are projected from the cell. The bud may free itself or remain attached and produce more buds, eventually forming a chainlike arrangement of individual cells.

133

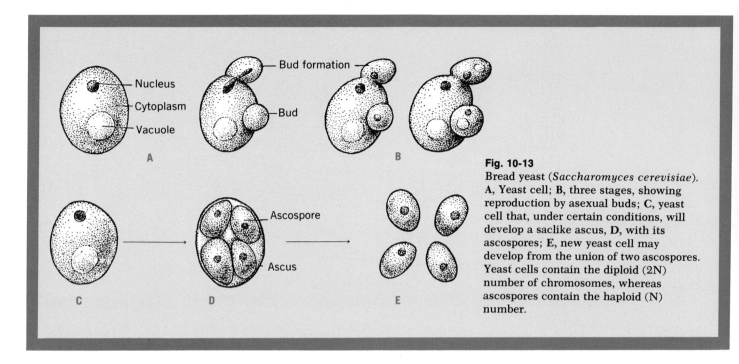

Fig. 10-13
Bread yeast (*Saccharomyces cerevisiae*). A, Yeast cell; B, three stages, showing reproduction by asexual buds; C, yeast cell that, under certain conditions, will develop a saclike ascus, D, with its ascospores; E, new yeast cell may develop from the union of two ascospores. Yeast cells contain the diploid (2N) number of chromosomes, whereas ascospores contain the haploid (N) number.

Under certain conditions a yeast cell may become an ascus, in which ascospores (usually four) are formed. In other instances two yeast cells may fuse before ascospores are formed. Certain yeasts produce an enzyme, zymase, which ferments sugars to form ethyl alcohol. In dough certain yeasts ferment sugars to release carbon dioxide whose bubbles cause the bread to "rise" (become porous). Some yeasts manufacture vitamins, while others synthesize protein from molasses and ammonia. The common yeast cake contains yeast cells and a quantity of starch. Yeasts may be classified as (1) *Saccharomyces* (sugar fungi), which are harmless or even beneficial, and (2) *Blastomyces* (germ fungi), which are pathogenic. Many sugar fungi are of commercial value, but many germ fungi produce a variety of common diseases (Table 10-3).

Pathogenic Ascomycetes

Pathogenic Ascomycetes produce a variety of diseases in plants, including the following: (1) In ergot of rye the fungus turns cereal grains and wild grasses purple and makes them poisonous to man. (2) In chestnut tree blight, the fungus destroys great numbers of chestnut trees by producing cankers and brown, shriveled leaves on them. The mycelium kills the cambium and inner bark, eventually stopping the food supply. (3) In Dutch elm disease the fungus (introduced to the United States in 1930) causes great damage to many elm trees. This disease is not to be confused with phloem necrosis of elms, caused by a virus. (4) In peach leaf curl the yeastlike fungus causes the leaves to curl and become yellow. (5) In apple scab the fungus reduces the yield of apples by millions of bushels by producing raised, discolored areas on the leaves and scabby spots on the fruits, which may be small and distorted. (6) In powdery mildews the parasitic hyphae appear as grayish, powdery areas on the leaves and stems of flowering plants, such as lilacs, roses, apples, grapes, cherries, cereals, clovers, and dandelions. Slender hyphae, the haustoria, absorb food from the host plant.

Class Basidiomycetes

Fungi in the class Basidiomycetes (ba sid i o mi -se′ tez) (Gr. *basidium*, club; *mykes*, fungus) bear basidiospores (usually four) on club-shaped basidia. Closely packed basidia are arranged in a

Table 10-3

Diseases of man and animals caused by pathogenic fungi

Disease	Causal agent and symptoms
North American blastomycosis (Gilchrist's disease)	Fungus causing chronic infection with suppurative and granulomatous lesions in body but especially in skin, lungs, and bones
Moniliasis	Fungus causing lesions in mouth, skin, nails, lungs, or vagina; creamy white, ulcerlike patches in mouth constitute disease called thrush
Coccidioidomycosis	Fungus causing acute but benign respiratory infection; or progressively chronic, malignant infection of skin, bones, or internal organs
Sporotrichosis	Fungus causing nodular lesions in skin and lymph nodes that may soften and break to form ulcers
Dermatomycoses (a) Ringworm of feet ("athlete's foot" or *Tinea pedis*)	Fungus causing infection of skin of feet
(b) Ringworm of body (*Tinea corporis*)	Fungus causing infection of skin of body
(c) Ringworm of scalp (*Tinea capitis*)	Fungus causing infection of scalp and hair, with scaly, red lesions and sometimes deep ulcers
Actinomycosis (lumpy jaw)	Anaerobic, pathogenic fungus causing chronic, systemic infection with lesions and abscesses

flat sheet called the hymenium. The filamentous hyphae have cross walls (septate), and the cells possess nuclei. In some types the hyphae are closely compacted and the interhyphal (L. *inter,* between) spaces are filled with rather solid materials, as in the bracket fungi. Each basidiospore is attached to the tip of a stalk called a sterigma (ste -rig′ ma) (Gr. *sterigma,* support). Basidia typically develop from binucleate, terminal cells of hyphae. As these cells enlarge, the two nuclei fuse to form one nucleus, which divides twice to form four nuclei, one of which passes into each basidium where it becomes a basidiospore. The mature basidiospores may be dispersed by winds, insects, or other agencies. When proper temperature, moisture, and foods are encountered, a spore germinates, forming a hypha with mononucleate cells. After some growth these cells become binucleate, or dikaryotic (Gr. *di-,* two; *karyon,* nucleus). This binucleate condition persists throughout the growth of these hyphae, the nuclei dividing as new cells are formed. Certain terminal cells of some hyphae produce the

135

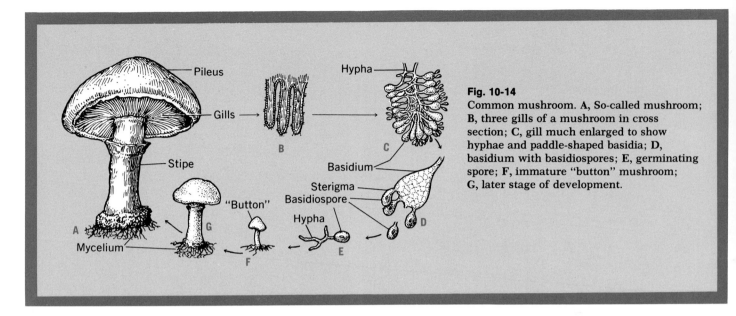

Fig. 10-14
Common mushroom. **A,** So-called mushroom; **B,** three gills of a mushroom in cross section; **C,** gill much enlarged to show hyphae and paddle-shaped basidia; **D,** basidium with basidiospores; **E,** germinating spore; **F,** immature "button" mushroom; **G,** later stage of development.

Mushroom

Hypha (+) (N)
(Dikaryotic)*
        ↘
      ↗Basidium (2N)→Basidiospores (N)→Hypha (N)
Hypha (−) (N) (By karyogamy)†
(Dikaryotic)*

Smuts

Hypha (N)→Chlamydospores→Basidium (2N)→Basidiospore (N)→Hypha (N)
     (Smut spores) (N) (By karyogamy)†

Fig. 10-15
Life cycles of certain true fungi of class Basidiomycetes, showing numbers of chromosomes of (N) or (2N). The asterisk (*) refers to two separate nuclei, each (N). The dagger (†) refers to the fusion of two nuclei.

basidia, in which two nuclei fuse by a process called karyogamy. As a group, Basidiomycetes differ from Ascomycetes by their larger, more conspicuous fruiting bodies and by fewer cases of sexual reproduction.

Mushrooms

Mushrooms are saprophytic fungi (Fig. 10-14) that derived their foods from decomposing organic materials in the soil, dead leaves, bard, and wood. The vegetative body is composed of masses of septate hyphae, some of which penetrate the substratum. Each sporophore (spor′ o for) (Gr. *spora*, spore; *phorein*, to bear) typically consists of a broad, caplike or umbrella-shaped pileus (pi′ le us) (L. *pileus*, cap) and a stalklike stipe (L. *stipes*, stalk). On the undersurface of the pileus are sheetlike gills. The gills are composed of plates of compacted hyphal tissues that bear the club-shaped basidia. The common, edible mushroom (*Psalliota campestris*) may produce nearly two billion spores, each of which may germinate to form a new hypha.

There are several hundred species of mushrooms, and only a few of them are poisonous. Most poisonous forms belong to the genus *Amanita*. Unless the collector is familiar with the mush-

The hyphae of tree-inhabiting shelf fungi secrete enzymes that digest the wood and bark and absorb organic substances from these tissues. In some species the bracket may be perennial, forming new, spore-forming hyphae in annual layers year after year. Shelf fungi cause the decomposition of wood, and parasitic species often attack living trees. Among the wood-rotting pore fungi is *Merulius lacrymans*, which causes the common "dry rot" of wood.

Smuts

Smuts are caused by basidiomycete fungi that parasitize flowering plants. Masses of septate hyphae penetrate the tissues of the host plant. They are called smuts because the fungi produce heavy-walled, dark-colored, ill-smelling smut spores called chlamydospores (klam′ i do spors) (Gr. *chlamys,* cloak; *spora,* spore) (Fig. 10-17). The resistant smut spores may lie dormant until the following spring, when each germinates into a tubelike basidium (three to four cells). Each basidium produces a basidiospore (sporidium). The basidiospores attack the host plant and form hyphae, which eventually form smut spores.

In some types of smuts chainlike groups of spores (conidia) may be formed on the parasitized plant. Thus in certain life cycles as many as three kinds of spores may be produced. Smuts primarily parasitize members of the grass family, including corn, oats, wheat, rye, rice, and barley. They can produce enormous crop losses.

Corn smut, *Ustilago zeae* (us ti -la′ go; ze′ a) (L. *ustilago,* thistlelike; *zea,* kind of grain), consists of tumorlike, black masses of smut on any part of the corn plant, but especially on the flower parts. When these tumors mature in the summer or fall, their hyphae contain masses of black chlamydospores that usually germinate the next spring or summer to infect new corn plants. The basidiospores, formed on basidia, produce germ tubes capable of infecting any part of the plant. The resulting hyphae mass together in certain areas and break out as smut tumors. The tumors are white at first but become black as the chlamydospores mature. Annual losses in the United States because of corn smut amount to millions of dollars.

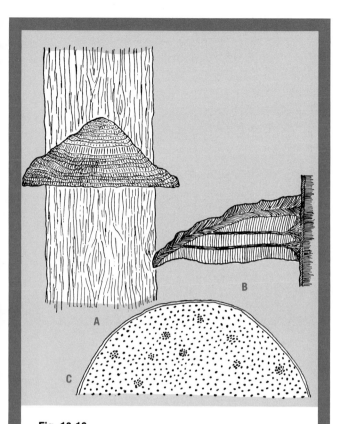

Fig. 10-16
Bracket (shelf) fungus. A, Attached to a tree; B, section of a sporophore, showing three annual layers of the porous hymenium, with the newest below; C, undersurface, showing the openings to pores, beneath which are borne basidiospores.

rooms, he should take no chances with the deadly species. It is better to forego the mushrooms than to suffer fatal illness.

Bracket (shelf) fungi

Bracket fungi belong to the group known as pore fungi because the underside of the brackets contain hundreds of tiny tubes that open as pores on the lower surface. This bracket is called the sporophore and is often tough and woody. The internal hyphae around the tubes form club-shaped basidia that bear basidiospores (Fig. 10-16) that escape through the pores.

Fig. 10-17
Corn smut (*Ustilago zeae*). Unbroken tumors are at right and broken and disseminating spores are at left; the inset shows chlamydospores of corn smut.
(From Hill, J. B., Overholts, L. O., and Popp, H. W.: Botany, New York, McGraw-Hill Book Company.)

Uredospores (Summer spores) ("Red rust" stage). Orange-red, rough, one-celled spores are borne at the tips of hyphae in blisterlike **uredinia** (u re din' i a) (Gr. *uredo*, blight) on the surface of WHEAT leaves and stems. In late summer the hyphae form pustules as described below.

Teliospores (Winter spores) ("Black rust" stage). Black, thick-walled, resistant, two-celled, stalked spores are borne in pustules called **telia** (Gr. *telios*, end), appearing as narrow black areas on WHEAT stems. Meiosis (reduction division) occurs here in which the chromosome number is reduced one-half.

Basidiospores During germination each teliospore develops a club-shaped **basidium** (Gr. *basis*, base) which bears 4 **basidiospores** on WHEAT. There are two kinds of basidiospores (plus and minus strains). Carried by the wind, a basidiospore may germinate to form pycnia as described below.

Pycniospores (Spring spores). Yellowish red spores are borne at the tips of hyphae in flask-shaped **pycnia** (pik' ni a) (Gr. *pyknos*, crowded) on the **upper surface** of the common BARBERRY leaf. Two kinds of pycniospores (plus and minus strains) are formed, which escape in a sweetish, insect-attracting liquid. When a plus pycniospore fuses with a minus pycniospore, a mycelium is formed which produces aeciospores.

Aeciospores (Spring spores). Orange-yellow spores are borne in chains in cuplike **aecia** (e' si a) (Gr. *aecium*, injury) on the **lower surface** of the common BARBERRY leaf. In the spring, aeciospores are carried to wheat where they germinate to form hyphae, which bear uredinia, thus completing the cycle.

Uredospores (Summer spores).

Aeciospores, uredospores, and teliospores represent the sporophyte generation (2N), whereas the basidiospores and pycniospores represent the gametophyte generation (N).

Fig. 10-18
Life cycle of the black stem rust of wheat (*Puccinia graminis*).

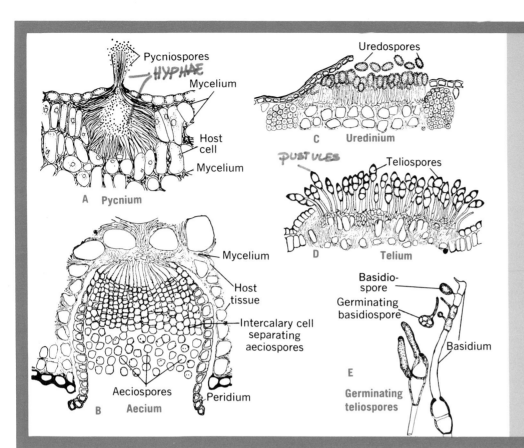

Pycniospores

HYPHAE

Mycelium

Host cell

Mycelium

A   Pycnium

Uredospores

C   Uredinium

PUSTULES

Teliospores

D   Telium

Mycelium

Host tissue

Intercalary cell separating aeciospores

Aeciospores

Peridium

B   Aecium

Basidio-spore

Germinating basidiospore

Basidium

E   Germinating teliospores

Fig. 10-19
Black stem rust of wheat (*Puccinia graminis*). **A,** Section of a leaf through the pycnium; **B,** section through the aecium in a barberry leaf; **C,** section through the uredinial sorus, showing unicellular, rough uredospores on slender stalks; **D,** section through the telial sorus, showing two-celled teliospores on long pedicels; **E,** germinating teliospores—left, both cells of the spore germinating; right, only an apical cell germinating; the germ tube has been transformed into a basidium, from each cell of which a basidiospore has been or is being formed. A germinating basidiospore is also shown.
(From Hill, J. B., Overholts, L. O., and Popp, H. W.: Botany, New York, McGraw-Hill Book Company.)

Table 10-4

Reduction of crop yield in the United States by plant diseases

Disease	Estimated reduction in yield
Wheat stem rust	85,452,000 bushels
Barley stem rust	12,046,000 bushels
Oat loose smut	32,728,000 bushels
Field corn smut	95,087,000 bushels
Field corn—all diseases	335,826,000 bushels
Potatoes—all diseases	58,673,000 bushels
Tobacco downy mildew	81,502,000 pounds
Cherry leaf spot	16,222,000 pounds
Apple scab	12,774,000 bushels
Strawberries—all diseases	746,000 crates
Grape black rot	31,830,000 pounds

Rusts

Rusts are parasitic fungi on flowering plants and ferns; they are so named because of reddish brown spores on the surface of leaves and stems. Hyphae penetrate the tissues of the host plant. A rust may parasitize two unrelated species of plants, alternating between the two hosts. Examples of two-host rusts include the "cedar-apple" rust, which alternates between cedars and apple trees; white pine blister rust, which alternates between white pine trees and wild gooseberries or currants; and the black stem rust of wheat, which alternates between wheat and the common European barberry (not the Japanese barberry). Other rusts may appear on corn, oats, rye, pears, plums, cherries, cone-bearing trees, garden vegetables, and cultivated flowers.

Black stem rust of wheat is a destructive disease caused by a basidiomycete fungus called *Puccinia graminis* (puk -sin′ i a; gram′ in is; after Puccini, and Italian anatomist; L. *graminis*, grass). The life cycle is described briefly in Figs. 10-18 and 10-19.

139

Some contributors to the knowledge of slime molds and true fungi

Clusius (Charles de la Cluse) (1529-1609)
He described and illustrated many edible and poisonous fungi (1601).

Gaspard Bauhin (1560-1624)
He classified about one hundred species of fungi and lichens.

Joseph P. de Tournefort (1656-1708)
He recognized six genera of fungi in his *Elements de Botanique* (1694).

Sebastian Vaillant (1669-1722)
He illustrated many fungi in his *Botanicon Parisiense* (1727)—studies that were not equaled for over 100 years.

Pier Antonio Micheli (1679-1737)
An Italian botanist who was an outstanding student of the microscopic anatomy of fungi and their classification.

Carolus Linnaeus (1707-1778)
A Swedish botanist who in his *Species Plantarum* recognized twenty-four classes of plants, one of which included fungi.

Christian H. Persoon (1755-1837)
He made great contributions to fungi classification and the roles of spores in this respect.

Elias M. Fries (1794-1878)
A Swedish mycologist who contributed greatly to fungi classification, especially the larger types. He recognized smuts and rusts as a natural group (1821). He is the "Father of Systematic Mycology." (*Historical Pictures Service, Chicago.*)

Fries

August C. J. Corda (1809-1849)
A Czechoslovakian mycologist who first recognized the sac fungi (Ascomycetes) as a group (1842) only a few years after the ascus had been described by **Joseph Henri Léveillé (1796-1870),** the French mycologist. Léveillé discovered the paddle-shaped basidium of Basidiomycetes (1837).

Louis Tulasne (1815-1885) and Charles Tulasne (1816-1884)
Two French mycologists who beautifully illustrated works on the structures of the ascus fungi (1861-1865). They contributed much to the knowledge of the morphology and classification of ascus fungi.

Louis Pasteur (1822-1895)
A French bacteriologist who demonstrated that yeasts cause fermentations (1871). He classified ferments into "organized ferments," such as yeasts and certain bacteria, and "unorganized ferments," such as digestive enzymes like pepsin.

Heinrich Anton De Bary (1831-1888)
A German mycologist who first recognized the algalike fungi (Phycomycetes) as a group (1873). He worked out many of the life cycles of fungi (1887), including the one for the black stem rust of wheat (*Puccinia graminis*). He thought that slime molds were more clsely related to animals than to plants.

Ernst A. Bessey (1877-1963)
An American who advanced the studies of fungi, especially those of the United States.

Pier Andrea Saccardo (1845-1920)
An Italian mycologist who published the *Sylloge Fungorum,* in which he reprinted in standardized form most of the original descriptions of species of fungi. The first complete set of eight volumes was published from 1882 to 1889, and seventeen supplements were issued up to 1931.

Eduard Buchner (1860-1917)
A German chemist and Nobel Prize winner who discovered that cell-free yeast extracts caused fermentation of sugars (1897). He called this water-soluble, heat-labile principle "zymase." (*Historical Pictures Service, Chicago.*)

Buchner

Bernard O. Dodge (1872-1960)
An American mycologist who pointed out the usefulness of the pink bread mold (*Neurospora*) as a tool for basic genetic studies (1930), and it has been widely used since.

G. W. Martin (1886-   )
Martin, of the State University of Iowa, an authority on slime molds, believes that fungi originated from a protozoanlike ancestor. He includes slime molds in his Division Fungi and calls them Myxomycetes (1949).

1 Define and distinguish between slime molds and true fungi.

2 Why are slime molds and true fungi classed in the subkingdom Thallophyta?

3 Why are slime molds considered to be plants? What animal characteristics do they possess?

4 Describe the life cycle of a typical slime mold.

5 List the distinguishing characteristics of the following classes of true fungi: Phycomycetes, Ascomycetes, and Basidiomycetes, including examples of each.

6 Give reasons why you consider fungi to be higher or lower types of plants than algae.

7 Discuss the ways in which fungi may be of importance to man.

8 Describe each of the asexual methods of reproduction that are found in the slime molds and in the true fungi, giving an example of each.

9 Describe the process of conjugation as found in certain fungi. In which fungi does conjugation occur? In what ways does this process resemble sexual reproduction?

10 Describe how sunlight and lack of moisture may be detrimental to fungi.

11 Explain how fungi obtain their nourishment and oxygen.

12 Diagram the life cycle of *Rhizopus nigricans*.

13 Explain how bread molds develop inside the loaf. How do they get into the loaf?

14 Diagram a typical life cycle of each class of true fungi.

15 Why are yeasts classed as Ascomycetes? In what ways do yeasts differ from other Ascomycetes?

16 Why is *Penicillium* classed as an Ascomycete? Of what economic importance is it?

17 Describe the life cycle of the black stem rust of wheat, including the host plants, various stages, types of spores, damages.

18 Which types of fungi possess septate hyphae? Which types possess filamentous hyphae?

Selected references

Alexopoulos, C. J.: Introductory mycology, ed. 2, New York, 1962, John Wiley & Sons, Inc.

Alexopoulos, C. J., and Bold, H. C.: Algae and fungi, New York, 1967, The Macmillan Company.

Batra, S. W. T., and Batra L. R.: The fungus gardens of insects, Sci. Amer. Nov., 1967.

Bonner, J.: The cellular slime molds, Princeton, New Jersey, 1959, Princeton University Press.

Bonner, J.: How slime molds communicate, Sci. Amer. 209:84-93, 1963.

Cochrane, V. W.: Physiology of fungi, New York, 1958, John Wiley & Sons, Inc.

Cook, A. H.: The chemistry and biology of yeasts, New York, 1958, Academic Press Inc.

Emerson, R.: Molds and men, Sci. Amer. Jan, 1952.

Gray, W. D.: The relation of fungi to human affairs, New York, 1959, Holt, Rinehart & Winston, Inc.

Lamb, I. M.: Lichens, Sci. Amer. Oct, 1959.

Maio, J. J.: Predatory fungi, Sci. Amer. July, 1958.

Pramer, D.: Nematode-trapping fungi, Science 144:382-388, 1964.

Rose, A. H.: Yeasts, Sci. Amer. 202:136-146, 1960.

Snell, W. H.: A glossary of mycology, Cambridge, Massachusetts, 1957, Harvard University Press.

Sparrow, F. K.: Aquatic phycomycetes, Ann Arbor, 1960, University of Michigan Press.

chapter eleven

Phylum Bryophyta— liverworts and mosses

Intermediate plants with chlorophyll in chloroplasts; without true leaves, stems, or roots; without true vascular (conducting) tissues; form multicellular embryos; subkingdom Embryophyta

KINGDOM Metaphyta

The members of the phylum Bryophyta (Gr. *bryon,* moss; *phyta,* plants) are terrestrial plants, although they require considerable moisture for growth and for the transmission of the sperm to fertilize the egg. Bryophytes possess chlorophyll-containing chloroplasts. In general the adult plant body is composed of blocks of rather thin-walled cells that form parenchymatous tissues in contrast to the simpler construction of the thallophytes. The mature plant is not filamentous, but the developmental protonema stage may be thread-like. None of the bryophytes grow to any great height.

The gamete-producing sex organs, the gametangia, are multicellular and possess a protective layer of sterile cells. With few exceptions the gametangia of thallophytes (mentioned in previous chapters) are unicellular. All bryophytes show alternation of generations between the multicellular, gamete-producing gametophyte generation and the multicellular, spore-producing sporophyte generation. The sporophyte is more or less dependent upon the gametophyte.

Bryophytes are placed in the subkingdom Embryophyta because they develop a multicellular embryo from the zygote. By cell division the sporophyte develops from the embryo. Spores are produced from spore mother cells within the sporangium. Asexual reproduction may occur by fragmentation of the plant or through special bodies known as gemmae (jem′ i) (L. *gemma,* bud). Bryophytes are without true vascular (conducting)

Table 11-1

Summary of some characteristics of certain groups of plants

Plants	Pigments*	Stored food	Cell wall
Bryophytes (liverworts and mosses)	Chlorophyll a and b Other pigments, depending upon the species	Starch	Cellulose
Tracheophytes (club "mosses," horsetails, ferns, gymnosperms, angiosperms)	Chlorophyll a and b Other pigments, depending on the species	Starch	Cellulose

*Many plants contain a variety of carotenoid pigments, carotenes, and xanthophylls not listed in this table. Carotenes vary from creamy white to yellow, orange red; xanthophylls produce yellow colors. Many plants also contain anthocyanin pigments that are not in plastids but free in the cytoplasm. Anthocyanins are known as glucosides and include deep reds (red beets), purples, and blues of certain plants.

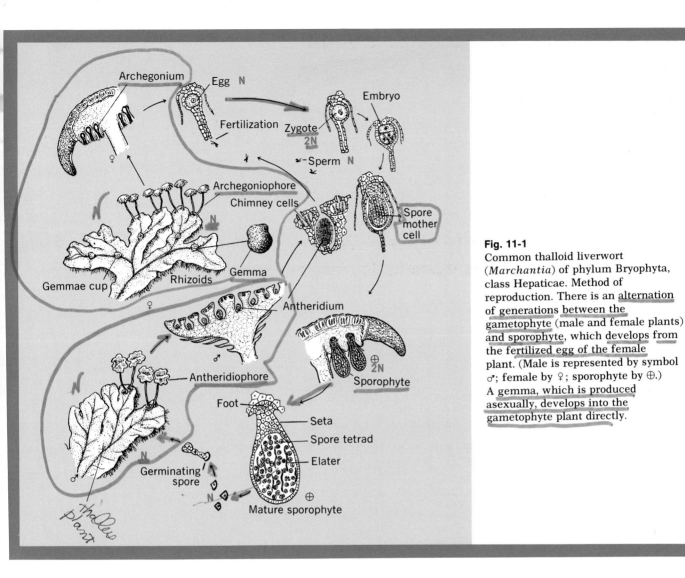

Fig. 11-1

Common thalloid liverwort (*Marchantia*) of phylum Bryophyta, class Hepaticae. Method of reproduction. There is an alternation of generations between the gametophyte (male and female plants) and sporophyte, which develops from the fertilized egg of the female plant. (Male is represented by symbol ♂; female by ♀; sporophyte by ⊕.) A gemma, which is produced asexually, develops into the gametophyte plant directly.

tissues, such as phloem and xylem, which are found in higher plants. Rootlike rhizoids anchor the plant and absorb materials from the substratum. Liverworts and true mosses possess similar methods of reproduction and life cycles and are much alike structurally and functionally despite differences that casual observation may indicate.

CLASS HEPATICAE (LIVERWORTS)

The members of the class Hepaticae (he -pat′ i se) (L. *hepaticus*, liver) usually grow flat on the substratum and have bodies that are flattened dorsoventrally with distinct upper and lower surfaces (Fig. 11-1). The thalloid liverworts have flat, lobed bodies that seem to resemble the lobes of a liver of higher animals, whereas the "leafy" liverworts have bodies with a stemlike axis upon which leaflike structures (not true leaves) grow. Many of the details can be learned from the following types.

Marchantia

Marchantia (mar -kan′ shi a) (after the French botanist Marchant) is the genus to which the flat, lobed thalloid liverworts belong. They are commonly found prostrate on moist rocks and soil along streams.

The surface of the branched thallus body has rhomboidal areas, each of which has a pore in its center. Internally, the thallus has air chambers and columns of cells (chimney cells) containing

143

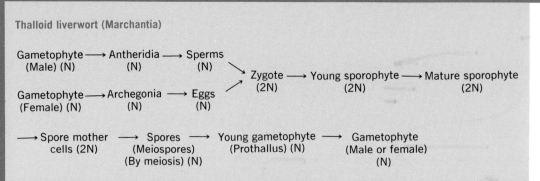

Thalloid liverwort (Marchantia)

Gametophyte ⟶ Antheridia ⟶ Sperms
(Male) (N)   (N)   (N)
                                    ⟶ Zygote ⟶ Young sporophyte ⟶ Mature sporophyte
Gametophyte ⟶ Archegonia ⟶ Eggs   (2N)   (2N)   (2N)
(Female) (N)   (N)   (N)

⟶ Spore mother ⟶ Spores ⟶ Young gametophyte ⟶ Gametophyte
   cells (2N)   (Meiospores)   (Prothallus) (N)   (Male or female)
                  (By meiosis) (N)   (N)

Fig. 11-2
Life cycles of liverworts, showing the alternation of generations. In another thalloid liverwort (*Riccia*) the gametophyte contains both antheridia and archegonia on the same plant.

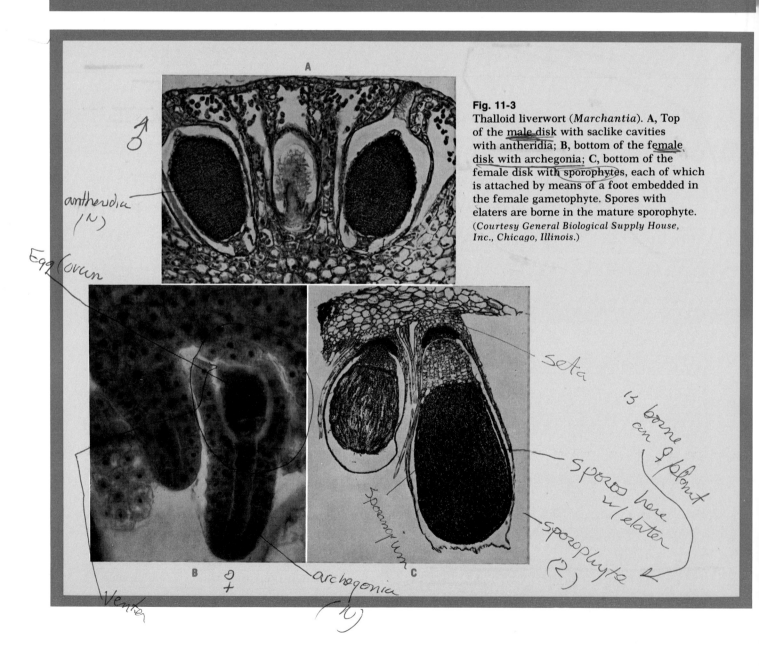

Fig. 11-3
Thalloid liverwort (*Marchantia*). **A,** Top of the male disk with saclike cavities with antheridia; **B,** bottom of the female disk with archegonia; **C,** bottom of the female disk with sporophytes, each of which is attached by means of a foot embedded in the female gametophyte. Spores with elaters are borne in the mature sporophyte. (*Courtesy General Biological Supply House, Inc., Chicago, Illinois.*)

chloroplasts. Rootlike rhizoids anchor the thallus and absorb materials from the substratum (see Figs. 11-1 and 11-3).

Marchantia is diecious, one thallus bearing antheridia and another thallus bearing archegonia. Stalklike antheridiophores, which have lobed disks at the tip, grow on the male thallus (male gametophyte). The male antheridia are borne in cavities that open on the upper surface of the disks. Each antheridium is an enlarged, oval structure that produces coiled, biflagellated sperms.

The stalklike archegoniophores, which have small, terminal disks bearing fingerlike rays, grow on the female thallus (female gametophyte). The female archegonia are borne on the undersurface of the disks. Each archegonium has a hollow, tubular neck and an enlarged venter at the base of which is a single egg.

The sperm swims through the water from the antheridium to the venter of the archegonium where the sperm and egg fuse. Through numerous cell divisions the fertilized egg forms a multicellular embryo from which the spore-producing sporophyte develops. The sporophyte consists of a foot embedded in the disk of the female gametophyte, a seta, and a sporangium. The sporangium produces numerous spores. Elongated, spiral-shaped, hygroscopic elaters (el' a ter) (Gr. *elater*, driver) are affected by moisture and expel the spores from the sporangium. The spores germinate to form new male or female gametophytes.

There is an alternation of generations between the gamete-producing gametophyte and the spore-forming sporophyte. The process is quite similar to that found in true mosses. Unlike the moss sporophyte, however, the *Marchantia* sporophyte does not possess stomata (stom' a ta) (Gr. *stoma*, opening). Also, it is usually smaller than that found in most mosses. *Marchantia* may also reproduce asexually by the formation of special bodies, known as gemmae, in little gemma cups or by the process of fragmentation.

Porella

Porella (por -rel' a) (L. *porus*, pore) is a common leafy liverwort (Fig. 11-4) that may form a green mass on moist soil, rocks, or rotten wood. Some species of leafy liverworts may grow on tree trunks

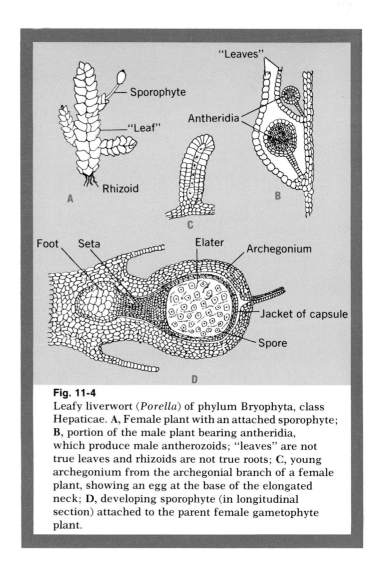

Fig. 11-4
Leafy liverwort (*Porella*) of phylum Bryophyta, class Hepaticae. **A**, Female plant with an attached sporophyte; **B**, portion of the male plant bearing antheridia, which produce male antherozoids; "leaves" are not true leaves and rhizoids are not true roots; **C**, young archegonium from the archegonial branch of a female plant, showing an egg at the base of the elongated neck; **D**, developing sporophyte (in longitudinal section) attached to the parent female gametophyte plant.

in damp forests. Some species may resemble true mosses, but the liverworts are prostrate on their substratum.

Porella has three rows of leaflike structures that grow on a stemlike axis. This axis may be branched and rhizoids are attached to its lower surface. The leaflike structures are much simpler than the gametophyte of *Marchantia*, consisting of one layer of cells without a midvein. The sporophyte of *Porella* is similar to that of *Marchantia*, consisting of foot, stalk, and sporangium. The sporangium bears spores and elaters, as in the thalloid liverworts. The life cycle is shown in Fig. 11-4.

Liverworts, mosses, and lichens, when growing on bare rocks, may mechanically and chemically

145

Fig. 11-5
Hairy cap moss (*Polytrichum*) growing in its natural habitat and showing antheridial, archegonial, and sporophyte plants.
(*Courtesy Carolina Biological Supply Co., North Carolina.*)

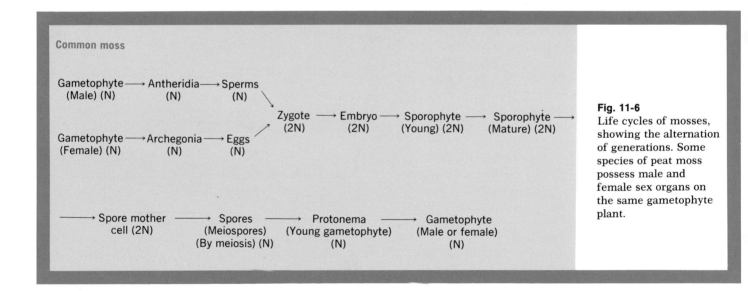

Common moss

Gametophyte ⟶ Antheridia ⟶ Sperms
(Male) (N)   (N)   (N)
                                    ⟶ Zygote ⟶ Embryo ⟶ Sporophyte ⟶ Sporophyte ⟶
                                        (2N)   (2N)   (Young) (2N)   (Mature) (2N)
Gametophyte ⟶ Archegonia ⟶ Eggs
(Female) (N)   (N)   (N)

⟶ Spore mother ⟶ Spores ⟶ Protonema ⟶ Gametophyte
   cell (2N)   (Meiospores)   (Young gametophyte)   (Male or female)
                   (By meiosis) (N)   (N)   (N)

Fig. 11-6
Life cycles of mosses, showing the alternation of generations. Some species of peat moss possess male and female sex organs on the same gametophyte plant.

convert the rocky surface into soil. Through their death they contribute valuable organic materials to the soil. Soon there may be sufficient soil formed to permit the growth of ferns and other simpler plants, and still later shrubs and trees may appear, thus creating what is called a plant succession.

CLASS MUSCI (TRUE MOSSES)

The members of the class Musci (mu' si) (L. *muscus*, moss) are small, green plants that usually grow upright and may be so abundant in some moist places that they form a mat of vegetation.

Although true mosses may possess structures that superficially resemble leaves and stems, they are not true leaves or stems because they lack the vascular tissues (phloem and xylem) of such true structures. Each individual plant consists of a stemlike axis with attached leaflike structures. The epidermal cells contain stomata (pores). Typical representatives of this class include the following.

Polytrichum

Polytrichum (po -lit' rĭ kum) (Gr. *poly-,* many; *thrix,* hair) is a common, true moss known as the hairy cap moss (Figs. 11-5 to 11-8).

Several antheridia are borne in a cluster at the tips of certain stemlike axes, and several archegonia are borne at the tips of other stemlike axes. *Polytrichum* has its sexes in separate plants and is therefore diecious. In species of mosses that are monecious (bisexual) the antheridia and archegonia are borne on the same plant.

In *Polytrichum* the antheridia are separated by multicellular, sterile hairs called paraphyses. Both antheridia and paraphyses are surrounded by a rosette of leaflike appendages that may be colored and resemble a "flower." Each antheridium consists of a short stalk and an enlargement that produces unicellular sperms. The sperms are coiled, bear two long, terminal flagella, and escape from the apex of the antheridium (Fig. 11-8).

The archegonia are separated by paraphyses, and each archegonium has a stalk supporting an enlarged venter that surrounds the egg. When mature, a canal leads through the long neck to the venter. During fertilization the motile sperm swims through water from the antheridium to the female plant. It travels down the canal to the venter where a sperm and egg fuse by the fertilization process known as oogamy. The fertilized egg is retained in the venter where it forms an embryo by numerous cell divisions. The embryo is parasitic on the female gametophyte plant, for it is given water, food, and protection. The embryo then grows to form a new plant, the sporophyte. A sporophyte consists of a foot, which is attached to the female plant, and a stalklike seta, which bears a sporangium at its tip. The sporangium is covered with a hairy cap, or calyptra (Gr. *kalyptra,* veil). When

Fig. 11-7
Hairy cap moss (*Polytrichum*), showing archegonial, sporophyte, and antheridial plants.
(*Courtesy Carolina Biological Supply Co., Elon College, North Carolina.*)

the calyptra is removed, a lidlike operculum (L. *operculum,* lid) is observed to cover the sporangium. Beneath the operculum is a ring of hygroscopic teeth, the peristome (Gr. *peri,* around; *stoma,* opening). The teeth are affected by moisture, and their movements expel the spores from the sporangium. When immature, the sporangium contains spore mother cells, each of which undergoes reduction division to produce four spores. The four spores of each tetrad are of two kinds. One kind contains a small Y chromosome (sex chromosome) and produces a male plant; the other kind contains a large X chromosome and produces a female plant. This method of sex determination

147

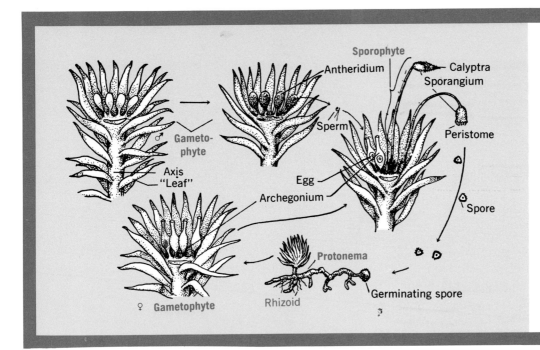

Fig. 11-8
Common moss of phylum Bryophyta, class Musci, showing the method of reproduction. There is an alternation of generations between the gametophyte (male and female plants) and the sporophyte, which develops from the fertilized egg of the female plant.

is similar to that in man, in whom there are also X and Y chromosomes.

Each spore germinates to form a threadlike, branched protonema. The cells of the protonema bear chloroplasts. Rhizoids anchor the young plant and absorb materials from the soil. Buds appear on the protonema, and these produce a new male or female moss plant by cell division. There is an alternation of generations between the gametophyte plant and the sporophyte plant. Under certain conditions some mosses may reproduce asexually by a fragmentation of the plant or by the formation of gemmae.

Sphagnum

Sphagnum (sfag' num) (Gr. *sphagnos,* moss) is the genus to which the peat or bog mosses belong. They commonly inhabit bogs, ponds, and other wet places. Their life cycles are similar to that of *Polytrichum.* The upright, branched axis may be 1 foot long and bears leaflike appendages that contain two types of cells—one stores water and the other contains chloroplasts. The water storage cells are large and empty and have openings to the outside. *Sphagnum* can absorb water up to twenty times its weight.

In certain species antheridia and archegonia

are present on the same plant (but on different branches). In others they are present on different plants. The fertilized egg develops into a sporophyte, which has a base embedded in the gametophyte, a short stalklike seta, and an enlarged sperangium. Because of the short seta, the gametophyte develops a structure known as the pseudopodium at the base of the foot in order to elevate the sporangium above the gametophyte. Alternation of generations occurs.

ECONOMIC IMPORTANCE OF SOME MOSSES

Sphagnum and other mosses grow on the edges of ponds and lakes where they may gradually fill in the entire body of water. During this process of filling in there may be masses of floating mosses.

Sphagnum is utilized in gardening to keep the soil porous and to increase its water-retaining capacity. Because of its water-holding quality, it is used by florists in packing cut flowers and in the development of seedlings. In the past *Sphagnum* and other mosses have accumulated in bogs and swamps where they have slowly decomposed and become compacted and carbonized. This process produces peat, a valuable fuel, which can be used in place of coal. Peat moss is also used for the bedding of farm animals. Some mosses are used commercially as packing material in shipping. Mosses may prevent soil erosion because of their abundant growth. Also, their natural water-absorbing qualities enable them to assist in flood control by preventing rapid runoff of water.

Review questions and topics

1 List the characteristics of Embryophyta, Bryophyta, Hepaticae, and Musci.
2 What are the general habitats for bryophytes?
3 Discuss the economic importance of liverworts and true mosses.
4 Diagram the life cycle of a liverwort, showing the stages in correct sequence.
5 Diagram the life cycle of a true moss, showing the stages in correct sequence.
6 Explain what is meant by alternation of generations.
7 Define gametophyte and sporophyte. Why may the sporophyte be considered to be a parasite upon the gametophyte? Contrast gametophytes and sporophytes in every way possible.
8 Describe the method of sex determination in a moss.
9 Give specific reasons why most bryophytes are rather short plants.
10 Contrast antheridia and archegonia, giving examples of each.
11 Describe the structure and functions of elaters. What is the function of a so-called foot?
12 Contrast what is meant by diecious and monecious (bisexual), giving examples of each. What advantages and disadvantages can you list for each phenomenon?

Selected references

Bodenberg, E. T.: Mosses, a new approach to the identification of common species, Minneapolis, 1954, Burgess Publishing Co.
Conard, H. S.: How to know the mosses, Dubuque, Iowa, 1956, William C. Brown Company, Publishers.
Grout, A. J.: Mosses with a hand lens, Newfane, Vermont, 1947, A. J. Grout.
Richards, P. W.: A book of mosses, Harmondsworth, England, 1950, Penguin Books, Ltd.
Smith, G. M.: Cryptogamic botany, vol. 2, New York, 1955, McGraw-Hill Book Company.
Steere, W. C.: Cenozoic and Mesozoic bryophytes of North America, Amer. Midl. Nat. 36:298-324, 1946.
Steere, W. C.: Bryology, in a century of progress in the natural sciences, 1859-1953, San Francisco, 1955, California Academy of Sciences.
Thieret, J. W.: Bryophytes as economic plants, Econ. Bot. 10:75-91, 1955.

Some contributors to the knowledge of liverworts and mosses

Marchant
A French botanist who studied the liverwort, *Marchantia* (1678), and the genus was named after him.

Johann Hedwig (1730-1799)
A German botanist who established the class Musci (1782) and is often called the "Father of Bryology."

Antoine Laurent de Jussieu (1748-1836)
He first established the class Hepaticae (1789).

Samuel F. Gray (1766-1828)
A British botanist who first recognized the bryophytes as a natural group (1821), although the name Bryophyta was not applied until nearly 50 years later.

Stephen L. Endlicher (1804-1849)
An Austrian botanist who classified all plants into two primary groups, the Thallophyta and Cormophyta (1836). The term Embryophyta is now used instead of his term Cormophyta (Gr. *kormos*, trunk; *phyta*, plants) and includes the Bryophyta (liverworts and true mosses).

William Starling Sullivant (1803-1873)
An American, who made significant contributions to the knowledge of liverworts and mosses, especially in their classification (1848).

John Merle Coulter, Sereno Watson, and Lucien Marcus Underwood
They are all Americans who contributed to the knowledge of classification of liverworts and mosses (1890).

chapter twelve

Phylum Tracheophyta— club "mosses," horsetails, and ferns

Prothallium – young plant [handwritten]

Higher plants with chlorophyll in chloroplasts; with true leaves, stems, and roots; with vascular tissues (phloem and xylem); without seeds; form multicellular embryos; subkingdom Embryophyta

GAMETOPHYTE IS DEPENDENT UPON SPOROPHYTE [handwritten]

The phylum Tracheophyta (Gr. *tracheia*, vessel; *phyta*, plants) contains plants with true vascular (conducting) tissues of varying degrees of complexity. The development of vascular systems is one of the great steps in the evolution of plants. In accordance with the system of classification being used, this phylum contains the club "mosses," horsetails, ferns, gymnospermous plants, and angiospermous (flowering) plants. The latter two will be considered in Chapters 13 and 14, respectively. Tracheophytes have true leaves, stems, and roots; supporting tissues for more or less upright growth; stomata or small openings for the exchange of gases; and a protective layer of waxlike cutin in certain parts of the plant (Table 12-1).

Tracheophytes undergo alternation of generations. The gamete-producing gametophyte is small and inconspicuous, whereas the spore-producing sporophyte is relatively large and independent when mature. All tracheophytes form multicellular sex organs of two kinds; antheridia and archegonia. They all form multicellular embryos that develop from a fertilized egg. The multicellular sporangia are usually borne on leaves known as sporophylls. Sporangia may (1) occur singly on the upper surface of the sporophylls to form a clublike strobilus

150

(Gr. *strobilos*, cone), as in club "mosses"; (2) occur in groups upon shield-shaped sporangiophores to form conelike strobili, as in horsetails; or (3) be borne in clusters known as sori (Gr. *soros*, heap), as in ferns. In certain species the sporophylls bear sporangia in which the spores are all alike (homosporous); in others there are two kinds of sporophylls, two kinds of sporangia, and two kinds of spores (heterosporous). Spores germinate to form different types of gametophytes: (1) colorless, rather bulky prothalli (Gr. *pro*, before; *thallos*, young shoot) with male antheridia and female archegonia, as in club "mosses"; (2) thin, green, irregular gametophytes with antheridia and archegonia, as in horsetails; (3) small, green, thin, heart-shaped prothalli with antheridia and archegonia, as in ferns. In general, club "mosses," horsetails, and ferns have somewhat similar methods of reproduction and life cycles.

Tracheophytes are primarily terrestrial, although some are aquatic.

SUBPHYLUM LYCOPSIDA (CLUB "MOSSES")

The subphylum Lycopsida (Gr. *lykos*, wolf; *opsis*, appearance) includes a variety of so-called club "mosses" that are also commonly called "ground pines," because they are principally perennial, creeping "evergreens." The sporophyte plants of club "mosses" have rather simple vascular tissues. The true green leaves are usually spirally arranged, and the true stems and true roots are usually branched dichotomously (Gr. *dicha*, in two; *tomein*, to divide).

Sporangia occur singly on the upper surface of specialized leaves known as sporophylls (spore leaves). Usually the sporophylls and their sporangia are arranged at the tips of the stems to form strobili (cones). The spore-bearing organs are often foot- or club-shaped. Additional details may be learned by studying such representative types as the following.

Lycopodium

Plants belonging to the genus *Lycopodium* (Figs. 12-1 and 12-2) are called club "mosses" because of their small, mosslike leaves and their club-shaped strobili. They are also called "ground pines" because of their prostrate, creeping habits

Table 12-1

Summary of Bryophyta and Tracheophyta

| | Phylum Bryophyta | | Phylum Tracheophyta | | |
| | Class Hepaticae | Class Musci | Subphylum Lycopsida | Subphylum Sphenopsida | Subphylum Pteropsida |
Characteristics	Liverworts	True mosses	Club "mosses"	Horsetails	Ferns
Gametophyte	Relatively large, conspicuous; independent; may live long time	Relatively large, conspicuous; independent; may live long time	Relatively small, inconspicuous, lumpy, and colorless; usually independent and subterranean	Relatively small, inconspicuous, cushionlike, and with numerous erect green lobes; independent on ground	Relatively small; independent; short lived
Sporophyte	Relatively small; dependent (parasitic); short lived	Relatively small; dependent (parasitic); short lived	Larger than gametophyte; parasitic for time on gametophyte (prothallus); independent when mature; may live some time	Larger than gametophyte; parasitic for time on gametophyte (prothallus); independent when mature; may live long time	Relatively large, conspicuous; independent; may live long time
True leaves, stems, and roots	Absent	Absent	Small, green leaves, usually spirally arranged; branched, underground stem (rhizome); true roots	Small leaves (sometimes scalelike) in whorls; underground rhizome; hollow, upright, jointed stems, usually roughened, and with silica; true roots	Large conspicuous leaves (fronds); true roots
True vascular system	Absent	Absent	Simple	Simple	Rather complex
Alternation of generations	Present	Present	Present	Present	Present

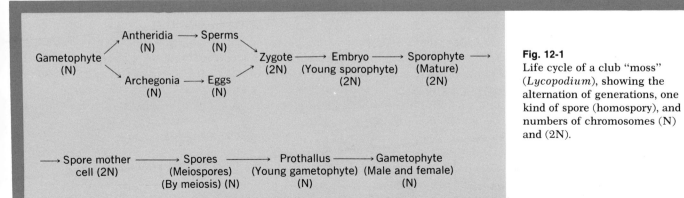

Fig. 12-1
Life cycle of a club "moss" (*Lycopodium*), showing the alternation of generations, one kind of spore (homospory), and numbers of chromosomes (N) and (2N).

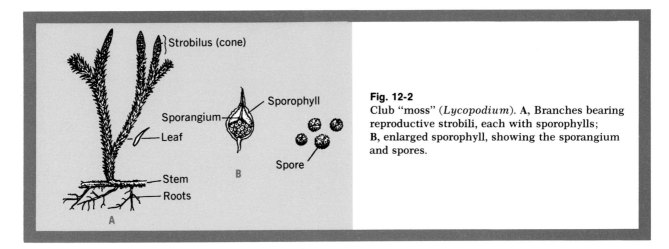

Fig. 12-2
Club "moss" (*Lycopodium*). A, Branches bearing reproductive strobili, each with sporophylls; B, enlarged sporophyll, showing the sporangium and spores.

and resemblance to miniature, evergreen pine trees.

The main stem (rhizome) is prostrate on the ground and is branched. It possesses true roots and sends up numerous upright stems, usually a few inches tall. *Lycopodium* has a simple vascular system consisting of a solid, central cylinder with alternate strands of phloem (sieve tubes and companion cells) and thick-walled xylem cells (tracheids). Stomata occur on the leaves and stems for the exchange of gases.

Sporangia-bearing leaves are called sporophylls. In some types the single sporangia are borne on leaves that may be located on any part of the stem. In other types the sporophylls are concentrated at the tips of the branches to form strobili. The spores are all alike (homosporous) and are wind disseminated. A spore germinates to form a colorless, rather bulky young gametophyte called a prothallus, which is usually in the soil. In some cases a part of the inconspicuous prothallus is above the ground and is green.

Antheridia and archegonia are present on the upper surface of the gametophyte, which is anchored by rootlike rhizoids. Antheridia produce biflagellated sperms that swim to the egg, which is produced in the archegonium.

After fertilization the egg divides to form two cells, one of which forms a suspensor that pushes the embryo into the food tissues of the gametophyte. The other cell develops into a multicellular embryo that grows to become a young, leafy sporophyte. This sporophyte remains as a temporary parasite on the gametophyte. The embryo forms a special organ, the foot, that acts as an absorbing organ. A root forms, and the gametophyte tissues eventually decay, thus leaving the sporophyte independent.

The various species of *Lycopodium* are common in forests and on mountains. They are widely used in the preparation of decorations, wreaths, and other articles where such evergreen materials are needed. Spores of *Lycopodium* are used as "lycopodium powder" in pharmacy. Club "mosses" have assisted in the formation of coal.

Selaginella

The genus *Selaginella* (L. *selago*, shrubby plant) contains the perennial, smaller club "mosses" (Figs. 12-3 and 12-4), which are widely distributed, especially in the tropics. One species, known as the "resurrection plant," can withstand the dry conditions in the southwestern United States by rolling into a ball. Other specimens are grown as ornamental plants in greenhouses. Most species of *Selaginella* are creepers, although a few are erect.

The branched stems of the mature sporophyte bear tiny, lanceolate, green, stomata-bearing leaves, usually in four rows, two of larger leaves and two of smaller ones. Roots anchor the plant and absorb materials. The vascular system is rather

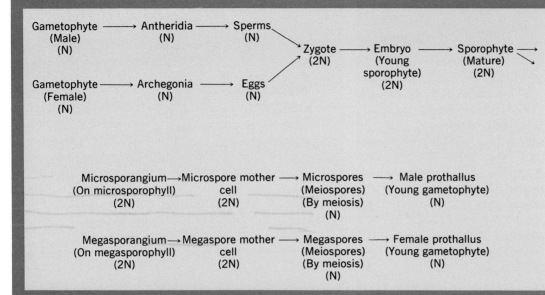

Fig. 12-3
Life cycle of a smaller club "moss" (*Selaginella*), showing the alternation of generations, two kinds of spores (heterospory), and numbers of chromosomes (N) and (2N).

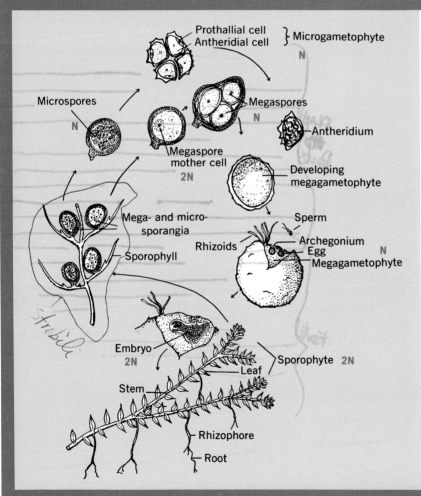

Fig. 12-4
Smaller club "moss" (*Selaginella*), showing its life cycle. The mature sporophyte is shown below: Note the reduction in the number of chromosomes from 2N to N. The biflagellated sperms are produced in the antheridium. A fertilized egg forms a zygote from which develops an embryonic sporophyte.

simple and varies with the species. The so-called vascular cylinder usually consists of a group of separate vascular bundles. Between these and the surrounding cortical tissue there is an air space when the stem is view in cross section. The different species of *Selaginella* may be distinguished from those of *Lycopodium* by the presence of a membranous ligule (L. *ligula*, little tongue), which has an unknown function at the base of each leaf. Strobili at the tips of the branches are composed of spore-bearing sporophylls. In a cone the upper surface of each microsporophyll (male) has in its axil (upper angle) a small microsporangium that produces many small microspores. The megasporophylls (female) produce four large, thick-walled megaspores in each megasporangium located in the axil. Frequently, the microsporophylls are located toward the tip of the cone, whereas the megasporophylls are located lower but on the same cone. Since *Selaginella* produces two kinds of spores (microspores and megaspores), it is heterosporous.

A megaspore develops to form a megagametophyte within the megaspore while the megaspore is still within the megasporangium. As the megagametophyte develops, it forms several egg-containing archegonia, stored foods, rhizoids, and chlorophyll.

A microspore develops within the microsporangium to form a small, parasitic microgametophyte, which is surrounded by the microspore wall, has no chlorophyll, and consists of one prothallial cell and one male antheridium. The antheridium produces a biflagellated sperm that swims to the female archegonium where the egg is fertilized to form a zygote. The zygote forms two cells by cell division, the upper one becoming a suspensor for pushing the developing embryo into the stored food of the megagametophyte. The other cell of the dividing zygote develops into the embryo, which forms a mass of cells, the foot, for absorbing food from the megagametophyte. Eventually, the embryo sporophyte forms stems, roots, and leaves. Later, the developing sporophyte becomes independent and photosynthesizes its own food. The formation of a suspensor by the zygote is somewhat similar to a phenomenon in higher, seed-forming plants. The megagametophyte, together

154

with its embryo, resembles a seed of a higher plant in some ways, but it lacks some of the tissues found in a true seed.

SUBPHYLUM SPHENOPSIDA (HORSETAILS)

The horsetails belong to the subphylum Sphenopsida (sfen -op' si da) (Gr. *sphen*, wedge; *opsis*, appearance), because in some species some of the leaves have a wedgelike appearance. The sporophyte plants of the horsetails have a simple vascular system; small leaves (sometimes scalelike or wedgelike) that are arranged in whorls at the joints (nodes) of the hollow stems; stems that are roughened by ribs and impregnated with silica, the ribs being opposite the vascular bundles and alternating with small, air-filled canals; and true roots. The roughened stems account for another name and use applied to them in pioneering America, that is, "scouring rushes."

In most species a horizontal, branched, underground stem known as a rhizome bears two types of aerial stems: (1) the sterile, green, branched, vegetative stems for food manufacture and (2) the colorless, unbranched, fertile, reproductive stems with a single, terminal strobilus.

Sporangia are borne on a shield-shaped sporangiophore whose stalk attaches it to a central axis where, with many other sporangiophores, they form the strobilus. The spores are alike, and each has ribbonlike elaters that are affected by moisture changes to assist in the movement of spores. A germinating spore forms a small, green, ribbonlike young gametophyte with rhizoids and usually with antheridia and archegonia. The spiral, multiflagellated sperms produced by the antheridia swim to the egg-producing archegonium where the egg is fertilized. The egg develops into a multicellular embryo, which in turn develops into a sporophyte plant. Hence there is an alternation of generations. More detailed information can be obtained by studying the common horsetail (*Equisetum*).

Equisetum

In addition to the characteristics given, the following apply to the members of the genus *Equisetum* (L. *equus*, horse; *seta*, tail). The horsetails are widely distributed and vary from a few inches

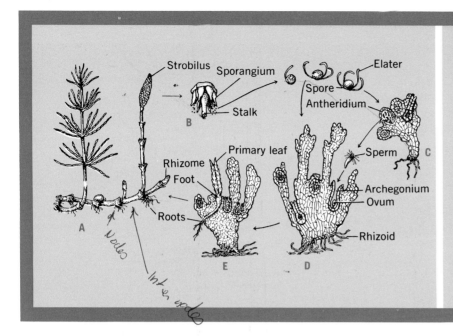

Fig. 12-5

Common horsetail or scouring rush (*Equisetum*) of class Equisetineae. A, Sterile, vegetative branch (left) and fertile, reproductive branch (right); B, one unit of the strobilus is much enlarged to show several sporangia; C, prothallus with antheridia for sperm production and an archegonium (upper right) for egg (ovum) production; D, gametophyte with archegonia for egg production and antheridia for sperm production; E, embryo of the horsetail plant developing from a fertilized egg in the gametophyte. The embryo consists of two primary leaves, a stalklike foot, and a primary root. The prothallus eventually disappears and the embryo develops into a mature plant with both kinds of aerial branches, as shown in A.

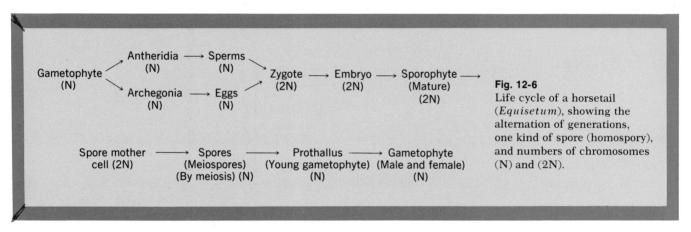

Fig. 12-6

Life cycle of a horsetail (*Equisetum*), showing the alternation of generations, one kind of spore (homospory), and numbers of chromosomes (N) and (2N).

to 3 feet in height, depending upon the species (Figs. 12-5 to 12-7). Photosynthesis occurs primarily in the stems. The vascular system consists of vascular bundles, each composed of phloem and xylem tissues. The rhizome has distinct nodes and internodes.

The spores, produced in tetrads, are all alike. Each spore has four elaters, or delicate bands with spoon-shaped tips. Moisture causes the elaters to uncoil. In so doing they assist in the dispersal of the spores. A spore germinates into a green, flat, ribbonlike gametophyte that consists of a thick mass of tissue from which grow thin, irregular vertical lobes that may be branched. Rhizoids attach the gametophyte to the substratum. Near the bases of the vertical lobes are the archegonia, although only their necks may be visible. The antheridia are also present on the gametophyte and produce coiled, multiflagellated sperms. The egg is fertilized by a sperm to form a zygote from which the young embryo develops. In contrast to some species of club "mosses," however, the embryo is without a suspensor.

Horsetails have helped to form coal in early geologic times. In certain places they may help to prevent soil erosion. Their growth on the shores of ponds and lakes may alter the shore lines and may assist in filling in the pond completely.

155

Fig. 12-7
Horsetail (*Equisetum*). A, Two fertile, reproductive branches; B, two sterile, vegetative branches. A sporangia-bearing strobilus is present on the tip of each reproductive branch.

SUBPHYLUM PTEROPSIDA (FERNS)

The subphylum Pteropsida (ter -op′ si da) (Gr. *pteris*, fern; *opsis*, appearance) includes the ferns, cone-bearing plants, and flower-bearing plants. The latter two groups will be considered in Chapters 13 and 14, respectively. The sporophyte plants of ferns have a rather well-developed vascular system; true leaves (fronds), which are usually conspicuous and large; underground stems (rhizomes); and true roots (Figs. 12-8 to 12-15). There is an alternation of generations between the conspicuous sporophyte and the small gametophyte. Multicellular sporangia are borne in clusters called sori on the lower surface of the leaves or on the margins of the leaves in certain species. In some ferns the sori are covered with a membranous indusium (L. *indusium*, tunic). When sporangia are young, they contain spore mother cells that divide to form spores. In ferns the sporophyte at maturity is independent, free-living, and larger than the gametophyte, which is also independent but small.

The leaves of the true ferns have a characteristic way of unrolling as they grow, giving rise to the popular name of fiddleheads. Usually the leaves are compound and pinnate (L. *pinna*, feather). Fern leaves continue to grow at their tips for a long time, whereas in higher seed plants the growth period of leaves is short. Ferns are worldwide in their distribution but usually grow in moist, shady places. They require sufficient water to permit the transfer of sperm to the egg for fertilization purposes. They do not produce seeds.

A typical sporangium consists of a stalk, a thin-walled capsule, and an annulus (L. *annulus*, ring). The annulus is a band of cells, in which portions of the cell walls are thick and other parts thin.

156

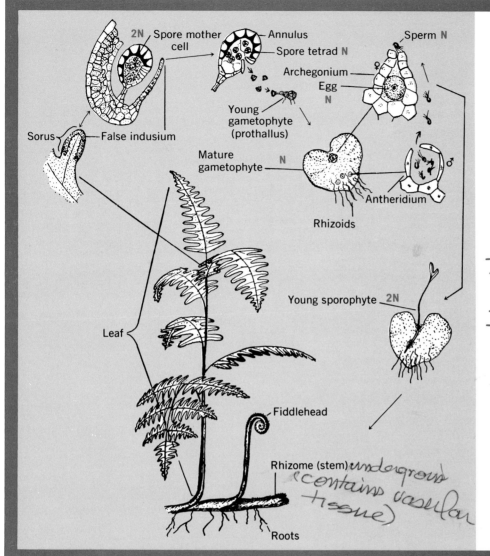

2N Spore mother cell

Annulus

Spore tetrad **N**

Sperm **N**

Archegonium

Egg **N**

Young gametophyte (prothallus)

Sorus — False indusium

Mature gametophyte **N**

Antheridium

Rhizoids

Leaf

Young sporophyte **2N**

Fiddlehead

Rhizome (stem) *underground (contains vascular tissue)*

Roots

Fig. 12-8
Common fern of class Filicineae. The adult sporophyte with its rhizome (underground stem) is shown below with the fiddlehead (young unrolling leaf or frond). Each sorus contains numerous sporangia (spore cases), which produce spores. The moisture-sensitive annulus is composed of a band of cells with thick inner walls and a thin outer wall. This structure responds to moisture changes by bending open and throwing spores outward. The young sporophyte develops from the fertilized egg into an adult sporophyte, and the mature gametophyte eventually disappears.

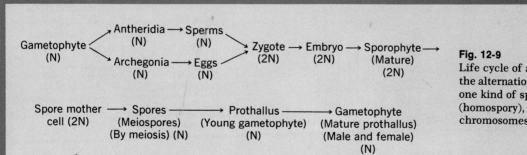

Gametophyte (N) → Antheridia (N) → Sperms (N) → Zygote (2N) → Embryo (2N) → Sporophyte (Mature) (2N) →

Gametophyte (N) → Archegonia (N) → Eggs (N) → Zygote (2N)

Spore mother cell (2N) → Spores (Meiospores) (By meiosis) (N) → Prothallus (Young gametophyte) (N) → Gametophyte (Mature prothallus) (Male and female) (N)

Fig. 12-9
Life cycle of a fern, showing the alternation of generations, one kind of spore (homospory), and numbers of chromosomes (N) and (2N).

Fig. 12-10
Fruiting fronds of various ferns, showing the ways in which spores are produced. 1, Sensitive fern; 2, cinnamon fern; 3, climbing fern; 4, common grape fern; 5, common polypody fern (*Polypodium*); 6, bracken fern (*Pteridium*); 7, maidenhair fern; 8, common chain fern; 9, Christmas fern; 10, spinulose shield fern; 11, common bladder fern; 12, obtuse woodsia fern; 13, boulder fern; 14, walking fern; 15, ebony spleenwort fern.
(*Courtesy General Biological Supply House, Inc., Chicago, Illinois.*)

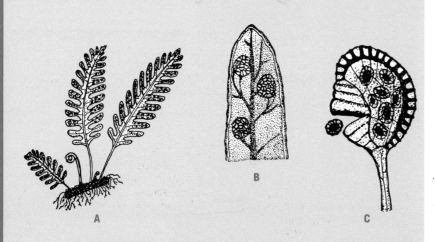

Fig. 12-11
Polypody fern (*Polypodium*). A, Horizontal rhizome, slender adventitious roots, and fronds with numerous sori on the lower surface; young frond (fiddlehead) is shown unrolling; B, lower surface of a frond, showing three sori (enlarged), each consisting of numerous sporangia, which produce spores; C, sporangium (much enlarged), showing the stalk, internal spores, capsule with thin walls, and moisture-sensitive annulus (a band of cells, each of which has three thick walls and an outer thin wall); the annulus responds to changes in moisture and bends back, thus throwing the spores to the outside.

Fig. 12-12
Fern prothalli growing in a
special liquid medium.
(*Courtesy General Biological
Supply House, Inc., Chicago,
Illinois.*)

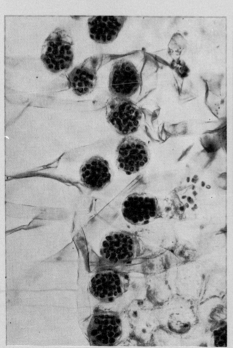 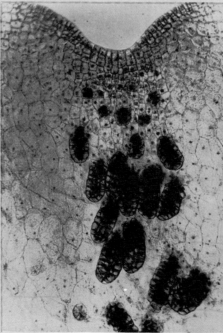

A B

Fig. 12-13
Fern prothalli (in part). **A**, Male
antheridia; **B**, female archegonia.
(*Courtesy General Biological
Supply House, Inc., Chicago,
Illinois.*)

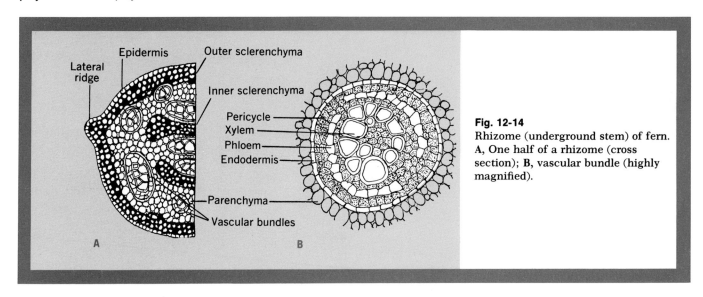

Fig. 12-14
Rhizome (underground stem) of fern. **A,** One half of a rhizome (cross section); **B,** vascular bundle (highly magnified).

Labels in figure: Lateral ridge, Epidermis, Outer sclerenchyma, Inner sclerenchyma, Pericycle, Xylem, Phloem, Endodermis, Parenchyma, Vascular bundles

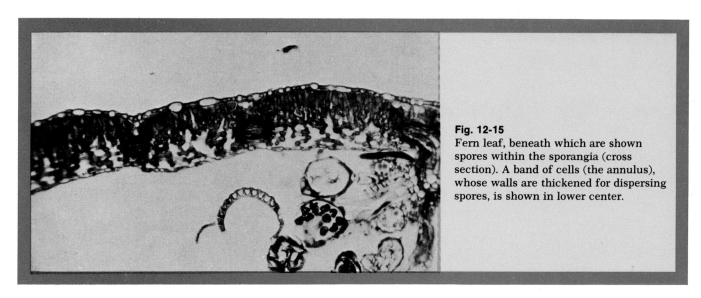

Fig. 12-15
Fern leaf, beneath which are shown spores within the sporangia (cross section). A band of cells (the annulus), whose walls are thickened for dispersing spores, is shown in lower center.

The annulus is affected by moisture (hygroscopic), which causes it to bend and spring forward, thus dispersing the spores.

A germinating spore forms a small, thin, green, heart-shaped, haploid gametophyte (also called a prothallus or prothallium), which will eventually develop antheridia and archegonia on its lower surface. The antheridia produce motile, flagellated sperms that contain the haploid number (N) of chromosomes. The sperms swim through water to fertilize the egg (N) in the archegonium, thus producing a diploid zygote (2N). By mitosis the zygote will develop eventually into a multicellular embryo (2N) from which a sporophyte will develop (2N).

Ferns are placed in the class Filicineae (fil i -sin' e e). The beautiful fronds are used extensively for decorating purposes. Because of the starch content, certain underground rhizomes are eaten despite their tannin content. A drug derived from

160

the so-called male fern *Dryopterus* is used medicinally as a worm-eliminating medicine. Fossil ferns also assist in the formation of coal.

■ Pteridium

The bracken ferns (*Pteridium*) are common in the woods or clear areas of the temperate regions. The rhizome of *Pteridium* is long, slender, rarely branched, and grows a few inches below the surface of the soil. It grows at its anterior end and may die at the posterior end. The rhizome may even separate into parts, each existing as an individual plant. In structure the rhizome (Fig. 12-14) consists of an outer layer of thick-walled epidermal cells covering a thick layer of rigid mechanical tissue called the outer sclerenchyma which gives rigidity. Much of the interior part of the stem consists of thin-walled parenchyma cells, which serve for storage. Two well-defined strands of mechanical tissue, called the inner sclerenchyma, are located near the center of the stem, and numerous vascular bundles (phloem and xylem) of various sizes are located in the center of the stem. As the rhizome elongates, it gives rise to many slender adventitious roots a short distance from the cap. Each root has a root cap and small root hairs. Roots that arise from stems or leaves are called adventitious (L. *adventicius*, foreign). The rhizome resumes its growth each spring and summer.

Each leaf begins its development as a coiled structure attached to the rhizome underground. Eventually it pushes into the air and uncoils, an action that is characteristic of nearly all ferns. A mature leaf consists of a slender petiole and a blade that is greatly divided. Two stalk rows of primary leaflets are borne upon the central axis of the blade. Several pairs of the lower primary leaflets may even be divided. Internally, the structure of the leaf resembles that of higher seed plants. Each leaflet has a lower and an upper epidermis, a palisade layer of elongated cells, spongy tissue in which are numerous spaces, and veins for the conduction of liquids. Stomata are numerous in the lower epidermis.

All leaves of the bracken fern are green, but only some of them are sporophylls and bear sporangia (see Fig. 12-15). Only the leaves without sporangia photosynthesize food. Along the edge of the underside of the sporophyll leaflets a narrow band of tissue is developed. Numerous sporangia grow from the surface of this band. The band may be partly covered by the curved margin of the leaflet. A sporangium consists of a stalk, a jacket-like capsule, and a row of thick-walled cells, the annulus. Within the jacket are formed numerous spore mother cells, each of which forms four spores, as in the mosses. When spores are formed, the number of chromosomes is reduced from the diploid (2N) number to the haploid (N) number by reduction division. The cells of the annulus are thickened on all sides except the outer one and are sensitive to changes in moisture. Hence the annulus straightens, breaks the capsule, snaps forward, and disperses the spores.

The spores are shed in late summer and are hard, brown, and irregular in shape. When a spore germinates, it forms a thin, green elongated plate (prothallus) with rhizoids (see Fig. 12-12). If moisture and light conditions are favorable, the prothallus develops typically into a flat, heart-shaped prothallus with a shallow notch at its anterior (apical) end. Prothalli may mature in a few months and usually are not more than $\frac{1}{4}$ inch in diameter. The prothallus is the gametophyte of the fern, just as in the mosses, and produces antheridia and archegonia. The small, dome-shaped antheridia are usually most numerous on the undersurface of the posterior (older) part of the prothallus where the rhizoids are most abundant. Numerous spiral-shaped, multiflagellated sperms (antherozoids) are produced within the antheridia. Archegonia are usually restricted to the undersurface of the prothallus just behind the apical notch and are constructed much like those of liverworts and mosses. The basal venter is embedded in the surrounding tissues and connects with a short neck to the outside. When mature, a passageway is developed to the egg. Although antheridia and archegonia are formed on the same gametophyte, most antheridia mature and discharge their sperms before the archegonia on that plant mature. Thus sperms from one gametophyte usually unite with the egg of another gametophyte.

Only one sperm units with an egg to form a zygote, which remains within the venter of the archegonium and develops into an embryo (young

161

sporophyte). The embryo becomes four lobed. One lobe develops into a foot that temporarily absorbs food from the gametophyte; another lobe produces a primary root that grows into the soil; a third lobe produces a primary leaf that develops a simple, green blade; and the fourth lobe develops the stem. At first the young sporophyte is parasitic upon the gametophyte, but later, as secondary leaves and secondary or adventitious roots develop from the stem, the gametophyte dies. Thus the mature sporophyte develops from only one lobe of the embryo, namely that which developed the stem. The structures derived from the other three lobes of the embryo function only in the early stages of young sporophyte development. The life cycle of the bracken fern, like that of other ferns, liverworts, and mosses, includes an alternation of generations between the gametophyte and sporophyte.

■

Polypodium

The common polypody fern has a thick rhizome and numerous adventitious roots. From the rhizome develops a series of rather large leaves on the underside of which are numerous sori consisting of several or many sporangia. Within the sporangia the spore mother cells undergo reduction division when they divide to form spores, as described for the bracken fern on p. 161. In *Polypodium* (pol i -po′ di um) (Gr. *poly-*, many; *podion*, small foot) and in most ferns spores are eventually produced on all the leaves, in contrast to the bracken fern. In general the structures and life cycles of many ferns are quite similar to those of the polypody and bracken ferns, although there are differences in structure, distribution of the sporangia, sori, and indusia. The embryologic developmental stages, the antheridia and archegonia on the undersurface of the gametophyte, and the life cycle are similar to those described for *Pteridium*.

Some contributors to the knowledge of club "mosses," horsetails, and ferns

Wilhelm Hofmeister (1824-1877)
A German botanist, who described the life cycles of horsetails, certain club "mosses," and many ferns (1851) and made a great contribution in this field. He discovered alternation of sporophytic and gametophytic generations.

Karl E. von Goebel (1855-1932)
A German botanist who divided ferns into two groups (1880) on the basis of the structure and developmental history of the spore cases.

Charles E. Bessey (1845-1915)
An American botanist who first recognized the club "mosses" and horsetails as distinct groups (1907).

Edward C. Jeffrey (1866-1952)
An American botanist, who pointed out that on anatomic grounds the vascular plants could be divided into two fundamental groups. One group, including ferns and modern seed plants, tends to have relatively large leaves with a branching vein system and leaves with a stalklike petiole and an expanded blade. The second group, including the club "mosses," tends to have relatively small, narrow leaves with one, unbranched midvein, and the leaves are not divided into petiole and blade.

Walter Zimmerman (1892-   )
A German botanist who is the author of the telome theory (Gr. *telos*, end), whereby he attempted to explain the origin of all leaves by modification of branch stems or branch-stem systems. He calls the ultimate branches of a dichotomously branching stem "telomes." Although certain parts of the theory are still controversial, the basic concepts are generally accepted.

1 List the characteristics of the phylum Tracheophyta.

2 List the characteristics of the subphyla Lycopsida, Sphenopsida, and Pteropsida.

3 Discuss the economic importance of club "mosses," horsetails, and ferns.

4 Explain how the sporophytes and gametophytes differ in club "mosses," horsetails, and ferns.

5 Diagram a fern life cycle, showing the stages in correct sequence.

6 Explain the significance of reduction division (meiosis).

7 Why are the leaves of ferns and their allies considered to be true leaves.

8 Why are ferns and their allies classed as embryophytes?

9 Contrast the sporophytes and gametophytes of ferns with those of mosses.

10 In what ways are the gametangia (sex organs) of tracheophytes similar to those of the bryophytes but different from those of the thallophytes?

11 Describe the structure and functions of stomata.

12 Describe the type of gametophyte formed from a germinating spore in club "mosses," horsetails, and ferns.

13 Describe how the multicellular sporangia are borne in club "mosses," horsetails, and ferns.

14 Why is water necessary for fertilization in ferns?

15 Describe the structure and function of annulus, suspensor, and foot.

16 Compare the alternation of generations in ferns with similar stages in the horsetails and club "mosses." Be specific in your comparisons.

17 Contrast the terms prothallus and protonema, with examples of each.

Selected references

Andrews, H. N.: Evolutionary trends in early vascular plants, Cold Spring Harbor Symposium, Quant. Biol. 24:217-234, 1960.

Andrews, H. N.: Studies in paleobotany, New York, 1961, John Wiley & Sons, Inc.

Delevoryas, T.: Morphology and evolution of fossil plants, New York, 1962, Holt, Rinehart & Winston, Inc.

Foster, A. S., and Gifford, E. M.: Comparative morphology of vascular plants, San Francisco, 1959, W. H. Freeman and Co., Publishers.

Sporne, K. R.: The morphology of the pteridophytes, London, 1962, Hutchinson & Co. (Publishers), Ltd.

White, R. A.: Tracheary elements of the ferns, Part I, Amer. J. Bot. 50:447-454, 1963.

chapter thirteen

Phylum Tracheophyta— gymnosperms

Higher plants with chlorophyll in chloroplasts;
with true leaves, stems, and roots; with vascular tissues;
with exposed ("naked") seeds; form multicellular embryos;
subkingdom Embryophyta

The gymnosperms are placed in the subphylum Pteropsida and the class Gymnospermae (Gr. *gymnos,* exposed; *sperma,* seed). Seeds are formed on the exposed surface of the megasporophylls and are not protected by an ovary wall, as in angiosperms (flowering plants). Gymnosperms are mainly evergreen (retain leaves for more than one growing season) and are usually rather large and woody perennial plants, although some types are short and shrubby. They possess true leaves, stems, and roots. A detailed treatment of the anatomy of these structures is given in Chapter 14. The leaves in certain cone-bearing evergreens are needlelike or scalelike. Conebearing evergreens (conifers) and their relatives are included in the class Gymnospermae.

The sporophyte generation is large, complex, and independent, whereas the gametophyte generation is much smaller (microscopic) and dependent (parasitic) upon the sporophyte. Two kinds of cones composed of sporophylls are usually present. In the conifers the male cones (composed of microsporophylls) and the female cones (composed of megasporophylls) are present on the same plant or on different plants, depending upon the species.

Immature, undeveloped seeds (ovules) and later the true seeds are borne exposed on the megasporophylls. Two kinds of spores (heterospores) are formed, namely microspores, which develop into microgametophytes, and megaspores, which develop into megagametophytes. The two kinds of spores may be the same size, or the microspores may be larger than the megaspores. Yet they produce different types of gametophytes.

Pollination is aided by wind dispersal, the pollen grains (microspores) landing near or in the small opening or micropyle of the ovule. The pollen grains are drawn into the ovule by the drying and shrinking of a sticky material secreted by the female cone. A pollen tube is developed through which the sperms travel to the egg.

In gymnosperms single fertilization occurs, in which one sperm is involved in fertilizing the egg (in contrast to double fertilization, which occurs in flowering plants). After fertilization the egg develops into the embryo, the rest of the female megagametophyte tissues form the endosperm (stored food), and the integument eventually forms the seed coat of the mature seed.

In pine trees the lapse of time between pollination and subsequent fertilization (actual union of sex gametes) is a distinct feature. For example, if pollination occurs in June, fertilization does not ordinarily occur until July of the next year. This time lapse varies with the species and locality, but usually a year elapses between pollination and fertilization. After fertilization the seed develops quickly, reaching maturity by the end of the year in which fertilization occurs.

Gymnosperms are considered to be higher plants than club "mosses," horsetails, and ferns because of (1) the transfer of the small microgametophyte (produced by a microspore) directly through a pollen tube to the area near the megagametophyte; (2) the retention of the megaspore within the megasporangium (nucellus) where the megaspore germinates to produce the megagametophyte, or embryo sac (in which the archegonia are formed); (3) the enclosure of the megasporangium and embryo sac by an integument (covering); (4) the development of the young sporophyte upon and at the expense of the mother sporophyte; and (5) the release of dormant seeds that have a reserve supply of food and are protected by a seed coat.

The gymnosperms are of great economic importance. The cone-bearing trees (conifers) produce much valuable timber, as verified by the use of yellow pines, redwoods, pitch pines, firs, cedars, hemlocks, and white pines. Billions of board feet of lumber are used in the United States annually. Pine lumber is durable because of its composition, is easily worked, and is quite resistant to the attacks of insects, probably because of its resin content. Certain conifers are used extensively in the manufacture of wood pulp. Red cedars are used in making pencils, cigar boxes, chests, trunks, and similar articles. Conifers yield large quantities of resins, oils, and amber products used in arts, industries, and medicine. Examples are turpentine, balsam, spruce gum (for chewing gum), oil of juniper, and oil of savin. The seeds of certain pines may be used as food, for example, the edible or nut pine of western United States and the sugar pine of California. The barks of conifers, such as hemlock and spruce, furnish important materials for use in tanning animal skins for leather. Millions of youngsters and adults enjoy the decorated conifers at Christmas. The wood of certain conifers is particularly resonant, so that it is used in making certain musical instruments. The remains of conifers are often found as fossils and as fossil-resin amber in which other fossils have been embedded.

ORDER CONIFERALES (CONIFERS)

The conifers are placed in the order Coniferales (L. *conus*, cone; *ferre*, to bear) because most of them bear cones composed of sporophylls (sporangium-bearing leaves). There are over 500 species of conifers, including the various species of pine, spruce, fir, juniper, cedar, hemlock, larch, yew, cypress, and redwood. Some of the oldest and largest plants are conifers, for example, the giant sequoia trees of California (Fig. 13-1) may be more than 30 feet in diameter, 300 feet tall, and 4,000 years old.

Pine trees

Pine trees (*Pinus*) have large, branched stems, and the needlelike leaves are borne in clusters. The number of leaves per cluster (two to five) and the

Fig. 13-1
General Sherman Tree, a giant sequoia, in winter. It is more than 270 feet high, more than 36 feet in diameter. Its size can be compared with nearby trees, which are about 35 feet high. Its age is estimated to be about 3,500 years. It is located in Sequoia National Park, California. (*Courtesy National Park Service, The United States Department of the Interior.*)

length of the leaves vary with the species (Figs. 13-2 to 13-5).

In the pines the male and female cones are borne on the same tree. In other conifers the male and female cones are borne on separate plants. The simple staminate (male) cones are smaller than the

165

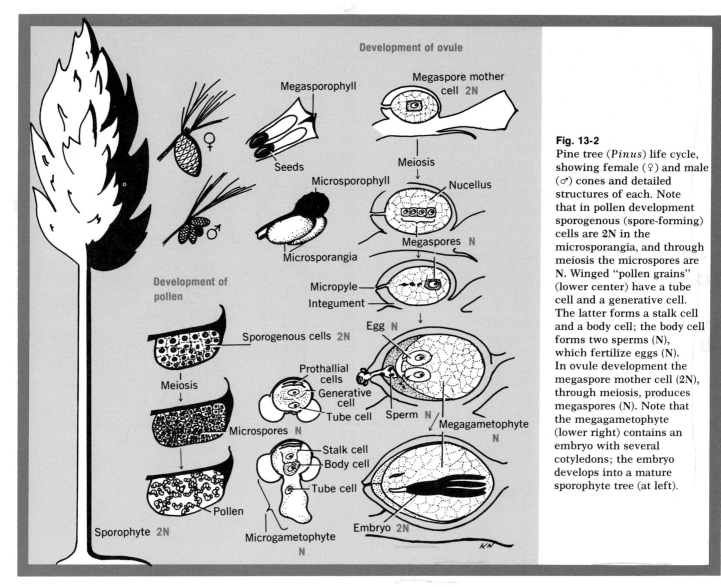

Development of ovule

Megasporophyll

Megaspore mother cell 2N

Meiosis

Seeds

Microsporophyll

Nucellus

Microsporangia

Megaspores N

Micropyle

Integument

Egg N

Development of pollen

Sporogenous cells 2N

Prothallial cells

Generative cell

Tube cell

Sperm N

Megagametophyte N

Meiosis

Microspores N

Stalk cell

Body cell

Tube cell

Pollen

Microgametophyte N

Sporophyte 2N

Embryo 2N

Fig. 13-2
Pine tree (*Pinus*) life cycle, showing female (♀) and male (♂) cones and detailed structures of each. Note that in pollen development sporogenous (spore-forming) cells are 2N in the microsporangia, and through meiosis the microspores are N. Winged "pollen grains" (lower center) have a tube cell and a generative cell. The latter forms a stalk cell and a body cell; the body cell forms two sperms (N), which fertilize eggs (N). In ovule development the megaspore mother cell (2N), through meiosis, produces megaspores (N). Note that the megagametophyte (lower right) contains an embryo with several cotyledons; the embryo develops into a mature sporophyte tree (at left).

female ones and are borne in a group. Each male cone is composed of microsporophylls that are spirally arranged and attached to a central axis. Each microsporophyll bears two microsporangia on the undersurface. Numerous microspore mother cells, each of which produces four microspores (pollen grains), are produced in the microsporangia. Each pollen grain develops into a microgametophyte by producing two prothallial cells

and an antheridial cell. The latter divides to form a generative cell and a tube cell. During this process a pair of "wings" forms on the four-celled pollen grain to assist in its dissemination by the wind—in some cases for hundreds of miles.

The ovulate (female) cones are larger than the male ones and are usually borne singly. Each female cone (Fig. 13-2) is composed of scalelike megasporophylls attached to a central axis. Each

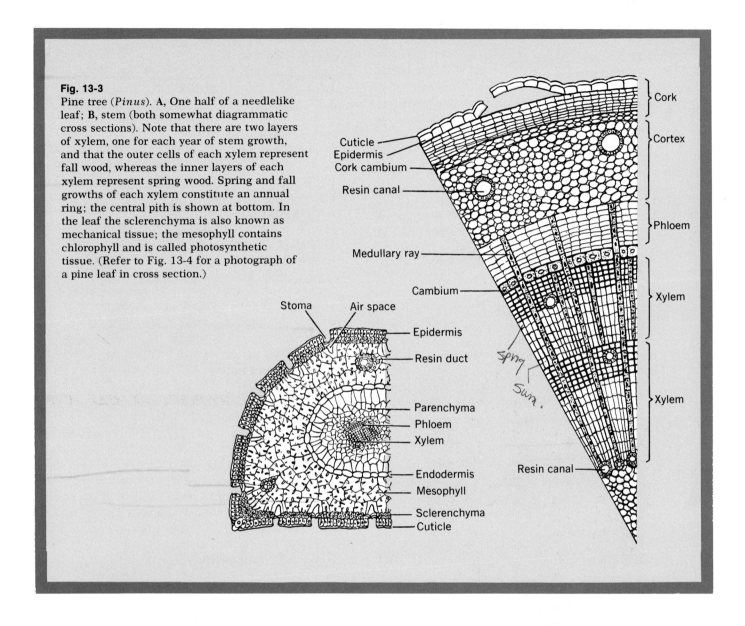

Fig. 13-3
Pine tree (*Pinus*). **A**, One half of a needlelike leaf; **B**, stem (both somewhat diagrammatic cross sections). Note that there are two layers of xylem, one for each year of stem growth, and that the outer cells of each xylem represent fall wood, whereas the inner layers of each xylem represent spring wood. Spring and fall growths of each xylem constitute an annual ring; the central pith is shown at bottom. In the leaf the sclerenchyma is also known as mechanical tissue; the mesophyll contains chlorophyll and is called photosynthetic tissue. (Refer to Fig. 13-4 for a photograph of a pine leaf in cross section.)

megasporophyll bears two ovules on its upper surface. An ovule consists of an external, protective integument that has a micropyle (small opening) for the entrance of a pollen grain and a central megasporangium (nucellus). When young, the megasporangium has one megaspore mother cell that produces four megaspores, three of which abort. The one megaspore develops into the female megagametophyte, which contains two or three

archegonia. Each archegonium contains an egg within a venter.

During pollination the wind carries the pollen grains to the female cones where pollen enters through the micropyle and contacts the megasporangium by means of a sticky liquid. The pollen grains form pollen tubes through the megasporangium toward the archegonia. The generative cell and the tube cell pass through the pollen tube,

167

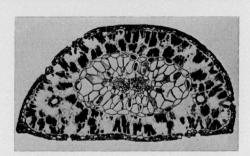

Fig. 13-4
Pine leaf (needle), showing the outer epidermis with stomata (openings) and, beneath, air spaces located in the mesophyll tissue (photographed in cross section). Refer to Fig. 13-3, A, for diagrammatic presentation of tissues.)
(*Courtesy Carolina Biological Supply Co.*)

Fig. 13-5
Cluster of cones on the stem of pitch pine (*Pinus rigida*).
(*Courtesy Jane Evelyn Richardson.*)

and the generative cell produces two nonmotile sperms (male nuclei).

About 1 year after pollination the fertilization process occurs and consists of the fusion of one sperm with the egg within the archegonium. The resulting zygote eventually produces an embryo and suspensor cells by cell division. The suspensor cells force the embryo into contact with the food endosperm (transformed female megagametophyte). Later, the embryo develops an epicotyl (ep i -kot′ il) (Gr. *epi*, upon; *kotyle*, seed leaf) and a hypocotyl (hi po -kot′ il) (Gr. *hypo*, under), which bear a number of primary, embryonic seed leaves known as cotyledons. The embryo is surrounded by the endosperm (food), which is covered by the seed coat (hardened integument). The seed thus formed was originally the ovule and contains a wing for wind dispersal. When a seed germinates, the embryo produces a young seedling that eventually develops into a pine tree (sporophyte generation). Since gymnosperms produce two different kinds of spores (microspores and megaspores), they are heterosporous.

ORDER CYCADALES (CYCADS)

The cycads ("sago palms") (Fig. 13-6) belong to the class Gymnospermae and the order Cycadales (sik a -da′ lez). They are palmlike trees or low shrubs with unbranched stems, terminating in a tuft of thick, fernlike leaves that are often spiny edged. The trunk of *Cycas* contains starch, hence the common name of "sago palm" (Malay *sagu*, starch of sago palm), although the true sago palm is native to the East Indies.

Zamia

In *Zamia* (L. *zamia*, fir cone) (Fig. 13-6), which occurs in Florida, the short, tuberous stem is not more than 4 feet tall and bears a crown of leathery, pinnate leaves. Sometimes much of the stem is underground. In general the cycads are inhabitants of the tropics or semitropics.

Zamia is diecious, since male and female cones are borne on separate plants. The carpellate cones (female) are composed of shield-shaped megasporophylls, each of which bears two ovules, with a micropyle in the enclosing integument. In the center of the ovule is the megasporangium. This

Fig. 13-6
Cycads. **A,** *Cycas* with a crown of young leaves unfolding; **B,** male plant of *Zamia* at the time of "flowering"; **C,** mature fruiting plant of *Zamia*. (*From Weatherwax, P.: Plant biology, ed. 3, Philadelphia, 1959, W. B. Saunders Company.*)

Some contributors to the knowledge of gymnospermous plants

Robert Brown (1773-1858)
A Scottish botanist who first observed (1827) that pine trees and their kin have ovules (undeveloped seeds) and mature seeds that are "naked" (not protected by an ovary). He classified gymnosperms and separated them from true, flower-bearing angiosperms (1825).

Charles J. Chamberlain (1863-1943)
An American botanist who was an outstanding student of cycads, those gymnosperms that are somewhat fernlike with large, pinnately compound leaves and seeds that are usually aggregated into a simple, terminal cone.

Charles E. Bessey (1845-1915)
An American botanist who first recognized cone-bearing plants as a distinct group (1907) and treated the cycads as a separate group. His ideas have been accepted by many in succeeding years.

Rudolph Florin (1894-   )
A Swedish botanist who is an outstanding authority on cone-bearing plants, particularly fossil forms.

contains one megaspore mother cell that produces four megaspores, three of which disintegrate. The nucleus of the remaining megaspore divides to form two nuclei, and further division results in numerous nuclei. Walls separate the nuclei, so this multicellular tissue becomes the female megagametophyte, which produces two to six archegonia. Each archegonium consists of one large egg and a neck.

The staminate cones (male) are smaller than the female ones and consist of numerous microsporophylls, each of which bears thirty to forty microsporangia on the lower surface. Each microsporangium contains many microspore mother cells, which produce many microspores. These microspores are carried by the wind to the female cone where they pass through the micropyle of the ovule to the megasporangium. About 6 months after pollination a motile sperm (antherozoid) fertilizes an egg in the archegonium, forming a zygote that produces a multicellular embryo, two cotyledons (embryonic leaves), and a coiled suspensor to push the embryo in contact with the endosperm (stored food) of the megagametophyte.

169

Review questions and topics

1 List the characteristics of the class Gymnospermae.
2 In what ways are gymnosperms considered to be higher plants than ferns, horsetails, and club "mosses"?
3 Discuss the differences and similarities between conifers and cycads.
4 Discuss the economic importance of gymnosperms.
5 Make a detailed diagram of the life cycle of the pine tree, including the various stages in correct sequence. Label each stage correctly.
6 Explain the phenomenon of alternation of generations in gymnosperms.
7 What is the sporophyte and what is the gametophyte in the pine? How do they differ concerning size and independence?
8 Explain what is meant by the term "evergreen."
9 Contrast pollination and fertilization, giving examples of each.
10 Explain the significance of heterospores.
11 Describe the structure and function of the micropyle.
12 What is the significance of the endosperm?
13 Discuss the structure and function of the pollen tube.
14 Contrast ovules and true seeds, giving an example of each.
15 What is the significance of the motile sperms of cycads and the nonmotile sperms of pine trees?

Selected references

Arnold, C. A.: An introduction to paleobotany, New York, 1947, McGraw-Hill Book Company.
Florin, R.: The systematics of the gymnosperms. In Kessel, E. L. (editor): A century of progress in the natural sciences, San Francisco, 1955, California Academy of Sciences.

chapter fourteen

Phylum Tracheophyta— angiosperms

[handwritten: OVARY OR CARPELS]

Higher plants with chlorophyll in chloroplasts; with true leaves, stems, and roots; with vascular tissues; with enclosed seeds; form multicellular embryos; subkingdom Embryophyta

Flowering plants are placed in the subphylum Pteropsida and the class Angiospermae (Gr. *angio-*, enclosed; *sperma,* seed). The seeds are enclosed within an ovary (carpels) that has probably been evolved by a folding together of the edges of the carpel (megasporophylls). At maturity the ovary constitutes the fruit. In some types of fruits there is more than one ovary. Angiosperms constitute the dominant, the most economically important, and the largest class in the plant kingdom, comprising about 200,000 species. These species are widely distributed and are primarily terrestrial, although a few, such as water lilies, are aquatic.

Angiosperms (see Figs. 14-2 to 14-4) possess true leaves, stems, and roots. Each of these structures contain conducting tissues (phloem and xylem). A unique characteristic of the angiosperms is the presence of flowers. Most angiosperms contain chlorophyll, but a few, like the Indian pipes (*Monotropa*), are colorless. The angiosperm plant is the conspicuous, independent, adult sporophyte part of the alternation of generations, while the gametophyte is very small, without chlorophyll, and dependent upon the sporophyte. The sporophyte produces two kinds of spores (heterospores), although, as in gymnosperms, the (male) microspores may be larger than the (female) megaspores. The cells of the sporophyte contain the diploid (2N) number of chromosomes, whereas the cells of the gametophyte contain the haploid (N) number.

In the evolution of plants from the lowest types to the angiosperms, there has been an increase in the size and independence of the sporophyte. Also, there has been a reduction in the size and independence of the gametophyte.

Angiospermae are divided into the subclasses Dicotyledoneae (Gr. *di-,* two; *kotyle,* seed leaf) and Monocotyledoneae (Gr. *mono-,* one), which may be differentiated as shown in Tables 14-1 to 14-3.

ROOTS

Roots are an adaptation to life on land and are typically cylindric in form. They arise embryologically from the hypocotyl region of the embryo of the seed. The radicle (L. *radix,* root), or basal part of the hypocotyl, emerges from the germinating seed to form the first or primary root of the new plant.

[handwritten: 1. Hypocotyl 3. Primary Root 4. 2nd Root. 2. Radicle]

The primary root may form many branches called secondary roots. In general the primary root grows downward, whereas the secondary roots grow somewhat horizontally.

The root systems of plants are commonly called (1) taproot systems, in which the single, main primary root is distinctly larger than the secondary roots; for example, radishes, carrots, beets, and dandelions, or (2) diffuse (fibrous) root systems, which are comprised of numerous slender roots, many of which are of equal size, and are usually without a single main root that is larger than the others, for example, grasses, corn, and potatoes. Sometimes the larger fibrous roots become enlarged with stored foods, as in sweet potatoes and dahlias. A dandelion plant that has forced its way through an asphalt driveway is shown in Fig. 14-1.

The principal functions of roots include (1) anchorage of the plant, thus binding the soil particles to reduce soil erosion by water or wind; (2) absorption of water and dissolved minerals and the conduction of substances through the roots; and (3) storage of foods, as in radishes, carrots, and turnips. Some roots are biennial (carrots, turnips, and beets). They grow and store foods the first season and then use this stored food to produce flowers and seeds during the second growing season.

Some roots perform specialized functions, such as the following: (1) Adventitious prop roots arise from the aerial stem and grow downward to the

171

HERBACEOU
1. SMALL
2. GREEN
3. SOFT

Fig. 14-1
Dandelion plant forcing its way through an asphalt driveway.

soil, thus giving added support, as in corn plants. (2) Adventitious climbing roots (ivies and certain vines) anchor plants to walls and trees. (3) Specialized suckers called haustoria are formed by certain parasitic seed plants (mistletoe and dodder) for absorbing food from the host plant. (4) Spongy, water-absorbing aerial roots of epiphytic plants, such as epiphytic orchids, take water from the atmosphere and rain, anchor the plants to objects (poles, trees, or wires), and absorb materials from the debris that collects about the roots. (5) Root outgrowths called "knees" or pneumatophores (niu -mat' o fors) (Gr. *pneuma*, air; *phorein*, to bear) are produced by certain swamp plants (bald cypress trees) and grow into the air above the water or swamp soil, possibly to conduct air to the roots. (6) Reproductive functions are performed by roots of certain plants (apple, cherry, or other trees) and by "suckers" that develop into new plants. The roots of potatoes and dahlias are used for propagation purposes. (7) Spices and other aromatic substances are formed by roots of plants, such as horse-radish, sassafras, and sarsaparilla. (8) Pharmaceutical preparations are obtained from

172

plants, such as ginseng, licorice, rhubarb, goldenseal, and gentian.

STEMS

Stems may be herbaceous or woody. Herbaceous stems are soft and usually small in diameter. Also, they are usually green, have few hard tissues, and are chiefly annuals. Woody stems are tough, usually larger in diameter than herbaceous ones, are not green, have well-developed fibers and other strengthening cells, and are chiefly perennials. The stems of all gymnosperms are woody, as are the stems of many angiosperms. Some examples of dicotyledons with woody stems are oaks, elms, maples, hickories, poplars, apples, and lilacs. Woody stems in monocotyledons are uncommon, but are present in palms and lilies. The hardened stems of bamboo, although used as wood, are fibrous in nature and not true wood (xylem). Woody stems are more primitive than herbaceous stems and probably developed from woody types. Table 14-1 illustrates the differences between the herbaceous stems of dicotyledons and those of monocotyledons.

The principal functions of stems are (1) the production and support of leaves, flowers, or cones; (2) the conduction of substances; and (3) the storage of foods. Substances are conducted upward in stems by the xylem; downward by the phloem; and horizontally by the wood rays, which are found in woody stems. Any living cell may store foods, but the pith, the cortex, and the wood and phloem parenchyma (particularly the parenchyma of the rays) primarily perform this function. Starch is the most commonly stored food, but fats and proteins may also be stored.

In addition to the general functions described above certain stems are of special importance. For example, in dicotyledonous trees and shrubs the great part of the stem consists of wood tissue. Through the activity of the cambium a new concentric layer of woods cells is added annually to the outside of the woody cylinder, thus forming annual rings. Spring growth is called springwood, and summer growth is known as summerwood. The dead annual rings in the centers of stems constitute the heartwood, whereas the functional, living part of the wood is called the sapwood. Wood

Table 14-1

Differences in herbaceous stems of dicotyledonous and monocotyledonous plants

Usually Annual

Herbaceous dicotyledonous stems	Herbaceous monocotyledonous stems
1. Examples: beans, sunflowers, clovers, potatoes, violets, snapdragons, and delphiniums	1. Examples: corn, wheat, other grasses, irises, daffodils, and cattails
2. Growth chiefly primary (from terminal bud) with little secondary growth	2. Growth chiefly primary with little secondary growth
3. Primary tissues present	3. Primary tissues present
(a) Epidermis, composed of single layer of protective cells	(a) Epidermis, composed of single layer of protective cells
(b) Cortex, composed of parenchyma cells and collenchyma cells, and generally much like cortical tissues of woody dicotyledons, but usually thinner	(b) Sclerenchyma tissue, just beneath epidermis and extending inward to vascular bundles; gives strength to stem (examples: corn and bamboo)
(c) Stele, composed of	(c) Vascular bundles composed of
(1) Endodermis, thin layer separating collenchyma and parenchyma cells of cortex from parenchyma cells of pericycle (fibers)	(1) Primary phloem, usually in outer part of each vascular bundle and not arranged in continuous layer around stem
(2) Primary phloem, usually in outer part of each vascular bundle	(2) No cambium (except in certain palms and lilies), hence no secondary growth
(3) Cambium, relatively inactive, thus producing little growth in diameter and separating phloem from xylem	(3) Primary xylem, forming inner part of each vascular bundle and not arranged in continuous layer around stem
(4) Primary xylem, forming inner part of each vascular bundle	(4) Pith (parenchyma) surrounding various vascular bundles scattered throughout stem and serving for support and storage of materials
(5) Pith, usually occupying large part of center of stem	
4. In some herbaceous dicotyledons cambium is continuous layer (foxgloves), extending as complete cylinder both through and between vascular bundles, whereas in other species cambium occurs only within separate vascular bundles (clovers and delphinums); in some cases (sunflowers) stele tissues distinct bundles in young stems, but continuous layer in older stems	4. In herbaceous monocotyledons individual vascular bundles (phloem and xylem) usually scattered throughout stem and not arranged in definite circular cylinder; vascular bundles commonly surrounded by strengthening cells and by sclerenchyma fibers; pith and cortex not sharply separated by vascular tissues, and difficult to distinguish one from other, entire ground tissue being called pith

Table 14-2
Comparison of gymnospermous and angiospermous plants

Gymnospermous plants	Angiospermous plants (flowering plants)
1. Ovules (immature seeds) and true seeds exposed (uncovered) on sporophylls	1. Ovules and seeds enclosed by ovary
2. No flowers present	2. Flowers (floral organs) present
3. Integument (covering of ovule) and seed coat present	3. Integument and seed coat present
4. Endosperm (food stored in seed for embryo development) present, but cells have single number (haploid) of chromosomes (N) and are actually female gametophyte	4. Endosperm present, cells have triple (triploid) number of chromosomes (3N); endosperm formed in different way from that in gymnosperms
5. Several cotyledons (embryonic seed leaves) present	5. Cotyledons present, one in monocotyledonous plants and two in dicotyledonous plants
6. Single fertilization (one sperm involved per embryo)	6. Double fertilization (two sperms used)
7. Stomata present in leaves	7. Stomata present in leaves
8. Vascular bundles usually without vessels, but xylem contains single-celled tracheids; no companion cells	8. Vascular bundles usually have long, true tubular vessels in xylem, companion cells present
9. Cambium (meristematic tissue) present	9. Cambium present
10. Two kinds of spores produced (heterospores)	10. Two kinds of spores produced (heterospores)
11. Sporophyte relatively large and independent	11. Sporophyte relatively large and independent
12. Gametophyte relatively smaller (microscopic) and dependent (parasitic) on sporophyte	12. Gametophyte relatively smaller (microscopic) and dependent (parasitic) on sporophyte
13. Pollen tube and pollination	13. Pollen tube and pollination
14. Many produce two kinds of cones; mostly evergreen, woody, perennial trees	14. Possess flowers (floral organs); mostly deciduous; woody or herbaceous; annual, biennial, or perennial trees, shrubs and vines

Table 14-3

Distinguishing characteristics of dicotyledonous and monocotyledonous plants

Dicotyledonous plants	Monocotyledonous plants
Two embryonic seed leaves (cotyledons) supply embryo with food during early development	One embryonic seed leaf (cotyledon) supplies embryo with food during early development
Flower parts usually in four's or five's, or multiples of these	Flower parts typically in three's or multiples of three
Leaves net veined	Leaves parallel veined, usually long and narrow
Some have woody stems, other have soft herbaceous stems	Most have herbaceous stems (few exceptions)
Vascular bundles of stems usually arranged in circular cylinder	Vascular bundles scattered throughout stem in no definite pattern
Cambium (meristematic tissue) present between phloem and xylem of vascular bundle	Usually no cambium between phloem and xylem of vascular bundle
Seem more ancient because of resemblance, particularly in vegetative structures, to primitive gymnospermous condition	Seem more advanced than dicotyledonous plants
Examples: beans, sunflowers, roses, violets, clovers, snapdragons, buttercups, potatoes, elms, oaks, apples, maples, hickories, poplars, and lilacs	Examples: corn, wheat, bluegrass, lilies, irises, daffodils, and cattails

serves a variety of purposes in construction industries, such as furniture and paper manufacture.

Certain parenchyma cells that are elongated at right angles to the long axis of the woody stem are arranged in horizontal, ribbonlike bands among the woody cells and constitute the wood rays that emanate from the central pith toward the periphery of the stem like the spokes of a wheel. These rays transfer materials horizontally within the stem and also store foods. The stems of many conifers have specialized resin canals that secrete "pitch," an economically valuable product. Tannin, various oils, and certain mucilaginous substances are secreted by special cells in the stems of numerous plants. The milky liquid, latex, is produced by various plants and is the source of natural rubber and other products.

175

Certain underground stems have lost their foliage-bearing abilities and have become shortened and thickened, constituting a tuber, as in the potato plant and Jerusalem artichoke. Such tuberous underground stems produce buds (a characteristic of stems) and store large quantities of food.

One of the important foods derived from typical plant stems is extracted from the sap of sugarcane and crystallized into sugar. Other sources of sugar are the stems of sorghums and the sap of maple trees. Other modified stems that serve as foods include the bulbs of onions and the swollen stems of kohlrabi. Among chemicals and commercial products derived from wood are dyes, tannin, tars, oils, wood alcohol, cellophane, and rayon. The liquid cellulose called "viscose" is forced through minute pores to produce tiny streams that harden into fibers that may be spun into threads. If forced through a narrow slit, transparent films of cellophane are formed.

Among the commercial fibers produced by plant stems are those of flax plants, hemp plants, jute, and ramie. The bark of the Spanish oak tree produces a thick layer of cork tissue that can be harvested as commercial cork every few years. The barks of oak and hemlock yield tannin which is used in tanning hides. The bark of the cinchona of South America and the East Indies yields quinine, an important medicine. Among the spices, cinnamon comes from the bark of the cinnamon tree of southeastern Asia.

LEAVES

Leaves arise from the terminal growing point of the stem. Unlike stems, leaves have no organized growth points and thus stop growing when mature. The blade of a leaf may be attached directly to the stem (sessile); most frequently it is attached to the stem by a stalklike petiole, which serves to transport materials and to expose the blade to light. Leaves may vary greatly in size, shape, texture, margin, and type of venation. They may vary from small scales (arborvitae) to needlelike leaves (pines), or to broad leaves (many deciduous trees). Leaves may vary from long and narrow (grasses) to circular forms (nasturtium). Detailed structures and functions of leaves are considered in Chapter 15.

FLOWERS AND REPRODUCTION

Flowers develop from buds. Some buds produce only flowers (elm, poplar, and morning glory), whereas others produce both flowers and leaves (apple and buckeye). There are many variations in flowers in the different species of angiosperms. The number of flower parts, the sizes and colors of the petals, the various degrees of fusion of parts, and the relative position and arrangement of the various floral organs are among the chief variations.

Flower colors usually result from the presence of pigments called anthocyanins or carotenoids. The anthocyanins are blue, purple, or reddish, and the carotenoids are yellow, orange, or sometimes reddish pigments. Plant pigments are considered in more detail in Chapter 15.

A flower is said to be complete when it consists of four floral organs: sepals, petals, stamens, and pistil (Fig. 14-2). Sepals (Gr. *skepe*, covering) are the outermost floral organs. They are usually green and may give some protection, especially when young. Petals (Gr. *petalon*, leaf) are the floral organs just inside the sepals and are usually colored. All the sepals together consitute the calyx (ka′ lix) (Gr. *kalyx*, cup), and all the petals together constitute the corolla (co -rol′ a) (L. *corolla*, crown). Sepals and petals together constitute the perianth (per′ i anth) (Gr. *peri*, around; *anthos*, flower). A stamen (microsporophyll) (Gr. *stemon*, thread) is composed of an enlarged, pollen-producing anther at the tip of a stalklike filament. A pistil (L. *pistillum*, pestle) is composed of a pollen-receiving stigma (Gr. *stigma*, mark) and a style (Gr. *stylos*, pillar) to conduct pollen to the enlarged ovary at the base.

Carpels (Gr. *karpos*, fruit) are the units of which a pistil is composed. They bear and enclose the ovules (immature seeds). Pistils may be composed of one or several carpels, depending upon the flower. An ovule is composed of an embryo sac (megagametophyte), a nucellus, and integuments (coverings). It develops into a seed after fertilization. A nucellus, or megasporangium, is the central part of an ovule. It encloses the megagametophyte and is covered by the integument.

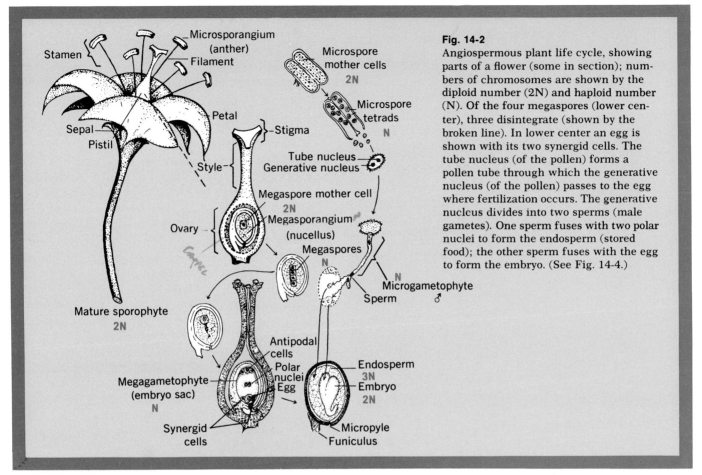

Stamen
Microsporangium (anther)
Filament
Petal
Sepal
Pistil
Stigma
Style
Tube nucleus
Generative nucleus
Megaspore mother cell 2N
Megasporangium (nucellus)
Megaspores N
Ovary
Mature sporophyte 2N
Antipodal cells
Polar nuclei
Egg
Megagametophyte (embryo sac) N
Synergid cells
Microspore mother cells 2N
Microspore tetrads N
Microgametophyte ♂
Sperm
Endosperm 3N
Embryo 2N
Micropyle
Funiculus

Fig. 14-2
Angiospermous plant life cycle, showing parts of a flower (some in section); numbers of chromosomes are shown by the diploid number (2N) and haploid number (N). Of the four megaspores (lower center), three disintegrate (shown by the broken line). In lower center an egg is shown with its two synergid cells. The tube nucleus (of the pollen) forms a pollen tube through which the generative nucleus (of the pollen) passes to the egg where fertilization occurs. The generative nucleus divides into two sperms (male gametes). One sperm fuses with two polar nuclei to form the endosperm (stored food); the other sperm fuses with the egg to form the embryo. (See Fig. 14-4.)

Pollination

If the angiosperm flower consists of both male and female parts, its structures and functions are as follows. Within the microsporangium (pollen sac) each diploid microspore mother cell produces four haploid microspores (pollen grains). The number of chromosomes is reduced by one half when the microspores are formed from the spore mother cells by meiosis. Before the pollen is shed the microspore nucleus divides to form a tube nucleus and a generative nucleus, which are not separated by a wall. If the pollen alights on the stigma of the proper plant, a pollen tube is formed that extends from the stigma through the style and ovary to the megagametophyte in the ovule. By the time the generative nucleus passes through the pollen tube it has divided into two microgametes (sperms). Hence, the microgametophyte is very small and is reduced to three nuclei (one tube nucleus and two sperms, or microgam-

etes). Each microgamete contains a haploid number of chromosomes.

Within the megasporangium each diploid megaspore mother cell forms four haploid megaspores by meiosis. However, three of the four megaspores degenerate. The nucleus of the remaining megaspore divides to form two, one moving to each end of the young megagametophyte. Each of these nuclei divides into two, and these two into four, so there are two groups of four nuclei each. One from each group moves to the center of the megagametophyte where the two polar nuclei fuse (Figs. 14-2 to 14-4). Thus at the microphyle end of the megagametophyte there are typically three cells, one egg and two nonfunctional synergid cells (Gr. *synergos*, working together). The synergid cells are probably vestigial remains of the archegonium, which has now disappeared. At the opposite end of the megagametophyte are three nonfunctional

177

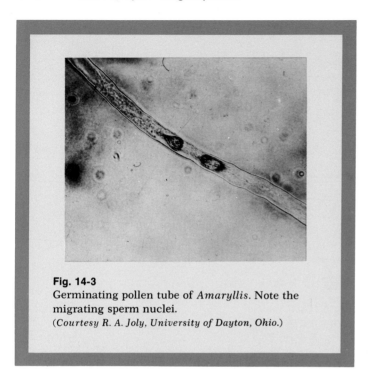

Fig. 14-3
Germinating pollen tube of *Amaryllis*. Note the migrating sperm nuclei.
(Courtesy R. A. Joly, University of Dayton, Ohio.)

antipodal cells (an -tip′ o dal) (Gr. *anti*, opposite; *pous*, foot), which are vestiges of the vegetative tissues of the megagametophyte. The endosperm nucleus, resulting from previous fusion of two polar nuclei, is in the center.

The pollen tube grows toward the ovule, being nourished by the tissues of the style through which it passes. The pollen tube usually enters the ovule through a small opening called the micropyle. The two sperms that come down the pollen tube are discharged into the megagametophyte (Fig. 14-3). One sperm fertilizes the egg to form a zygote, which will develop into the embryo, and the other sperm fertilizes the endosperm nucleus (two polar nuclei) to form the unique triploid number of chromosomes (3N). This triploid nucleus gives rise to the nutritive endosperm tissue, by mitosis.

Hence, double fertilization occurs. One sperm (N) fuses with one egg (N) to form a zygote (2N) and eventually the embryo. The other sperm (N) fuses with the endosperm nucleus (2N) to form the endosperm tissue (3N).

Pollination of flowers may be accomplished by wind, insects, or birds. Many flowers that are pollinated by insects are brightly colored or scented or both. Their pollen is heavy, sticky, and not readily carried by wind. Such flowers are often provided with nectaries (Gr. *nektar*, sweet) that secrete a fluid that has a rather high sugar content. Nectaries vary greatly in form and location, depending upon the species. Insect pollination is accomplished chiefly by bees, wasps, butterflies, and moths, although other insects may frequent flowers to gather pollen and nectar.

The sense of smell of honeybees is sufficiently sensitive to enable them to detect flower odors. Honeybees cannot distinguish red from dark gray or black. They can distinguish yellow, blue-green, and blue but not violet or purple. Butterflies frequently pollinate red flowers. Many types of cultivated plants depend upon insect pollination.

FRUITS AND SEEDS

The commonly accepted idea of what constitutes a fruit may differ from the botanical meaning of the term. To the average person a fruit probably means an apple, peach, or pear, or something similar at the food market. After fertilization occurs a ripened ovary of a flower, its seed, and sometimes the adjacent parts that may be associated with it develop into a fruit. Sometimes a group of ovaries may develop into a multiple or aggregate fruit.

After fertilization an ovary is composed of a thickened ovary wall called the pericarp (per′ i karp) (Gr. *peri*, around; *karpos*, fruit) and one or more ovules, which develop into mature seeds. The pericarp is derived from one or more carpels, which are modified sporangia, called megasporophylls. Fruits are important to plants for seed dissemination and they also serve as sources of food for animals and man. A fruit always develops from a flower and is always composed of at least one ripened ovary that may be fused with other floral parts or structures associated with the flower. Any edible part of a plant that does not conform to this definition of a fruit is classed as a vegetable.

Successful reproduction requires the formation of seeds and their dispersal. Seeds and fruits may

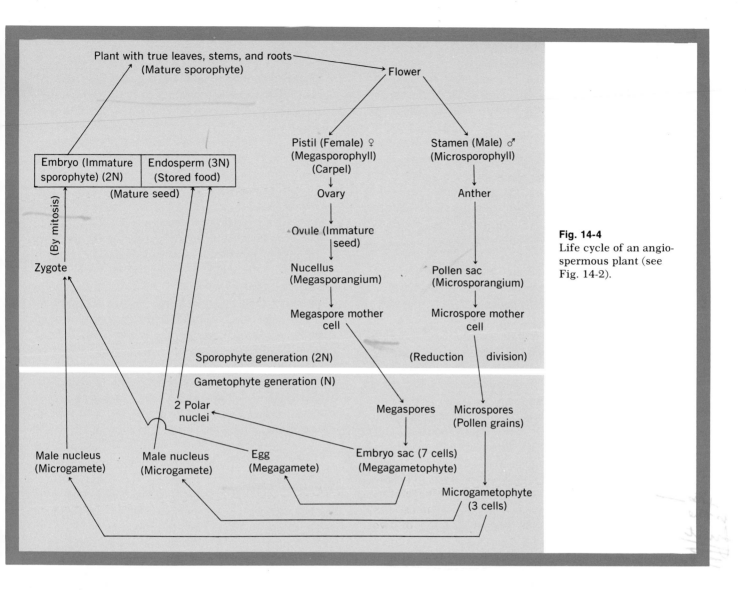

Fig. 14-4
Life cycle of an angio-
spermous plant (see
Fig. 14-2).

Contents of the diagram:

Plant with true leaves, stems, and roots (Mature sporophyte) → Flower

Flower → Pistil (Female) ♀ (Megasporophyll) (Carpel)
Flower → Stamen (Male) ♂ (Microsporophyll)

Embryo (Immature sporophyte) (2N) | Endosperm (3N) (Stored food)
(Mature seed)

(By mitosis)

Zygote

Pistil (Female) ♀ (Megasporophyll) (Carpel) → Ovary → Ovule (Immature seed) → Nucellus (Megasporangium) → Megaspore mother cell

Stamen (Male) ♂ (Microsporophyll) → Anther → Pollen sac (Microsporangium) → Microspore mother cell

Sporophyte generation (2N)   (Reduction   division)

Gametophyte generation (N)

2 Polar nuclei   Megaspores   Microspores (Pollen grains)

Male nucleus (Microgamete)   Male nucleus (Microgamete)   Egg (Megagamete)   Embryo sac (7 cells) (Megagametophyte)   Microgametophyte (3 cells)

---

be dispersed by wind (winglike structures of maples and ashes). In fleshy fruits, such as cherries,
grapes, and tomatoes, the soft pulp is often colored
and tasty, so that animals may disperse the seeds
by eating them. In some instances spines and hooks
(as in the burdock) permit dispersal by attachment
to animals.

As an ovary matures or ripens into a fruit, its
wall, the pericarp, may thicken and be differentiated into layers of tissues: (1) the outer exocarp,
consisting of one or several layers of epidermal
cells; (2) the middle mesocarp, consisting of one
layer of cells, or a large mass of tissues, several
inches thick; and (3) the inner endocarp, which

varies greatly in thickness, structure, and texture
in the fruits of different species.

Fruits may be classified on the basis of their
structure and the number of ovaries from which
they are derived. In some species flower parts other
than the ovaries may adhere to or enclose the mature ovaries to form so-called accessory fruits,
which are considered later in this chapter. Fruits
may be simple, aggregate, or multiple (compound),
as described in the following.

Simple fruits

Simple fruits are derived from one ripened ovary
(pistil) plus, in some species, such adherent parts   179

as sepals and stamens. In some species the ovary is composed of more than one carpel (a member of a compound pistil). The fruits of most angiosperms are simple fruits.

Dry fruits

When mature, the ovary wall of these fruits (pericarp) is dry and may become paperlike, leathery, or woody. The wall does not become fleshy.

Dehiscent fruits

At maturity dehiscent fruits dehisce (open) along definite seams or at definite points and may contain several to many seeds.

Legume. Legumes (L. *legare,* to gather) develop from a single carpel and usually dehisce along two sides. The so-called pod is a single carpel and contains the seeds. However, in some legumes (peanut) the fruit does not open at maturity. Examples are string bean, lima bean, pea, locust tree, and peanut.

Follicle. Follicle (L. *folliculus,* small sac) dry fruits consist of a single carpel and dehisce along one seam only. Examples are milkweed, larkspur, peony, spiraea, and columbine.

Capsule. Capsule (L. *capsula,* little box) dry fruits develop from a compound pistil (two or more fused carpels). The capsule may dehisce in various ways. In the iris it may split along the back of the individual carpels. In the poppy it may open by forming a circular row of pores near the top of the fruit, seeds coming out through the pores. In the horse chestnut the capsule splits into three valves, thus releasing the seed. In the Brazil "nut" the capsule is a thick shell, often 6 inches in diameter, and may contain twenty seeds, each with a hard seed coat. A lidlike cover at the tip of the fruit opens at maturity but does not permit the exit of seeds. Upon germination the seedlings grow out through the capsule openings. Examples are iris, lily, tulip, poppy, pansy, azalea, horse chestnut, snapdragon, violet, plantain, jimsonweed, Brazil "nut," and cotton.

Indehiscent fruits

Indehiscent fruits do not split at definite seams or points and usually contain one or two seeds.

Achene. Achenes are (Gr. *a-,* not; *chainein,* to open) rather small dry fruits contain a single seed that nearly fills the cavity of the fruit. The seed coat does not adhere to the pericarp, except at the point of attachment of the seed. Examples are sunflower, buttercup, buckwheat, and dandelion.

Grain or caryopsis. Grains or caryopsis (Gr. *karyon,* nut; *opsis,* appearance) are one-seeded fruits and are similar to achenes, but the thin seed coat is fused to the inner surface of the pericarp and is thus not easily removed. These fruits serve as cereals. Examples are corn, wheat, oats, barley, rice, and most grasses.

Samara (key or wing fruit). Samara (L. *samara,* seed of elm) are dry, indehiscent winged fruits and are usually single seeded, the wings being formed from the pericarp for the purpose of wind dispersal. Examples: elms, ashes, basswood, and maples.

True nut. These one-seeded fruits have a pericarp that is rather thick, hard, stony, or woody when mature. The edible inner part is the seed. Examples: filbert (hazelnut), beechnut, acorn (of oaks), chestnut, and hickory nut.

Fleshy fruits

When mature, the pericarp, or some part of it, is filled with sap and becomes fleshy. The pericarp has three parts: (1) an outer exocarp, which is usually thin, (2) a middle mesocarp, which is a fleshy pulp, and (3) an inner stony or fleshy endocarp. Fleshy fruits may be classified as true berry, false berry, drupe, and pome.

True berry

The botanist's definition of a berry is somewhat different from that of a layman. In these fruits the entire pericarp ripens, becoming fleshy and often edible and juicy. In tomatoes the ovary becomes large and fleshy and bears many seeds. Citrus fruits are modified berries, although the peel, or rind (pericarp), is not usually edible. Sections of young citrus fruits develop multicellular outgrowths from the surface of the carpel walls, and these outgrowths become large, juicy, and edible. Each wedgelike segment of oranges and grapefruits represents a carpel filled with the multicellular outgrowths. In citrus fruits the fleshy part with its seeds is endocarp, whereas the peel is exocarp and mesocarp. In grapes the thin skin is exocarp, and the fleshy material with its seeds is mesocarp and endocarp. Examples are tomato, grape, date, avocado, red pepper, citrus fruits (orange, lemon, and grapefruit), blueberry, and gooseberry.

False berry

Some fruits may include not only a matured ovary but also other flower parts or closely related structures. When such structures constitute an important part of the fruit, it is said to be an accessory fruit. Most accessory fruits are simple fruits derived from both the pericarp and the floral tube (composed of basal parts of sepals, petals, and stamens). This tube is fused with the pericarp and becomes fleshy when ripe. Since the entire fruit is fleshy when ripe, as in true berries, such accessory fruits may be termed "false" berries. In contrast to true berries "false" berries have remnants of the flower persisting at the part of the fruit that is opposite the stem. The apical end of the cranberry has remnants of sepals, whereas the tip of the banana fruit has a large scar where flower parts have fallen away. When the berry has a hard rind (watermelon, squash, and cucumber), it is called a pepo. Examples are cucumber, pumpkin, watermelon, gourd, cantaloupe, cranberry, and banana.

Drupe or "stone fruit"

In drupes (Gr. *dryppa*, berry) the pericarp is divided into an outer, thin skin, the exocarp; a fleshy, thick mesocarp; and a stony (or woody) endocarp. The endocarp ("pit" or "stone") encloses one seed (rarely two to three).

In olives and apricots the so-called pit is endocarp with the seed inside; the edible, fleshy part is mesocarp; and the skin is exocarp. In almonds the fleshy exocarp and mesocarp are leathery, inedible, and removed when the fruit is harvested. The stony shell (endocarp) contains the edible almond seed. The shell of the coconut is composed of fibrous exocarp and mesocarp, so that the coconut is often called a dry, or fibrous, drupe. The stiff fibers that compose the husk are used commercially in making coco mats, brushes, and cordage. The hard, woody endocarp that we purchase has ger-

mination pores and encloses a large seed. Examples are olive, apricot, coconut, peach, plum, almond, cherry, and walnut.

Pome

Pomes (L. *pomum*, apple) are fleshy, indehiscent accessory fruits; the pericarp has parts that are somewhat like drupes. In the pome the true fruit (ripened ovary) constitutes the central, inedible "core." Thus in apples and pears the edible parts are not the matured ovary but the stem and floral tube tissues in which the true fruits (ovaries) are embedded. The outer fleshy part is exocarp and mesocarp, and the inner endocarp is leathery (papery). In apples and pears the inner leathery endocarp contains seeds. A group of vascular bundles, sometimes called the "core line," marks the outer boundary of the exocarp. Between the inner leathery endocarp and the exocarp is the mesocarp. The part of the apple extending from the "core line" to the outside of the fruit is derived from the floral tube. Apples and pears are accessory fruits because the remains of sepals and stamens are present at the tip of the mature fruit. Examples are apples, pears, quinces, and hawthornes (haws).

Aggregate fruits

An aggregate (L. *ad*, to; *gregare*, to collect together) fruit develops from a cluster of several to many ripened, simple ovaries (pistils) produced by one flower and borne on the same basal receptacle (axis of the flower stalk bearing floral organs).

In raspberries the simple fruits develop into tiny drupelets (drupes) that adhere to one another but separate as a unit from the dome-shaped receptacle. Blackberries are like raspberries, but the elongated receptacle is also fleshy and forms part of the fruit, thus making it an aggregate-accessory fruit. In strawberries the numerous simple ovaries develop into tiny achenes that are visible on the exterior of the fruit and are commonly called seeds. Each ovary contains one ovule that develops into a seed within the achene. The edible, accessory part of the fruit is the enlarged, fleshy receptacle (modified stem), thus making the strawberry an aggregate-accessory fruit. Examples are raspberry, boysenberry, blackberry, and strawberry.

Multiple (compound) fruits

Multiple fruits are formed from a cluster of several to many ripened ovaries produced by several flowers associated closely together on the same inflorescence rather than from a single flower. The ovary of each flower forms a fruit, and at maturity all these fruits remain in a mass. As in aggregate fruits, the small fruitlets of a compound fruit may be berries, drupes, or nutlets.

In the pineapple from 100 to 200 sessile flowers are attached to the elongated, fleshy axis of the inflorescence, which is leafy at the top. The flowers are fused with each other, and together with their fleshy bracts (modified leaf associated with a flower) they ripen at the same time. Each of the units visible on the surface of the mature fruit represents a flower together with its bract, thus making it a multiple-accessory fruit. The edible part is mostly the thickened, pulpy central stem in which the individual fruits are embedded. The edible part of a fig is the mature axis of the inflorescence, which is hollow and encloses many tiny achenes, each formed from a single flower. Examples: pineapple, Osage orange ("hedge apple"), mulberry, and fig.

DICOTYLEDONOUS PLANTS

Legume family

The podlike fruits (legumes) of leguminous plants are frequently gathered for various purposes. The family includes foods, such as peas, beans, and peanuts and forage crops, such as clovers and alfalfa, and plants, such as sweet pea, lupine, wisteria, Judas tree and locust tree. Leguminous plants add nitrogenous materials to the soils through the fixation of free nitrogen by the actions of special types of bacteria that inhabit the enlarged nodules on the roots.

Garden bean (DICOT)

The common bean plant *Phaseolus* (L. *fabaceus*, from *faba*, bean) may be a short, bushlike, or vinelike annual, depending upon the variety (Fig. 14-5). The stems (stalks) may be long and slender, and because of unequal rates of growth on opposite sides, they have a tendency to twine spirally around objects with which they come in contact. The ex-

ternal epidermis is thin and affords limited protection. In certain parts of the stem and leaves there are hairs (outgrowth of epidermal cells). The leaves of the bean plant are usually trifoliate (three leaves arising from one point) and net veined (frequently branched). The thin epidermal layer contains stomata for the exchange of gases. Guard cells control the size of the opening of the stoma. Below the stomata are air spaces surrounded by cells that contain chlorophyll in green chloroplasts. Part of the manufactured foods is used by the plant, and the remainder is stored as carbohydrates and proteins in the developing seeds. The veins (vascular bundles) of leaves conduct materials throughout the leaf and are continuations of the vascular bundles of the leaf petiole and stem.

The flowers of the common bean are irregular (bilaterally symmetrical with petals that are various sizes and unequally spaced), perfect (both stamens and pistils), and complete (all four sets of flower parts). They are usually small, whitish purple, and racemose (flowers are along an elongated axis). The calyx is composed of four to five green sepals, which are more or less united. The corolla is papilionaceous (butterfly-like) and consists of four to five petals, some of which may be fused. In those species with definitely irregular flowers the large, recurved, somewhat contorted upper petal is called the standard, the two lateral petals are called wings, and the two lower (anterior) petals are fused to form the keel, which may be spirally coiled. There are usually ten stamens, nine of which may be united into a thin sheath around the pistil. One stamen is free. Each stamen bears a pollen-producing anther at its tip.

The pistil is composed of an elongated ovary (one carpel) that contains several ovules, a filamentous style, and a pollen-receiving stigma. Pollen tubes are formed through the style and extend from the stigma to the ovary. Fertilization occurs in the ovary, and the fertilized ovules develop into the true seeds (Fig. 14-5). There is no endosperm (food), as in the case of corn, but its place is taken by the two cotyledons. Each seed is attached to the pod by a stalklike funiculus (fu -nik′ u lus) (L. *funiculus*, small cord). The seeds, like those of other legumes, are developed in a

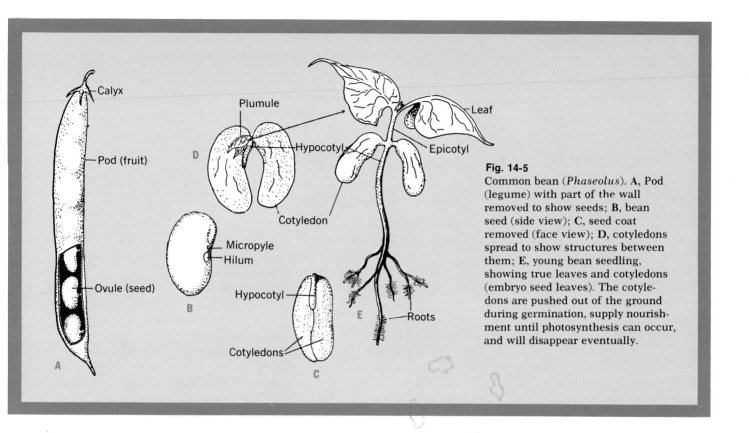

Fig. 14-5
Common bean (*Phaseolus*). **A,** Pod (legume) with part of the wall removed to show seeds; **B,** bean seed (side view); **C,** seed coat removed (face view); **D,** cotyledons spread to show structures between them; **E,** young bean seedling, showing true leaves and cotyledons (embryo seed leaves). The cotyledons are pushed out of the ground during germination, supply nourishment until photosynthesis can occur, and will disappear eventually.

bivalved, multiseeded legume (pod). Since a ripened ovary is known as a fruit, the legume is a rather special type of fruit.

Each individual seed consists of (1) a small, prominent scar, the hilum (hi' lum) (L. *hilum*, small), where it was attached to the pod; (2) a prominent ridge above the hilum, the raphe, which is formed by the ovule beneath; (3) the micropyle, which is a small opening in the seed coat near the hilum for the entrance of pollen; (4) seed coats, which form the protective covering; (5) two cotyledons, which are the two fleshy halves of the bean for the storage of food; (6) the plumule (or epicotyl) with its true leaves folded over the growing tip; (7) the hypocotyl; and (8) the radicle, which is continuous with the hypocotyl and forms the embryonic root. All of these structures may best be seen in seeds that have been soaked to initiate germination.

During germination the food of the cotyledons is digested and transferred to the plumule, hypocotyl, and radicle. The embryonic primary root is formed from the radicle, which bends downward under the influence of gravity. The hypocotyl elongates and carries the plumule and two cotyledons with it out of the soil. The two cotyledons spread to allow the foliage leaves of the plumule to develop. The cotyledons may develop chlorophyll for carrying on photosynthesis for a time, but eventually they shrivel. The roots absorb water and nutrients from the soil and conduct them to the stalk (stem).

Crowfoot family (Ranunculaceae)

Marsh marigold, hepaticas, anemones, columbine, and larkspurs belong to the Crowfoot family, along with the common buttercup, *Ranunculus* (Fig. 14-6). The roots or leaves of certain species of the family are poisonous when eaten.

Buttercup

When a thin cross section of a buttercup root (see Fig. 15-7) is studied, it will be noted that the xylem is shaped like a four-pointed cross and is composed of thick-walled cells for conducting materials upward toward the stem. Alternating

183

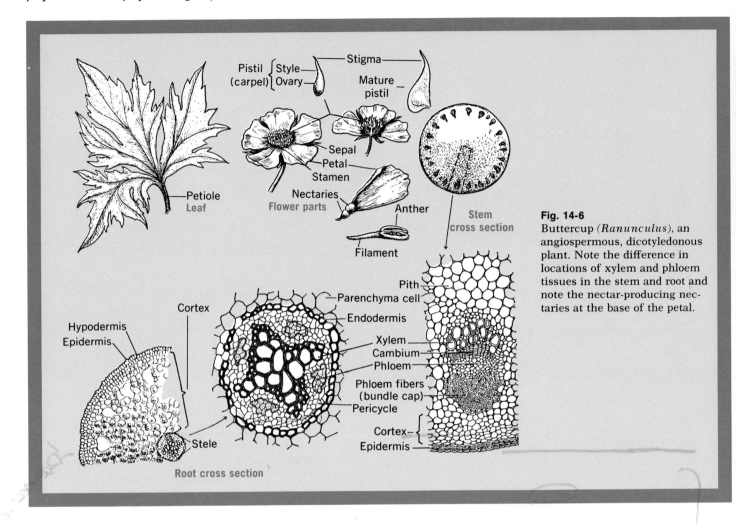

Pistil {Style — Stigma
(carpel) {Ovary — Mature pistil

Sepal
Petal
Stamen

Nectaries
Flower parts

Anther

Filament

Petiole
Leaf

Stem cross section

Fig. 14-6

Buttercup (*Ranunculus*), an angiospermous, dicotyledonous plant. Note the difference in locations of xylem and phloem tissues in the stem and root and note the nectar-producing nectaries at the base of the petal.

Pith
Parenchyma cell
Endodermis

Xylem
Cambium
Phloem

Phloem fibers
(bundle cap)
Pericycle

Cortex
Epidermis

Cortex

Hypodermis
Epidermis

Stele

Root cross section

with the xylem points are four areas of phloem, composed of thin-walled cells for conducting materials downward from the stem. Surrounding these vascular bundles (xylem and phloem) is a pericycle of thin-walled cells, and external to this is the ringlike endodermis, composed of thick-walled cells, although thin-walled cells may be present. The xylem, phloem, pericycle, and endodermis constitute the small, centrally located stele (Gr. *stele*, column). External to the endodermis is a very wide cortex (L. *cortex*, bark) of thin-walled cells, which is surrounded by a hypodermis and an epidermis.

When a thin cross section of a buttercup stem is studied (Fig. 14-6) the following will be noted.

An external, protective epidermis is present, beneath which is a cortex of thin-walled cells. Numerous vascular bundles are arranged in a circle toward the periphery of the stem, and each bundle contains thin-walled phloem cells, thick-walled xylem cells, and a band of cambium (L. *cambiare*, to exchange) between them (Fig. 14-7). Internally, there is a large pith composed of thin-walled cells.

The leaves have typically lobed blades attached to a petiole and contain palmately netted veins (main veins branch out from the petiole like the fingers of a hand). The veins (vascular bundles) connect with the vascular bundles of the petiole and stem. Some variations in the leaves will be found in different plants.

184

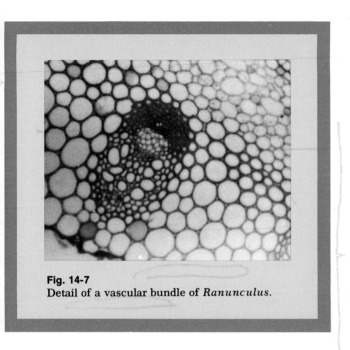

Fig. 14-7
Detail of a vascular bundle of *Ranunculus*.

The flowers of the buttercup (Fig. 14-6) are typically regular (floral parts of each whorl are similar in shape) and perfect (bearing both male stamens and female carpels). The numerous pollen-producing stamens (anther and filament) and the many free, one-ovuled carpels are spirally arranged. There are five green sepals and five yellow petals, the latter having a nectar gland at the base of each. The ovule develops into an achene, a small, dry fruit that does not open at maturity and contains a single seed.

Composite family (Compositae)

Plants of this family have many closely compacted individual flowers (florets) that form a head that is commonly mistaken for a flower (Fig. 14-8). They include dandelions, ragweeds, goldenrods, daisies, asters, zinnias, dahlias, marigolds, and sunflowers. Their enormous seed production and efficient dispersal devices have distributed

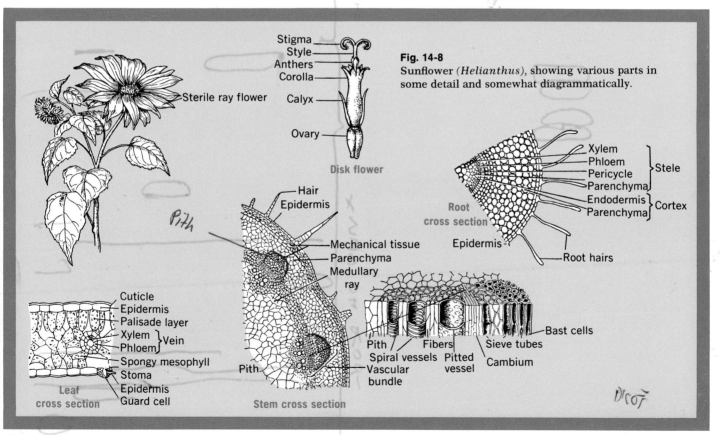

Fig. 14-8
Sunflower (*Helianthus*), showing various parts in some detail and somewhat diagrammatically.

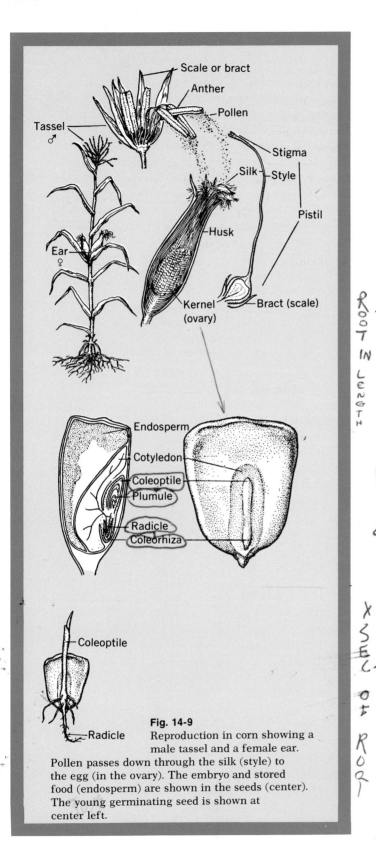

Scale or bract
Anther
Pollen
Tassel
♂
Stigma
Silk — Style
Pistil
Husk
Ear
♀
Pistil
Kernel — Bract (scale)
(ovary)

Endosperm
Cotyledon
Coleoptile
Plumule
Radicle
Coleorhiza

Coleoptile

Radicle

Fig. 14-9
Reproduction in corn showing a
male tassel and a female ear.
Pollen passes down through the silk (style) to
the egg (in the ovary). The embryo and stored
food (endosperm) are shown in the seeds (center).
The young germinating seed is shown at
center left.

them widely. The pollen of many composites, including ragweeds and goldenrods, causes pollen ("hay") fever.

Sunflowers

An older sunflower plant, *Helianthus* (Gr. *helios*, sun; *anthos*, flower), has a primary root with lateral branch roots that may have an extensive branching root system for anchorage and absorption.

The regions of a mature root (Fig. 14-9), beginning at the tip, include (1) the cup-shaped root cap that protects the root as it grows through the soil; (2) the meristematic region, or growing region, which is covered by the root cap and is composed of small, similar, closely packed, rapidly dividing cells; (3) the elongation region, just behind the growing region, which is composed of cells that are increasing in length; (4) the maturation region, in which the cells are differentiated and are taking on mature characteristics; and (5) the mature region, in which most cells have completed their development. These regions can be observed in a longitudinal section.

A microscopic examination of a thin cross section of a mature region (root hair area) of a root (Fig. 14-8) shows the following tissues (from the outside to the center): (1) An external, protective epidermis is composed of one layer of cells, some of which may bear projections known as root hairs for increased absorption. Growth occurs near the root tip rather than in the region of the root hairs, so that the hairs are not injured as the root pushes through the soil. (2) A cortex is composed of (a) large, thin-walled, spherical or ovoid parenchyma cells with numerous intercellular spaces for storing foods and transporting water and minerals, which are absorbed by root hairs, to the conducting tissues in the center of the root and (b) the endodermis, a layer of nearly rectangular cells, some of which have walls thickened by suberin, which is fatty and impermeable to water, whereas other endodermal cells, without suberin, pass water toward the central stele. The endodermis forms the outer boundary of the centrally located stele, which is the conducting and strengthening part of the root. (3) The stele is composed of (a) a pericycle, which is one or more layers of thin-walled parenchyma cells just inside the endodermis; (b) the xylem, which is in the shape of a cross (+) and is composed of thick-walled cells (tracheids and vessels) for strengthening the root and transporting liquids and salts toward the stem; (c) the phloem, which contains clusters of thin-walled cells (sieve tubes and companion cells) for conducting materials brought downward from the stem and located between the strands of the xylem cross or in short bands alternating with the radially arranged xylem groups; and (d) thin-walled parenchyma cells surrounding the xylem and phloem tissues within the stele.

The stem of a sunflower plant contains nodes (joints) (L. *nodus*, knot), at which leaves are borne, and internodes, which are present between the successive nodes. Growth occurs by an elongation process at the internodes.

A thin cross section of a mature stem of a dicotyledonous plant, such as a sunflower (Fig. 14-8), shows the following structures: (1) A central pith region is composed of thin-walled parenchyma cells. (2) Vascular bundles are arranged in a circle toward the periphery of the stem and are composed of (a) xylem, which is composed of thick-walled cells (single-celled tracheids and vessels), (b) phloem, which is composed of nonnucleated sieve tubes (with perforated sieve plates), and elongated, nucleated companion cells, and (c) a layer of cambium, which separates the xylem and phloem. (3) The pericycle is a cylinder of mechanical tissue that is external to the vascular bundles and is composed of thick- or thin-walled cells. Individual bundles may be separated by radial strands, medullary rays, which are composed of parenchyma cells to conduct materials across the stem. This entire central core of the stem constitutes the stele. (4) A layer of cortex is external to the stele and is composed of large, thin-walled parenchyma cells. (5) A layer of mechanical tissue is external to the cortex and is composed of thick-walled cells. (6) The external epidermis is composed of elongated cells whose outer walls contain a waxy cutin to make them impermeable to water. Certain epidermal cells may produce extensions known as hairs. A connection between the vascular bundle of a stem and a leaf is called a leaf trace.

A mature leaf from a dicotyledonous plant, such as the sunflower, consists of a broad blade, which is attached to the stem by a slender petiole. Net veins conduct materials throughout the leaf.

A thin cross section of a leaf blade shows (1) an external, protective cuticle; (2) an epidermis, whose outer cell walls contain a waxy cutin; (3) a palisade layer, which is composed of column-shaped cells closely arranged and containing many chloroplasts for photosynthesis; (4) a spongy mesophyll layer (Gr. *mesos*, middle; *phyllon*, leaf) beneath the palisade layer, which is composed of irregularly shaped, loosely packed cells with many intercellular spaces for the storage of materials (chloroplasts are also present in the spongy tissue cells); (5) veins, which are vascular bundles composed of xylem and phloem, as in the petiole and stem; (6) a lower epidermis, which is similar to the upper epidermis except that it contains stomata for the exchange of gases. Each stoma is bordered by two guard cells that contain chloroplasts and also regulate the size of the stoma. The stomata open into the intercellular spaces (substomal cavities) where gases may be exchanged.

A composite flower, such as the sunflower, is composed of many individual flowers grouped together on a disklike head that resembles a single flower in a general way. At the edge of the head are two or more spirals of overlapping, flat, green bracts (L. *bractea*, thin plate).

There are two types of flowers on the head—ray flowers and disk flowers. Ray flowers form one or two circles at the edge, each in the axil of a small bract (modified leaf). These flowers consist of a strap-shaped corolla (petals), one side of which is modified into a broad, flat structure. The male stamens and the female style may be abortive (poorly developed). These marginal ray flowers may be sterile, or they may contain only female pistils. Disk flowers, in the center of the head, consist of a basal ovary, containing one functional ovule (immature seed). The pollen-producing anthers are united at their edges to form an anther tube around the female style. The fused petals (corolla) surround the style, and their tips have blunt teeth. The style extends beyond the surrounding anther tube and corolla and terminates in two pollen-receiving stigmas. After fertilization the ovary enlarges and eventually contains one mature seed.

187

MONOCOTYLEDONOUS PLANTS

Indian corn

Indian corn (*Zea mays*) (Gr. *zea*, corn) has an erect stem from which adventitious roots may form at the nodes. These roots assist the fibrous true roots in absorbing water and dissolved materials from the soil and in anchoring the plant (Fig. 14-9). Such unusual adventitious roots are called "brace roots," or "prop roots."

Corn has a typical monocotyledonous stem, composed of parenchyma cells of various shapes and sizes, with numerous vascular bundles scattered throughout it. When a thin cross section of a corn stem is studied, it will be noted that the external, protective epidermis is composed of relatively small, thick-walled cells (Fig. 14-10). Beneath the epidermis is a narrow layer of (mechanical) sclerenchyma tissue whose cells are small and thick-walled with lignin (lig′ nin) (L. *lignum*, wood). Each vascular bundle is surrounded by a sheath or layer of thick-walled sclerenchyma tissues. Internally, each bundle consists of phloem and xylem. There is no meristematic cambium separating the phloem and xylem, as in dicotyledonous stems, so there can be no secondary increase in size after the primary tissues are mature. Bundles lacking cambium are called "closed" bundles because of their inability to grow indefinitely.

The phloem of a mature bundle conducts liquids both upward and downward and consists of regularly arranged, nonnucleated sieve tubes and companion cells. The sieve tubes have their adjacent end walls supplied with a perforated sieve plate. Often a narrow, thin-walled, elongated, nucleated companion cell lies parallel to the sieve tube (Fig. 14-11).

The xylem conducts liquids primarily upward and consists of large vessels with pitted walls, which are located next to the phloem. Between these vessels are a few hollow, one-celled tracheids. The innermost part of the xylem contains vessels whose walls have ring-shaped or spiral thickenings. Between these vessels and the sheath of sclerenchyma tissue is a large hollow intercellular (air) space.

188

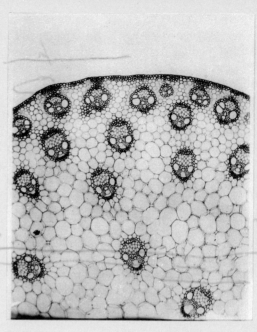

Fig. 14-10
Corn stem (cross section). Note the typical vascular bundles.
(*Courtesy General Biological Supply House, Inc., Chicago, Illinois.*)

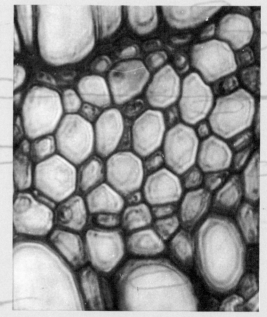

Fig. 14-11
Detail of phloem in the vascular bundle of a corn stem.

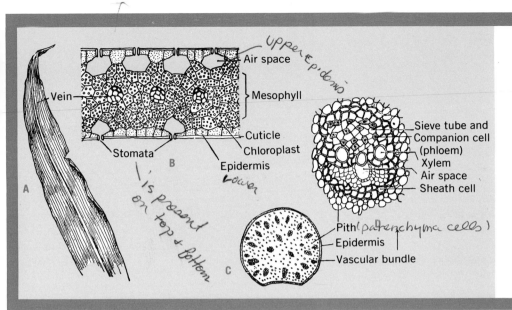

Fig. 14-12
Corn (*Zea mays*). **A,** External view of a leaf with parallel veins; **B,** leaf (somewhat diagrammatic cross section); **C,** stem (cross section); **D,** one vascular bundle (much enlarged).

Labels in figure: Air space, Mesophyll, Cuticle, Chloroplast, Epidermis, Vein, Stomata, *upper epidermis*, *lower*, *is present on top & bottom*, *B*, *A*, *C*, Sieve tube and Companion cell (phloem), Xylem, Air space, Sheath cell, Pith (*parenchyma cells*), Epidermis, Vascular bundle, *vein*

The leaves of corn are characterized by numerous main veins running parallel to the long axis and all connected by a network of fine, inconspicuous branches (Fig. 14-12). The veins are actually vascular bundles that are connected with the vascular bundles of the stem. The broad portion of a leaf is called the blade. The tissues of the corn leaf are as follows: (1) An epidermis, which is composed of one layer of cells, is present on the upper and lower surfaces. Openings on the surfaces, stomata, are for the exchange of gases. Each stoma is bordered by guard cells to regulate the size of the opening. Just beneath each stoma is an irregularly shaped, intercellular (substomal) air space for the storage of gases. (2) A mass of compactly arranged cells contain chloroplasts for photosynthesis. (3) Veins, which are vascular bundles, are composed of phloem and xylem.

The flowers of the corn plant are incomplete and are on different parts of the same plant. The tassel at the tip of the stem consists of pollen-bearing stamens (male flowers). Each stamen consists of a stalklike filament at the tip of which is the enlarged pollen-producing anther (Fig. 14-9).

The female flowers (pistils) consist of a series of enlarged ovaries ("kernels") arranged on the corncob to form the corn "ear." A long style (the "silk" of corn) is attached to each ovary, and the tip of the style, the stigma, is sticky for receiving the wind-disseminated pollen. A pollen tube, for the conduction of pollen, grows through the style to the ovary. Fertilization takes place within the ovary.

A grain of corn is really a fruit because it consists of a ripened ovary. A mature grain of corn consists of an outer pericarp firmly fused to the seed coat beneath. On the concave side of the grain, beneath the pericarp, is the embryo, which is embedded in the extensive endosperm (food). The endosperm is composed of three parts: (1) a single layer of cells next to the nucellus, called the aleurone layer; each cell in the layer being filled with grains of protein known as aleurone (a lu' ron) (Gr. *aleuron*, flour); (2) an inner, starchy endosperm; (3) an outer, horny endosperm containing proteins.

The embryo (Fig. 14-9) consists of (1) one broad cotyledon for absorption of food from the endosperm; (2) a well-developed plumule consisting of a stem and one or more foliage leaves; (3) a radicle, which forms the primary root of the seedling; (4) a sheathlike coleoptile (ko le -op' til) (Gr. *koleos*, sheath; *ptilon*, feather), which encloses the plumule and (5) a sheathlike coleorhiza (ko le o -ri' za) (Gr. *koleos,* sheath; *rhiza,* root), which encloses the radicle.

Upon germination the radicle breaks through the coleorhiza and forms a temporary primary root. The plumule breaks through the protective coleoptile to form true leaves.

189

Review questions and topics

1 List the distinguishing characteristics of the class Angiospermae, subclass Dicotyledoneae, and subclass Monocotyledoneae, including examples of each.

2 Describe the sporophyte and gametophyte generations in angiosperms, including their relative sizes and independence.

3 Compare the pollination process in angiosperms with that in gymnosperms, such as the pine tree.

4 Describe the structures and functions of the various parts of a complete flower.

5 Describe how seeds are produced and protected in angiosperms.

6 List the chromosome number in gametes, sporophyte, and endosperm. How can you explain the number of chromosomes in each?

7 Describe the formation and function of the pollen tube in angiosperms.

8 Contrast the phenomena of pollination and fertilization, giving examples.

9 Discuss the so-called double fertilization process and its significance. Contrast this with fertilization in gymnosperms.

10 Define heterospory in angiosperms, including examples.

11 In a botanical sense contrast: ovary and fruit, ovule and mature seed, giving examples.

12 Discuss the embryologic origin of roots.

13 Discuss the various functions of roots.

14 Discuss the principal external differences between roots and stems.

15 Describe the structure of typical stems, including the functions of different tissues.

16 Contrast herbaceous dicotyledonous stems with herbaceous monocotyledonous stems, including the significance of the differences.

17 Discuss the functions of various plant stems.

18 Define such terms as herbaceous, woody, vascular bundle, annual rings, wood rays, heartwood, sapwood, latex, and "viscose" (liquid cellulose).

19 Describe leaves, including the functions of the different tissues.

20 Describe the various kinds of leaves, including examples of each kind that you may be able to collect.

21 Define such terms as stomata, guard cells, palisade layer, spongy layer, veins, xylem, phloem, leaf trace, and chloroplast.

22. Describe the structure of a typical flower.

Selected references

Cronquist, A.: The evolution and classification of flowering plants, Boston, 1968, Houghton Mifflin.

Esau, K.: Plant anatomy, New York, 1965, John Wiley & Sons, Inc.

Evans, L. T. (editor): Environmental control of plant growth, New York, 1963, Academic Press Inc.

Foster, A. S., and Gifford, E. M.: Comparative morphology of vascular plants, San Francisco, 1959, W. H. Freeman & Co., Publishers.

Grant, V.: The fertilization of flowers, Sci. Amer. June, 1951.

Gundersen, A.: Families of dicotyledons, Waltham, Massachusetts, 1950, Chronica Botanica Co.

Johansen, A.: Plant embryology, Waltham, Massachusetts, 1950, Chromica Botanica Co.

Mangelsdory, P. C.: Wheat, Sci. Amer. July, 1953.

Swain, T. (editor): Chemical plant taxonomy, New York, 1963, Academic Press Inc.

chapter fifteen

Biology of higher plants

As in all living things, the various metabolic activities of plants are in a state of constant change. The rate and kind of activity vary with internal and external factors, such as nutrition, growth, age, light, and water supply. All these factors form the subject matter of that area of biology called plant physiology. Some of these activities are considered in this chapter.

PLANT COLOR

Plant colorations are caused by a number of pigments that perform a variety of functions. In general they may be grouped as (1) the chlorophylls; (2) the carotenoids; and (3) the anthocyanins. Some of the characteristics of these pigments are shown in Table 15-1.

Plastids are present in the cytoplasm of many plant cells and may be classified as (1) chloroplasts, which are bodies that contain chlorophyll; (2) chromoplasts, which are bodies of various shapes that contain nonchlorophyll pigments and are usually yellowish; and (3) leukoplasts (Gr. *leukos*, white), which are colorless, widely distributed, and associated with starch production.

Carotenoids are present in chromoplasts, either with chlorophyll or without it. The most abundant

Table 15-1
Pigments in higher plants

Pigment	Formula	Present in plastids	Color	Location
Chlorophyll a	$C_{55}H_{72}O_5N_4Mg$	Yes	Bluish green	Leaves and other plant parts
Chlorophyll b	$C_{55}H_{70}O_6N_4Mg$	Yes	Yellowish green	Leaves and other plant parts
Carotenes	$C_{40}H_{56}$	Yes	Creamy white to orange red	Associated with chlorophyll
Xanthophylls	$C_{40}H_{56}O_2$	Yes	Pale yellow	Associated with chlorophyll
Anthocyanins	All glucosides $C_3H_6O_5$	No (dissolved in cell sap)	Red, blue, purple, depending on type of anthocyanin	Flower petals, certain leaves (wandering jew and purple cabbage), roots (red beets), fruits (grapes and plums)

carotenoids are carotenes ($C_{40}H_{56}$) and xanthophylls (Gr. *xanthos*, yellow; *phyllon*, leaf) ($C_{40}H_{56}O_2$). Carotenes vary from creamy white to orange red and are present in carrots, sweet potatoes, pumpkins, and leafy vegetables. They are valuable because they are converted into vitamin A by the body. Xanthophylls produce yellow colors and are present in a variety of organisms.

The development of yellow color during the ripening of fruits may involve the formation of carotenoid pigments and the disappearance of chlorophyll, as in the ripening of bananas. The anthocyanin pigments, such as red, blue, and purple, may be present in many parts of plants, for example, in red beets, red radishes, cherries, grapes, plums, asters, geraniums, tulips, and poinsettias. Both anthocyanins and carotenoids may be present in the same flower. In the nasturtium there are yellow carotenoids and red anthocyanins.

The chromoplasts and their pigments play important roles in the coloration of autumn leaves. One group of autumn colors, the purples, blues, and reds, are influenced by anthocyanins, whereas the yellows are produced by carotenoids and xanthophylls. As the quantity of chlorophyll diminishes in the autumn, the associated pigments are revealed. In many woody plants the anthocyanins are so abundant in the autumn that the carotenoids are concealed, forming shades of red and purple.

The production of anthocyanins is determined by the heredity of the particular plant and by such environmental factors as temperature and light. Many woody plants, such as lilacs and privets, develop no anthocyanins, hence their coloration is affected accordingly. Lower temperatures generally influence anthocyanin formation, although actual frosts are not necessary to produce autumn colors. Light stimulates anthocyanin formation, as revealed by the brighter shades on the outer leaves of trees in contrast with the yellow leaves within the same tree. Poison ivy colors may be bright if leaves have been subjected to strong light but yellow if the leaves have been exposed to less light.

The red and purple colors of the leaves of hard maples, oaks, sweet gums, and sumacs are caused by anthocyanins. This coloration is correlated with the presence of sugars. If the main vein of a leaf is experimentally cut in early autumn, the blade beyond the cut may become a deeper red because sugars cannot leave that part of the leaf, thus promoting anthocyanin formation.

RADIANT ENERGY

Radiant energy is the energy possessed by the sun's rays (solar energy). When the electromagnetic waves of sunlight are passed through a prism, a spectrum (L. *spectrum*, image), or bands of different colors, is produced (Fig. 15-1). The visible part of the spectrum extends from the violet, whose wavelengths start at about 400 mμ (1 mμ, that is, 1 millimicron, is approximately 1 one-millionth of a millimeter), through the red, with wavelengths up to 760 mμ. The shorter, visible violet rays at one end of the spectrum grade through the blue, green, yellow, and orange to the longer red rays at the opposite end of the spectrum. At each end of the visible spectrum there are no visible wavelengths or colors, but the rays continue to become shorter beyond the violet end and are called ultraviolet rays, whereas the rays beyond the red end become longer and are called infrared rays. The red rays of the visible spectrum and those of the infrared, are known as heat rays. The various rays of the spectrum produce different effects on living protoplasm. Of the various regions of the visible spectrum, the violet, blue, and red wavelengths are absorbed by chlorophyll, although lesser amounts of the orange and yellow may be absorbed as well. Chlorophyll is prevalent particularly in the palisade layers of leaves, where maximum light is available. The green color of leaves is caused by the transmission of certain light rays to our eyes, whereas other rays are absorbed or transmitted through the leaves. The absorption of certain rays by chlorophyll is essential in manufacturing foods. Certain lower plants, such as the blue-green, the brown, and the red algae, as well as such higher plants as coleus and red cabbage, possess other pigments that mask the chlorophyll.

The color of an object depends on which wavelengths are reflected to our eyes to produce the sensation of color. When all wavelengths of the visible spectrum are reflected, the object appears white; when all wavelengths are absorbed, it appears black (the absence of all color). The specific color sensation produced in our eyes depends on the

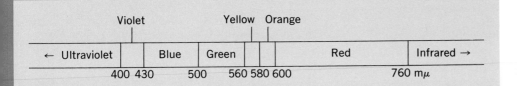

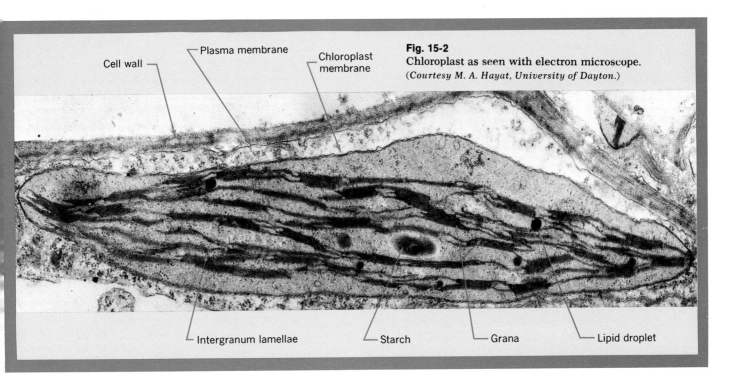

Cell wall — Plasma membrane — Chloroplast membrane

Fig. 15-2
Chloroplast as seen with electron microscope.
(*Courtesy M. A. Hayat, University of Dayton.*)

Intergranum lamellae — Starch — Grana — Lipid droplet

specific wavelengths or combination of wavelengths that are reflected from the object. Green leaves absorb blue and red wavelengths. Green is reflected principally and yellow and orange are reflected partially so that leaves appear greenish to our sense of vision.

PHOTOSYNTHESIS

Most life depends upon the manufacture of foods by chlorophyll-bearing plants through the process of photosynthesis (Gr. *phos*, light; *syntithenai*, to build). The chlorophyll-bearing cells are the world's primary transformers of radiant energy into chemical energy, and thus are the source of foods for the producing plants and also for animals that consume the plants. The immensity of the overall process can be seen in the estimate that a total of

100 billion tons of carbon goes through photosynthesis every year.

Chlorophyll-bearing plants form the essential base of a food chain and are followed in succession by herbivorous animals (plant eating) and carnivorous animals (flesh eating). Omnivorous animals feed on a mixed diet.

Photosynthesis takes place in the chloroplasts, located principally in the inner mesophyll tissues of leaves. The chlorophyll is concentrated in small bodies known as grana (Fig. 15-2) within the chloroplasts. The grana are about 0.5μ in diameter (1μ, 1 micron, is 1/1,000 of a millimeter). Approximately fifty grana all present in each chloroplast in spinach cells. It is thought that each granum contains several million molecules of chlorophyll.

193

In higher green plants chlorophyll consists of two pigments: a blue-green chlorophyll a ($C_{55}H_{72}O_5N_4Mg$) and a yellow-green chlorophyll b ($C_{55}H_{70}O_6N_4Mg$). Chlorophyll is quite similar chemically to hemoglobin, the red pigment of blood, although the latter contains iron instead of magnesium. When chlorophyll is dissolved from a leaf by alcohol or acetone, the solution exhibits a red fluorescence, that is, red by reflected light but green by transmitted light.

The source of energy necessary for photosynthesis is light, either solar or artificial. Only 1 or 2% of the absorbed energy is converted into chemical energy, depending on the species of plant, the intensity of the light, and other factors.

The overall process of photosynthesis is usually discussed under the following three headings: (1) photolysis, or the light reaction, in which hydrogen is split from water and held in readiness for incorporation in a carbohydrate molecule; (2) carbon dioxide fixation, or the dark reaction, in which the interaction of normal cell components, carbon dioxide (CO_2), and hydrogen (H) from photolysis produces the carbohydrate food; and (3) photophosphorylation, in which light energy acts directly with chlorophyll to produce high-energy ATP (adenosine triphosphate).

■ Photolysis

In the light-requiring process of photolysis the absorption of energy by the chlorophyll molecule causes it to become more energetic or "excited." What do we mean by "excited chlorophyll"? When light strikes a chlorophyll molecule, it affects primarily the electrons of the atoms that make up the chlorophyll molecule. All atoms normally have a given number of electrons that orbit around the atomic nucleus at given distances. If such an atom absorbs light of sufficient energy, one of the electrons may be displaced from its normal orbit to a new orbit farther away from the atomic nucleus. This atom is then said to be in an "excited" or more energetic state.

For example, when blue-violet light is absorbed by chlorophyll, some of the energy excites the chlorophyll, and the excess energy of the light dissipates as heat. The atoms of the chlorophyll that trap the light energy remain excited for an incredibly short time. When an electron of an atom has attained a more distant orbit, it quickly goes back to its original orbit. When it reverts from the excited to the normal state, an atom releases the energy that it had absorbed originally. The atom may undergo repeated excitation–de-excitation cycles of this type.

Some unknown mechanism exists that prevents the useless loss of most of the excitation energy, a mechanism that in some manner transfers the available energy from chlorophyll into the chemical process that splits water.

The neatly layered arrangement of chlorophyll in the grana (Fig. 15-2) may be of importance in the splitting of water. The crystal-like structure of intact grana seems to be uniquely associated with the efficient energy transfer in the water-splitting reaction. If the internal structure of the grana is destroyed, it cannot mediate photolysis. If chlorophyll is extracted from the grana, it cannot act as an effective photosynthetic agent.

Thus chlorophyll seems to function as an energy carrier. As the released excitation energy of chlorophyll splits water, de-excited chlorophyll reappears, again available to trap and carry light energy.

As stated previously, the photolytic splitting of water yields hydrogen and oxygen, the latter escaping from the chloroplast into the cytoplasm of cells where it is used in plant respiration or passed into the atmosphere as a by-product. The hydrogen remains in the grana where it combines eventually with carbon dioxide to produce carbohydrates.

Up to the time of this combination the hydrogen is not actually "loose," but it joins temporarily with hydrogen acceptors, which are normal constituents of protoplasm. In photolysis these hydrogen acceptors unite with hydrogen as soon as it is produced when water is split. These acceptors then pass the hydrogen on to other acceptors, and after a series of such transfers they deliver the hydrogen into the carbon dioxide–fixing reactions. This last acceptor is a derivative of a nucleotide and is called NADP (nicotinamide-adenine-dinucleotide-phosphate). As hydrogen is delivered into the carbon dioxide–fixing process, free NADP reappears, again available to accept newly liberated hydrogen. With the formation of NADP · H_2 the photolytic phase of photosynthesis is completed.

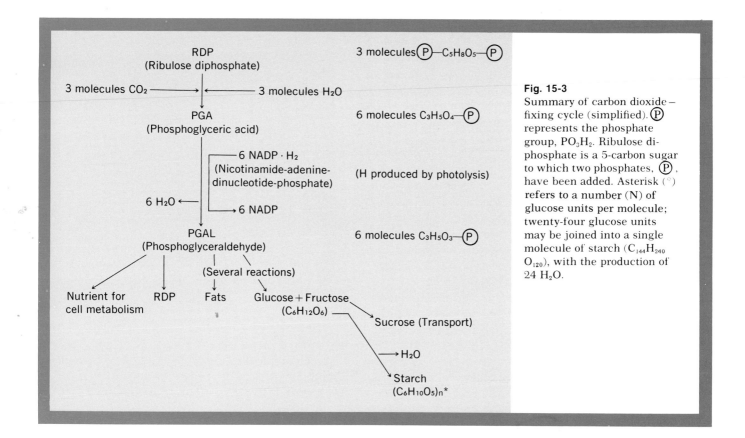

Fig. 15-3
Summary of carbon dioxide–fixing cycle (simplified). P represents the phosphate group, PO_3H_2. Ribulose diphosphate is a 5-carbon sugar to which two phosphates, P, have been added. Asterisk ($*$) refers to a number (N) of glucose units per molecule; twenty-four glucose units may be joined into a single molecule of starch ($C_{144}H_{240}O_{120}$), with the production of 24 H_2O.

We may summarize the activity of photolysis by the following:

$$H_2O + NADP \xrightarrow[\text{Chlorophyll}]{\text{Light energy}} NADP \cdot H_2 + O \uparrow$$

This of course is not a chemical equation, but it does indicate the result of photolysis, namely, gaseous oxygen and hydrogen in a form in which it may enter the next sequence of activity.

Carbon dioxide fixation

Carbon dioxide fixation is often called the "dark reaction," since it is the process in photosynthesis that does not require light. Carbon dioxide fixation is a cycle of reactions to which a steady stream of raw materials is channeled and processed by a series of reactions. A stream of finished products emerges at another point, and some of the reactions along the way return products to the starting point, where they pick up new supplies of raw materials.

The reaction cycle of carbon dioxide fixation consists essentially of three stages or parts: (1) carbon dioxide is brought in; (2) hydrogen enters; and (3) carbohydrate is produced. Each stage or part consists of a series of interconnected reactions, and each reaction of each part must be catalyzed by a very specific enzyme.

The chemicals that form the cycle are all normal constituents of protoplasm and are called phosphorylated carbohydrates. Since they are carbohydrates, their molecules contain carbon atoms linked into chains with hydrogen and oxygen attached to them. Since they are phosphorylated, their molecules also possess one or more phosphate groups, which are derivatives of phosphoric acid (H-O-PO_3H_2). Here, the —PO_3H_2 is the phosphate group. The phosphate group is often abbreviated by —P, and in phosphorylated carbohydrates a —P occupies the place normally filled by a hydrogen atom. Whenever a carbohydrate gains a phosphate group, it loses a hydrogen atom at the same time; likewise, in the loss of a phosphate group a hydrogen atom is gained.

For convenience the carbon dioxide–fixation

195

process can be summarized briefly as shown in Fig. 15-3. Three molecules of ribulose diphosphate (RDP) react with three molecules of carbon dioxide and three molecules of water. The formation of phosphoglyceric acid (PGA) ends the first stage of the carbon dioxide–fixing cycle.

In the next stage hydrogen, which has been produced by photolysis, has been accepted by NADP, making it NADP · H_2. This is the first point in the cycle that requires energy. The transfer of hydrogen from NADP · H_2 to PGA requires energy. Twelve hydrogen atoms, produced earlier by the splitting of water molecules, participate in this reaction. The formation of PGAL (phosphoglyceraldehyde) ends the second stage of the cycle.

Some of the six molecules of PGAL are used for cell metabolism; others undergo a long, complicated series of reactions in which the atoms are rearranged, a process which requires energy. In some cases an additional phosphate group is added (to replace a hydrogen) to form RDP, thus regenerating the starting point of the cycle; that is to say, five 3-carbon chains (PGAL) change into three 5-carbon chains (RDP). Thus three individual carbon atoms supplied as raw inorganic material (CO_2) form a single organic molecule of PGAL.

As before, we may summarize the activity of carbon dioxide fixation by the following:

$$RDP + CO_2 \xrightarrow[H_2O]{NADP \cdot H_2} PGAL$$

Again, this is not a chemical equation, but it does indicate the result of the process, namely, PGAL (phosphoglyceraldehyde), the basic food molecule.

Photophosphorylation

It has been established that a second source of energy-yielding molecule is formed during the light sequence. This involves the action of light energy, inorganic phosphates, cellular nucleotides, a type of molecule similar to chlorophyll called a cytochrome, and ADP (adenosine diphosphate).

The generalized summary is as follows:

$$ADP + -\textcircled{P} \xrightarrow[\text{Nucleotides, Cytochromes}]{Energy} ATP$$

ATP (adenosine triphosphate) will be recognized

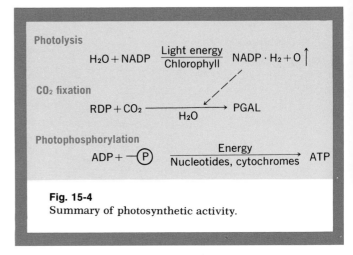

Fig. 15-4
Summary of photosynthetic activity.

as the principal form of energy available to cells for all activities. The activities involved in photosynthesis are shown in Fig. 15-4.

Study of photosynthesis

Plant physiologists have been investigating activities involving carbohydrate formation for many years. While it was known that the final product was a carbohydrate, the sources of atoms in the hypothetical carbohydrate $(CH_2O)_x$ were not fully known. Evidently, the carbon (C) must come from carbon dioxide (CO_2), and the source of hydrogen must be water (H_2O). Where does the oxygen (O) come from to manufacture CH_2O? Does it come from H_2O, from CO_2, or from both?

Through the use of artificially prepared water, in which the oxygen contains an isotope of ordinary oxygen, namely ^{18}O, this question can perhaps be answered. This isotopic oxygen (^{18}O) has an atomic nucleus of mass 18 units, whereas every atom of ordinary oxygen has a mass of 16 units. Thus each ordinary oxygen atom (^{16}O) is sixteen times heavier than a hydrogen atom (which is the lightest of atoms and is given the arbitrary mass 1 unit).

When such artificially prepared water is used, the fate of its labeled oxygen can be traced. The isotopic oxygen (^{18}O) of the prepared water appears only in the oxygen by-product and does not appear in the carbohydrate. Hence, the oxygen of water does not enter into the formation of carbohydrate but is liberated as a by-product of free oxygen.

Therefore, the oxygen source must be from CO_2.

In other words, of the atoms in CH_2O, the C and O come from CO_2, and H_2 comes from water in the input. The other atoms of the input form such by-products as water and molecular oxygen. In normal photosynthesis water is an input raw material as well as an output by-product, but twice as much water must be applied as is actually used. Thus, if water supplies only its hydrogen (and not its oxygen) to carbohydrate production, the hydrogen atoms must be split from the oxygen atoms at some point during photosynthesis.

The hydrogen and oxygen of water are joined together very firmly, so that a great amount of energy is required to split them. In order to find the source of this energy leaves were dried and powdered, and the chloroplasts in the powder were isolated and cleaned. The pure chloroplast preparation was then suspended in water to which certain iron salts were added. These iron salts serve merely as acceptors (absorbers) of hydrogen. The mixture was then illuminated. As long as the light was on, the chloroplasts actively evolved bubbles of oxygen.

Isotopes have also been used to investigate the mechanics of CO_2 fixation. Artificially prepared CO_2 was experimentally given to photosynthesizing plants. The carbon atom of the prepared CO_2 was an isotope of ordinary carbon. The latter is ^{12}C, and the isotope used was ^{14}C, which was 2 mass units heavier. Plants using the isotope CO_2 as raw material were killed at successive times during photosynthesis, and their cells were analyzed for the isotope ^{14}C. Since the plants manufactured substances from the isotope CO_2, this procedure could reveal the sequence of reactions leading to the formation of the end product as well as identify the photosynthetic end product.

ABSORPTION AND TRANSLOCATION

Many materials necessary for the various activities of plants must be absorbed by the roots and particularly by the root hairs. The absorption of water and of substances dissolved in water involves several complex phenomena, including imbibition (L. *imbibere*, to drink in), osmosis, and diffusion.

Imbibition is the process by which solid particles, chiefly colloidal particles, absorb liquids and increase in volume. Plant cell walls absorb water by this process, much like a dry sponge soaks up water and swells. Colloidal materials of plants, such as cellulose, the proteins of the protoplasm, pectic substances, and other organic compounds, imbibe large quantities of liquids.

Diffusion is the spreading out of the molecules of a substance from places of greater molecule concentration to places of lower concentration. When the cork is removed from a bottle of perfume, the molecules of the perfume, because of their ceaseless movements, tend to diffuse throughout the air in the room. Likewise, the molecules of a particle of dye in a beaker of water tend to diffuse throughout the water, and the molecular actions of materials assist in their diffusion in plant roots.

Osmosis is the diffusion of water through a differentially permeable membrane from a region of high water concentration to a region of lower water concentration. Osmotic pressure is the maximum pressure that can be developed in a solution separated from pure water by a rigid membrane that is permeable only to water. Ordinarily, the osmotic concentration of the cell liquids of root epidermal cells is higher than that of the soil solution. Thus osmosis seems to permit continued absorption by osmotic action, for water usually diffuses from a region of low osmotic concentration of solutes (or high concentration of water) toward a region of high concentration of solutes (or low concentration of water).

Ion exchange occurs between soil particles and the adhering root hairs (Fig. 15-5) without entering into solution in the soil. Root cells absorb solute ions from the soil solution and also from the surfaces of soil particles as a result of the exchange of cations (electropositive ions).

Both osmosis and imbibition assist in the transfer of water and solutes from cell to cell across the cortex of the root to the xylem cells of the central stele. This process is accomplished in part by simple diffusion and in part by active absorption of solute molecules and ions. In the region of the root hairs the cells are constructed to permit the transverse passage of liquids, but in other regions the presence of a waxy substance, suberin, partly prevents the free passage transversely, so that the liquids seem to be guided upward toward the stem. Water and solutes are conducted through the xylem

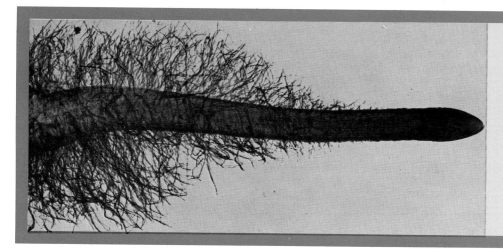

Fig. 15-5
Tip of a root, showing the root cap and root hairs.
(Courtesy General Biological Supply House, Inc., Chicago, Illinois.)

Fig. 15-6
Detail of a sieve tube with material moving through a pore in a sieve plate.
(Courtesy D. R. Geiger and D. Cataldo, University of Dayton.)

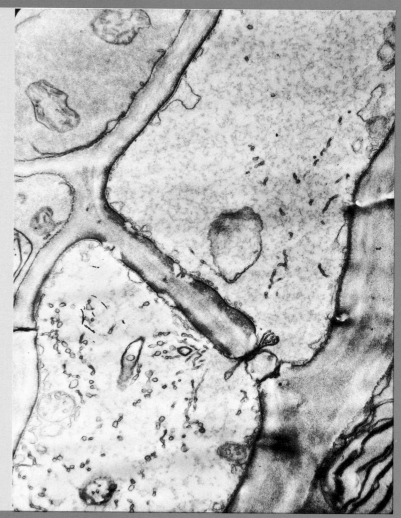

tissues of the root, stem, and veins of the leaves. Materials in the leaves, including manufactured foods, are conducted downward through the phloem tissues of the leaves, stems, and roots. In woody stems wood rays conduct materials horizontally (radially) through bands of parenchymatous tissue, hence, they are often called vascular rays.

Water is important to plants because (1) the colloidal constituents of protoplasm are dispersed in water; (2) water and carbon dioxide are used to manufacture food by photosynthesis; (3) solid materials are usually dissolved in water before they can enter or leave a cell or be moved from one part of a plant to another; (4) it provides internal pressure to maintain form and give support to the plant; and (5) it is the chief medium in which most of the chemical reactions of living protoplasm occur.

Water and organic solutes, such as sugars and amino acids, move through the vascular tissues of the phloem and xylem. The transport of these organic compounds is called translocation.

In general the direction of solute movement is governed by the metabolic requirements of the plant rather than by the strict morphology of the conducting system. Prior to the full expansion of a growing leaf, sugars and amino acids move into its tissues. After maturity the leaf is an exporting organ. Most of its photosynthetic product is translocated in the form of sucrose to such regions as young leaves, shoot tips, root tips, and young fruits. This movement is primarily through the sieve tubes of the phloem (Figs. 15-6 and 15-7). At any one time different phloem elements of the same vascular bundle may be transporting solutes in different directions.

During translocation sucrose may move across the sieve tubes into the parenchyma or cortex cells, where it is stored as starch. Analysis of the sieve tube contents may reveal a sucrose content of up to 25% as well as much smaller quantities of sugar alcohols and amino acids.

Xylem sap contains significant amounts of nitrogenous compounds, especially amino acids. The main movement of xylem sap is from the roots to stems, leaves, and fruits.

The translocation (conduction) of liquids and organic solutes through the stem is a complex phenomenon that is incompletely understood.

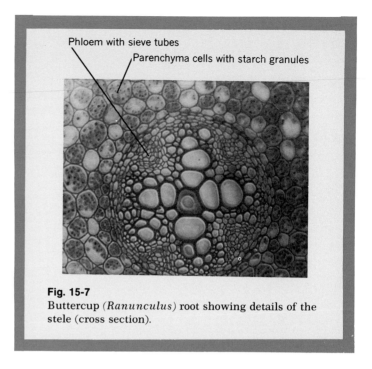

Phloem with sieve tubes

Parenchyma cells with starch granules

Fig. 15-7
Buttercup (*Ranunculus*) root showing details of the stele (cross section).

Water and nitrogenous compounds are conducted upward largely by the xylem tissues. Proof of this can be seen by experimentally cutting away the phloem and other tissues external to the cambium without appreciably decreasing the upward flow of the water and solutes. Likewise, by experimentally severing the xylem of the stem the leaves wilt, and the aerial parts of the stem usually die. Great quantities of water evaporate from the aerial parts of plants. In addition water is used in food manufacture and growth so that a supply must be conducted upward almost constantly. The rate of ascent is influenced by many external environmental factors, including humidity, air temperature, and light intensity. Water absorbed by roots may ascend in the wood of a stem sometimes as fast as 4 feet per hour.

Among the various theories that have been proposed to account for the ascent of liquids in stems are the following examples. It can be observed that liquids will exude from the stumps of some plants and from the cut branches or tapped sapwood of such trees as sugar maples. The force that causes this so-called bleeding is root pressure. Root pressure may assist in the ascent, but it is present in great amounts only during the spring and thus is lacking or only a minor factor when the ascent is greatest. Likewise, cut stems, when

199

placed in water, are able to translocate water upward even though no roots are present.

Another theory to explain the ascent of liquids in plant stems involves the force of atmospheric pressure. Atmospheric pressure can force up and maintain a column of water approximately 33 feet high (at sea level). It can do this only if the column of water is unbroken, if the atmospheric pressure is exerted directly on the surface of the water at the base of the column, and if there is a vacuum at the tip of the column. Evidence against this explanation includes the absence of an actual vacuum in the leaves and the fact that the water column is actually broken by cross walls in the conducting channels of stems. Since liquids rise for hundreds of feet in some trees, atmospheric pressure is at best only a minor factor in the ascent.

When glass tubes of very small diameter are placed vertically with their lower ends in water or similar liquids, the liquid will rise in the tubes to a level above that in the container in which they are standing. The smaller the diameter of the tube, the higher the level to which the liquid rises. This rise is caused by capillarity, or surface attraction, between the molecules of the liquid and those of the tube substance. This force of capillarity has been thought to be involved in the ascent of liquids in stems because the hollow tracheids and xylem vessels are somewhat comparable to the tubes just described and function in a similar way. Possibly, capillary forces may assist in the ascent of liquids, but they could cause a rise of only a few feet at the most.

Another important explanation for the ascent of liquids is the cohesion theory, based on the presence of several known factors and forces in stems. As water evaporates from the cells of leaves and other aerial parts of plants, the colloidal materials of their protoplasm are somewhat dried and an increase of osmotic concentration in those cells occurs. Because of the powerful attraction of the colloidal particles in protoplasm for water, the partially dried protoplasm of these cells absorbs water, chiefly by colloidal imbibition and partly by the osmotic forces of the leaf cells, from adjacent cells that have a higher water content. Thus cells with less water pull water from cells with more water, so a pull on the water forms in the tracheids and vessels of the leaf veins. This pull is extended through the xylem cells of the leaf stalk through the xylem tissues of the stem and roots. It is transmitted through the long columns of water because of the enormous cohesive power of water molecules, which remain together with such great mutual attraction for each other that enormous power is required to separate them. Thus a tremendous tension or "negative pressure" in the water columns is started by water evaporation in the leaves, resulting in definite decreases in the diameter of stems. A strong upward osmotic pull in the leaf cells will raise a tall column of water, such as that found in the conducting xylem of wood, because of the high cohesive power of a thin column of water.

Thus the absorption of water and solutes from the soil into the xylem tissues of roots is performed by the living cells of the root epidermis and root cortex so that water in roots is available for translocation upward through the stem to the leaves. If air bubbles enter the conducting vessels, the water column is broken and the upward rise is stopped. If the air bubble is eventually dissolved, the water may resume its ascent. This is the reason for cutting off, under water to prevent the entrance of air, an inch or more of the lower ends of cut flower stems. This cutting eliminates air bubbles that may have entered the stems, and it allows the ascent of water to prevent wilting of the cut flowers.

TRANSPIRATION AND GUTTATION

Transpiration (L. *trans*, beyond; *spirare*, to breathe) is the loss of water vapor from aerial parts of plants, particularly through the stomata of leaves. Guttation (L. *gutta*, a drop) is the exudation of water from plants, usually in the form of drops. The early morning dew is often water from guttation. Of the amount of water absorbed from the soil, a plant uses only a fraction for its various activities, but much escapes from the leaves and stems as water vapor. A very high percentage of transpiration occurs through the leaves.

The structures of leaves are directly or indirectly involved in transpiration. Often leaves are broad, thus exposing large surfaces to the air. The presence of stomata in the lower and upper surfaces permits the escape of water vapor. The number of stomata per unit area of leaf varies with the species and is influenced by the age of the leaf and certain environmental factors. Usually, the lower surface has many more stomata than the upper. In fact, some plants, such as apple, oak, and orange, have no stomata in the upper surface but have thousands in the lower. The water lily, on the other hand, has stomata only in the upper surface. The bean leaf has 4,000 stomata per square centimeter in the upper surface and approximately six times that number in the lower. Corn has a ratio of 6,000 to 10,000, and the sunflower has a ratio of about 8,000 to about 16,000.

The mesophyll tissues inside the leaf are surrounded by many intercellular spaces that connect with large spaces just beneath the stomata. Hence large areas are exposed and therefore can pass much water into the air. Water lost from the cells is replaced by the movement of water from adjacent cells and eventually from the leaf veins.

Since much of the transpiration occurs through the stomata and since the rate of water loss even in the same plant varies from time to time, the relative sizes of the stomata are of significance, even though these openings occupy less than 1% of the leaf surface. In general the stomata are closed at night and open in the day, although in low light intensities they may close during the day.

The most influential environmental factors that affect the transpiration rate include the humidity and temperature of the air, air currents, the intensity of the light, and the amount of water in the soil. If the humidity of the surrounding air is high, the amount of water evaporation will be reduced. Air movements may also remove the water vapor from around the leaf, thus affecting the rate of transpiration. Because of a lack of water in the soil, if there is a minimum of water circulating in a plant, the amount of transpiration might be affected.

Although most water is passed from plants as a vapor, it may leave in liquid form by the process of guttation, in which the drops of water are confined to the margins and tips of the leaves. Guttation usually occurs when water loss by transpiration is less than the amount being absorbed rapidly by the roots. Such is the case when warm days are followed by cool nights. Probably osmosis plays an important role in guttation. This phenomenon can be demonstrated experimentally in such plants as corn, potato, tomato, cabbage, nasturtium, and grass, especially if environmental conditions are right and there is sufficient absorption by the roots.

RESPIRATION AND DIGESTION

Respiration

Respiration is a chemical oxidative process whereby all living protoplasm breaks down certain organic substances with the release of energy necessary for various activities. Through respiration the common sugar, glucose, is oxidized into simpler substances with the release of energy as summarized by the following:

$$C_6H_{12}O_6 + 6\ O_2 \rightarrow 6\ CO_2 + 6\ H_2O + Energy$$

From the standpoint of energy loss or gain the reaction in the process of respiration is the opposite of photosynthesis. The process of respiration involves a series of complex stages, the use of a number of enzymes, and the formation of several intermediate compounds. Only the basic principles are summarized here.

The fundamental differences between photosynthesis and respiration are shown in Table 15-2.

Energy is the capacity to do work. The energy liberated in respiration is used in part to form heat and in part for such plant functions as the flowing of living protoplasm; the movement of cell parts; the process of assimilation, whereby food is converted into living protoplasm or its products; the synthesis of fats and proteins from sugars; the accumulation of solutes from the soil; and the synthesis of pigments, vitamins, enzymes, oils, resins; organic acids, and tannins.

Perhaps all chemical reactions within living cells are controlled by organic catalysts called enzymes. In respiration the action of certain enzymes results in the oxidation of glucose with the liberation of energy and the production of carbon dioxide and water. During the day photosynthesis usually proceeds at a rate several times that of respiration, so that foods and energy are stored. At night only respiration occurs, and the plant uses food and oxygen and liberates carbon dioxide.

Table 15-2

Fundamental differences between photosynthesis and respiration

Photosynthesis	Respiration
1. Occurs only in chlorophyll-bearing cells of plants	1. Occurs in all living cells of plants and animals
2. Occurs only in presence of energy-supplying light	2. Occurs throughout life of cell, without light
3. Releases O_2 as by-product	3. Releases CO_2 and H_2O as end products
4. Uses CO_2 and H_2O	4. Uses O_2 and food
5. Solar (radiant) energy converted into chemical energy and stored	5. Energy released by oxidation, chemical energy being converted into heat and useful energy for various plant activities
6. Results in weight gain	6. Results in weight loss
7. Foods manufactured	7. Foods broken down

The type of respiration considered so far occurs when a sufficient supply of free oxygen is available and is termed aerobic respiration (Gr. *aer*, air; *bios*, life). Under certain conditions respiration may occur in the absence of free oxygen and is called anaerobic respiration (Gr. *an*, without; *aer*, air; *bios*, life). Both types of respiration are controlled by specific enzymes and release energy. Anaerobic respiration does not release as much energy or form the usual end products of carbon dioxide and water, but it produces a number of compounds, including alcohol, carbon dioxide, and organic acids. In fact the commercial productions of certain products are obtained from anaerobic respiration. An example of anaerobic respiration is shown by the following equation, in which several stages are omitted:

$$C_6H_{12}O_6 + \text{Specific enzymes} \rightarrow$$
Sugar (in the absence of free oxygen)

$$2\ C_2H_5OH + 2\ CO_2 + \text{Energy}$$
Alcohol

Digestion

Insoluble stored foods, such as starch or protein, can neither be used in the cells in which they are stored nor translocated to other parts of plants until they are changed into a soluble, diffusible form through enzymatic actions. This process is known as digestion (L. *digestio*, from *dis*, apart; *gerere*, to carry).

One of the most common digestive activities in plants is the change of starch into sugar through the catalytic activity of a specific enzyme called amylase (am' i lase) (L. *amylum*, starch) with the production of maltose. Nearly all digestive enzymes change a complex compound into one or more simpler compounds through a reaction with water, which is known as hydrolysis (Gr. *hydro*, water; *lysis*, dissolve). Digestion of starch may be shown as follows, with several stages and intermediate carbohydrates being omitted:

$$2(C_6H_{10}O_5)_n + n(H_2O) + \text{Amylase} \rightarrow n(C_{12}H_{22}O_{11})$$
Starch Maltose

In plant tissues maltose is hydrolyzed by another enzyme, maltase, to form glucose, as shown by the following equation:

$$C_{12}H_{22}O_{11} + H_2O + \text{Maltase} \rightarrow 2\ C_6H_{12}O_6$$
Maltose Glucose

Many other enzymes digest various types of carbohydrates in living plants.

Oils and fats in plants are digested by the enzyme, lipase, to form glycerin (glycerol) and fatty acids, such as palmitic, oleic, and linoleic.

Proteins are hydrolyzed into amino acids by enzymes known collectively as proteases. The process occurs in several stages, and intermediate products are formed. Proteases also catalyze the conversion of amino acids into proteins. There are many plant proteases, including bromelin in fresh pineapple, which liquefies (digests) the gelatin in desserts, and papain in papaya, a tropical fruit. Papain digests proteins and is used in tenderizing meats, in surgery, and in the tanning and brewing industries.

PLANT GROWTH FACTORS

There are many internal factors in plants that independently or in conjunction with each other influence their growth and development. These are so numerous and complex that only a general outline can be given here.

Inherited traits

The genetic traits of an organism basically determine its structures and activities. Changes in the number and construction of the chromosomes have important effects. Plants with tetraploid chromosomes (four sets) often have larger organs with larger cells and wider leaves and fruits than diploid plants of the same species. Sometimes the differences in size and shape of a fruit (such as disk and spherical squashes) may result from a single gene difference, whereas in other plant phenomena the interaction of several genes may be responsible. Such internal structures as cell shape and size, the character of the cell wall, and the plane of cell division are influenced by heredity. Traits that result from inheritance may influence the development and activities of that plant.

Hormones

Growth substances may be formed in one part of an organism and in some cases transported to other parts where they produce their effects. Such substances have a variety of names, including auxins, gibberellins, phytohormones, growth regulators, and florigens ("flower-forming" substances). So-called wound hormones are produced at the sites of injuries, where they stimulate mitosis and the formation of protective tissues. In all cases the hormones are effective in exceedingly small quantities.

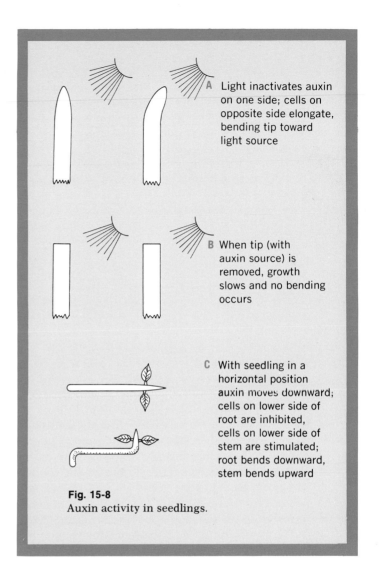

A Light inactivates auxin on one side; cells on opposite side elongate, bending tip toward light source

B When tip (with auxin source) is removed, growth slows and no bending occurs

C With seedling in a horizontal position auxin moves downward; cells on lower side of root are inhibited, cells on lower side of stem are stimulated; root bends downward, stem bends upward

Fig. 15-8
Auxin activity in seedlings.

Auxins

Auxins (Gr. *auxein*, to increase) are growth substances of plants that have been studied extensively. They are formed chiefly in young, physiologically active parts of plants.

Auxin is concerned with phototropic (light) as well as geotropic (gravity) changes. In young oat seedlings or seedlings of other grains, the young shoot is surrounded by a light-sensitive, sheath-like coleoptile. This coleoptile will bend toward light because the cells on the darker side elongate more rapidly than those on the lighter side. If about 1 mm. of the tip of the coleoptile is experimentally removed, no bending toward light results (Fig. 15-8). When the cut tip is replaced on the stump,

203

Fig. 15-9
Demonstration of the effects of applying indoleacetic acid to the plant *Kalanchoe*. The control plant at left had only lanolin applied, whereas the plant at right had indoleacetic acid in lanolin applied. The indoleacetic acid stimulated the plant to grow toward the right, to be followed by normal upright growth several hours later.

light sensitivity is restored. This growth substance can be extracted from the tip and applied to the stump with similar results. The auxin passes in only one direction from the tip toward the base of the coleoptile, showing what is called physiologic polarity.

Several types of auxins are organic acids of the indoleacetic acid group and are produced by living protoplasm. Certain artificially synthesized substances, such as indolebutyric acid and naphthaleneacetic acid, are able to produce effects similar to those of the natural growth substances (Fig. 15-9).

Some types of plants produce fruits even though pollination and fertilization do not occur, a phenomenon known as parthenocarpy (par' the no kar pi) (Gr. *parthenos*, virgin; *karpos*, fruit). Such a fruit is usually seedless, and the fruit seems to maintain an auxin content that is sufficiently high to continue growth. Parthenocarpy is found in seedless grapes, in some kinds of citrus fruits, and in cultivated varieties of banana and pineapple.

Seedless fruits of tomato, squash, cucumber, and watermelon may be produced artificially by using synthetic auxinlike growth substances. The use of plant hormones to increase fruit yield when plants are not easily pollinated may be of practical value.

Root formation is also stimulated by auxin. More roots will form on a cutting if leaves or buds are present because of the production of growth substances. If the basal end of a cutting is treated with the proper growth substances for a few hours before planting, more abundant roots will form.

In some plants the development of shoots is caused by an auxin deficiency rather than by any shoot-forming substance. The phenomenon of dominance and inhibition is associated with growth substances. When a terminal shoot or bud is present, the buds below may often fail to develop, but they will do so after the inhibiting terminal tip is removed. Evidently the tip secretes a substance that moves downward to inhibit bud development below.

Auxin is also influential in certain geotropic changes. Roots placed in a horizontal position tend to bend downward. Auxin checks the elongation of root cells, so that the upper surface of such roots with less auxin grows faster. On the other hand, shoots placed in a horizontal position tend to bend upward because gravity causes auxin to accumulate on their lower side, where it stimulates elongation. The differences between root and shoot reactions to auxin probably result from roots being stimulated by weak auxin concentrations but being inhibited by the stronger ones that stimulate the growth of stem cells.

Gibberellins

Other plant growth hormones, accidentally discovered in the last decade of the nineteenth century, are the gibberellins. Japanese rice farmers observed some greatly elongated seedlings, which never lived to maturity and only rarely produced flowers. The condition was named "foolish seedling" disease. It was discovered in 1926 that these seedlings were all infected with *Gibberella*, a fungus that could be transferred to healthy plants, thereby producing the disease. If grown in a nutrient medium in a flask, the fungus produces a substance that causes typical overgrowth symptoms when applied to a receptor plant. When this gibberellic acid is applied to plants, enormously elongated stems result, and in some instances the leaf area is reduced. The fungus also produces prompt stimulation of flowering in those plants known as "long day plants." In the grape industry applications of gibberellins to seedless grape clusters results in the retention and development of a greater number of large-sized grapes. Applied to celery, gibberellins produce larger, more succulent plants in a shorter time.

Other hormones

Many other substances, most of which are not completely understood, have been discovered and are being investigated. These include the kinins, which appear to promote cell division, and the florigens, which promote flower formation.

ABSCISSION

Abscission (L. *abscissus*, cut off) is the separation of plant parts, such as leaves, stems, flowers, and fruits, usually by the natural dissolution of certain cell walls of the so-called abscission layer. The leaves of deciduous plants fall at the end of their growing season. The pectins in the cell walls of the abscission layer are dissolved through the action of enzymes, partially detaching the petiole from the stem and leaving only the vascular bundles to hold the leaf until it eventually falls. A layer of cork tissue develops at the petiole base for the protection of the exposed tissues. Because of tough vascular bundles and lack of an abscission layer, leaves may remain for some time, as in the case of red oak leaves. Pectic substances are gelatinous and viscous and tend to hold cells together. In evergreens the leaves remain for several seasons (4 years or more), but new leaves are produced to replace those that are shed periodically.

Abscission may be demonstrated experimentally by cutting away the blade portion of a leaf. In a few days the remaining part of the leaf will fall. However, if a plant hormone, such as indoleacetic acid, is applied to the cut surface of the petiole, the abscission layer will not form, and the petiole will remain on the plant. The abscission of flowers and fruits is also controlled in part by auxins. If certain immature fruits are sprayed by growth substances, their fall is often delayed.

Among probable causes of abscission are decreased amounts of light available, decreased production of plant growth hormones, and decreased soil moisture.

LIGHT AND FLOWERING

Light affects different plants in various ways. Nonchlorophyll-bearing plants do not require light, some even being affected adversely by it. On the other hand, chlorophyll-bearing plants require certain quantities and qualities of light for photosynthetic purposes. Some lower plants, such as algae and certain types of bacteria, have pigments (other than chlorophyll) that are affected in different ways by light.

In addition to its role in photosynthesis light influences the development and functions of many parts of plants. The direction and intensity of light influence the amount and direction of the growth of stems and leaves. In darkness plants tend to have long, weak stems, poorly developed leaves, and tissues without chlorophyll, a condition called etiolation (Fr. *étioler*, blanch). Leaves growing in relatively bright light tend to have thicker epidermis and palisade layers and smaller intercellular spaces. Very intense light, particularly the ultraviolet rays, affect the protoplasm so that the structure and position of plant parts may be modified to subject as

205

little of the plant as possible to the harmful rays. In general, light of longer wavelengths (toward the red end of the spectrum) seems to affect plants in much the same manner as if they were grown in the dark. Blue light (at the opposite end of the spectrum) seems to have the opposite effect, increasing tissue differentiation.

A plant hormone, florigen (L. *flos*, flower; *gignere*, to produce), develops in the leaves of some plants and controls their flowering. In some asters, if the leaves of one branch are covered for a few hours each day during the early part of the summer, the entire plant will come into flower and will continue to flower through the autumn when the control plants normally do so. The black covering must be in place for the number of hours that would reduce the length of the longer summer day to the same length as the shorter autumn day when the plant ordinarily begins to flower.

On the basis of the effect of light on flower production, plants may be divided into (1) "long day plants," such as red clover and spinach, which form flower buds as soon as the day length exceeds 12 hours; (2) "short day plants," such as poinsettia, aster, dahlia, cosmos, and chrysanthemum, which form flower buds only when the day length is fewer than 12 hours; and (3) "indifferent plants," including many types, which appear to be unaffected by day length.

By withholding light for part of each day "short day plants" can be induced to flower entirely out of season. By providing artificial light for an extra period at the end of the day they can be prevented from flowering, even though they are mature. "Long day plants" respond in the opposite way to artificially controlled periods of illumination. If poinsettias are desired at Christmas, they are placed in the dark until noon each day, starting in November, and then return to the light all afternoon to allow for sufficient photosynthesis. Such responses of plants to periodic light and dark conditions are phenomena of photoperiodism (Gr. *phos*, light; *peri*, around; *odos*, way).

Temperature

Plants are constantly influenced by variations in the temperature of the air and soil. Most of them grow best with an optimum temperature of 70° to 90° F. Plant activities tend to decrease as temperatures approach freezing or 100° F. The maximum, optimum, and minimum temperatures for various plants and their activities vary with species, age, water content, and inherited abilities. In general, seeds with low water content tend to resist extremes of temperature better than actively growing plants. Certain seeds and spores have existed for months in a temperature as low as −226° F. and for shorter periods at temperatures as high as 220° F. (boiling point of water is 212° F.). Certain bacteria and blue-green algae live in the water of hot springs just below the boiling point.

If germinating seeds of winter wheat are exposed to low temperatures for a time, the developing plants will flower earlier than normal. By exposing an embryo to a low temperature during early seedling development the early stages of development seem to be accelerated and the development time shortened. When such a young seedling emerges, it is in a more advanced stage of development than if it had developed at a higher temperature. The explanation and use of this phenomenon may have far-reaching effects in future plant developments.

Growth movements

Unequal growth rates in different parts of a plant organ may occur. These include the following. (1) Nutations (L. *nutare*, to nod) are nodding movements of the growing apices of certain organs. Different growth rates are internally induced on different sides of the growing apex, as shown in the movements of the epicotyls. (2) Twining movements are discernible, spiral movements of the growing apex of stems, for example, the twining of morning glories, hops, and beans. (3) Nastic movements (Gr. *nastos*, pressed close) are movements of such flattened organs as petals, leaves, and bud scales, in which one surface grows faster than its opposite, thereby resulting in the closing or opening of an organ.

Tropisms (Gr. *trope*, a turning) are rather slow bending movements of cylindrical organs, such as stems, leaf petioles, flower stalks, and roots, in response to external stimuli. Growth occurs more rapidly on one side than on the opposite, which results in a bending that usually tends to place the structure in a more favorable position as far as that stimulus is concerned. Tropisms are common in fungi, bryophytes, ferns, and seed plants.

Among the common tropisms caused by an unequal distribution of auxin because of a specific stimulus are (1) phototropism (Gr. *phos*, light), which is a light-induced growth movement that is usually positive for aerial parts, such as leaves and stems, and places them in advantageous positions; (2) chemotropism (Gr. *chemeia*, chemical), which is a growth reaction to chemical substances; (3) hydrotropism (Gr *hydro*, water), which is a growth reaction to water whereby roots obtain necessary supplies; (4) thigmotropism (Gr. *thigma*, touch), which is a growth reaction to contact stimuli; and (5) geotropism (Gr. *ge*, earth), which is a growth response to gravity that is positive when roots grow toward soil and negative when stems bend away from a horizontal position.

Turgor movements

Turgor movements are caused by a change in the water pressure (turgor pressure) of certain tissues of an organ and do not result from growth rate differences, as previously described. In general, turgor movements are temporary and often are of little significance to the plant. Many are rapid, in some cases occurring in fractions of a second.

Some of the more common types of turgor movements are contact movements, rapid movements, and so-called sleep movements. Contact movements are induced by contact and pressure stimuli, such as displayed by leaves and flower parts. The contact turgor movements of the two parts of a Venus's flytrap leaf result in a closing and possible trapping of an insect. The rapid turgor movements of certain flower parts are beneficial in efficient insect pollination, in some cases covering the stimulating insect with pollen.

Another rapid turgor movement, displayed by the reactions of the leaves of the sensitive plant *Mimosa*, may be initiated by such stimuli as contact and temperature change. Other adjacent leaves react successively as the excitation chemicals (possibly auxin) are transported from the original point of stimulation. The leaves return to normal after several minutes (see Fig. 5-2).

The closing, opening, and change in size of the stomata of plants are caused by changes in the surrounding paired guard cells. The internal water pressure in cells is referred to as turgidity (L. *turgere*, to swell). When guard cells are filled with water, the stomata open, and a decreased water content causes them to close. This is possible because the differences in the thickness of their cell walls cause the guard cells to change their shapes.

The changes in the turgor of guard cells are usually correlated with the ratio of starch to sugar in them. Starch grains are formed in the chloroplasts of the guard cells. When light strikes the guard cells, the starch molecules are converted into many sugar molecules through enzymatic action, thereby increasing the number of particles in the liquid. This decreases the water concentration of the guard cells, thus permitting an osmosis of water from surrounding cells. This additional water increases the guard cells' turgor, causing the stomata to open. In the reverse process, when light is less intense, the sugar in the guard cells is converted into starch. This causes a greater concentration of water than in adjacent cells, thus permitting water to move out of the guard cells with a loss of turgor, which causes the stomata to close.

The so-called sleep movements are turgor reactions induced in many plant leaves because of changes in light intensity (nyctitropism). Such changes in leaf positions are shown by white clover leaflets and garden beans, which assume a somewhat horizontal position in bright light and a vertical position in diminished light.

LIFE CYCLES IN PLANTS

Life cycles are characteristic of living organisms. An organism starts at a certain stage in its development and, after passing through successive stages, eventually reaches that same stage again. Organisms that reproduce sexually go through a cycle, such as egg, embryo, adult, and egg. In most plants that reproduce sexually this cycle is complicated by having two parts, a gamete-producing gametophyte generation and a spore-producing sporophyte generation. The alternation between these two generations is the alternation of generations and corresponds with the change in the number of chromosomes in the two stages of the life cycle, as described earlier.

Alternation of generations is found in certain

spores = specilized cells for reproduction
proces called : sporugenosis

Table 15-3

Showing relative traits of certain sexually reproducing plants that possess an alternation of generations

Plant	Sporophyte (2N)	Gametophyte (N)
Liverworts	Less conspicuous; dependent on gametophyte	Usually prominent
True mosses	Usually less conspicuous; dependent on gametophyte	Usually prominent
Ferns	Prominent and large when mature; independent of gametophyte	Small and less conspicuous; independent of sporophyte
Gymnospermous plants	Prominent and large when mature; independent of gametophyte	Small and inconspicuous; dependent on sporophyte
Angiospermous (flowering) plants	Prominent and large when mature; independent of gametophyte	Small and inconspicuous; dependent on sporophyte

gametes = sex cells
♀ + ♀ = zygote

= size gametes = isogametes
≠ un = size gametes = heterogametes
when heterogametes unite — gametogenesis happens

algae, liverworts, true mosses, horetails, club "mosses," ferns, gymnosperms, and angiosperms (flowering plants). In many colonial green algae the plant body is a haploid (N) gametophyte that gives rise to haploid gametes that fuse to form the sporophyte generation. In certain brown algae the diploid generation is rather conspicuous and the haploid one is small. In most red algae and some brown algae the two generations are much alike and independent.

The relative traits of the sporophyte and gametophyte generations of certain plants that reproduce sexually are given in Table 15-3.

Gametogenesis and sporogenesis in plants

Many plants reproduce by specialized sex cells called gametes. Fusion of pairs of gametes is called fertilization, and the resulting cell is a zygote. When gametes are similar in size and structure, the process is called isogamy; if gametes are unlike, it is called heterogamy. Sometimes the smaller gamete is called a sperm and the larger, an egg. The formation of gametes is called gametogenesis.

In many plants minute specialized cells called spores are produced for reproduction by a process called sporogenesis.

There are two phases in the life cycle of most plants, each of which produces the other, a process called alternation of generations. One generation, the gametophyte, bears male and female sex organs and produces gametes that fuse to effect fertilization. The fertilized egg does not develop into another gamete-producing individual (as it does in most animals) but into another generation, known as the sporophyte. The sporophyte produces spores, each of which develops eventually into a new gametophyte again (without sexual fusion).

In seed-producing angiospermous plants, as in a few of the more advanced spore-bearing lower plants, two kinds of spores are produced: the microspores, which produce male gametophytes, and the megaspores, which produce female gametophytes (Fig. 15-10). In higher plants the megaspore is re-

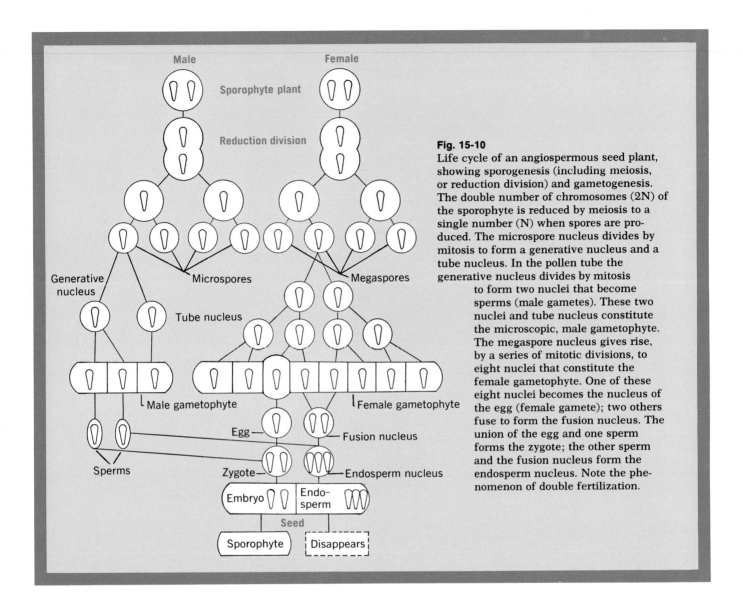

Fig. 15-10
Life cycle of an angiospermous seed plant, showing sporogenesis (including meiosis, or reduction division) and gametogenesis. The double number of chromosomes (2N) of the sporophyte is reduced by meiosis to a single number (N) when spores are produced. The microspore nucleus divides by mitosis to form a generative nucleus and a tube nucleus. In the pollen tube the generative nucleus divides by mitosis to form two nuclei that become sperms (male gametes). These two nuclei and tube nucleus constitute the microscopic, male gametophyte. The megaspore nucleus gives rise, by a series of mitotic divisions, to eight nuclei that constitute the female gametophyte. One of these eight nuclei becomes the nucleus of the egg (female gamete); two others fuse to form the fusion nucleus. The union of the egg and one sperm forms the zygote; the other sperm and the fusion nucleus form the endosperm nucleus. Note the phenomenon of double fertilization.

tained within the spore case, closely attached to the mother plant, where it germinates into a small female gametophyte. This whole structure, with the protective integument (covering), is known as an ovule (immature seed). Usually only a few ovules are formed, in contrast to the great numbers of spores produced by lower plants.

Microspores (pollen grains) are liberated into the air, but instead of falling to the ground and germinating there, they are carried to the ovule (or near it), where each microspore produces two sperms (male gametes). One sperm fertilizes the egg to form an embryo, which obtains its nourishment from the mother plant, thus acting like a parasite. Around the embryo is placed a supply of stored food in the endosperm, which results from the union of another sperm and the female fusion nucleus. The embryo grows and forms an embryonic root and one or more embryonic leaves. The embryo and endosperm are enclosed by the protective integu-

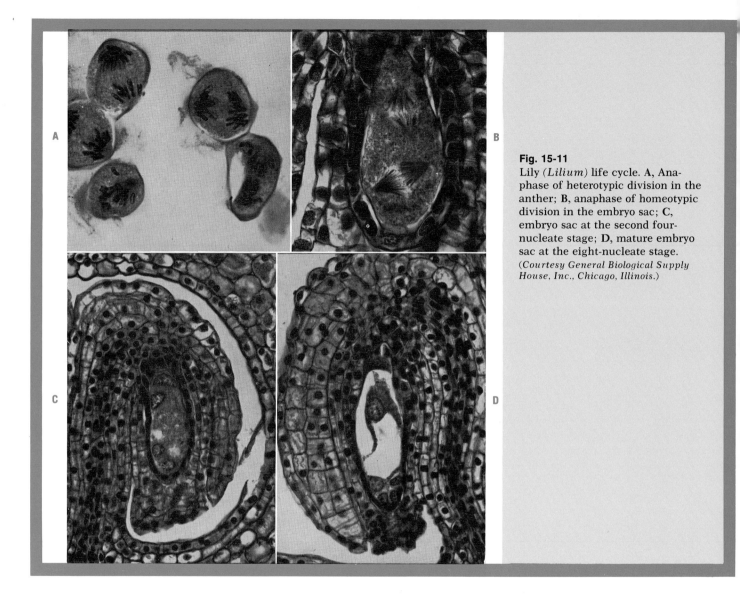

Fig. 15-11
Lily *(Lilium)* life cycle. **A,** Anaphase of heterotypic division in the anther; **B,** anaphase of homeotypic division in the embryo sac; **C,** embryo sac at the second four-nucleate stage; **D,** mature embryo sac at the eight-nucleate stage. *(Courtesy General Biological Supply House, Inc., Chicago, Illinois.)*

ment of the ovule, thus forming a seed. This seed eventually detaches from the parent and may remain dormant but viable for a long time, sometimes for years.

Under favorable conditions the seed germinates, and the embryo starts growing by absorbing stored food, forming leaves and roots, and developing into a new plant. Reproduction by seeds is advantageous and helps to make seed-bearing plants successful and beneficial. Thus most plants possess an alternation of generations between the asexual sporophyte (with the double, or diploid [2N], number of chromosomes) and the sexual gametophyte (with the single, or monoploid or haploid [N], number of chromosomes).

Meiotic division (meiosis) in plants

In all organisms that reproduce sexually there must be a reduction of the number of chromosomes by one half, so that when two sex cells unite at fertilization, only the normal diploid number (2N) will result. This reduction division process is called meiosis, and through it the number of chromosomes is reduced from the diploid (2N) number to the haploid number.

The time at which meiosis occurs in the life cycle of plants varies in different groups of plants. In flowering, angiospermous plants (Figs. 15-10 and 15-11) it immediately precedes the formation of microspores (pollen grains) and megaspores (embryo sacs). Such spores are produced in fours, by two divisions of a so-called spore mother cell.

In meiosis each chromosome does not split and then separate as in mitotic division (mitosis), but instead the two members of each of the homologous pairs fuse temporarily by the process of synapsis. During synapsis there may be an exchange of segments of chromosomes (crossing-over), which is followed by a separation of chromatids (half chromosomes). The process of meiosis has been considered in Chapter 4 and should be reviewed in connection with gametogenesis and sporogenesis.

Plant embryogenesis

A zygote (fertilized egg) divides repeatedly by mitosis to form groups of cells that eventually differentiate to produce various tissues and organs. Associated with the embryo are one or two cotyledons, or embryonic seed leaves (sometimes more than two in gymnospermous seeds), which are attached to an elongated embryo axis. Cotyledons rarely resemble mature leaves structurally, and they function in the absorption, digestion, and storage of food from the endosperm (if present), which lies adjacent to the embryo. The broad, flat cotyledons of bean seeds may persist temporarily after germination, become green, and even photosynthesize food.

Endosperm tissue contains water-insoluble stored foods that are to be used by the embryo before or during seed germination. The seeds of many plants contain large quantities of carbohydrates (chiefly starches), as in beans, corn, wheat, and rice. Smaller quantities of sugars may be present. Carbohydrates supply energy for growth and materials for forming the cellulose of cell walls.

Stored fats and oils, used primarily for energy, are present in the seeds of such plants as castor beans, peanuts, coconuts, and sunflowers. Proteins are stored in seeds and are used in forming new protoplasm. Large quantities of proteins are stored in the seeds of such plants as beans, peas, and soybeans.

The conversion of the water-insoluble foods into usable water-soluble foods through the process of digestion is necessary because the embryo uses foods that dissolve in water. Seeds of angiospermous, flowering plants may be divided into (1) monocotyledonous seeds with one embryonic seed leaf (cotyledon), such as corn, grasses, lilies, and irises, and (2) dicotyledonous seeds with two cotyledons, such as garden beans, peas, castor beans, sunflowers, geraniums, and oak trees.

Embryology of corn plants

A germinating seed of Indian corn (see Fig. 14-9) shows the embryonic stages of a monocotyledonous, angiospermous plant. Beneath an external seed coat is a large amount of endosperm. Toward the basal tip of the seed is the small embryo with its disklike scutellum (sku -tel' um) (L. *scutellum*, small shield) and probably one cotyledon. The scutellum absorbs food from the nearby endosperm for the developing embryo.

The part of the embryo axis above the point where the cotyledon is attached is called the epicotyl, sometimes called the plumule (L. *pluma*, feathery), and it is somewhat enclosed by the scutellum, which is enclosed by a sheathlike coleoptile and points upward. The growing tip of the epicotyl forms the young stem of the plant upon which mature leaves and floral organs eventually develop.

The part of the embryo axis below the point of cotyledon attachment is called the hypocotyl, the lower end of which is the radicle, or root primordium. The radicle is enclosed by a sheathlike coleorhiza. The radicle forms the primary (first) root of the seedling, and it points downward.

As the embryo develops, the true leaves, stem, and roots are formed from their respective embryonic areas. The embryonic development of other flower-bearing plants, in general, resembles the stages found in corn, although there are some differences, as might be expected. For example, in dicotyledonous plants, such as beans and peas, there are two cotyledons instead of one, as in corn.

211

Some contributors to the knowledge of the biology of certain higher plants

Theophrastus (370-285 B.C.)
A student of Aristotle and "Father of Botany" who observed the association of certain fungi and the roots of oak trees. Such an association of filamentous fungi and roots is known as mycorhiza (Gr. *mykes*, fungus; *rhiza*, root), in which, from a food standpoint, both types of plants may benefit.

John Ray (1628-1705)
An English Botanist who studied plant physiology and described experiments explaining the rise of sap. He recognized differences between dicotyledonous and monocotyledonous plants.

Stephen Hales (1677-1761)
An English physiologist who conducted critical experiments on the manufacture of foods by plants and on the transportation of materials within plant bodies. He is the "Father of Modern Plant Physiology."

Robert Brown (1773-1858)
An Englishman who discovered the minute "areola" (spaces) in pollen grains, which he regarded as points for the production of pollen tubes.

N. T. de Saussure (1767-1845)
A Swiss who published the first quantitative treatment of photosynthesis expressed in somewhat modern terms (1804).

René Dutrochet (1776-1847)
A French physiologist who recognized that chlorophyll was necessary for photosynthesis (1837).

Julius Sachs (1832-1897)
A German who proposed experimental methods for studying photosynthesis, respiration, and transpiration in plants.

Wilhelm Pfeffer (1845-1920)
A German botanist who observed the role of fungi in the mycorhiza of nonchlorophyll-bearing plants (1877). Such plants, including the Indian pipe (*Monotropa*), depend on the fungi for food, water, and minerals.

W. J. Beal
An American botanist who experimentally showed the longevity of seeds of common plants (1879), some of which retain their ability to germinate after 90 years.

H. H. Dixon (1869-1953)
A British botanist who proposed the theory of cohesion, which is generally accepted as a principal explanation of the rise of liquids in the xylem tissues of plants. Root pressures are also of importance, especially in shorter plants.

W. W. Garner (1875-1956)
An American botanist who discovered photoperiodism, or the effect of natural day length on flowers (1920), and noted that natural day length could be extended by artificial light or shortened by placing plants in the dark, or by at least covering them.

Richard Willstäter (1872-1942)
He and co-workers suggested the chemical nature of chlorophyll (1913), although many investigators since that time have made additional contributions.

Peter Boysen Jensen (1883-1959)
A Danish botanist who performed experiments that suggested that stem growth resulted from a stimulus transmitted by a chemical that was formed in the tip and migrated down into the region of bending.

Frits W. Went (1903-   )
An American botanist who proved the existence of plant hormones (1926) by growing decapitated coleoptile tips of oats on blocks of agar. (*Wide World Photos, Inc.*)

Went

Robert Woodward (1917-   )
Woodward, of Harvard University, won a Nobel Prize (1965) for his synthesis of chlorophyll, cortisone, cholesterol, and quinine. The synthesis of chlorophyll required fifty-five separate and very complicated steps. (*Wide World Photos, Inc.*)

Woodward

1 Explain how materials are absorbed and translocated by plants, including the roles of imbibition, osmosis, diffusion, root pressure, atmospheric pressure, capillarity, and cohesion.

2 Which theory, or theories, seems to explain the ascent of plant liquids in the most logical manner? Give specific reasons why you believe this.

3 Explain the phenomena of transpiration and guttation, including the functions of both.

4 Explain how chlorophyll-bearing plants photosynthesize foods. What foods are made? Might we consider photosynthesis as a chemicophysical process? Explain the origin of fats, proteins, and such special products as cellulose and aleurone grains.

5 Discuss the raw materials, source of energy, end products and role of chlorophyll in photosynthesis. Give reasons for dividing photosynthesis into such stages as photolysis and carbon dioxide fixation.

6 List all the characteristics of chlorophyll that you can. How many kinds of chlorophyll are there? How many carotenoids function in photosynthesis?

7 Explain why all living plants and animals must respire. Contrast aerobic and anaerobic respiration.

8 Contrast photosynthesis and respiration in as many ways as possible.

9 Explain the process of digestion in plants.

10 Explain how leaves and other plant parts may fall and why this is desirable in certain regions of the world.

11 List ways in which plants, or plant products, may be of economic importance, giving specific examples to prove your points.

Selected references

Arnon, D. T.: The role of light in photosynthesis, Sci. Amer. 203:105-18, 1960.

Bold, H. C.: The plant kingdom, Englewood Cliffs, New Jersey, 1964, Prentice-Hall, Inc.

Clevenger, S.: Flower pigments, Sci. Amer. June, 1964.

Crofts, A. S.: Translocation in plants, New York, 1961, Holt, Rinehart & Winston, Inc.

Galston, A. W.: The life of the green plant, Englewood Cliffs, New Jersey, 1964, Prentice-Hall, Inc.

Jensen, W. A., and Kavaljian, L. G. (editors): Plant biology today, Belmont, Calif., 1963, Wadsworth Publishing Co., Inc.

Lehninger, A. L.: How cells transform energy, Sci. Amer. 205:62-73, 1961.

Meyer, B. S., Anderson, D. B., and Bohning, R. H.: Introduction to plant physiology, Princeton, New Jersey, 1960, D. Van Nostrand Co., Inc.

Rabinowitch, E. J., and Govindjee: The role of chlorophyll in photosynthesis, Sci. Amer. 213:74-83, 1965.

Salisbury, F. B.: Plant growth substances, Sci. Amer. 196:125-134, 1957.

Salisbury, F. B.: The flowering process, New York, 1963, The Macmillan Company.

Steward, F. C.: Plants at work, Reading, Mass., 1964, Addison-Wesley Publishing Co., Inc.

Whittingham, C. P.: The chemistry of plant processes, New York, 1964, Philosophical Library, Inc.

ANIMAL BIOLOGY

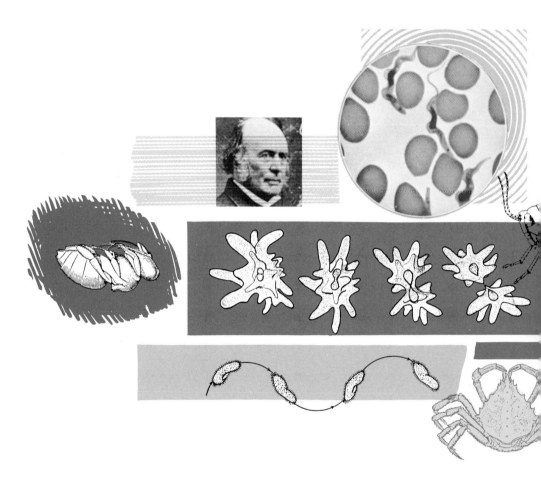

chapter sixteen

Protozoa

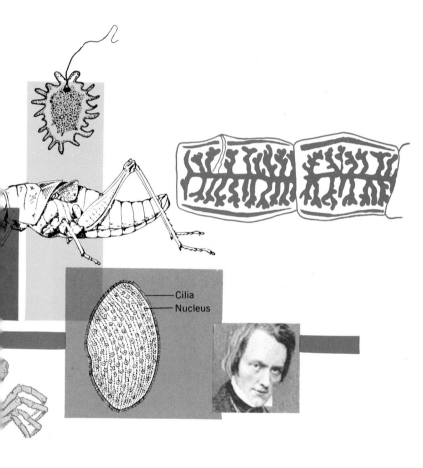

Cilia
Nucleus

CLASSIFICATION OF PROTOZOA

Class Sarcodina (Gr. *sarx*, protoplasm or flesh)
Locomote by protoplasmic projections called pseudopodia ("false feet"). Example: *Amoeba*.

Class Mastigophora (Flagellata) (Gr. *mastix*, whip; *phorein*, to bear); (fla jel -la' ta) (L. *flagellum*, whip)
Locomote by one or more whiplike flagella. Examples: *Peranema* and *Trypanosoma*.

Class Ciliata (L. *cilium*, hairlike cilia)
Locomote by hairlike cilia. Example: *Paramecium*.

Class Sporozoa (Gr. *spora*, spore; *zoon*, animal)
Have no locomotor organelles but may change body shapes. Reproduce by forming spores. Example: *Plasmodium*.

The organisms considered in this chapter are extremely varied.* Although often described as simple, single-celled animals, they are in fact very complex and have many differences as well as similarities. Many biologists call attention to this diversity by placing the protozoa in a separate subkingdom.

At any rate, protozoa are single-celled animals that are found in a wide variety of microenvironments. They may be free-living or may exist as partners in every type of symbiotic relationship. Also, protozoa function through the use of many kinds of organelles.

Most people become aware of their existence only by hearing of such problems as the increase of malaria among servicemen in Viet Nam and the outbreak of amebic dysentery among the athletes at the Olympic Games. However, in addition to their place in nature, protozoa are important research organisms. Our knowledge of cell metabolism has been greatly increased by studying them.

CLASS SARCODINA

Amoeba

Amoeba (Gr. *amoibe*, change) is a common freshwater protozoan, averaging about 500μ (1/50 of an inch) in length (Figs. 16-1 to 16-4). It appears to be

*Before beginning this chapter, refer to Chapter 7 for general information about animals.

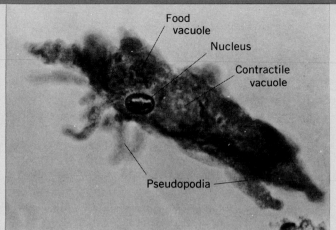

Food
vacuole

Nucleus

Contractile
vacuole

Pseudopodia

Fig. 16-1
Amoeba proteus. Note the nucleus, contractile vacuole,
food vacuole, and pseudopodia.

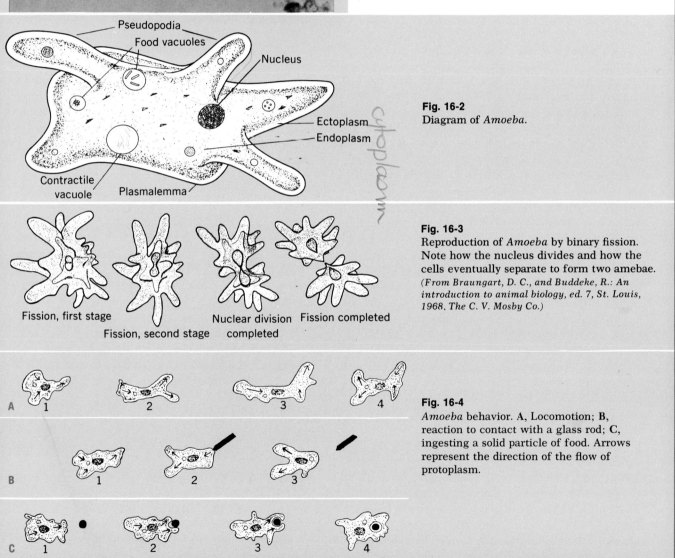

Pseudopodia
Food vacuoles

Nucleus

Ectoplasm
Endoplasm

Contractile
vacuole

Plasmalemma

cytoplasm

Fig. 16-2
Diagram of *Amoeba.*

Fission, first stage

Fission, second stage

Nuclear division
completed

Fission completed

Fig. 16-3
Reproduction of *Amoeba* by binary fission.
Note how the nucleus divides and how the
cells eventually separate to form two amebae.
*(From Braungart, D. C., and Buddeke, R.: An
introduction to animal biology, ed. 7, St. Louis,
1968, The C. V. Mosby Co.)*

A   1   2   3   4

B   1   2   3

C   1   2   3   4

Fig. 16-4
Amoeba behavior. **A,** Locomotion; **B,**
reaction to contact with a glass rod; **C,**
ingesting a solid particle of food. Arrows
represent the direction of the flow of
protoplasm.

VISCOUS- thick, ~~etc.~~ sirupy & sticky

a colorless, jellylike, granular mass that frequently changes shape by forming fingerlike pseudopodia. A disk-shaped nucleus is present but is not easily seen in living specimens. The protoplasm displays a streaming movement. Some details are observed more easily in stained specimens. There is an outer, delicate plasma membrane (plasmalemma), which has a clear ectoplasm beneath it. Internally, there is a granular endoplasm, which is made up of a rather viscous plasmagel and a more fluid plasmasol, which exhibits the flowing movements.

Locomotion is accomplished by forming temporary pseudopodia. At the beginning a pseudopodium is a fingerlike projection of ectoplasm; then the granular plasmasol flows into the projection as it lengthens. Mast's theory, based upon the colloidal nature of protoplasm, states that the movement is based upon the reversible change from the fluid sol state to the gel state. In this process the posterior part of the moving *Amoeba* is changed from the plasmagel to the plasmasol, while just the reverse is occurring at the anterior end where the pseudopodium is forming. In the gel state the protoplasm may contract somewhat and squeeze the sol, thus pushing it into the pseudopodium, the tip of which is not a gel. When the fluid sol comes near the tip of the pseudopodium, it moves to the sides and is converted into the gel state. Hence, at the point where a pseudopodium forms, the ectoplasm liquefies as a result of local chemical action, thus changing to the sol state. The endoplasm flows through this liquefied area, and as it does, the surface gels to form a tube of ectoplasm. Elsewhere the ectoplasm is converted to endoplasm. An *Amoeba* may form pseudopodia from any part of the cell. This type of activity, known as ameboid movement, is also found in some cells of complex animals, such as in the white blood cells of man.

Nutrition of an *Amoeba* is maintained by the ingestion of food particles. Amebae feed on other protozoans, rotifers, bacteria, and algae. Ingestion may occur at any part of the *Amoeba* by a thrusting out of pseudopodia to engulf the food. By this process the ectoplasm surrounds the food, and the structure eventually becomes a food vacuole that flows along in the endoplasm. Enzymes from the cytoplasm move into the food vacuole, and the

process of digestion begins. The digestive enzymes assist in breaking down complex foods, such as proteins and fats, into simpler molecules that can pass through the food vacuole boundary into the cytoplasm. Here these simple molecules are used as food or recombined into more complex substances. Other enzymes assist in recombining these simpler molecules into complex foods. Food vacuoles, when first formed, give an acid reaction but later become alkaline. As digested foods pass from the food vacuole into the protoplasm, the vacuole becomes smaller, and finally only the indigestible materials are egested (eliminated) through the cell surface.

Respiration occurs by diffusing oxygen, which is dissolved in the water, through the body surface. In order to secure necessary energy an *Amoeba* oxidizes foods, which results in such waste materials as water, urea, and carbon dioxide. Wastes may be eliminated through the body surface, but some are excreted by a special spheroid organelle, called a contractile vacuole. As this vacuole slowly collects water, it increases in size, contacts the cell surface, and expels its contents. It then repeats the process.

An *Amoeba* responds to various stimuli in different ways, as do other animals, but its behavior pattern is simpler than that of higher organisms (Fig. 16-4). The responses of amebae and other protozoans are often called taxes (Gr. *taxis*, locomotion), which are orientations of the organism, either toward the source of a stimulus (positive taxis) or away from the stimulus (negative taxis). Responses are influenced by the quality and intensity of the stimulus, as well as by the physiologic status of the organism at that particular time.

Reproduction occurs asexually by binary fission, in which the nucleus divides by mitosis, and the cytoplasm elongates and divides to form two daughter amebae. Usually an *Amoeba* will divide every two or three days, although this rate may be influenced by many factors.

Several kinds of amebae are parasitic on such things as *Hydra*, annelids, and vertebrates. One of these, *Entamoeba histolytica*, is an intestinal parasite in man, which causes the disease amebic dysentery and leads to intestinal abcess, liver damage, and other complications.

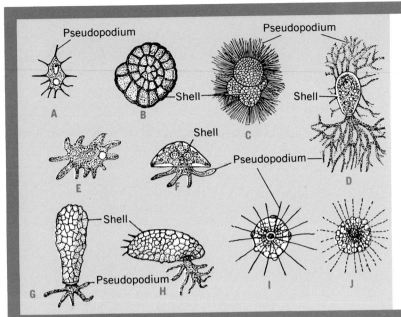

Fig. 16-5
Representative protozoans of class Sarcodina, showing various types of pseudopodia. **A**, *Protomonas*; **B**, *Rotalia*; **C**, *Globigerina*; **D**, *Allogromia*; **E**, *Amoeba*; **F**, *Arcella*; **G**, *Difflugia*; **H**, *Centropyxis*; **I**, *Actinophrys*; **J**, *Thalassicola*. (All enlarged and somewhat diagrammatic.)

Several other protozoans of the class Sarcodina are shown in Fig. 16-5.

CLASS MASTIGOPHORA

Peranema

Peranema (Gr. *pera,* sac; *nema,* thread) is a single-celled, flagellated protozoan found in stagnant fresh water and in ponds. When not swimming, the body form may assume many different sizes (25 to 100μ long by 10 to 15μ wide) and temporary shapes, which suggest that its external pellicle is flexible. One large, stiff flagellum and a smaller one are present on the anterior end. The posterior end of the body is blunt but tapers anteriorly. An oval nucleus contains chromatin. Nutrition is holozoic (ingestion and digestion of solid organic materials) and saprozoic (absorption of soluble organic materials). Chlorophyll is absent so food cannot be photosynthesized. A contractile waste vacuole is located near a saclike reservoir that is connected to a gullet. The gullet is supported by rods so that it opens and closes. A slitlike mouth (cytosome or pharynx) opens into the cytoplasm near the base of the flagellum. Food vacuoles are formed in which foods are digested by enzymes. Foods are stored as paramylum (starchlike ma-

terial). Asexual reproduction occurs by longitudinal binary fission.

Trypanosoma

Trypanosoma (tri pan o -so' mah) (Gr. *trypanon,* awl; *soma,* body) is a genus containing elongated, ribbonlike, mononucleated, monoflagellated protozoans that are parasitic in certain vertebrate animals. The flagellum is attached to the trypanosome body by a thin undulating membrane, and the actions of the two propel the organism.

Trypanosoma gambiense (Fig. 16-7) (Gambia, a country on the west coast of Africa) causes a form of African sleeping sickness. The trypanosomes are injected into the human bloodstream by blood-sucking, trypanosome-carrying tsetse flies in their search for food. Normally, the trypanosomes may live in the blood plasma of certain vertebrates, where they absorb foods and oxygen. The tsetse flies bite infected animals and suck up the trypanosomes, which multiply and migrate to the salivary glands in about 3 weeks. If an infected fly bites a man or other mammal, the saliva and reorganized trypanosomes are injected. The multiplication of trypanosomes in the human blood liberates fever-producing toxins. Soon, these toxins may invade the nervous system, resulting in a

219

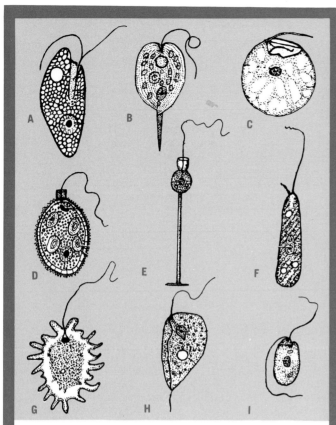

Fig. 16-6

Representative flagellated protozoans of class Mastigophora. **A**, *Chilomonas;* **B**, *Phacus;* **C**, *Noctiluca;* **D**, *Trachelomonas;* **E**, *Monosiga;* **F**, *Peranema;* **G**, *Mastigamoeba;* **H**, *Cercomonas;* **I**, *Bodo.* (All enlarged and somewhat diagrammatic.)

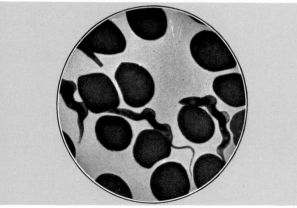

Fig. 16-7

Trypanosoma gambiense, a protozoan of class Mastigophora, is the cause of tropical sleeping sickness in central and western Africa.

(Courtesy General Biological Supply House, Inc., Chicago, Ill.)

lethargy that becomes sleeping sickness and ends in death.

Flagellated protozoa in termites and roaches

The digestive tracts of certain termites and wood roaches contain many kinds of flagellates. These protozoans perform an essential function in the digestion of wood. The insect host eats the wood, the flagellates digest the cellulose it contains, and both absorb the digested products. This type of mutual collaboration between two different kinds of organisms in which each depends on the other for the accomplishment of a process beneficial to both is called mutualism (L. *mutuus*, exchanged).

CLASS CILIATA

Paramecium

Paramecia (par ah -me′ se a) (Gr. *paramekes*, oblong) (Figs. 16-8 to 16-12) are ciliated protozoans, which are sometimes called "slipper animalcules" because they seem to resemble a slipper. Paramecia are common in fresh water that contains an abundance of decaying organic matter. Paramecia have holozoic nutrition. They live on bacteria, algae, and other organisms. A common type, *Paramecium caudatum* (ko -dah′ tum) (L. *cauda*, tail), has a tuft of tail-like cilia on the posterior end and may be up to 0.3 mm. long, whereas *Paramecium aurelia* (o -re′ lia) (L. *aurum*, brown or gold) may be less than 0.2 mm. long. A depression on the ventral side is called the oral groove and runs obliquely backward, ending just posterior to the middle of the body.

Covering the entire surface is a clear, elastic membrane, the pellicle, which is covered with fine cilia. The pellicle is divided by tiny elevated ridges into many small hexagonal areas. Just beneath the pellicle is the clear ectoplasm, which encloses the granular endoplasm. Embedded in the ectoplasm, just below the surface, are spindle-shaped cavities called trichocysts (trik′ o sist) (Gr. *thrix*, hair; *kystis*, sac), which are filled with a semifluid that may be discharged as long trichocyst threads presumably for defense purposes. There are two kinds of nuclei, the larger macronucleus

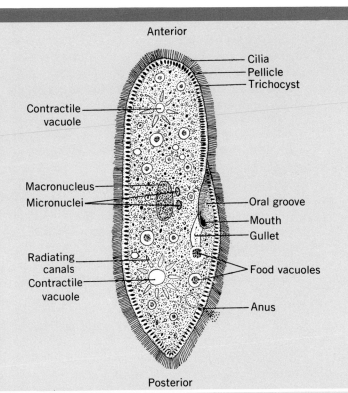

Anterior

Cilia
Pellicle
Trichocyst

Contractile vacuole

Macronucleus
Micronuclei

Oral groove
Mouth
Gullet

Radiating canals
Contractile vacuole

Food vacuoles

Anus

Posterior

Fig. 16-8
Paramecium aurelia, a ciliated protozoan of class Ciliata.

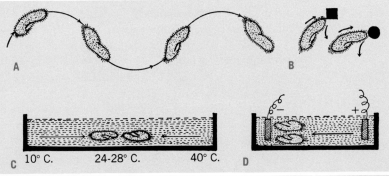

A

B

C   10° C.   24-28° C.   40° C.

D

Fig. 16-9
Paramecia showing various behaviors. **A**, Locomotion (helical movement); **B**, reaction to contacts with solid objects (avoiding reaction); **C**, reaction to temperatures; **D**, reaction to electric current. (All enlarged and somewhat diagrammatic.)

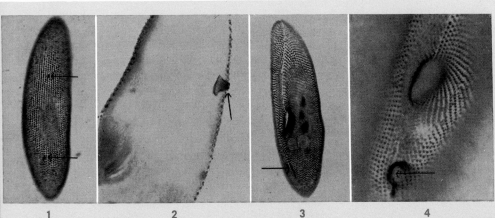

1   2   3   4

Fig. 16-10
Paramecium, showing certain detailed structures (arrows). 1, Contractile vacuole pores; 2, contractile vacuole pore, enlarged; 3, anal pore, or cytopyge (Gr. *kytos*, cell; *pyge*, posterior or rump); 4, anal pore, enlarged. (*Courtesy Dr. Elizabeth E. Powelson, Wittenberg University.*)

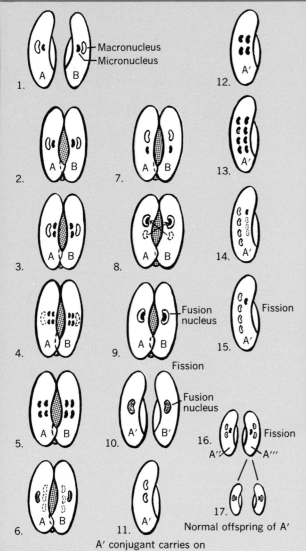

Fig. 16-11

Paramecium caudatum conjugation. 2, Two partners meet and fuse at the oral groove. 3, 4, and 5, Macronucleus in each disappears and micronucleus divides twice. 6 and 7, Three of the four micronuclei disappear and the fourth one divides unequally. 8 and 9, Paramecia exchange nuclei, which then fuse. 10, Partners separate and the nucleus in each now has genes from two sources. 11, 12, and 13, Fusion nucleus divides three times, resulting in a total of eight. 14, Three of the new nuclei disappear, four of them enlarge, and one remains a micronucleus. 15, Micronucleus *and organism* divide. 16, New micronucleus and new organism(s) divide again. 17, Process is now complete. Each of the eight new paramecia (four from each original parent) has a micronucleus and a macronucleus.

(Modified from Braungart, D.C., and Buddeke R.: An introduction to animal biology, ed. 7, St. Louis, 1968, The C. V. Mosby Co.)

and a smaller micronucleus. In *Paramecium aurelia* there are typically two micronuclei.

Cilia in the oral groove sweep food in the water into the mouth (cytostome). Along the gullet (cytopharynx) are rows of cilia fused into an undulating membrane to pass food inwardly from the mouth to the posterior part of the gullet where the food is collected into a food vacuole. This, when of proper size, disconnects from the gullet to circulate through the endoplasm. Enzymes digest the food in the food vacuoles, which gradually become smaller. Food vacuoles at first have an acid reaction and later become alkaline. Indigestible fecal materials are ejected through the anus, which may be seen as a pore posterior to the oral groove (Fig. 16-10).

Respiration occurs when the oxygen dissolved in the water diffuses into the cell and the carbon dioxide passes out.

Toward each end, near the dorsal surface, is a contractile vacuole, composed of a central hollow area and several radiating canals. These canals conduct excess water and possibly wastes to the vacuole, which increases in size. When a certain size is reached, the contractile vacuole contracts and discharges its contents to the outside through a pore. The effect of this is to maintain water and ion balance.

By special fixation and staining it can be observed that the cilia are attached to basal granules that are connected to each other by interciliary fibrils. This comprises the noncellular neuromotor mechanism for the coordination of ciliary action. The cilia vibrate either forward or backward, so that the animal can move in either direction. By beating obliquely the cilia cause a rotation on the long axis. The longer cilia of the oral groove vibrate more rapidly so that the anterior end swerves. A combination of such actions causes the animal to move along a helical path.

When a *Paramecium* encounters an undesirable stimulus, such as certain chemicals, it performs an avoiding reaction by reversing its cilia to move backward. The anterior end is then swerved aborally (away from the mouth), thus pivoting on the posterior end. This permits a sampling of new areas by the oral groove, which continues until a condition is encountered that again permits for-

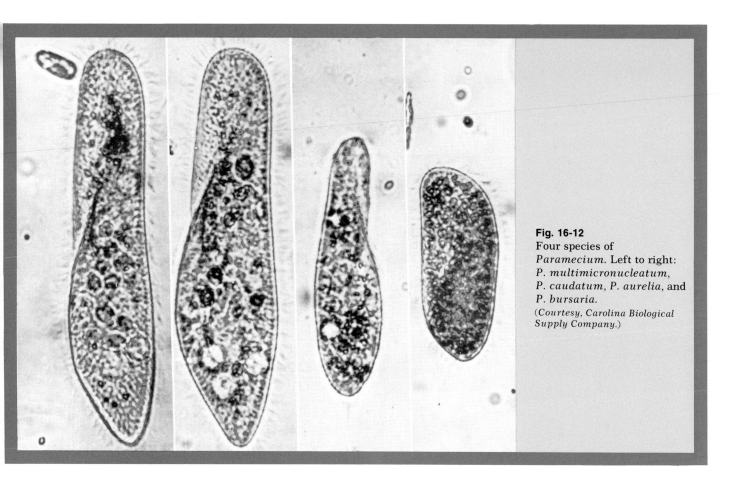

Fig. 16-12
Four species of
Paramecium. Left to right:
P. multimicronucleatum,
P. caudatum, P. aurelia, and
P. bursaria.
*(Courtesy, Carolina Biological
Supply Company.)*

ward movement. This might be called a "trial-and-error" type of adjustment.

The specific type of response of a *Paramecium,* even to the same stimulus, varies from time to time, being influenced by a great number of conditions. The responses of protozoans are attempts to orient themselves, either toward or away from a stimulus, and are called taxes, which were mentioned earlier in this chapter. The names of taxes are based on the type of stimulus suggested by the following: (1) thermotaxis (Gr. *therme,* heat; *taxis,* locomotion), response to heat; (2) chemotaxis (Gr. *chemo,* juice or chemical), response to chemicals; (3) phototaxis (Gr. *phos,* light), response to light; (4) thigmotaxis (Gr. *thigema,* touch), response to contact; (5) geotaxis (Gr. *ge,* earth), response to gravity; (6) rheotaxis (Gr. *rhein,* to flow), response to water or air currents; and (7) galvanotaxis (L. Galvani, Italian physiologist), response to electric currents.

Paramecium has an average optimum temperature of about 26° C. and tends to be negatively thermotaxic to temperatures much above or below this temperature; it reacts negatively to most chemicals, yet may react positively to certain concentrations of mild acids; it avoids either bright light or no light at all; it usually avoids objects; it responds negatively to gravity; it usually swims against water currents and will point its anterior end toward the negative pole (cathode).

A *Paramecium* reproduces by transverse binary fission, whereby the micronucleus divides by mitosis into two micronuclei, which migrate to opposite ends of the cell. The macronucleus divides into two parts, without mitosis, and the two take positions in opposite ends of the cell. The gullet buds off a new one, and two new contractile vacuoles are formed. A transverse constriction deepens across the middle of the cell, thus dividing the cytoplasm. Hence, two smaller paramecia are visible in about 20 minutes. Over 600 generations may be produced annually by this process. All in-

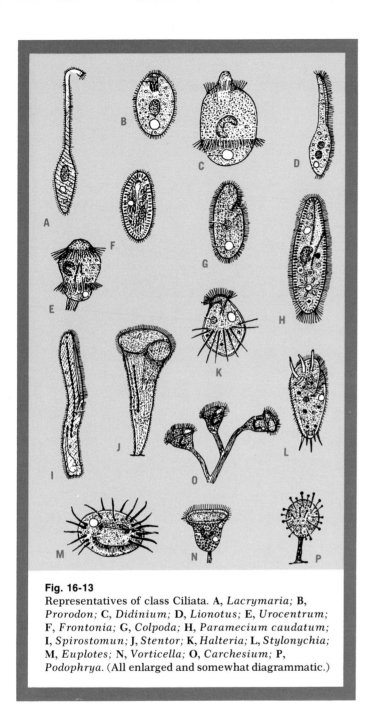

Fig. 16-13
Representatives of class Ciliata. A, *Lacrymaria;* B, *Prorodon;* C, *Didinium;* D, *Lionotus;* E, *Urocentrum;* F, *Frontonia;* G, *Colpoda;* H, *Paramecium caudatum;* I, *Spirostomun;* J, *Stentor;* K, *Halteria;* L, *Stylonychia;* M, *Euplotes;* N, *Vorticella;* O, *Carchesium;* P, *Podophrya.* (All enlarged and somewhat diagrammatic.)

At certain times two paramecia may attach themselves temporarily at their oral groove regions. They then build between them a protoplasmic conjugating tube through which they will mutually exchange micronuclear materials by a process called conjugation (Fig. 16-11). The behavior of the micronucleus and macronucleus is quite complex during this process. After the mutual exchange of the micronuclear materials the two paramecia separate again and may continue to divide by fission. Conjugation may not be necessary for rejuvenation, since cultures have been kept for years without it taking place. However, in those cases a related process called autogamy usually took place. Conjugation is more common in *Paramecium caudatum* than in *Paramecium aurelia.*

Autogamy (Gr. *autos,* self; *gamos,* marriage) is another method of reproduction in which there is a nuclear reorganization within a single *Paramecium.* The macronucleus disintegrates, and the micronuclei divide. Two of the micronuclei (each with one half the normal chromosome number) then fuse and restore the original chromosome number. The *Paramecium* may then divide by fission. This process involves nuclear reorganization but differs from conjugation in that only one individual is involved.

There are several species of paramecia, each with its inherited characteristics. Certain paramecia have two mating types, neither of which will conjugate with another of its own type but only with the complementary type.

Certain particles that have the ability to self-duplicate have been discovered in the cytoplasm of paramecia. They constitute a cytoplasmic mechanism for the transmission of certain hereditary traits and are often called plasmagenes (cytoplasmic genes) to distinguish them from the genes within the nucleus. Valuable information in the field of heredity has been obtained by studies of the plasmagenes of paramecia.

Sonneborn observed that certain paramecia, called killers, release a substance into the water that will cause the death of other paramecia, which are designated as sensitives. He called the gene responsible for the transmission of the killer characteristic from parent to offspring the "K" gene. He and his co-workers also have demon-

dividuals produced from a single individual by binary fission have a similar heredity and constitute a group called a clone (klon) (Gr. *klon,* twig). Such groups of organisms can be used experimentally where numbers of individuals with similar heredity are required.

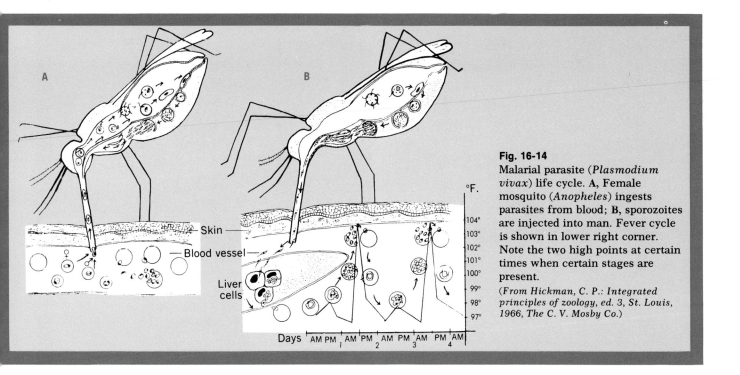

Fig. 16-14

Malarial parasite (*Plasmodium vivax*) life cycle. **A,** Female mosquito (*Anopheles*) ingests parasites from blood; **B,** sporozoites are injected into man. Fever cycle is shown in lower right corner. Note the two high points at certain times when certain stages are present.

(*From Hickman, C. P.: Integrated principles of zoology, ed. 3, St. Louis, 1966, The C. V. Mosby Co.*)

strated what they call kappa particles in the cytoplasm, which are responsible for the secretion of the actual killing chemical substance, known as paramecin. Hence, the dominant "K" gene in the chromosome and the kappa particle in the cytoplasm must both be present in order for a *Paramecium* to produce paramecin and thus be a killer. Without one or the other the *Paramecium* is not a killer. These kappa particles are similar to genes (of the nucleus) because they are transmitted from parent to offspring, have the ability to reproduce themselves, and are able to mutate.

CLASS SPOROZOA

Plasmodium

Plasmodium (plas -mo′ di um) (Fig. 16-14) is a protozoan belonging to the class Sporozoa. There are no well-defined organelles for locomotion, although certain immature stages may be motile. There are no contractile vacuoles and no digestive organelles, since food is absorbed. Several species of *Plasmodium* infect man and other vertebrates, producing different types of malaria. The parasites are transmitted by the female mosquito of the genus *Anopheles* (a -nof′ el ez) (Gr. *anopheles*, harmful).

When an infected female *Anopheles* mosquito bites (actually this "bite" is really a puncture) a human, she injects minute, slender sporozoites along with saliva from the salivary glands. These sporozoites pass through the bloodstream and are filtered out in the liver. They undergo multiple divisions in the liver and finally enter the red blood cells of the general circulatory system.

Inside the red blood cells they grow and divide into many merozoites, which are released as the red blood cells burst.

Because of the great numbers of parasites and the toxins that they produce, the characteristic fevers and chills develop. The time between the fever and the chill varies with the type of malaria.

After the asexual stages some merozoites become sexual forms called gametocytes. When the gametocytes are sucked into the mosquito stomach, they become microgametocytes and macrogametocytes. Two gametes of opposite sex unite to form a zygote, which enters the stomach walls. Later, these zygotes enlarge to form oocysts, which produce thousands of sporozoites. The sporozoites escape from their enclosing cyst and travel to the salivary glands from which the female mosquito transfers them by biting a human being. The developmental stages in the mosquito require from

225

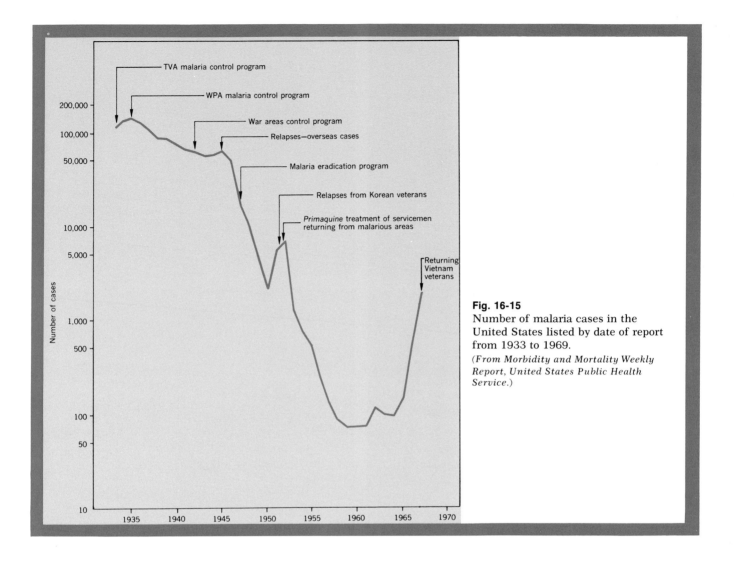

Fig. 16-15
Number of malaria cases in the United States listed by date of report from 1933 to 1969.
(From Morbidity and Mortality Weekly Report, United States Public Health Service.)

7 to 18 days or longer. In man the symptoms usually appear in 10 to 14 days after the parasites enter his body.

Although malaria has been eliminated as a major disease in the United States, its incidence has increased tremendously among servicemen in the Far East.

ECONOMIC IMPORTANCE OF PROTOZOA

Many protozoans are neither harmful nor beneficial to man. Some types contribute to the formation of soils and earth deposits, such as chalk and rocks formed by *Radiolaria* and *Foraminifera*. When they die, certain shelled forms form an ooze at the bottom of the ocean that may solidify into rocky formations. Water supplies may become un-

desirable because of bad tastes or odors produced by such protozoans as *Bursaria*, *Uroglena*, *Dinobryon*, and *Synura*.

The cellulose eaten by termites is digested in their intestines by certain flagellated protozoans, thus changing it into a form that is usable by the termites. In turn the termites give protection and distribution to the protozoans. Many types of protozoans live in the intestines of animals and man without causing harm. *Entamoeba gingivalis* and *Trichomonas tenax* may live in the mouth. Many types of protozoans have been studied in laboratories, and much has been learned about living protoplasm, the inheritance of traits, nutritional requirements, behaviors of living organisms, osmosis, and permeability.

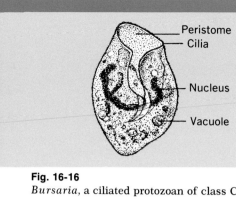

Fig. 16-16
Bursaria, a ciliated protozoan of class Ciliata.

Labels: Peristome, Cilia, Nucleus, Vacuole

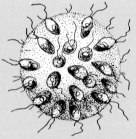

Fig. 16-17
Uroglena, a flagellated, spherical, colonial protozoan of class Mastigophora.

Fig. 16-18
Branched colony of *Dinobryon*, a protozoan of class Mastigophora.

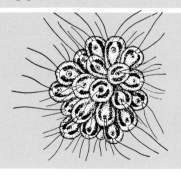

Fig. 16-19
Synura, a colonial protozoan of class Mastigophora.

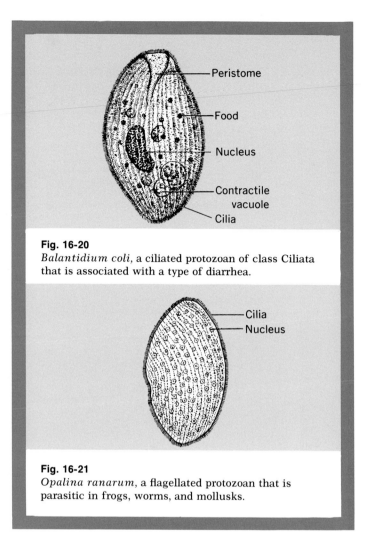

Fig. 16-20
Balantidium coli, a ciliated protozoan of class Ciliata that is associated with a type of diarrhea.

Labels: Peristome, Food, Nucleus, Contractile vacuole, Cilia

Fig. 16-21
Opalina ranarum, a flagellated protozoan that is parasitic in frogs, worms, and mollusks.

Labels: Cilia, Nucleus

Among the many human protozoan diseases are intestinal diarrhea, which may be caused by a variety of protozoans, including *Balantidium coli;* African sleeping sickness, caused by *Trypanosoma gambiense* and transmitted by the tsetse fly; amebic dysentery, caused by *Entamoeba histolytica;* and several types of malaria, caused by different species of *Plasmodium.*

Among protozoan diseases of animals are the parasitism of frogs, worms, and mollusks by *Opalina;* Nagana diseases of game animals of Africa, caused by *Trypanosoma brucei;* Texas fever of cattle, caused by *Babesia* and transmitted by a tick; coccidiosis of poultry, rabbits, and other animals, caused by *Coccidia;* and destruction of sperms and seminal vesicles in earthworms, caused by *Monocystis.*

227

Some contributors to the knowledge of protozoa

Konrad von Gesner (1516-1565)
A Swiss naturalist who described a foraminifera type of protozoan in 1565.

Antonj van Leeuwenhoek (1632-1723)
A Dutch microscopist who studied many types of cells and organisms. Through his rather accurate accounts and observations he discovered free-living and parasitic protozoans (1674-1703) and is known as the "Father of Protozoology."

Christian G. Ehrenberg (1795-1876)
A German microbiologist who was an enthusiastic student of protozoans, especially infusorians. He postulated that they are highly organized unicellular organisms and classified them. (*Historical Pictures Service, Chicago.*)

Ehrenberg

Felix Dujardin (1801-1860)
A French microscopic zoologist, recognized living substance in protozoans and applied the name "sarcode" to them (1835).

Carl T. E. von Siebold (1808-1896)
A German parasitologist who recognized protozoans as single-celled animals (1845) and classified them. (*Historical Pictures Service, Chicago.*)

Siebold

Émile François Maupas (1842-1916)
A French protozoologist who demonstrated sexuality in protozoans, which later stimulated much work in protozoology and parthenogenesis (reproduction without fertilization).

Lösch
He described a protozoan (*Entamoeba histolytica*) in the feces and intestinal ulcers in a fatal case of dysentery (1875).

tion, (6) excretion, (7) coordination, nervous, and sensory equipments, and (8) reproduction.

4 Why do unicellular (acellular) organisms not have organs and tissues? What is the function of organelles?

5 What characteristics of living protoplasm are displayed by protozoans?

6 Do you consider the various types of protozoans to be simple or complex? Explain.

7 Do individual protozoans die of old age? Of what significance is the fact that protozoans simply divide when they reach a certain stage in their life? Explain the relationship between the environmental conditions and the death of protozoans.

8 Discuss the economic importance of protozoans from beneficial and detrimental standpoints.

9 Explain the complex life cycle of *Plasmodium*, including the stages in proper sequence, the different hosts, and the detrimental effects.

10 How many studies of certain types of protozoans contributed to our knowledge of cytoplasmic inheritance? Explain what is meant by "killers" and "sensitives" in paramecia.

Selected references

Barnes, R. D.: Invertebrate zoology, ed. 2, Philadelphia, 1969, W. B. Saunders Company.

Calkins, G. N., and Summers, F. M. (editors): Protozoa in biological research, New York, 1941, Columbia University Press.

Chandler, A. C., and Read, C. P.: Introduction to parasitology, ed. 10, New York, 1961, John Wiley & Sons, Inc.

Cheng, T. C.: The biology of animal parasites, Philadelphia, 1964, W. B. Saunders Company.

Corliss, J. O.: The ciliated Protozoa, New York, 1961, Pergamon Press, Inc.

Honigberg, B. M., and others: A revised classification of the phylum Protozoa, J. Protozool. 11:7-20, 1964.

Hunter, S. H., and Lwoff, A.: Biochemistry and physiology of the Protozoa, 2 vol., New York, 1951 and 1955, Academic Press Inc.

Manwell, R. D.: Introduction to protozoology, New York, 1961, St. Martins Press, Inc.

Meglitsch, P. A.: Invertebrate zoology, New York, 1967, Oxford University Press, Inc.

Tartar, V.: The biology of Stentor, New York, 1961, Pergamon Press, Inc.

Vickerman, K., and Cox, F. E. G.: The protozoa, Boston, 1967, Houghton Mifflin Company.

Review questions and topics

1 From your studies of protozoans list the characteristics that they have in common.

2 From your studies of protozoans make a table listing the differences among them.

3 Describe the following for each protozoan studied: (1) integument (covering), (2) locomotion, (3) ingestion and digestion, (4) respiration, (5) circula-

chapter seventeen

Sponges and coelenterates

All multicellular animals are collectively known as the Metazoa. This classification infers that these higher animals have both a specialization of cells and a multicellular structure.

The two groups to be considered in this chapter, the sponges and the coelenterates, are Metazoan. However, there are some qualifications that must be made in describing them.

The sponges are a rather unusual group of organisms in many ways. While they have different types of cells, the cells are not clearly arranged into tissues. They have a body cavity, but it is more of a water filtering system than a digestive site.

Sponges were once thought to be plants because they lack movement and have an indefinite structure. Even now they are not considered to be on the direct line of animal evolution and, therefore, are placed in the branch Parazoa (Gr. *para*, beside) as an offshoot of the Metazoa.

The coelenterates, by contrast, appear in the evolutionary line as the lowest animals with tissues. They have a digestive cavity (an enteron) in which food is broken down by enzymes. Several cellular aggregates are present and function as tissues. A nerve network is also present.

CLASSIFICATION OF SPONGES (PHYLUM PORIFERA)

Class Calcarea (L. *calcarius*, lime)

Calcareous spicules, which may be straight or branched; marine types. Examples: *Sycon* and *Leucosolenia*.

Class Hexactinellida (Gr. *hex*, six; *aktin*, rays)

Six-rayed siliceous spicules arranged in three planes, which in some types are fused into a skeleton resembling spun glass; marine types. Example: Venus's flower-basket (*Euplectella*).

Class Demospongiae (Gr. *demos*, people; *spongos*, sponge)

Network of fibers of proteinlike spongin alone or associated with siliceous spicules; complex system of canals; mostly marine types. Example: Commercial sponges.

Canal systems

There are three kinds of canal systems that are found in the various types of sponges (Fig. 17-1).

Ascon type

The ascon type of canal passes directly from the external ostia (L. *ostia*, pores) to the spongocoel (Gr. *spongos*, sponge; *koilos*, cavity). This type of canal is lined with collar cells. *Leucosolenia*, a freshwater sponge, is a sponge with this type of canal system.

Sycon type

The sycon type of canal system has an incurrent system that ends blindly near the central spongocoel and an excurrent system of radial canals that ends blindly near the body surface. Both canals are connected by small openings. Only the radial canals are lined with collar cells. *Sycon* is a sponge with this type of canal system.

Leucon type

The leucon type of canal system is complex with many branches. These multibranched canals contain numerous chambers that are lined with collar cells. Commercial (bath) sponges have this type of canal system.

Class Calcarea

Sycon

Sycon is a vase-shaped, sessile, radially symmetrical, marine sponge with calcareous (limy) skeletal spicules. It may be up to 1 inch long and often forms a colony by budding (see Fig. 17-3). The surface has a fuzzy appearance and has numerous incurrent pores, or ostia, which open into the

229

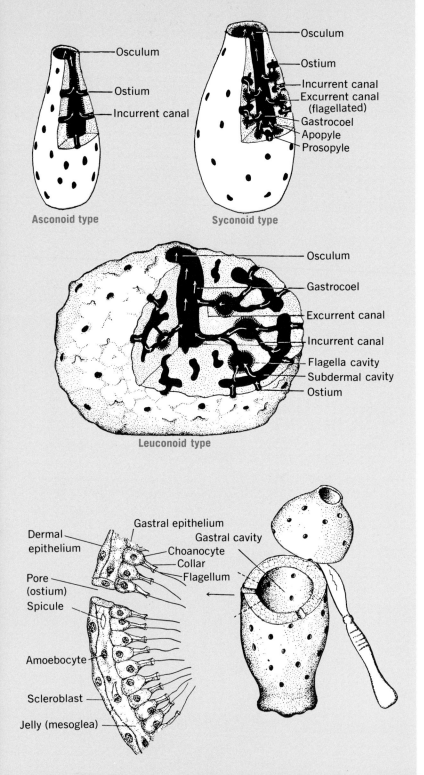

Asconoid type

- Osculum
- Ostium
- Incurrent canal

Syconoid type

- Osculum
- Ostium
- Incurrent canal
- Excurrent canal (flagellated)
- Gastrocoel
- Apopyle
- Prosopyle

Leuconoid type

- Osculum
- Gastrocoel
- Excurrent canal
- Incurrent canal
- Flagella cavity
- Subdermal cavity
- Ostium

- Dermal epithelium
- Gastral epithelium
- Gastral cavity
- Choanocyte
- Collar
- Flagellum
- Pore (ostium)
- Spicule
- Amoebocyte
- Scleroblast
- Jelly (mesoglea)

Fig. 17-1

Sponges, showing types of canal systems. Simple asconoid type, such as *Leucosolenia;* syconoid, such as *Sycon;* leuconoid, such as commercial (bath) sponge. Central cavity, or gastrocoel, is also called spongocoel. Flagellated cells are known as choanocytes. Observe the increase in complexity, particularly of the number of chambers lined with flagellated choanocytes.

Fig. 17-2

Leucosolenia, showing some of the structures in detail (cross section). The jellylike mesoglea is also called mesenchyme.

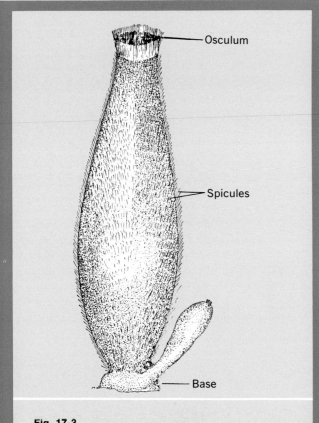

Fig. 17-3

Simple sponge with a young bud.

(From Braungart, D. C., and Buddeke, R.: An introduction to animal biology, ed. 7, St. Louis, 1968, The C. V. Mosby Co.)

incurrent canals just beneath. These canals end blindly near the central cavity, the spongocoel (Fig. 17-1). Openings in the spongocoel wall are called apopyles (Gr. *apo*, away from; *pyle*, gate) and are surrounded by perforated cells known as porocytes (Gr. *poros*, pore; kytos, cell). The apopyles lead into radial (excurrent) canals that run toward the outer surface and end blindly. The incurrent canals and radial canals are connected by small openings called prosopyles (Gr. *proso*, forward; *pyle*, gate). The incurrent canals are lined with dermal epithelium. The spongocoel and radial canals are lined with gastral epithelium. Radial canals are well supplied with flagellated collar cells, the choanocytes (Gr. *choane*, funnel; *kytos*, cell), to propel water and food. Water enters the spongocoel and eventually goes out through an opening, the osculum (L. *osculum*, small "mouth"), located at the free end of the sponge. Cells called myocytes (Gr. *myo-*, muscle; *kytos*, cell) around the ostia contract and expand to regulate the size of the opening and thus the flow of water.

The body wall of the *Sycon* is composed of an outer dermal epithelium and an inner gastral epithelium. A noncellular jellylike layer, the mesenchyme (Gr. *mesos*, middle; *en*, in; *chein*, to pour), is present between them (Fig. 17-2 and 17-4). This middle layer contains wandering ameboid cells, the amebocytes, for ingesting foods, forming

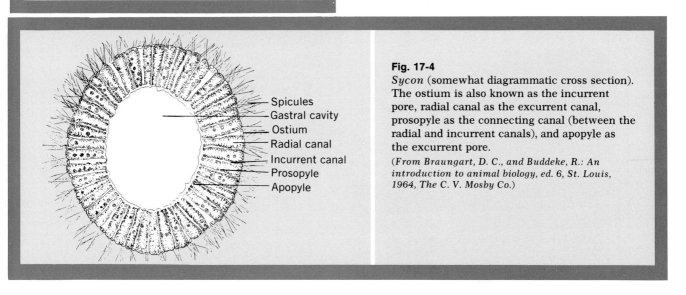

Fig. 17-4

Sycon (somewhat diagrammatic cross section). The ostium is also known as the incurrent pore, radial canal as the excurrent canal, prosopyle as the connecting canal (between the radial and incurrent canals), and apopyle as the excurrent pore.

(From Braungart, D. C., and Buddeke, R.: An introduction to animal biology, ed. 6, St. Louis, 1964, The C. V. Mosby Co.)

reproductive cells, and producing skeletal materials.

The body wall contains many calcareous spicules (calcium carbonate), which are interlaced to give protection and support. *Sycon* has four types, namely, short monaxon (straight), long monaxon, triradiate (three rays in one plane), and polyaxon (many rays). All spicules originate from specialized ameboid cells called scleroblasts (Gr. *skleros*, hard; *blastos*, bud or origin).

Sycon ingests microscopic organisms and small pieces of organic materials, drawing them in through the incurrent canals by the beating action of the flagella of the collar cells. Digestion is intracellular (within the collar cells). The collar cells engulf the foods by means of pseudopodia and form food vacuoles in which the foods are digested by digestive enzymes. Digested foods are circulated from cell to cell by diffusion and by the wandering amebocytes of the mesenchyme.

Respiration occurs by absorbing oxygen from the water and giving off carbon dioxide to the water as it circulates. Wastes are carried with the water and out through the osculum; hence the osculum functions more as an anus than as a mouth. *Sycon* reproduces asexually by budding, and many may be produced to form a cluster, or colony, of individuals.

Sexual reproduction occurs by sperm and eggs, both of which are formed by special ameboid cells and are present in the same animal (monecious). The fertilized egg (zygote) forms an embryo that becomes a flagellated larval stage, called the amphiblastula. One half of the amphiblastula bears flagella, whereas the other half does not. The motile larva escapes from the parent, swims freely, and finally attaches and grows to become an adult.

Class Hexactinellida
Venus's flower-basket

Venus's flower-basket, a glass sponge, belongs to the genus *Euplectella* and is composed of siliceous (silicon dioxide), six-rayed spicules interwoven into a beautiful cylindrical network that resembles spun glass. The spicules surround the numerous pores and canals and form a complex architectural pattern. Many of the spicules are star shaped. The chambers on the canals are lined with collar cells. *Euplectella* is found in the deeper regions of the ocean near the Philippines where it is attached by a mass of long glasslike threads.

Class Demospongiae
Commercial (bath) sponges

Commercial bath sponges have a skeleton of spongin fibers, which are composed of a proteinlike material. The fibers are branched and form a net-

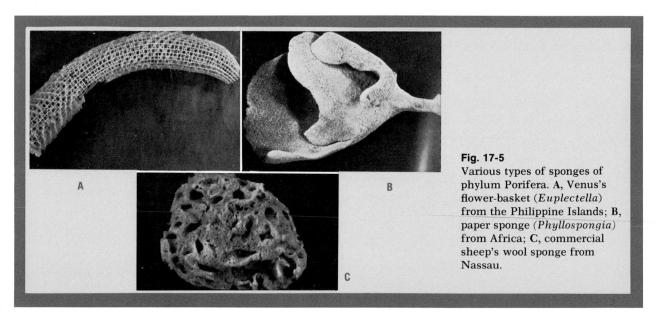

Fig. 17-5
Various types of sponges of phylum Porifera. A, Venus's flower-basket (*Euplectella*) from the Philippine Islands; B, paper sponge (*Phyllospongia*) from Africa; C, commercial sheep's wool sponge from Nassau.

work that supports and protects other parts of the sponge. Bath sponges, when alive, are frequently black, but some types are yellow, brown, or grayish. The canals are complex, very branched, and have numerous chambers lined with collar cells.

Commercial sponges are collected by diving, hooks, and dredging. They are killed by being placed in shallow water in order to rot the tissues, thus leaving the skeleton. They are then beaten, washed, and bleached in the sun. They may be farmed artificially by fastening small pieces of suitable species in the proper type of water and waiting several years for them to reach a usable size.

Economic importance of sponges

The skeletons of sponges have a great number of uses for bathing, washing, absorbing fluids during operations, packing materials, and soundproofing. Even though artificial sponges made of cellulose and other materials are widely used today, natural sponges are still in great demand. The best commercial sponges are found in the warm waters of the Gulf of Mexico, the West Indies, and the Mediterranean Sea. Over one million dollars' worth of sponges have been collected off the Florida coast in a single year.

Sponges are usually not used as food by other animals because of the skeletal materials, their bad odors or tastes, and their relatively small supply of nutritive materials. Some smaller animals may seek protection in or among sponges. Certain boring sponges attach themselves to the shells of clams and oysters and, by means of a secretion, bore the limy shells so full of holes that they destroy the animal inside. Siliceous sponges and protozoans of the order Radiolaria may initiate the process of flint formation. It is believed that beds of flint may be made from a mass of sponge skeletons in a few years.

Freshwater sponges may attach themselves to water pipes, water filtration equipment, and reservoirs and, together with other living organisms, form a feltlike mass that interferes with the water system.

Sponges may starve oysters and other shelled mollusks by attaching themselves to the shells and taking food from the mollusks. Fossil sponges that are similar to present-day forms have been found in chalk and flint formations that are estimated to be one-half billion years old.

CLASSIFICATION OF COELENTERATES (PHYLUM COELENTERATA)

Class Hydrozoa (Gr. *hydor*, water; *zoon*, animal) Marine and freshwater types; solitary or colonial; gastrovascular cavity is not held in position with membranous mesenteries; medusae (when present) with a velum (membrane on the undersurface); usually with alternation of generations of asexual polyps and sexual, umbrella-shaped medusae, although one or the other may be absent. Examples: *Hydra*, *Obelia*, *Gonionemus*, and *Physalia*.

Class Scyphozoa (Gr. *skyphos*, cup; *zoon*, animal) Marine types; solitary; gastrovascular cavity is held in position with mesenteries; polyp stage is absent or inconspicuous; umbrella-shaped medusa stage is large, without a velum, and typically with eight notches on the margin and supplied with sense organs. Examples: large jellyfishes.

Class Anthozoa (Gr. *anthos*, flower; *zoon*, animal) Marine types; solitary or colonial; polyp stage but no medusa stage; radially arranged mesenteries (eight or more) extend from the body wall to the central gullet. Examples: sea anemone (*Metridium*) and corals.

Class Hydrozoa

Hydra

Hydra (Figs. 17-6 to 17-8) is a common, cylindrical, radially symmetrical, solitary, freshwater animal, which may be about one fourth inch long when extended.

Two common species of *Hydra* are the green *Hydra* (*Chlorohydra viridissima*), which is green because algae live within its body symbiotically, and the brown *Hydra* (*Pelmatohydra oligactis*).

The free end (oral end) contains a mouth in the center of a hypostome (Gr. *hypo*, under; *stoma*, mouth), which is surrounded by a circle of about six slender, extendable tentacles. The mouth opens into a gastrovascular cavity (enteron) that connects with the hollow tentacles.

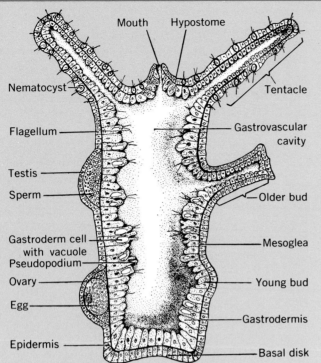

Fig. 17-6
Hydra, a freshwater coelenterate, in section (much enlarged and somewhat diagrammatic).

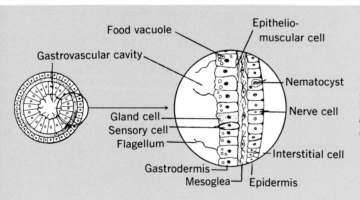

Fig. 17-7
Hydra, showing the nerve net (diagrammatic). Special staining techniques show the netlike arrangement of the nervous system and its concentration in the mouth region.

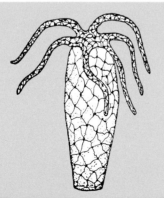

Fig. 17-8
Hydra (cross section and somewhat diagrammatic). The central cavity is the gastrovascular cavity (enteron).

The body wall consists of an outer epidermis (ectoderm) and an inner gastrodermis (endoderm). A thin, noncellular mesoglea is present between them (Fig. 17-10).

The epidermis cells are typically cube-shaped and include such types as (1) epitheliomuscular cells, which possess contractile fibers; (2) cnidoblasts, which contain the saclike nematocysts; (3) interstitial cells, which give origin to nematocysts, buds, and reproductive gonads; (4) gland cells, which secrete mucus on the basal disk for attachment; and (5) nerve and sensory cells, which form part of the so-called nerve net (Fig. 17-7).

The gastrodermis cells are typically column shaped and line the gastrovascular cavity. Among such gastrodermal cells are (1) digestive-muscular cells, which digest foods (intracellular) and possess flagella for moving water and fibers for contraction; (2) gland cells, which secrete mucus and albumin; (3) interstitial cells, which are thought to give rise to other types of cells; and (4) sensory and nerve cells.

The mesoglea consists of a jellylike substance and contractile fibers from the two cellular layers.

A characteristic of *Hydra* and of coelenterates in general is the presence of stinging cells, or nematocysts (Gr. *nema*, thread; *kystis*, sac), which are saclike, filled with fluid, and contain a coiled thread. A triggerlike cnidocil (Gr. *knide*, nettle; L. *cilium*, hair) projects from the cell and, when stimulated (by acetic acid), causes the nematocyst to explode, thus uncoiling the thread. Probably increased fluid pressure in the sac is responsible for this phenomenon. There are several kinds of nematocysts in *Hydra*, which perform such functions as paralyzing organisms by a poisonous secretion, producing an adhesive secretion for attachment, and coiling around their prey. When a nematocyst is discharged, the surrounding cnidoblast cell is destroyed and must be replaced. Most nematocysts do not harm man, but those of large jellyfishes and the Portuguese man-of-war may be dangerous.

Hydra feeds on small animals and plants, overpowering its prey with its tentacles and nematocysts. Food is forced into the gastrovascular cavity, where certain gastrodermal cells secrete digestive enzymes and the food is digested. Some food particles are engulfed by the pseudopodia of the endo-dermal epitheliomuscular cells, where digestion occurs within food vacuoles.

Indigestible materials are forced out through the mouth. Digested materials may diffuse from cell to cell. The flagella and body movements may assist in circulating the contents of the gastrovascular cavity. Respiration and excretion are performed by individual cells, since there are no special organs. Necessary energy is obtained by oxidizing foods within cells. Wastes such as urea and carbon dioxide pass from the cells into the water.

Hydra may attach itself by its basal disk, or it may move in different ways, such as by: (1) gliding on the basal disk; (2) a type of "somersaulting," whereby the basal disk is suddenly detached and the animal turns over completely and reattaches itself; and (3) a "measuring-worm" type of movement, in which *Hydra* bends over and attaches its tentacles, then detaches the basal disk and slides it along toward the tentacles, which are then released, and the *Hydra* assumes its upright position.

The nervous and sensory equipments of coelenterates consist of a plexus of nerve cells, considered to be a nerve net, which is connected with nerve fibers within the epidermis. The nerve net is connected to sensory cells, in order to receive stimuli, and to epitheliomuscular cells, in order to perform movements. The nervous system functions especially well in the hypostome and basal disk regions, and sensory cells are most numerous in these areas. Sometimes the sensory-nerve net is referred to as the neuromuscular system.

Hydra responds to external and internal stimuli. The body and tentacles contract and expand, the movements being slower, as a rule, when *Hydra* is well fed. The movements caused by the contractile fibers in the body wall result from the proper transmission of impulses through the nerve net. The degree of response depends partly on the kind and intensity of the stimulus. Usually, if the stimulus is strong, *Hydra* will respond negatively. Mild stimuli may cause only localized responses. *Hydra* tends to seek moderately intense light and avoids very strong light. It reacts negatively to strong chemicals. Cooler water seems to be most acceptable, for *Hydra* has a tendency to move away when temperatures approach 20° C. When subjected to electric current, the mouth end tends to orient toward the anode (positive charge) and the basal end toward the cathode (negative charge). By trial-and-error, *Hydra* seems to seek conditions best fitted for its existence.

Under natural conditions, particularly in the autumn when there is a reduction of water temperature, *Hydra* may produce several testes, which are rounded outgrowths near the mouth end. A single, larger, somewhat conical ovary is produced nearer the basal end. Numerous sperm, which are set free in the water, swim (each has one long flagellum) to the ovary where one will fertilize the egg. In the egg the normal number of chromosomes has been reduced from twelve to six. A similar reduction has occurred in the sperm, so that the fertilized egg (zygote) again has the normal number of twelve. Within the ovary the zygote undergoes divisions to form a hollow blastula, which is followed by a solid gastrula. The embryo passes from the parent and is surrounded by a protective, shell-like cyst. When favorable conditions are encountered, the shell breaks and the young *Hydra* emerges. Both male and female gonads may be present on the same individual, in which case self-fertilization may occur. When gonads are formed at different times, cross-fertilization must occur.

Asexual buds may project from the sides as outpocketings of the entire body wall, even with an internal extension of the parent gastrovascular cavity. The bud develops a mouth and tentacles, eventually constricts at its base, and detaches to live separately.

Hydra has a great ability to restore lost parts through regeneration. When cut into pieces, each part may regenerate a new individual. If the hypostome region is cut off, a new individual is formed. If the mouth region is split, an individual with two "heads" is produced. Parts of individuals may be grafted together experimentally, even between different species. The ability of animals to regenerate lost parts was first discovered in 1740 in the *Hydra*. Even if a *Hydra* is turned inside out, the cells of the epidermis and gastrodermis may migrate past each other through the mesoglea to their original positions. Regeneration occurs to a greater or lesser extent in all animals, being correlated with the increase in complexity of animal types and influenced by many external and internal factors.

Obelia and Gonionemus

Obelia (Fig. 17-9) is a marine, colonial animal that illustrates the principle of alternation of generations, in which the budding, asexual generation is represented by a conspicuous, polyp-type, hydroid stage, and the sexual generation is represented by a small, inconspicuous, jellyfish-type, medusoid stage. *Gonionemus*, another marine hydrozoan, also represents alternation of generations because its asexual generation is represented by a tiny, inconspicuous, hydroid stage (polyp) and its sexual generation by a conspicuous, jellyfish-type, medusoid stage. The two most conspicuous generations of these different animals will be studied, and the two inconspicuous stages will not be considered.

The colonial, hydroid stage of *Obelia* is attached to a substrate such as rock and possesses many branches that contain numerous polyps of two types: (1) the nutritive hydranths (Gr. *hydor*, water; *anthos*, flower) and (2) the reproductive gonangia (Gr. *gone*, offspring; *angios*, case). The hydranths of *Obelia* somewhat resemble *Hydra*, for they possess a hypostome and mouth, which are surrounded by many tentacles with nematocysts for capturing prey.

The vase-shaped gonangia contain an internal blastostyle on which are borne asexual buds that develop into umbrella-shaped medusae with tentacles around the margin. The medusae give rise to sperm and eggs, the sperm being produced in testes in one individual and the eggs being produced in ovaries in another individual (diecious).

The entire animal has an outer, horny, protective layer, the perisarc (Gr. *peri*, around; *sarx*, flesh). Inside the perisarc is the hollow, tubular coenosarc (Gr. *koinos*, common), consisting of epidermis, mesoglea, and gastrodermis. The internal cavity, called the gastrovascular cavity, is continuous throughout the colony.

The medusoid stage of *Gonionemus* is umbrella-shaped and about $\frac{1}{2}$ inch in diameter (Fig. 17-10). The numerous tentacles, which bear nematocysts, fringe the margin. Each tentacle bends sharply near its tip and bears an adhesive pad for attachment. The convex, or aboral, side is called the exumbrella, whereas the concave, or oral, side is called the subumbrella. A shelflike membrane, the velum (L. *velum*, covering), partly covers the con-

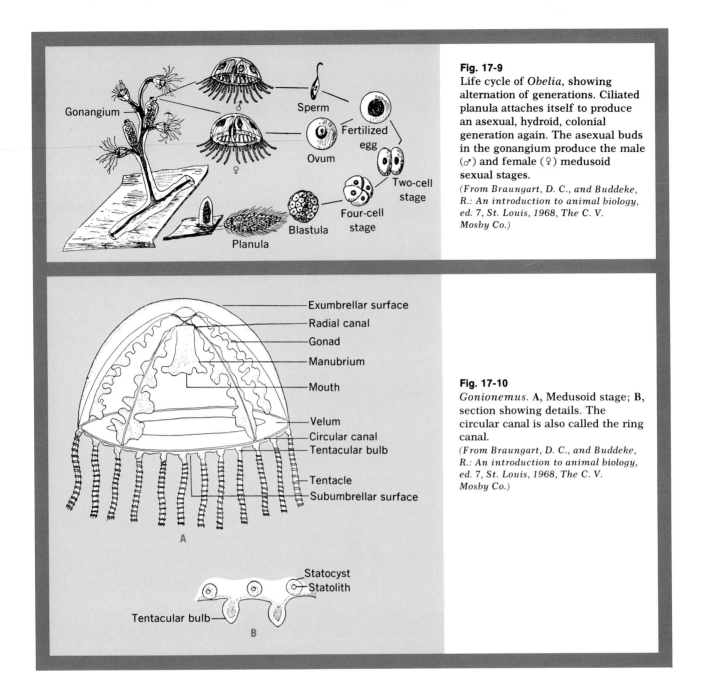

Fig. 17-9
Life cycle of *Obelia*, showing alternation of generations. Ciliated planula attaches itself to produce an asexual, hydroid, colonial generation again. The asexual buds in the gonangium produce the male (♂) and female (♀) medusoid sexual stages.
(From Braungart, D. C., and Buddeke, R.: An introduction to animal biology, ed. 7, St. Louis, 1968, The C. V. Mosby Co.)

Labels in Fig. 17-9: Gonangium, Sperm, Fertilized egg, Ovum, Two-cell stage, Four-cell stage, Blastula, Planula

Fig. 17-10
Gonionemus. **A**, Medusoid stage; **B**, section showing details. The circular canal is also called the ring canal.
(From Braungart, D. C., and Buddeke, R.: An introduction to animal biology, ed. 7, St. Louis, 1968, The C. V. Mosby Co.)

Labels in Fig. 17-10: Exumbrellar surface, Radial canal, Gonad, Manubrium, Mouth, Velum, Circular canal, Tentacular bulb, Tentacle, Subumbrellar surface, A, Statocyst, Statolith, Tentacular bulb, B

cave side. There is a circular opening in the velum through which water is forced by body contractions propelling the animal in a type of jet propulsion.

Hanging downward from the subumbrella is a tubular manubrium (L. *manubrium*, handle) with a mouth at its tip. The mouth leads into the gastrovascular cavity, and four radial canals lead to a ring canal, near the margin. Attached to the radial canals are testes in some individuals and ovaries in others. Hence, the medusoid stage of *Gonionemus* is diecious.

The egg fertilized by the motile sperm forms a zygote, which by repeated cell divisions produces a multicellular blastula. This becomes a free-swimming ciliated planula, which eventually settles down, attaches itself, and starts a new colony of the hydroid type again.

237

Physalia (Portuguese man-of-war)

The Portuguese man-of-war *(Physalia)* is a marine colony composed os several different types of polyps suspended in the water from a gas-filled, colored flat (pneumatophore), which may be as much as 1 foot long (Fig. 17-11). The chief polyps include (1) nutritive zooids that capture prey, which is digested in a digestive sac, thus nourishing the colony; (2) long defensive zooids with powerful, poisonous, stinging nematocysts; and (3) reproductive zooids. A sail-like crest on the float assists in carrying the animal along on the surface of tropical seas. The float and crest are beautiful, having different iridescent colors. The defensive zooids may be more than 50 feet long, and the nematocysts may be strong enough to injure or kill a man. Certain kinds of fish are captured for food, but other species of fish are immune to attack and even live in association with the tentacles, thereby obtaining protection.

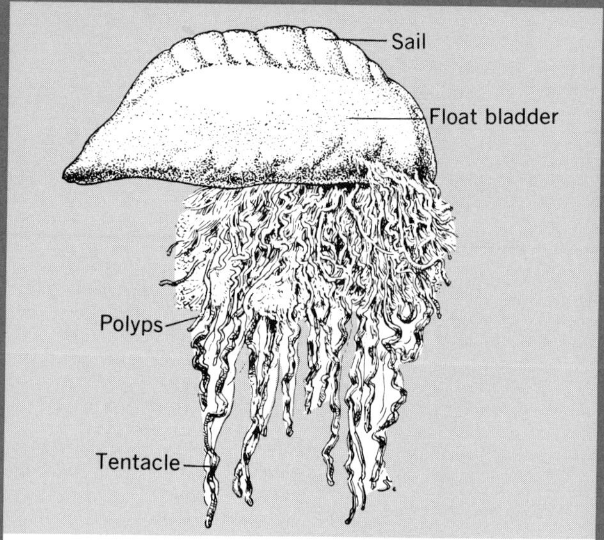

Fig. 17-11

Portuguese man-of-war *(Physalia)*. This colonial coelenterate floats on the surface of the sea. The male and female zooids, vegetative polyps, and long tentacles with nematocysts are suspended from the float bladder.

Class Anthozoa
Sea anemones

The sea anemone *(Metridium)* is a marine coelenterate with a cylindrical body, soft tough skin, and circles of hollow tentacles arranged around a mouth (Fig. 17-13). A pedal disk attaches it to rocks and wood. Sea anemones may move by gliding along on the pedal disk. Foods consist of smaller animals that they capture with their tentacles and nematocysts. Respiration and excretion occur by diffusion through the cells and body wall. Sea anemones are conspicuous, beautiful, and when expanded, resemble a garden with flowerlike

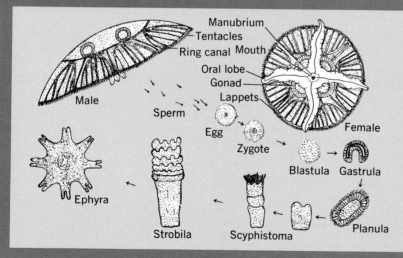

Fig. 17-12

Jellyfish *(Aurelia)* of class Scyphozoa life cycle. The ciliated, larval planula settles down and forms a small polyp, the scyphistoma, which divides by transverse fission to form a series of small jellyfishes. Each free-swimming ephyra stage is produced by the strobila stage.

(From Goodnight, C. J., and Goodnight, M. L.: Zoology, ed. 1, St. Louis, 1954, The C. V. Mosby Co.)

Fig. 17-13
Sea anemone (*Metridium*) showing numerous sensitive tentacles that surround the mouth.
(Courtesy Carolina Biological Supply Co., North Carolina.)

crowns of various colors. In nature they are not so flowerlike, being deathtraps for many smaller animals.

The mouth opens into a gullet (stomodeum), on either side of which is a ciliated groove called the siphonoglyphe (si′ fon o glif) (L. *siphon,* tube; *glyphein,* to carve). The gullet leads to the gastrovascular cavity, which is divided into six radially arranged chambers by six pairs of primary mesenteries (septa), which extend from the body wall to the gullet. Water passes from chamber to chamber through ostia in the mesenteries, which are open below the gullet. Numerous smaller mesenteries extend inwardly from the body wall, partially subdividing the larger chambers, but they do not reach the gullet. The free edges of the mesenteries below the gullet are thickened into digestive filaments, which secrete digestive enzymes. Near the base these filaments bear long threadlike acontia (a -kon′ ti a) (Gr. *akontion,* dart), which are armed with nematocysts and gland cells. The acontia capture prey by being extended through the mouth and from pores in the body wall.

Sex organs (gonads) are present on the margins of the mesenteries, the sexes being separate (die-

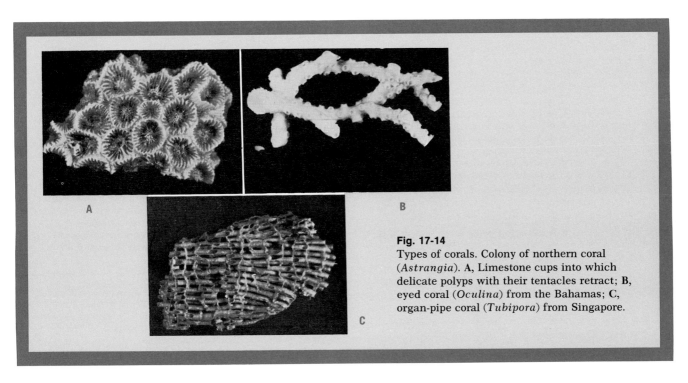

Fig. 17-14
Types of corals. Colony of northern coral (*Astrangia*). **A,** Limestone cups into which delicate polyps with their tentacles retract; **B,** eyed coral (*Oculina*) from the Bahamas; **C,** organ-pipe coral (*Tubipora*) from Singapore.

239

cious). The zygote develops into a ciliated larva that later forms an adult. No medusa stages have ever been discovered in any Anthozoa. Asexual reproduction may occur by budding.

Corals

The coral (Fig. 17-14) is a polyp with short tentacles and lives in a stony cuplike structure. The polyp is constructed somewhat like the sea anemone, but it secretes a calcareous skeleton (coral) around itself and between its mesenteries. Thus the cup is divided radially by ridges. The living polyps can withdraw into these cups. A number of individuals usually live in a colony, although some are solitary. They may display many branches or be rounded masses. Varieties of colors are found in different species. Corals live in the shallower waters of the sea. The polyps are alive only on the surface of coral masses, and they are rarely found below 100 feet. Many characteristics and physiologic processes of coelenterates are found in corals. The great coral reefs of southern seas and smaller, circular atolls that enclose lagoons are of coral origin. The Great Barrier Reef off the Australian coast is more than 1,000 miles long.

Economic importance of coelenterates

Coelenterates are of limited economic importance. Some animals utilize certain types as food. Some flatworms and mollusks may eat hydroids and use the stinging nematocysts for their own defense. Sea anemones may be used for food in certain areas where better foods are unavailable. Corals are important because, through years of growth, they may build calcareous materials that can be used for building purposes, including roadway construction. Certain types of corals are used in making ornaments and jewelry.

Review questions and topics

1 What are the characteristics of sponges and coelenterates?
2 What advances in structures and functions do we find in sponges and coelenterates in each of the following: (1) methods of obtaining and digesting foods; (2) excretion of wastes; (3) coordination of body parts; and (4) sensory equipment and defense mechanisms?
3 What advancement is made in the middle layer in these animals?
4 What is the significance of the different types of symmetry in these animals?
5 Contrast intracellular and extracellular digestion, giving examples of each.
6 What advancement is noted in the gastrovascular cavity?
7 Of what significance is an alternation of generations as shown in certain coelenterates?
8 What is meant by a nerve net? Is such a mechanism, with some modification, found in man?
9 Is the organization of an animal into oral and aboral ends an improvement over lower animals?
10 Discuss the types of skeletal materials found in sponges and certain coelenterates.
11 What is the improvement in the nervous mechanism as we proceed from the sponges to the coelenterates?
12 Discuss the economic importance of sponges and coelenterates.

Selected references

Barnes, R. D.: Invertebrate zoology, ed. 2, Philadelphia, 1969, W. B. Saunders Company.

DeLaubenfels, M. W.: A guide to the sponges of Eastern North America, Miami, 1953, University of Miami, Marine Laboratory Special Publication.

Hyman, L. H.: The invertebrates; vol. I, Protozoa through Ctenophora, New York, 1940, McGraw-Hill Book Company.

Hyman, L. H.: Coelenterata. In Edmondson, W. T. (editor): Fresh-water biology, New York, 1959, ed. 2, John Wiley & Sons, Inc.; 1st ed. by Ward, H. B., and Whipple, G. C., pub. 1918.

Jewell, M.: Porifera. In Edmondson, W. T. (editor): Fresh-water biology, New York, 1959, ed. 2, John Wiley & Sons, Inc.; 1st ed. by Ward, H. B., and Whipple, G. C., pub. 1918.

Lenhoff, H. M., and Loomis W. F. (editors): The biology of Hydra and of some other coelenterates, Coral Gables, Florida, 1961, University of Miami Press.

Meglitsch, P. A.: Invertebrate zoology, New York, 1967, Oxford University Press, Inc.

Moore, H. F., and Goltsoff, P. S.: Sponges. In Tressler, D. K., and Lemon, J. M.: Marine products of commerce, New York, 1951, ed. 2, Reinhold Publishing Corp.

Runcorn, S. K.: Corals as paleontological clocks, Sci. Amer. Oct., 1966.

chapter eighteen

Helminths—flatworms and roundworms

PHYLUM PLATYHELMINTHES (FLATWORMS)

Class Turbellaria (L. *turbella*, a stirring)
 Usually free living; flattened dorsoventrally; external epidermis is ciliated and contains secretory cells and rodlike rhabdites, mouth usually on the ventral surface; reproduction is usually monecious. Example: *Dugesia tigrina.*

Class Trematoda (Gr. *trematodes*, pore or sucker)
 Parasitic; leaflike or cylindrical in shape; flattened dorsoventrally; thick external cuticle without cilia; ventral suckers, and sometimes hooks; usually monecious. Examples: flukes, such as *Fasciola hepatica* and *Clonorchis (Opisthorchis) sinensis.*

Class Cestoda (Gr. *kestos*, girdle or ribbonlike)
 Parasitic; tapelike in shape; flattened; thick external cuticle without cilia; body divided into a series of proglottids (pro -glot' id); an anterior scolex with suckers or hooks, or both, for attachment; no digestive or sensory systems; usually monecious and self-fertilizing. Examples: tapeworms, such as *Taenia.*

In early systems of classification all wormlike animals were collectively known as Helminthes (Gr. *worm*). These organisms have both bilateral symmetry (that is, two sides are mirror images) and at least a semblance of a head and tail.

The groups considered in this chapter, the Platyhelminthes and the Nemathelminthes, are members of different and distinct phyla.

The Platyhelminthes have a distinct middle or third germ layer, called a mesoderm, in place of the mesoglea of the coelenterates. In addition to other organs, this tissue produces distinct muscle layers, which permit more varied movement. Because they have a compact mesoderm instead of an open body cavity, these organisms are often called acoelomates, that is, without a coelom, or lined body cavity. This coelom is not to be confused with the enteron, or digestive cavity.

The second group considered in this chapter, the roundworms or nematodes, have a body cavity in which the intestine and the reproductive system are supported more or less loosely. However, neither these organs nor the body wall are lined with mesodermal tissue. Therefore, roundworms are referred to as pseudocoelomates.

Class Turbellaria (free-living flatworms)

The free-living flatworms of the class Turbellaria are commonly called planarians. They and related forms occur in freshwater ponds and streams throughout the world. Planarians avoid light, and sometimes hundreds of them may be found on the underside of a submerged leaf or rock.

Dugesia (planaria)

Dugesia tigrina, a common type of planarian, will be discussed here to illustrate the characteristic structure and activities of the freeliving flatworms. The animal is a small, flat, wormlike form, which is usually about 10 to 20 mm. (about $\frac{1}{2}$ inch) in length. The broadly triangular anterior end has two lateral projections called auricles. A pair of dorsal, light-sensitive eyespots is located between the auricles. The structure of the eyespots is such that they give the animal a "cross-eyed" appearance. The body surface of *Dugesia* often has a brownish, mottled appearance. The ventral surface has two openings, the mouth and the genital pore, which are not easily observed.

An epidermis, which is formed by a single layer of cells resting on a basement membrane, covers the body. Most of these cells are alike and have numerous embedded rhabdites whose function is not known. Numerous mucous-secreting gland cells are also present, especially on the lower surface. Many cilia are also located on the ventral

241

surface, which aid in locomotion over a mucous surface formed by gland cells.

The interior of *Dugesia* is filled with an unorganized parenchyma, which is a spongy mass with many undifferentiated spaces. Many individual ameboid cells are also present. There is no distinct body cavity, since the various organs are surrounded by the parenchyma.

Circular, longitudinal, and dorsoventral muscles control the contractions that are necessary for crawling and wriggling movements. In addition to helping the organism perform obvious movement, muscles are important in the protusion of the pharynx to grasp or to tear food. Other muscles aid in the movement of food through the digestive system and throughout the parenchyma.

Dugesia's nervous system consists of a "brain," which is composed of two thickened cerebral ganglia and two longitudinal nerve cords, which extend toward the posterior end of the organism. Transverse connectives run between the nerve cords, and many peripheral nerves connect the surface and the main nerves.

Sense organs and several kinds of receptors are present. The most obvious of these are the eyes. The eye consists of an inverse pigment cup and light-sensitive neurosensory cells. In *Dugesia* these neurosensory cells are stimulated only by parallel light rays. This sensitivity to light rays provides a means of orientation for the organism.

Ciliated pits at the base of the auricles serve as chemoreceptors. *Dugesia*, like all planarians, gives immediate and highly positive responses to such substances as meat juice and blood.

Touch receptors, consisting of single neurosensory cells with bristles, are common on the external surface. Other cells may also serve to orient the animal to food, water currents, and the oxygen concentration of the water.

The excretory system consists of two longitudinal, multibranched canals with numerous terminal, hollow flame cells, which contain a tuft of vibrating cilia to propel wastes. The two main canals open through minute dorsolateral excretory pores. Respiration occurs through the body surface.

Dugesia reproduce asexually by dividing transversely behind the pharyngeal region, each resulting part reorganizing into a complete animal.

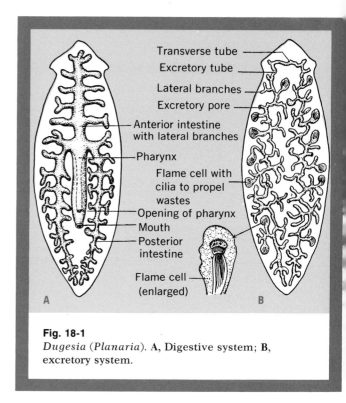

Fig. 18-1
Dugesia (Planaria). A, Digestive system; B, excretory system.

Sexual reproduction also occurs, as each organism has both testes and ovaries, hence it is monecious. Each of the many spherical testes has a tubular vas efferens. All the vasa efferentia on one side of the body join to form a larger tubular vas deferens. The right and left vasa deferentia enter a seminal vesicle where sperms are stored until discharged through a muscular penis.

Two ovaries, near the anterior end, are connected to two oviducts that carry eggs to the one vagina. Several yolk glands along the oviducts supply yolk (food) to the eggs. Connected to the vagina is a rounded uterus (seminal receptacle) that stores sperms from the mating partner. Although *Dugesia* is monecious, cross-fertilization is practiced.

During copulation each organism inserts its penis into the other, and there is a mutual exchange of sperms from the seminal vesicles of each animal. The sperms then pass through the oviducts to fertilize the eggs, forming a zygote. As the zygotes move down the oviducts, yolk (food) is added, and finally shells are formed around the eggs. A number of eggs, encased in

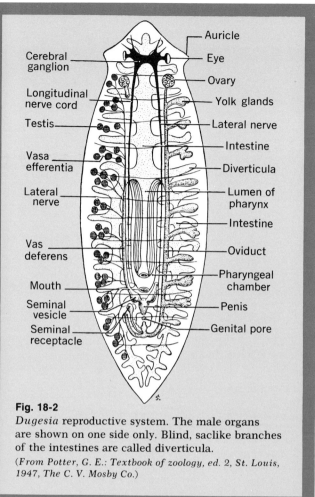

Fig. 18-2
Dugesia reproductive system. The male organs are shown on one side only. Blind, saclike branches of the intestines are called diverticula.
(From Potter, G. E.: Textbook of zoology, ed. 2, St. Louis, 1947, The C. V. Mosby Co.)

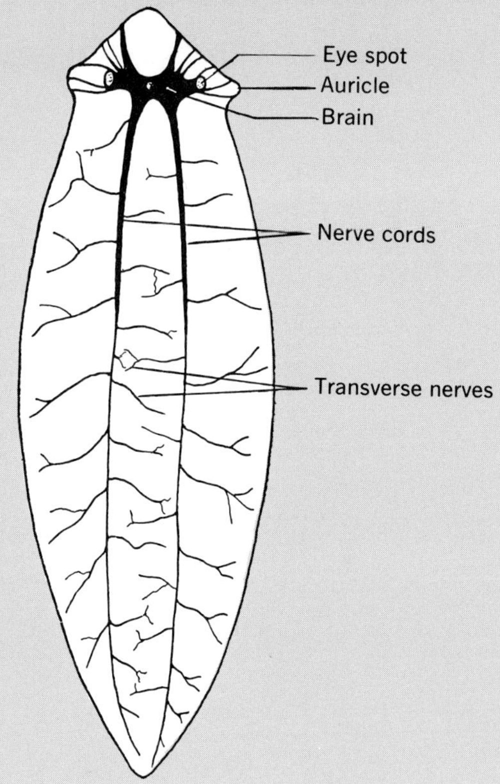

Labels in Fig. 18-2:
Cerebral ganglion — Auricle — Eye — Ovary — Yolk glands — Longitudinal nerve cord — Testis — Lateral nerve — Intestine — Vasa efferentia — Diverticula — Lateral nerve — Lumen of pharynx — Intestine — Vas deferens — Oviduct — Mouth — Pharyngeal chamber — Seminal vesicle — Penis — Seminal receptacle — Genital pore

Fig. 18-3
Dugesia nervous system. Note the nerves leading from the brain to the auricles.

Labels in Fig. 18-3:
Eye spot — Auricle — Brain — Nerve cords — Transverse nerves

a single cocoon, are passed into the water. After development, immature planaria emerge from the cocoon.

Dugesia has the remarkable ability to regenerate lost parts. When cut in two, the anterior part will regenerate a new tail, and the posterior part will regenerate a new head. The organism illustrates the theory of the axial gradient. Metabolic activity is measured by the amount of oxygen consumed and the amount of carbon dioxide given off. The metabolic rate is greatest at the anterior end of the animal and gradually and progressively decreases toward the posterior end. The metabolic rate of any part will depend on its position with respect to this axial gradient. In any fragment a head will develop where metabolic activity is greatest and a tail where it is lowest. *Dugesia* may

be grafted in many ways to produce animals with two heads and two tails.

Class Trematoda (Flukes)

Flukes are the flatworm parasites of a wide variety of vertebrates. Many of them are large enough to be seen without a microscope. The fact that they occur in parts of animals used for food has been known for hundreds of years. In spite of this it was not until 1883 that the complete life cycle of a fluke was described.

As will be shown, flukes produce a fantastic number of offspring during a rather complex life cycle. Since some stages of this cycle occur in man, flukes are of great medical importance in some areas of the world.

243

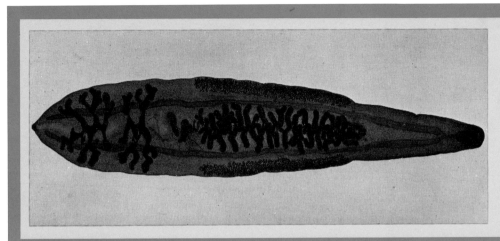

Fig. 18-4
Oriental liver fluke, *Clonorchis*
(*Opisthorchis sinensis*).
(*Courtesy, General Biological Supply
House, Inc., Chicago. Ill.*)

Table 18-1
Some common flukes

Fluke	Adult stages in primary (definitive) host	Immature stages in intermediate host	Where found
Oriental liver fluke (*Clonorchis* [*Opisthorchis*] *sinensis*)	Man, cat, dog, fish-eating mammals	Snails (different species) and freshwater fish	Very common in China, Thailand, Japan, Southeast Asia, Korea, Philippines
Sheep liver fluke (*Fasciola hepatica*)	Sheep, cow, pig	Snail	Europe, United States, Cuba
Blood fluke (*Schistosoma hematobium*)	Man (blood vessels of urinary system), also monkey	Snail (*Bulinus*)	Africa, Near East, Australia, Portugal
Blood fluke (*Schistosoma japonicum*)	Man (blood vessels), also cat, dog, pig, etc.	Snail (*Oncomelania*)	Japan, China, Philippines
Lung fluke (*Paragonimus westermani*)	Man, cat, dog, pig, tiger	Snails, crabs (freshwater), crayfish	China, Japan, Korea, Philippines, America
Skin fluke or "swimmer's itch" fluke (*Trichobilharzia*)	Aquatic birds	Snails (many species)	Widespread in northern lakes and coastal regions

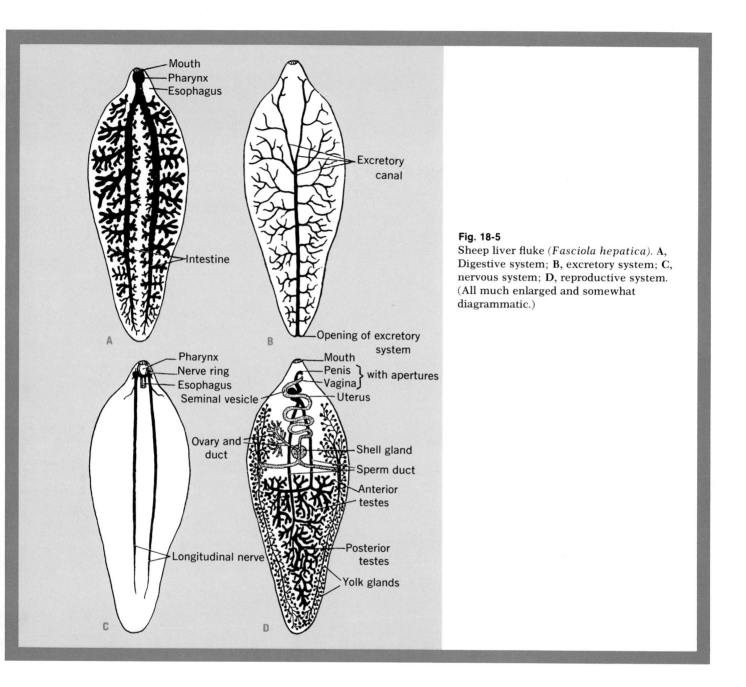

Fig. 18-5
Sheep liver fluke (*Fasciola hepatica*). **A**, Digestive system; **B**, excretory system; **C**, nervous system; **D**, reproductive system. (All much enlarged and somewhat diagrammatic.)

Oriental liver fluke

The average size of *Clonorchis (Opisthorchis) sinensis* is about 3 by 15 mm. It is covered with a rough cuticle (Fig. 18-4). It has an anterior (oral) sucker around the mouth and a ventral sucker below the oral one.

The digestive system consists of a muscular pharynx attached to two long, unbranched intestines. There is no anus; the excretory system consists of two tubes whose branches terminate in flame cells to propel wastes. One posterior excretory pore exists. The nervous system, as in planarians, consists of two cerebral ganglia connected to two ventral, longitudinal nerve cords that are transversely connected.

Like the planarians, this fluke is monecious, but cross-fertilization is normal. The reproductive system generally resembles that of *Dugesia*.

245

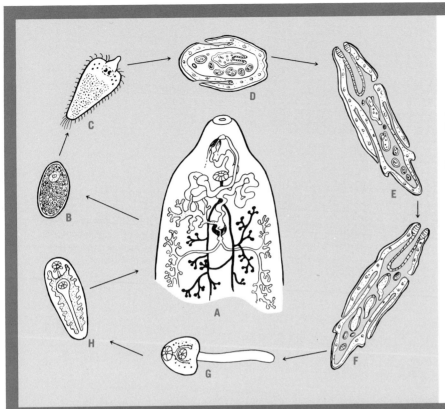

Fig. 18-6
Life cycle of *Fasciola hepatica.* **A,** Adult, showing the anterior part of the reproductive system, oral sucker, and ventral sucker (found in sheep); **B,** egg and yolk cells within a shell (found in feces); **C,** miracidium, showing cilia and awl-like proboscis for boring (found in water); **D,** saclike sporocyst with germ cells that develop into rediae I (in freshwater snail); **E,** redia, showing the gut, birth pore, and germ cells that develop into rediae II (in snail); **F,** redia, showing the gut, birth pore, germ cells, and cercariae (in snail); **G,** free-swimming cercaria with oral and ventral suckers, branched gut, and tail (in water); **H,** metacercaria (encysted cercaria) with suckers, branched gut, and cyst (on grass). Sheep (or man) eats grass, and adult fluke develops.

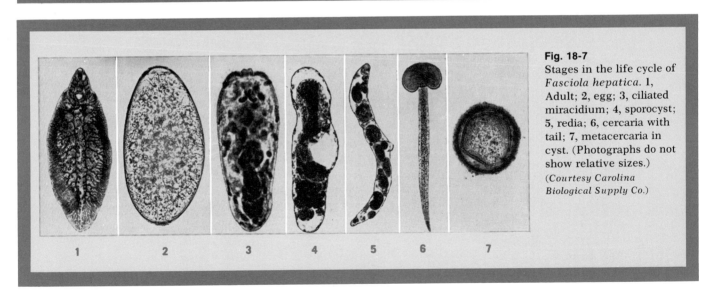

Fig. 18-7
Stages in the life cycle of *Fasciola hepatica.* 1, Adult; 2, egg; 3, ciliated miracidium; 4, sporocyst; 5, redia; 6, cercaria with tail; 7, metacercaria in cyst. (Photographs do not show relative sizes.) (*Courtesy Carolina Biological Supply Co.*)

Life cycle

The adult stage of the oriental liver fluke is normally spent in the bile ducts of man. The eggs, each containing a ciliated miracidium (mi ra -sid′ i um), are shed with the feces into water. When the eggs are ingested by certain species of snails, the miracidium is transformed into an elongated sporocyst, which produces one generation of rediae (after Redi, an Italian scientist). The rediae pass into the snail's liver where they produce tadpolelike cer-

246

cariae that escape into water where they bore into the muscles of certain fishes. In the fish the cercariae lose their tails and encyst to form metacercariae (Gr. *meta*, after). When man eats incompletely cooked fish, the protective cyst is dissolved, and the metacercariae move up the bile ducts to become adults. They may live in the bile ducts for many years, and serious infections may cause a severe liver disease and death. Additional information about flukes is given in Table 18-1.

Sheep liver fluke

Fasciola hepatica (L. *fasciola*, small bandage; Gr. *hepar*, liver) is parasitic in the liver of sheep, cows, and pigs, where it may cause great damage. The general characteristics and life cycle are shown in Figs. 18-5 to 18-7. There may be as many as 200 adults in a sheep liver, and if all live, they may produce a total of 100 million eggs. The larval stage develops after the eggs are passed into water. The ciliated larvae (miracidia) bore into the body of a certain freshwater snail where a complex series of developmental stages occurs. Finally, they pass from the snail and become encysted on grass. When this grass is eaten by a sheep, the cysts are dissolved to liberate the young fluke, which finally enters the liver to form an adult, thus completing the complex life cycle. Few cases of human liver infections are known, except in certain areas.

Blood flukes

Blood flukes are the cause of very serious diseases in many parts of Africa, Asia, and other areas of the world. For example, it is reported that in many areas of Egypt, up to 90% of the population may suffer from the effects of these flukes.

An event of significant historical interest occurred in the 1950s concerning blood flukes. The armies of mainland China were about to launch an attack on the Nationalist China forces on Taiwan. The mainland troops were massed, and supplies were gathered. Yet, nothing happened. It was later revealed that thousands of cases of schistosomiasis had occurred among the mainland soldiers. As we know, a sick man cannot fight.

Blood flukes are placed in the genus *Schistosoma* (Gr. *schistos*, divided; *soma*, body) because the

male has a ventral, splitlike fold in his body in which he carries the smaller female. Unlike most flukes, the sexes are in separate individuals (diecious) in blood flukes and the two branches of the digestive tube are united into one at the posterior part of the body. The life cycles of the various species of blood flukes are similar and generally resemble those of other flukes. However, the cercariae bore directly into the skin.

Class Cestoda (Tapeworms)
Pork tapeworm

The adult *Taenia solium* (Figs. 18-8 and 18-9) lives in the human alimentary canal, where it may become 10 feet long. Larval stages are found in the muscles of pigs. An enlarged anterior scolex has suckers and numerous hooks. The body surface is covered with a thick, protective cuticle. The animal is composed of a linear series of proglottids that may number 1,000. New proglottids are produced in the neck region by a process of transverse budding, so that the most immature proglottids are nearest the scolex, and the broader, mature ones are toward the posterior end.

There is no digestive system, and foods are absorbed from the surroundings. Two lateral excretory canals run throughout all the proglottids and possess flame cells to propel wastes. A transverse excretory canal connects the lateral canals in each proglottid. Two lateral nerve cords run throughout the proglottids and terminate in a nerve ring in the scolex. Each mature proglottid has a complete set of male and female sex organs (monecious). Fertilization of eggs may occur by sperms of the same proglottid (self-fertilization), from adjacent proglottids, or from another tapeworm (cross-fertilization). As the posterior, mature proglottids break off, they pass with the feces and disintegrate, thus depositing eggs and embryos on grass and soil, where they may be ingested by pigs.

The male reproductive system consists of many spherical testes, all of which are connected to a sperm duct (vas deferens) leading to the genital pore. The female reproductive system consists of a multibranched ovary; an oviduct; a yolk gland; a shell gland; a saclike, egg-containing uterus; and

247

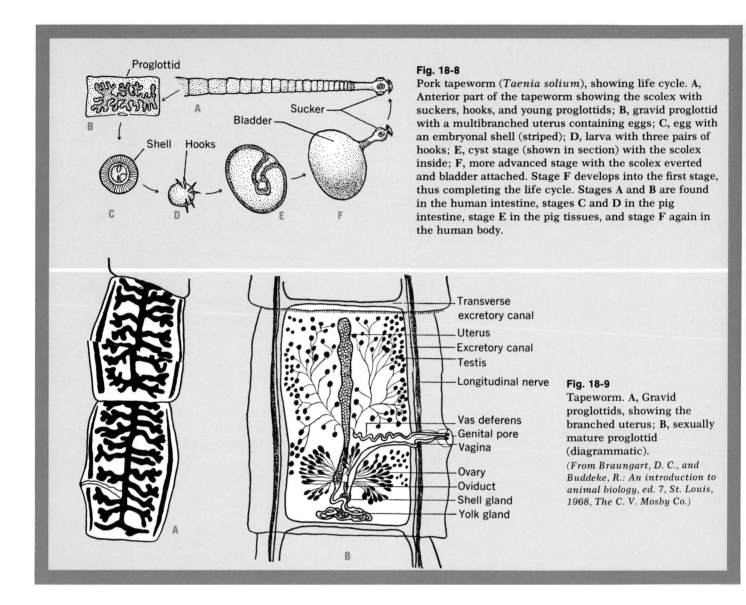

Fig. 18-8
Pork tapeworm (*Taenia solium*), showing life cycle. **A,** Anterior part of the tapeworm showing the scolex with suckers, hooks, and young proglottids; **B,** gravid proglottid with a multibranched uterus containing eggs; **C,** egg with an embryonal shell (striped); **D,** larva with three pairs of hooks; **E,** cyst stage (shown in section) with the scolex inside; **F,** more advanced stage with the scolex everted and bladder attached. Stage F develops into the first stage, thus completing the life cycle. Stages A and B are found in the human intestine, stages C and D in the pig intestine, stage E in the pig tissues, and stage F again in the human body.

Fig. 18-9
Tapeworm. A, Gravid proglottids, showing the branched uterus; B, sexually mature proglottid (diagrammatic).
(From Braungart, D. C., and Buddeke, R.: An introduction to animal biology, ed. 7, St. Louis, 1968, The C. V. Mosby Co.)

a tubular vagina leading to the genital pore. As the uterus matures with its developing eggs, it may become branched and almost fill the proglottid.

Life cycle

When pigs swallow a proglottid or an egg, the shell is dissolved. The six-hooked embryo that develops from the egg burrows through the intestinal wall to enter the blood and lymph vessels. It finally encysts in skeletal muscles and becomes a saclike bladder worm, called a cysticercus (sis te -ser' kus). In 10 to 20 weeks each of the cysticerci forms an invaginated (turned in) scolex with suckers and hooks. If such infested meat is improperly cooked and eaten by man, the cyst wall is dissolved by the digestive juices. The bladder disappears and the scolex evaginates (turns inside out), attaches itself to the intestine, and forms new proglottids that mature in 2 to 3 weeks.

Table 18-2

Some common tapeworms

Tapeworm	Site of adult stages (primary host)	Site of immature larval stages (secondary host)
Pork tapeworm (Taenia solium)	Man (intestine)	Pig (muscle)
Beef tapeworm (Taenia saginata)	Man (intestine)	Cattle (muscle)
Broad tape worm (Diphyllobothrium latum)	Man, dog, cat, fox (intestine)	Freshwater fish, certain crustaceans (cyclops)
Dog tapeworm (Dipylidium caninum)	Dog, cat, sometimes children	Dog lice, fleas
Hydatid tapeworm, or dog tapeworm (Echinococcus granulosus)	Dog, wolf (intestine)	Many mammals, including man, monkey, cattle, sheep, cat (in liver, lungs, brain, etc.)
Sheep tapeworm (Moniezia expansa)	Sheep, goats	A small mite (Galumna)

Beef tapeworm

Taenia saginata is much like the pork tapeworm, but the scolex does not contain hooks. It may be more than 30 feet long and may contain 20,000 proglottids. The adult worm lives in the human alimentary canal, and the larval stages are primarily associated with the muscles of cattle where it is called *Cysticercus bovis*. The greatest incidence of bovine cysticercosis in the United States occurs on the West Coast (Table 18-3).

Man becomes infested with *Taenia saginata* if he eats inadequately cooked beef that is infected with cysticercosis. The infection can be spread to cattle by indiscriminate defecation by humans in feed pens or pastures or by allowing sewage and septic tank effluent to pollute pastures where cattle graze.

The case described on the following pages illustrates the complexity of the problems involved in dealing with *Taenia saginata*.

Table 18-3

Numbers of beef carcasses infected with *Cysticercus bovis*, United States, 1967*

State	Cases
Arizona	303
California	10,455
Colorado	334
Texas	777
All other states	2,538
Total United States	14,407

*Source: Livestock Slaughter Inspection Division, U.S.D.A. (from Morbidity and mortality weekly reports, United States Public Health Service, April, 1968).

Economic importance of flatworms

The free-living types of flatworms are of little direct importance to man, but many of the flukes are dangerous parasites, especially in the Far East, and the tapeworms cause damage in all parts of the world. The various types of flukes may affect such organs and tissues as liver, intestine, blood, lung, and skin. Flukes spend one part of their life cycle in one animal host and the other part of the cycle in another animal host, thereby affecting two different animals. Tapeworms also alternate between two hosts in their life cycle. The particular tissues or organs involved depend on the species of tapeworm. In the dog tapeworm (*Echinococcus granulosus*) a larval stage, known as a hydatid cyst (Gr. *hydatis*, watery sac), may occur in the liver and other organs of man, cattle, horses, and sheep, where the fluid-filled sac may assume large proportions.

PHYLUM NEMATODA (ROUNDWORMS)

In contrast to flatworms, roundworms have a pseudocoel, which is a body cavity that is not lined by a special layer of epithelial cells, the mesenteries, as it is in higher animals. The interior organs lie free in this cavity. Because their body structure consists of an intestine in a cavity, which is surrounded by a body wall, roundworms are often referred to as having a "tube within a tube" structure.

A great many species of roundworms are known. While many of them are free-living, many others are predatory or parasitic. The latter are of much interest because they cause damage to crops and diseases in man and other animals.

Roundworms are often placed in the phylum (sometimes superphylum) Aschelminthes, along with many other groups of worms. Since these other worms are not considered in this textbook, we are using the phylum Nematoda, as well as the class Nematoda, for classification purposes.

Class Nematoda

Nematoda are slender elongate animals with a complex cuticle lacking cilia. They have a complete digestive system. The sexes usually separate, and the males are generally smaller than the females. Examples of nematodes are the human roundworm (*Ascaris lumbricoides*), the hookworm (*Necator americanus*), and the vinegar eelworm (*Turbatrix aceti*).

BOVINE CYSTICERCOSIS—Texas*

An epizootic of bovine cysticercosis has been reported from northern Texas. During the period from March 15 to April 1, 1968, 771 cattle from two large commercial feedlots were slaughtered in packing plants under the United States Department of Agriculture Inspection Program; 346 cattle (45 percent) were found to be infected with *Cysticercus bovis*. Infected cattle are known to have been shipped [to] plants in Oklahoma, Nebraska, Colorado, Missouri, Kansas, Iowa, Texas, and Florida.

An investigation is now in progress to determine the mode of spread of this zoonosis and the extent of the movement of infected cattle into retail channels.

TAENIASIS—Rhode Island*

On May 9, 1968, a 40-year-old female X-ray technician from Rhode Island recognized tapeworm proglottids in her stool. For approximately 2 months the patient had experienced mild abdominal cramps, borborygmi, and a change in her bowel habits from relative constipation to bowel movements on arising each morning. Because she believed that she was infected with pinworms, she had mistakenly been looking at her stools each day until May 9 when she first sighted the tapeworm segments.

The patient gave no history of recent travel. She ate rare beef, often sampling raw hamburger during its preparation, but she rarely ate pork. She purchased all her meat in a single Rhode Island supermarket.

*From Morbidity and mortality weekly reports, United States Public Health Service, April, May, June, 1968.

The patient's stool was examined and found to contain *Taenia* eggs. She was treated with oral Atabrine, but this was not successful. She was then treated with Niclosamide and she stated that within a day her bowel habits returned to the previous normal pattern. Her stools will be examined periodically to see if the entire worm was removed.

The commercial sources of beef for the single supermarket from which the [patient] purchased her meat were traced. It was found that two of the sources were slaughter houses in Nebraska and Iowa that had processed Texas cattle infected with *Cysticercus bovis* during the epizootic that first appeared in mid-March (MMWR, Vol. 17, No. 16). An investigation is now underway to determine if there are more cases of taeniasis in this region.

At the same time that this autochthonous case was found, two imported cases of taeniasis, one in an Ethiopian and the other in a Lebanese, were also reported in Rhode Island.

FOLLOW-UP BOVINE CYSTICERCOSIS — Texas

Intensive investigation of the epizootic of bovine cysticercosis which occurred in March 1968 in cattle from feedlots in northern Texas (MMWR, Vol. 17, Nos. 16 and 23) has recently been completed. The following information summarizes the epidemiologic findings at the feedlots near Gruver and Hereford, Texas, where the infected cattle originated.

Investigation near Gruver, Texas, at Feedlot A

Feedlot A is a commercial feedlot with a capacity for 8,000 cattle. The cattle are shipped to the feedlot from many sources for intensive feeding. On arrival, cattle are put in one of 34 pens at the feedlot. While at the feedlot one pen of animals is never mixed with another pen. Cattle are fed on consignment and careful records are maintained of the amount and type of feed given to cattle in each pen.

From January 1 to March 15, of 1,398 cattle from Feedlot A that were sent to slaughter, only one was found infected with *Cysticercus bovis*. However, from March 15 to June 12, of 5,870 cattle originating in Feedlot A, 743 were infected (overall infection rate 12.7 percent). Investigation into the sources for all cattle slaughtered since January 1, 1968, revealed that the cattle were assembled from a variety of sources by several commercial buyers.

In the infected animals, cysts were found in the masseter muscles, esophagus, liver, heart, and diaphragm, and the majority of cysts were degenerated and caseous, indicating that the infections were probably several months old.

The personnel at the feedlot were investigated for *Taenia saginata*. In addition to the manager, seven employees worked at Feedlot A at the time of the inquiry. Two former employees, who worked at the feedlot between the time the cattle entered the feedlot and were slaughtered, were also questioned. None of the present employees gave a history suggestive of taeniasis (*T. saginata*), but a former employee stated that 1 1/2 years ago he had passed motile, flat worms, approximately 2 cm long, in his stool. He had not been treated for his condition. The likelihood that he was the source of infection is also increased by temporal considerations. Since the cattle were not infected in pasture and since the majority of cysts were degenerated and caseous, the most likely time of infection would have been mid-October when the cattle first entered the feedlot. The former employee had started working in the feedlot at that time.

Human roundworm

Ascaris lumbricoides is one of the largest parasites of man. Females may reach a length of 12 to 14 inches with a diameter of $1/4$ inches. Males are much smaller and can be recognized by their strongly curved posterior. Females taper smoothly at both ends (Fig. 18-10).

The body wall is covered with a smooth cuticle, which is secreted by epidermal cells. The anterior mouth opening is modified by one dorsal and two ventral lips.

The body wall itself is relatively simple in structure. From the outside inward it consists of the

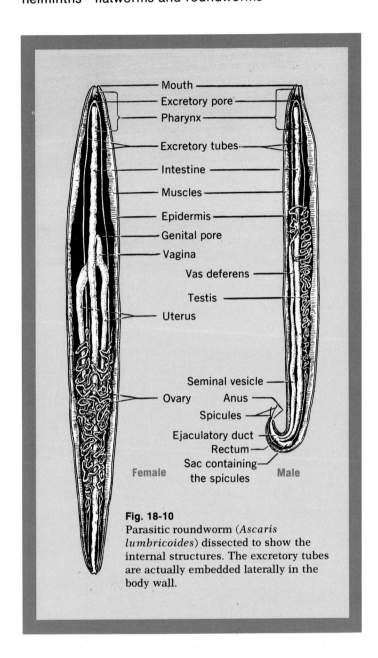

Mouth
Excretory pore
Pharynx

Excretory tubes

Intestine

Muscles

Epidermis
Genital pore
Vagina

Vas deferens

Testis

Uterus

Seminal vesicle
Ovary   Anus
Spicules
Ejaculatory duct
Rectum
Sac containing
the spicules

Female   Male

Fig. 18-10
Parasitic roundworm (*Ascaris lumbricoides*) dissected to show the internal structures. The excretory tubes are actually embedded laterally in the body wall.

The muscles contain contractile fibers and are innervated by extensions that pass directly to the nerve cords. The muscles are separated into four groups by the projection of the epidermal ridges. Movement occurs by the alternate contraction of four muscle groups.

The sensory system is apparently poorly defined. There is a nerve ring around the pharynx, and from it nerves pass anteriorly to the lips and posteriorly to other areas. In addition to the dorsal and ventral cord there are longitudinal nerves along the length of the lateral ridges.

A pseudocoel is present, which is lined by mesoderm only on the surface toward the body wall. No mesoderm covers the digestive tract. The digestive system consists of a mouth; a short, muscular, sucking pharynx; a long, nonmuscular intestine for absorption; a short rectum; and an anus at the posterior end. There are no respiratory or circulatory systems. An excretory tube without flame cells is present in each of the two lateral lines. The two tubes unite and empty wastes through a ventral excretory pore just posterior to the mouth.

In the male one coiled, threadlike testis produces sperms that are conducted by a vas deferens to an enlarged seminal vesicle where they are stored. A muscular ejaculatory duct empties into the cloaca. Two bristlelike, chitinous spicules attach the male during copulation.

In the female two coiled, tubular ovaries produce eggs. Two coiled oviducts carry the eggs to two larger, tubular uteri that unite and empty into a short vagina that opens through the genital pore about one third of the way behind the anterior end. Fertilization occurs in the uteri, and the egg is enclosed by a thick, rough shell and passed out through the genital pore.

Life cycle

A female may produce eggs at the rate of 200,000 daily, and one worm may contain over 25 million eggs at one time. The eggs pass with the feces, are deposited in soil, and develop into immature embryo worms inside their shells. Eggs usually develop embryos in 3 weeks, and when taken into the intestine in this state, they will produce small larvae (0.2 mm.) that burrow through the intestinal wall into the veins and lymph vessels. From here

cuticle, an epidermis that is largely syncytial (that is, nuclei are not separated by definite cell boundaries), and a muscle layer of large longitudinal cells.

The epidermis projects inward in four definite ridges on the top, bottom, and both sides. The lateral ridges surround longitudinal excretory canals that run the length of the body. The dorsal and ventral ridges contain longitudinal nerve cords.

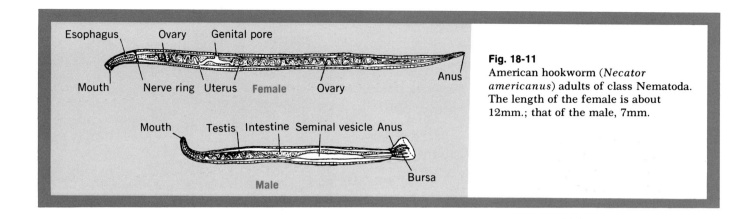

Fig. 18-11
American hookworm (*Necator americanus*) adults of class Nematoda. The length of the female is about 12mm.; that of the male, 7mm.

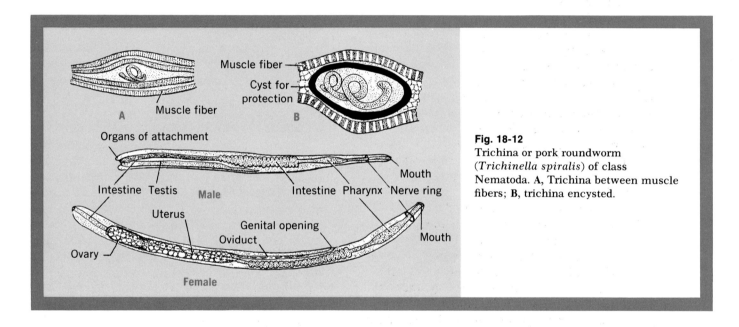

Fig. 18-12
Trichina or pork roundworm (*Trichinella spiralis*) of class Nematoda. **A**, Trichina between muscle fibers; **B**, trichina encysted.

they may pass through the heart to the lungs, break into the air passages, travel into the trachea (windpipe) and down the esophagus, and finally go to the intestine where they mature in 2 to 3 months. Great damage may be done by the migrating embryos, especially in the lungs. When very numerous, they may migrate to the bile ducts, appendix, sinuses, and nose.

Hookworm (Necator)

The American hookworm, *Necator americanus* (L. *necare*, to kill) (Fig. 18-11) is common in the southern United States. Adult worms are about 12 mm. long, males being slightly shorter than females. They are called hookworms because of a bend in the anterior of the male. Cutting, platelike teeth in the mouth cut holes in the human intestine. They suck blood and other fluids by means of a sucking pharynx and digest them in the intestine. An anticoagulant in the mouth prevents blood clotting in the host.

The methods of reproduction and the life cycle are somewhat similar to those of *Ascaris*. Larvae (0.5 mm.) may live for weeks or months. When infested soil contacts the skin, the larvae burrow into the blood, thus causing an irritation known as "ground itch." Although the larvae usually enter human beings through the skin of the feet, they may also be taken orally. Their journey within the human body is similar to that described for *Ascaris*,

253

finally reaching the intestine. They may remain in a host for years, where their presence may cause anemia, loss of energy, and retarded physical and mental growth. Hookworm disease is common in many parts of the world, especially in rural areas. Preventing the entrance of larvae through the skin and proper disposal of feces may reduce possible infestations. Proper medication will eliminate them from the human body.

Pork roundworm (trichina)

The human parasite *Trichinella spiralis* (Fig. 18-12) is about 2 to 4 mm. long and may cause a serious disease called trichinosis. Man becomes infested by eating improperly cooked pork. The adults live in the small intestine, where the female burrows into the intestinal wall and produces larvae (0.1 mm. long), which burrow into veins or lymph vessels and are carried to the skeletal muscles, especially those of the neck, eyes, tongue, and diaphragm. In the muscles they coil and are surrounded with a limy cyst. In the encysted stage they may live for years. When cysts are ingested with infested meat, the larvae are liberated in the intestine and reach maturity in a few days.

In addition to man, mammals such as pigs, bears, dogs, cats, and rats, may become infested by eating meat containing trichina cysts. Many of the people of the United States have trichina infestations, but most cases are mild and are not reported. Some human symptoms include intestinal disturbances, fever, muscular pains, and edema. Precautionary measures must include thorough cooking of pork. See p. 255 on cases of trichinosis from bear meat.

Vinegar eelworm

The free-living worm *Turbatrix aceti* (L. *turbatio*, disturbance; *acetum*, vinegar) (Fig. 18-13) lives in cider vinegar. The sexes are separate (diecious), the female being about 2 mm. long and the male about 1.5 mm. long. This difference illustrates sexual dimorphism. There is a false body cavity (pseudocoel) and a cuticle in the eelworm. In general the various systems are rather typical of nematodes.

The eggs are fertilized within the female worm where they develop into active larvae. Sometimes eggs develop into embryos within the mother's

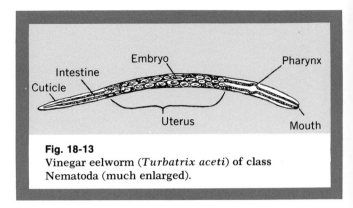

Fig. 18-13
Vinegar eelworm (*Turbatrix aceti*) of class Nematoda (much enlarged).

body but are not nourished by her bloodstream. Such a condition is called ovoviviparous (o vo vi -vip′ a rus) (L. *ovum*, egg; *vivus*, alive; *parere*, to bear).

Economic importance of roundworms

In general, the free-living roundworms do not seem to be of great direct importance to man. The parasitic nematodes are very numerous and affect large numbers of animals and plants. Some species of these nematodes live in plants where they may be of great economic importance because they cause crop loss. It is thought that there are thousands of unclassified nematodes, and when some of them are studied, their economic importance may be ascertained. Filarial worms (*Wuchereria bancrofti*) (Fig. 18-14) live in the lymphatic system of man, where they obstruct the flow of lymph and may cause a severe condition known as elephantiasis. Enormous swelling and growth of connective tissue, especially in the arms and legs, occur, thus giving the appearance of the limb of an elephant.

Adult pinworms may live in the cecum, appendix, and large intestine of children. Females with eggs may migrate to the anal regions at night to deposit eggs, thus causing an irritation. Scratching may transfer them to the mouth if sanitary precautions are not taken. No intermediate host is necessary, and eggs taken into the intestine hatch into adults. Thus a reinfestation occurs. Each generation lasts 3 to 4 weeks and will die out unless reinfestation occurs. In 1969, pinworm eggs were found in fossilized human feces estimated to be 10,000 years old.

TRICHINOSIS FROM BEAR MEAT — Alaska*

In Alaska, in July and August 1968, three persons developed clinical trichinosis and five others had serologic evidence of infection after ingestion of bear meat. Symptoms for the clinically ill persons included fever, myalgia, vomiting, diarrhea, periorbital edema, and muscle ache. In addition, one patient had a generalized maculopapular rash. All three recovered after therapy with thiabendazole.

The incriminated meat was from two black bears killed on June 6. The meat from Bear 1 was frozen in two home freezers that could maintain a minimum temperature of 0° F. Then it was served at meals on June 28, July 7, and July 11 (see figure). The meat served on June 28 was "well-cooked" on a charcoal grill while that served on July 11 was cubed and sauteed in butter and was described as "pink in the center." All five guests of G. K., Sr., who ate the meat on June 28 had clinical and/or serologic evidence of *Trichinella spiralis* infection. Three of these persons ate meat from Bear 1 again on July 11. One other person at the July 11 meal who had not eaten the June 28 meal also be-

came ill. No persons at the July 7 meal became ill.

The meat from Bear 2 was cut into stew meat and chops, aged for 5-6 days, flash-frozen in a commercial locker at −25° F., and then stored at 0° F. It was served at five meals. Although no one at these meals developed clinical illness, two had serologic evidence of *T. spiralis* infection. The charcoal card flocculation test for trichinosis was used to identify the symptomatic and asymptomatic infections.

Microscopic examination of meat samples from both bears demonstrated heavy *T. spiralis* infections (see table). Pepsin hydrochloric acid digestion in the laboratory of the meat 81 days after the death of each animal showed definitive sluggish movement of larvae in the meat from Bear 1 and no larval movement in the meat from Bear 2. The larvae in Bear 2 showed no signs of calcification.

All 30 persons who had consumed the bear meat were contacted, symptomatic persons were treated, and the other persons maintained under surveillance for illness for 30 days. All the remaining meat was destroyed under the supervision of the USDA in Anchorage.

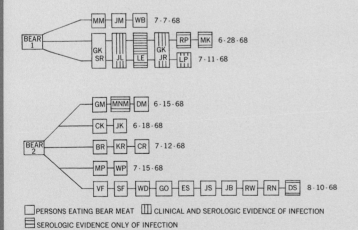

PERSONS EATING BEAR MEAT ☐   CLINICAL AND SEROLOGIC EVIDENCE OF INFECTION
SEROLOGIC EVIDENCE ONLY OF INFECTION

Incidence of trichinosis in persons eating infected bear meat by date of meal.

Larvae per gram of bear meat

Animal	Origin of sample	Larvae per gram
Bear 1	Hind quarter	110
Bear 2	Chops	80
Bear 2	Stew meat	190

*From Morbidity and mortality weekly reports, United States Public Health Service, Sept. 1968.

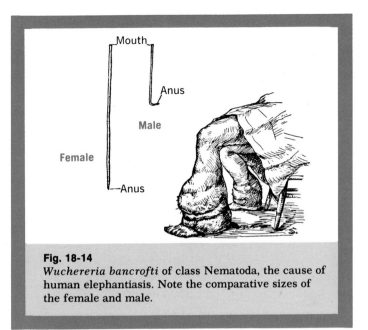

Fig. 18-14
Wuchereria bancrofti of class Nematoda, the cause of human elephantiasis. Note the comparative sizes of the female and male.

Damage produced by ascaris worms, hookworms, and trichina worms have been considered earlier. Guinea worms (*Dracunculus*) are found in Africa, India, South America, and places in the Orient. The females are long, some being $\frac{1}{25}$ inch in diameter and up to 4 feet long. They are present in dogs and minks in the United States. The female lies beneath the skin where it may appear like an enlarged vein. The anterior end of the worm protrudes through a sore, and immature stages may be discharged. If the larvae are taken into the body of a certain crustacean (*Cyclops*), they will develop there. If water containing *Cyclops* is taken by man, the worms pass to the intestine and then migrate to areas beneath the skin to become mature in about a year. Stoll (1947) estimated that there are over two billion human roundworm infections.

Review questions and topics

1 What are the general characteristics that describe the flatworms as a group?
2 What are the characteristics that distinguish planarians, flukes, and tapeworms from each other?
3 Discuss the status and significance of the head (cephalization) and sensory organs in certain flatworms, including specific examples.
4 Discuss the importance of polarity and axiate organization in planarians.
5 What is the significance of bilateral symmetry?
6 Name the advantages and disadvantages of monecious reproduction?
7 What are the effects of prolonged parasitism on the host and upon certain structures and systems of the parasite itself? Give specific examples.
8 Of what importance are meat inspections and the proper cooking of meats?
9 Discuss the two types of hosts that certain flatworms commonly use in their complete life cycles, including specific examples.
10 What are the general characteristics that describe the roundworms as a group?
11 Discuss the significance of a complete digestive system that extends from a mouth to an anus, including the improvement over a gastrovascular cavity.
12 Discuss the status and significance of cephalization and sensory organs in certain roundworms, including specific examples.
13 Discuss parasitism as found in certain roundworms, including the number of hosts usually involved in the life cycle.
14 Discuss the role of proper inspections and cooking of meats to curtail the spread of roundworm parasites.
15 Explain how serious infestations by hookworms may affect human beings.

Selected references

Chandler, A. C., and Read, E. C.: Introduction to parasitology, New York, 1961, ed. 10, John Wiley & Sons, Inc.

Cheng, T. C.: The biology of animal parasites, Philadelphia, 1964, W. B. Saunders Company.

Faust, E. C., Beaver, P. C., and Jung, R. C.: Animal agents and vectors of human disease, Philadelphia, 1968, Lea & Febiger.

Faust, E. C., and Russell, P. F.: Clinical parasitology, Philadelphia, 1964, ed. 7, Lea & Febiger.

Hyman, L. H.: The invertebrates; vol. II, Platyhelminthes and Rhynchocoela, New York, 1951, McGraw-Hill Book Company.

Hyman, L. H., and Jones, E. R.: Turbellaria. In Edmondson, W. T.: Fresh-water biology, New York, 1959, John Willey & Sons, Inc.; 1st ed. by Ward, H. B., and Whipple, G. C., pub. 1918.

Meglitsch, P. A.: Invertebrate zoology, New York, 1967, Oxford University Press, Inc.

Rogers, W. P.: The nature of parasitism, New York, 1962, Academic Press Inc.

Sasser, J. N., and Jenkins, W. R. (editors): Nematology, Chapel Hill, North Carolina, 1960, University of North Carolina Press.

Thorne, G.: Principles of nematology, New York, 1961, McGraw-Hill Book Company.

Wardle, R. A., and McLeod, J. A.: The zoology of tapeworms, Minneapolis, 1952, University of Minnesota Press.

Yamaguti, S.: Systema helminthum; vol. 3, The nematodes of vertebrates, New York, 1961, Interscience Publishers, Inc.

Some contributors to the knowledge of flatworms and roundworms

Avicenna

Avicenna (981-1037 A.D.)

A Persian physician who described four kinds of worms, apparently including several types of tapeworms and roundworms. (*Historical Pictures Service, Chicago.*)

Jehan de Brie

He discovered (in 1379) the first fluke, *Fasciola hepatica*, the cause of sheep liver rot. It was described more accurately by **Gabucinus** in 1547.

Francesco Redi (1626-1697)

An Italian scientist who recognized the larval, cysticercus stage of the tapeworm *Taenia* as an animal. Much later, **E. Küchenmeister** proved by feeding experiments that these bladder worms represented the immature stages in the life cycle of the tapeworm, and that as a rule, they required a different host from that of the adult worm (1851). Küchenmeister also worked out the life cycle of the pork tapeworm, *Taenia solium* (1885).

J. F. McConnell

He discovered the Oriental liver fluke, *Clonorchis*, in 1874.

Carolus Linnaeus (1707-1778)

A Swedish biologist who recognized the genus *Ascaris* in 1758.

Carl T. E. von Siebold (1808-1884)

A German parasitologist who extended the idea of worms alternating between hosts. He proved that intestinal parasites did not arise spontaneously.

R. Leuckart and A. P. Thomas

They discovered (in 1882 and 1883, respectively) the complete life cycle of the sheep liver fluke, involving a snail as a required host.

Patrick Manson (1844-1922)

An English pathologist who proved that the blood-sucking mosquito served as the larval host for Bancroft's filaria (1878) and that the periodicity of these microfilaria in the human peripheral blood seemed to be related to the life cycle. (*Historical Pictures Service, Chicago.*)

Manson

C. J. Davaine

He first discovered that ascaris larvae hatched from eggs in the small intestine of experimental rats (1863).

Eduard van Beneden (1846-1910)

A Belgian cytologist who studied the formation and maturation of sperms and eggs of the roundworm ascaris (1883).

A. Looss

He first demonstrated the life cycle of hookworms, including the infectiveness of larval stages that usually entered through the skin (1896-1911).

Theodor Boveri (1862-1915)

A German zoologist who did pioneer work on ascaris, especially with the chromosomes in the cell nucleus.

N. A. Cobb (1859-1933)

An American who made valuable contributions to nematode biology.

B. H. Ransom and E. B. Cram

They traced the course of migration of *Ascaris* larvae (1921).

257

Phylum Annelida— segmented worms

Segmented worms are often called "real" worms because, unlike all other wormlike organisms, they have distinct ringlike segments composing their bodies. Annelids and all animals considered in the following chapters have a distinct body cavity, the coelom, which is lined by epithelium.

CLASSIFICATION OF SEGMENTED WORMS

Class Oligochaeta (Gr. *oligos*, few; *chaite*, bristle) Internal and external segments; relatively few bristlelike setae; no fleshy parapodia; brandlike clitellum around the body; both sexes in the same animal (monecious). Example: earthworm (*Lumbricus*).

Class Polychaeta (Gr. *poly-*, many; *chaite*, bristle) Internal and external segments; fleshy parapodia with numerous bristlelike setae; distinct head with eyes, tentacles, and fingerlike sensory palps; no clitellum; sexes are usually separate (diecious). Example: sandworm (*Nereis*).

Class Hirudinea (L. *hirudo*, leech) Body flattened; fixed number of segments (usually thirty-four), each of which may be subdivided into several false segments; no setae or parapodia; anterior and posterior suckers; coelom is reduced by connective tissues and muscles; both sexes in the same animal (monecious). Examples: leeches (*Hirudo*).

CLASS OLIGOCHAETA

Earthworm

The common earthworm is *Lumbricus terrestris*. Its body is elongated, soft, and segmented. It bur-

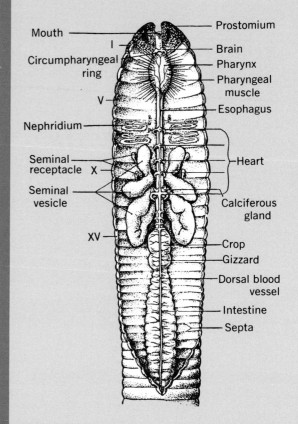

Fig. 19-1

Earthworm (*Lumbricus terrestris*) of class Oligochaeta, dissected from the dorsal side to show the internal organs of the anterior end (somewhat diagrammatic). **I, V, X,** and **XV,** Number of segments or somites. Nephridia are shown only in a few segments; there may be some variations in the location of certain structures in different earthworms.

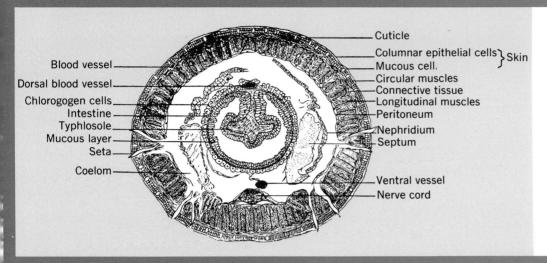

Blood vessel
Dorsal blood vessel
Chlorogogen cells
Intestine
Typhlosole
Mucous layer
Seta
Coelom

Cuticle
Columnar epithelial cells } Skin
Mucous cell
Circular muscles
Connective tissue
Longitudinal muscles
Peritoneum
Nephridium
Septum

Ventral vessel
Nerve cord

Fig. 19-2
Earthworm cross section posterior to the clitellum (cut between the septa). (*From Braungart, D. C., and Buddeke, R.: An introduction to animal biology, ed. 7, St. Louis, 1968, The C. V. Mosby Co.*)

rows in the soil, forcing debris through its alimentary tract and passing this debris on the surface as "castings." A conspicuous, saddlelike clitellum (kli -tel' um) is present on segments XXXI to XXXVII. A true body cavity, or coelom, is present. Membranous, internal septa separate adjacent segments (Fig. 19-1).

Integument

A thin, noncellular, transparent cuticle contains pores of glands located in the epidermis just beneath. Four pairs of chitinous, bristlelike setae are located in each segment (Fig. 19-2).

Motion and locomotion

An outer layer of circular muscles and an inner layer of longitudinal muscles in the body wall aid in locomotion. Setae are moved by protractor and retractor muscles at their bases. The setae are set at certain angles to provide friction or to reduce it as desired.

Ingestion and digestion

The foods of earthworms consist of dead plant and animal materials and soils rich in organic substances. Living materials are rarely molested. A fleshy prostomium (Gr. *pro*, before; *stoma*, mouth) projects over the mouth at the anterior end. The mouth leads into a buccal pouch that connects with a thick, muscular, sucking pharynx. Muscles attached to the outside of the pharynx contract and expand it to produce suction. A narrow esophagus connects the pharynx with the large, thin-walled crop, which is used for storage. Posterior to the crop is the thick, muscular gizzard used for grinding food by grains of sand and similar materials. The gizzard leads to the long intestine, which has a deep, dorsal fold, the typhlosole (tif' lo sole) (Gr. *typhlos*, blind; *solen*, channel). The typhlosole increases the absorbing surface of the intestine and is filled with chlorogogen cells which probably aid in the digestion of foods and the elimination of wastes. The anus is at the posterior end of the earthworm.

Three pairs of calciferous glands (L. *calx*, lime; *ferre*, to bear) secrete calcium carbonate into the nearby esophagus to neutralize acid foods. The calcium carbonate also lines the tunnels (burrows), thus preventing their collapse. These plastered tunnels, through which the worm crawls, allow moisture and air to penetrate the soil. In this way the earthworms cultivate the soil.

Enzymes from the digestive juices act on foods in a manner similar to that of higher animals. Absorption in the earthworm occurs through the walls of the intestine, being assisted by the ameboid action of some of the lining epithelial cells. Some absorbed foods are placed in the coelomic cavity where they are circulated by the coelomic fluid. Other absorbed foods are placed in the circulatory system to be taken to various parts of the body.

259

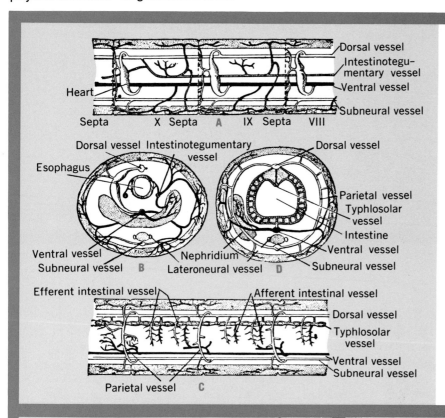

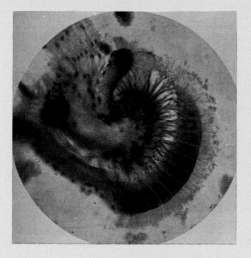

Fig. 19-3
Earthworm circulatory system (somewhat diagrammatic). **A,** Longitudinal view in segments **VIII, IX,** and **X; B,** cross section of the same region; **C,** longitudinal view in the region of the intestine; **D,** cross section of the same region.
(*From Hegner, R. W.: College zoology, New York, The Macmillan Company.*)

Fig. 19-4
Ciliated, funnel-shaped nephrostome of the earthworm from a cross section just posterior to the clitellum.
(*Courtesy General Biological Supply House, Inc., Chicago, Illinois.*)

Circulation

The blood vessels form a so-called closed type of circulatory system (Fig. 19-3), in which the blood flows, more or less continuously, in the vessels and only a limited amount of the blood's constituents pass in and out through the vessel walls. The more important parts of the system are (1) a dorsal blood vessel above the digestive tract; (2) a ventral vessel below the digestive tract; (3) five pairs of pulsating, looplike heart arches in segments VII to XI, which connect the dorsal and ventral vessels; (4) a subneural vessel beneath the ventral nerve cord; (5) two lateral neural vessels on either side of the nerve cord; and (6) numerous branches from the vessels with their thin-walled capillaries (L. *capillaris*, hair) to supply blood to all body parts. The blood is propelled through the vessels by the peristaltic contractions of the hearts and dorsal blood vessel, thus forcing it from the posterior part of the dorsal vessel toward the anterior end. Valves in the hearts and dorsal vessel prevent backward flow. Blood is

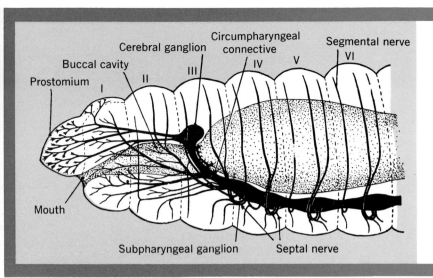

Fig. 19-5
Earthworm nervous system. A side view of the anterior end with the cerebral suprapharyngeal ganglion and larger nerves.
(After Hess, from Hegner, R. W.: College zoology, New York, The Macmillan Company.)

returned from the body wall to the lateral neural vessels, in which it flows posteriorly, and eventually reenters the dorsal blood vessel.

The blood of the earthworm consists of (1) liquid plasma with hemoglobin, an oxygen-carrying red pigment, dissolved in it and (2) numerous colorless white blood cells, which resemble those of human blood. The blood carries absorbed foods from the digestive tract to all parts of the body, transports wastes rapidly from the tissue to the organs of elimination, and exchanges oxygen and carbon dioxide by flowing near the body surface.

Respiration

There is no respiratory system, but oxygen is obtained and carbon dioxide eliminated through the moist body surface. Many thin-walled capillaries just beneath the cuticle make the exchange of gases possible. Excess water around the animal interferes with respiration, which partly explains why earthworms are "rained out" after a rain.

Excretion and egestion

A pair of nephridia (Gr. *nephros*, kidney) is present in each segment, except in the first three and the last one. The internal, free end of each is a ciliated, funnel-shaped nephrostome that filters the wastes from the coelomic fluid (Fig. 19-4). The nephridia select the wastes and pass them through the ciliated tubes to the exterior through openings called nephridiopores. These are located on the ventral side of the body just posterior to the segment in which their particular nephridia are lo-

cated. Chlorogogen cells, covering the intestinal wall and filling the typhlosole, may act to eliminate wastes. Solids are eliminated through the anus.

Coordination and sensory equipment

A bilobed brain (suprapharyngeal ganglion) is located dorsally from the pharynx near segment III (Fig. 19-5). The circumpharyngeal ring, or commissure, encircles the pharynx and connects the brain with the subpharyngeal ganglion below the pharynx. The ventral nerve cord extends posteriorly from the subpharyngeal ganglion and has an enlarged ganglion (Gr. *ganglion*, a swelling) that gives origin to three pairs of nerves in each segment. These ganglia serve as subordinate "brains" where nerve impulses are received and redirected. Nerves connect the body segments to coordinate their various activities. The muscles of the setae are controlled to make them perform their functions properly.

There are epidermal sense organs in the peripheral tissues that, when stimulated, send impulses over nerves. Sensory hairs penetrate the cuticle and are connected with the nervous system. Earthworms react to light, contact, moisture, chemicals, and sound.

Reproduction

Both male and female sex organs are present in the same earthworm (monecious) (Fig. 19-6). The female organs include one pair of small ovaries (segment XIII), which are not visible from the dorsal side; one pair of small oviducts, which are

261

modified nephridia (segment XIII); one pair of egg sacs connected with the oviducts (segment XIV); one pair of oviduct openings on the ventral side (segment XIV); two pairs of seminal receptacles (spermatheca) in segments IX and X; and two pairs of seminal receptacle openings between segments IX and X and X and XI. The male organs include two pairs of testes (segments X and XI), which are covered by the seminal vesicles and not visible from the dorsal surface; one pair of vasa deferentia (sperm ducts) with ciliated funnels (segments X to XV); one pair of vasa deferentia openings on the ventral surface (segment XV); and three pairs of large, conspicuous seminal vesicles (segments IX to XII). The bases of these vesicles are attached in segments IX to XII, although they may extend beyond them.

During copulation the ventral surfaces of two earthworms are in contact and the anterior ends point in opposite directions. A slimy band secreted by the clitellum encircles the two worms. A pair of temporary seminal channels is formed on the ventral surface of each worm, so that sperm expelled from the vasa deferentia of one worm travel to the openings of the seminal receptacles of the other, within which the sperms are stored. Copulation (Fig. 19-7) results in a mutual exchange of sperms. No discharge of eggs or fertilization occurs at this time. One earthworm cannot fertilize its own eggs, but there is a mutual cross-fertilization in the cocoon.

After copulation the worms pull away from each other, and half of the slimy band is slipped over the anterior end of each worm. In doing so the eggs from the oviducts are discharged into the slimy tubes, which also receive sperms from the seminal receptacles (segments IX and X). The elastic ends of the cocoons close to imprison sperm, eggs, and a liquid food for the developing embryos. A young worm eventually breaks from the cocoon and shifts for itself in the soil. After a few weeks the embryo becomes an adult.

CLASS HIRUDINEA

(Leeches)

Leeches, or "blood suckers," have anterior and posterior suckers, which are used to attach them-

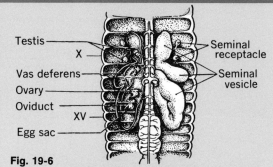

Fig. 19-6
Earthworm reproductive system. The seminal vesicles on left are dissected to show the male and female organs.

Fig. 19-7
Copulation in the earthworm. The anterior ends point in opposite directions, and the ventral surfaces are held together by mucous bands secreted by the clitellum of segments XXXI to XXXVII. Male sperms pass out of the vasa deferentia of each worm, through the seminal channels on the ventral surface, to the openings of the seminal receptacles of opposite worms. After a mutual exchange of sperms the worms separate. (*Courtesy General Biological Supply House, Inc., Chicago, Illinois.*)

Fig. 19-8
Earthworm cocoons, in which young earthworms develop in the soil. Each cocoon is about $\frac{1}{8}$ inch long.

Fig. 19-9
Common leech of class Hirudinea, showing segments and suckers.

selves to the host, and chitinous jaws to obtain blood from their victims (Fig. 19-9). They usually have thirty-four true segments, and each true segment is subdivided into several false grooves. Leeches have definite annelid characteristics, such as segmented bodies, a series of nephridia for excretion, a series of ganglia on the ventral nerve cord, and reproductive organs (gonads) in the coelom, although they do not have bristlelike setae.

The medicinal leech (*Hirudo medicinalis*) is 4 to 5 inches long, and its anterior sucker surrounds the mouth, which is supplied with three jaws of chitinous teeth. The salivary glands secrete an anticoagulant (hirudin) into the wound to prevent blood clotting in the attacked animal. Great amounts of blood are sucked up by a muscular pharynx and are stored in a large crop that is supplied with many pairs of pouchlike ceca. Digestion occurs at intervals for months.

Respiration, circulation, and excretion occur in much the same way as in other annelids. The nervous system is annelidlike, and there are sense organs of taste, touch (tactile), and "sight" (light sensitive). The muscular system consists of circular, oblique, and longitudinal bands.

Leeches have both sexes in one animal (monecious), but there is cross-fertilization.

Because excess blood was believed to cause certain abnormal bodily conditions, the medicinal leech was used for years in medical practice for extracting blood from human beings. During those times pharmacies kept a supply of living leeches for this purpose.

ECONOMIC IMPORTANCE OF ANNELIDS

Earthworms are nature's great cultivators of the soil because they redistribute soil ingredients as they move along. By their migrations the help to aerate it and redistribute the chemicals that are present in the various layers. They bring such food elements as potassium and phosphorus from the subsoils toward the surface. They carry leaves and other organic materials into their burrows, thus bringing them closer to plant roots. Earthworms can ingest their own weight in soil every 24 hours. In his classic work *The Formation of Vegetable Mould Through the Action of Worms*, Charles Darwin (1809-1882) showed the great value of worms in the mixing and aerating of soils. He estimated that up to 18 tons of earth per acre pass through their bodies annually and are thus mixed, which is of great importance in agriculture.

Leeches are widely distributed as parasites on a great many kinds of animals, including man.

Review questions and topics

1 What general characteristics describe the segmented worms as a group?

2 How must a body cavity be constructed in order to qualify as a coelom?

3 Discuss the value of ganglia on the ventral nerve cord, especially if present in a series of segments.

4 What is meant by a so-called closed circulatory system? What are some advantages of such a construction?

5 What improvements are to be found in the digestive system of an earthworm? Explain the structure and functions of the typhlosole and calciferous glands.

6 Of what value is monecious reproduction when we consider the habitats of earthworms?

7 Because earthworms obtain their oxygen through their body surface, might this explain the erroneous statement that "it rains earthworms"? Give an explanation for the emergence of earthworms from their earthen tunnels during excessive rains.

Selected references

Dales, R. P.: Annelids, London, 1963, Hutchinson & Co. (Publishers) Ltd.

Edmondson, W. T. (editor): Fresh-water biology, New York, 1959, John Wiley & Sons, Inc.; 1st ed. by Ward, H. B., and Whipple, G. C., pub. 1918.

Hyman, L. H.: The invertebrates; vol. II, Platyhelminthes and Rhynchocoela, New York, 1951, McGraw-Hill Book Company.

Laverack, M. S.: The physiology of earthworms, New York, 1963, The Macmillan Company.

Mann, K. H.: Leeches (Hirudinea); their structure, physiology, ecology and embryology, New York, 1962, Pergamon Press, Inc.

Meglitsch, P. A.: Invertebrate zoology, New York, 1967, Oxford University Press, Inc.

Pennak, R. W.: Freshwater invertebrates of the United States, New York, 1953, The Ronald Press Company.

Prosser, C. L.: The nervous system of the earthworm, Quart. Rev. Biol. 9:181-200, 1934.

chapter twenty

Phylum Mollusca

The name Mollusca (L. *soft*) refers to the fact that the animals placed in this phylum have a soft, unsegmented body, which is often not apparent in those forms that have distinct shells or leathery skins.

A number of very distinct forms are included in the phylum that do not seem to be related to the common types of mollusks until they are studied carefully. In general mollusks have an anterior head, a dorsal visceral mass, and a ventral foot that may be modified. The presence of the foot is indicated in the "poda" of several of the classes of mollusks. (See Table 20-1 for details of classification and examples.)

CLASS PELECYPODA

■ Clams

A freshwater clam, such as *Lampsilis* (Figs. 20-1 to 20-3), has a pair of limy valves held together by internal muscles and an elastic, ligamentous hinge. Each valve has concentric lines that indicate successive growth stages, several being formed each season. The shell has an outer horny layer, a middle layer of calcium carbonate, and an inner, limy, iridescent layer of nacre (na′ kir), or mother-of-pearl. Concentric layers of nacre may form natural pearls.

Internally, the soft body consists of the visceral mass and a muscular foot. The foot may be extended forward between the valves into sand or mud. The thicker, distal end of the foot anchors the animal. One lobe of a mantle adheres to the inner surface of each valve, and together they form a mantle cavity that encloses the enitre body mass. The mantle secretes the calcium carbonate used in forming the shell.

Posteriorly, the mantle is modified to form a short, tubular, dorsal excurrent (exhalant) siphon and a short, ventral, incurrent (inhalant) siphon, both of which regulate the output and intake of water. Food and oxygen in the water are drawn into the mantle cavity through the ventral, incurrent siphon by the action of cilia. Water and feces go out the excurrent siphon. In contaminated waters this siphon system may result in the buildup of harmful bacteria. (See page 2 for an example.) The digestive system consists of (1) the mouth,

located between two pairs of fleshy, ciliated, flap-like labial palps, which carry food to the mouth; (2) a short esophagus leading to (3) an enlarged stomach, which receives digestive enzymes from the paired liver (digestive gland) surrounding the stomach; and (4) a coiled intestine followed by a rectum (surrounded by the pericardium), which leads to an anus. The anus opens into the mantle

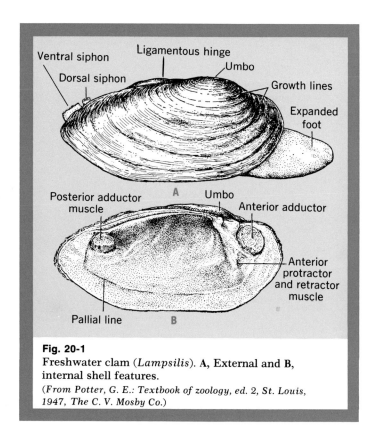

Fig. 20-1
Freshwater clam (*Lampsilis*). **A,** External and **B,** internal shell features.
(*From Potter, G. E.: Textbook of zoology, ed. 2, St. Louis, 1947, The C. V. Mosby Co.*)

Table 20-1
Distinguishing characteristics of mollusks

Characteristic	Amphineura (Gr. *amphi-*, both; *neuron*, nerve)	Pelecypoda (Gr. *pelekys*, hatchet; *pous*, foot)	Gastropoda (Gr. *gaster*, belly; *pous*, foot)	Scaphopoda (Gr. *skaphe*, boat; *pous*, foot)	Cephalopoda (Gr. *kephale*, head; *pous*, foot)
Shell	Eight limy, dorsal plates	Two limy, lateral valves (bivalved)	Usually limy, coiled (univalved); may be flat, or absent (in some species)	Limy, tubular, tusk shaped	May be present or absent (octopus); if present, may be external or internal (as in squid)
Mantle	Present	Sheetlike, bilobed; muscular siphons present	Present	Tubelike around body	Thick, muscular; muscular siphon present
Foot	Flat, ventral (absent in some)	Hatchet-shaped for burrowing	Flat, ventral for creeping	Conical, trilobed for boring	Modified into a "head-foot"
Head	Small, with rasping, tonguelike radula; no eyes; no tentacles	Absent; no radula; no eyes; no tentacles; labial palps present	Distinct, with radula, eyes, and tentacles	Rudimentary; mouth with tentacles	"Head-foot" with radula, paired eyes; eight to ten long tentacles with suckers
Respiration	Row of gills (between foot and mantle); all marine	Paired, platelike gills on either side of mantle cavity; aquatic or marine	Lungs (in terrestrial types); gills (marine and aquatic)	No gills; all marine	Paired gills (squids); two pairs of gills (nautili); all marine
Sexes	Diecious	Diecious (usually)	Diecious or monecious (depends on species)	Diecious	Diecious
Symmetry	Bilateral	Bilateral	Bilateral (head and foot); asymmetry (visceral hump)	Bilateral	Bilateral
Examples	Chitons	Clams (*Unio*, *Anodonta*, *Lampsilis*); oysters; scallops; shipworm (*Teredo*)	Snails (*Helix*, *Polygyra*, *Physa*) slugs (*Limax*); abalones	Tooth shells (*Dentalium*)	Squids (*Loligo*); octopuses; nautili; cuttlefish (*Sepia*)

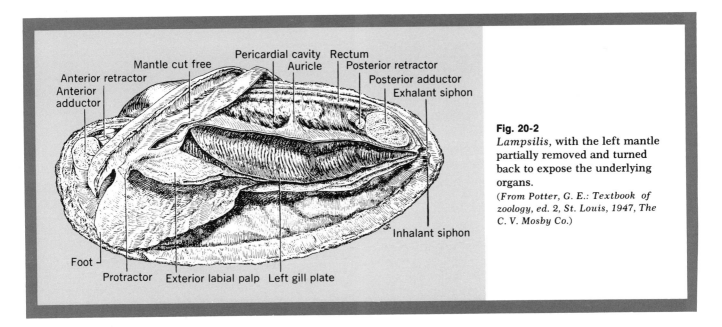

Fig. 20-2
Lampsilis, with the left mantle partially removed and turned back to expose the underlying organs.
(From Potter, G. E.: Textbook of zoology, ed. 2, St. Louis, 1947, The C. V. Mosby Co.)

Labels (clockwise from top): Mantle cut free, Pericardial cavity, Rectum, Auricle, Posterior retractor, Posterior adductor, Exhalant siphon, Inhalant siphon, Left gill plate, Exterior labial palp, Protractor, Foot, Anterior adductor, Anterior retractor

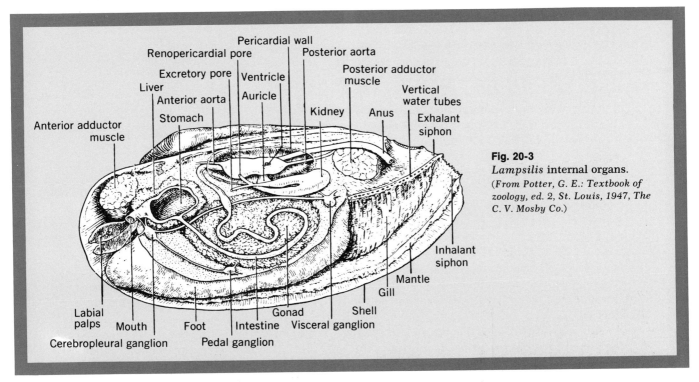

Fig. 20-3
Lampsilis internal organs.
(From Potter, G. E.: Textbook of zoology, ed. 2, St. Louis, 1947, The C. V. Mosby Co.)

Labels: Pericardial wall, Renopericardial pore, Posterior aorta, Excretory pore, Ventricle, Posterior adductor muscle, Liver, Anterior aorta, Auricle, Kidney, Anus, Vertical water tubes, Stomach, Exhalant siphon, Anterior adductor muscle, Inhalant siphon, Mantle, Gill, Shell, Labial palps, Mouth, Foot, Intestine, Gonad, Visceral ganglion, Cerebropleural ganglion, Pedal ganglion

cavity near the dorsal excurrent siphon. Internally, the rectum has a longitudinal fold, the typhlosole, to increase the surface area.

Foods consist of small plants and animals carried by water inwardly through the incurrent siphon. Mucous secretions on the mantle and gills catch the food particles, and cilia move the masses of mucus and food toward the mouth. Foods are digested within cells of the digestive gland ("liver"), which surrounds the stomach.

The circulatory system is an "open" one, consisting of a heart within a pericardial cavity, arteries, sinuses (L. *sinus*, cavity), and veins. The heart consists of two auricles and one ventricle. Blood is pumped from the ventricle through an anterior aorta (artery) to the foot and viscera and backward through a posterior aorta to the mantle and rectum. Some of the blood picks up oxygen in the mantle (becoming oxygenated) and returns directly to the auricles. Other blood circulates through sinuses, travels through a vein to the kidneys, to the gills to be oxygenated and back to the auricles. From the auricles it goes to the ventricle. Wastes and carbon dioxide are carried to the kidneys and gills for elimination.

There are two pairs of thin gills, one pair on either side of the foot. Each gill is formed by two platelike lamellae (L. *lamella*, small plate). Each lamella consists of many vertical, filamentous gill bars, strengthened by chitinous rods. Cross partitions between the lamellae divide the gill into numerous vertical water tubes. The gill surface has many small water pores (ostia) through which water passes by the action of cilia.

The excretory system consists of two U shaped kidneys located just below the pericardium. Each kidney empties into a ciliated tube leading to a bladder that empties into the suprabranchial chamber.

The nervous system consists of three pairs of ganglia (cerebropleural, pedal, and visceral) that are connected by nerve connections known as commissures. Sensory organs include tactile organs along the mantle margins; light-sensitive organs on the siphon margins; a pair of hollow statocysts, each containing a limy statolith and located in the foot for equilibrium purposes; and a yellowish osphradium (Gr. *osphradion*, strong scent) located over each visceral ganglion, which the clam may use to test the water chemically.

Reproduction is diecious. Gonads (testis or ovary) are located among the intestines within the foot. The sperm ducts (vasa deferentia) in the male and the oviduct in the female open into the supra-branchial chamber near the openings from the kidneys.

Sperms pass to the outside through the excurrent siphon where water carries them to a female, which they enter through the incurrent siphon. Eggs are fertilized internally. The resulting

Fig. 20-4
Clam glochidium.

zygote undergoes cleavage and forms a hollow mass of cells, the blastula. The blastula forms a gastrula, which in turn produces a bivalved glochidium (glo -kid′ i um) (Fig. 20-4). This is discharged from the excurrent siphon of the female and attaches itself to a fish by closing the two valves. The fish tissues grow around it forming a "blackhead" that lives parasitically for 2 to 3 months. The miniature clam then breaks out, sinks to the bottom, and continues development. Information on clams and other mollusks is given in Table 20-1.

CLASS GASTROPODA

Snails

There are many kinds of land snails. The imported European snail (*Helix*) is a typical one. The well-developed head has two pairs of retractile, sensory tentacles and a pair of light-sensitive eyes on the

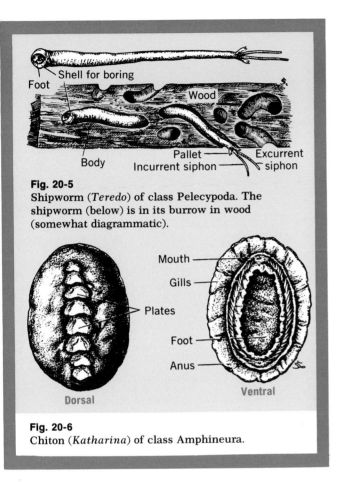

Fig. 20-5
Shipworm (*Teredo*) of class Pelecypoda. The shipworm (below) is in its burrow in wood (somewhat diagrammatic).

Fig. 20-6
Chiton (*Katharina*) of class Amphineura.

hollow, longer tentacles. A muscular foot is attached to the head and is supplied with mucous glands to assist in locomotion. The internal organs (visceral mass) form a hump surrounded by a mantle, which secretes the external, coiled shell. A muscle can pull the body within the shell.

The digestive system consists of a mouth; a pharynx with a radula of chitinous teeth for rasping green vegetation; an esophagus; a thin-walled crop; a stomach; a long, coiled intestine; and an anus, located in the margin of the mantle at the edge of the shell. A pair of salivary glands pours their secretions into the pharynx to assist in the digestion process. A large liver, high up in the spiral shell, pours its secretions into the stomach. Since *Helix* is a land snail, it has a membranous lung, well supplied with blood vessels (veins) to aerate the blood. Air is drawn into and forced out of the lungs through the respiratory pore, which is located in the margin of the mantle at the edge of the shell.

The circulatory system consists of one heart (one auricle and one ventricle), arteries to carry blood to various organs, and veins to return blood. The excretory system consists of a kidney (nephridium) to secrete wastes, which are carried by a tubular ureter to the mantle cavity.

The nervous system consists of a concentration of ganglia (cerebral, buccal, pedal, and visceral) that encircle the pharyngeal region and nerves leading to various organs. Sensory organs include light-sensitive eyes; olfactory organs on the tentacles; chemical and tactile (touch) organs on the head and foot; and a pair of statocysts (Gr. *statos*, stationary; *kystis*, sac) near the pedal ganglia for equilibrium.

Helix is monecious, but cross-fertilization is common. The gonad is called an ovotestis (producing both eggs and sperms) and is located high in the visceral hump and surrounded by the liver. A hermaphroditic duct (Gr. *hermaphroditos*, containing both sexes) leads from the ovotestis and connects with the albumin gland. A tubular vas deferens connects with the hermaphroditic duct and carries sperms to the penis at the genital pore on the right side near the mouth. Another tube, the oviduct, also connects with the hermaphroditic duct and leads to the enlarged vagina, which also empties through the genital pore. To the vagina are connected the duct from the seminal receptacle, the dart sac, and the fingerlike oviducal glands. The penis has a slender, tubular flagellum in which sperms are formed into spermatophores (bundle of sperms).

During copulation each snail inserts its penis into the vagina of its partner and transfers a spermatophore. After separating each snail deposits its fertilized eggs into shallow burrows in moist soil. Young, small snails develop directly.

When dry weather exists, snails may form a temporary covering, or epiphragm, of limy mucous secretions, which can close the opening in the shell.

CLASS CEPHALOPODA

Squids

The common Atlantic coast squid (Fig. 20-10), known as *Loligo pealei*, has a torpedo-shaped body, which may be up to 1 foot long. The skin contains

269

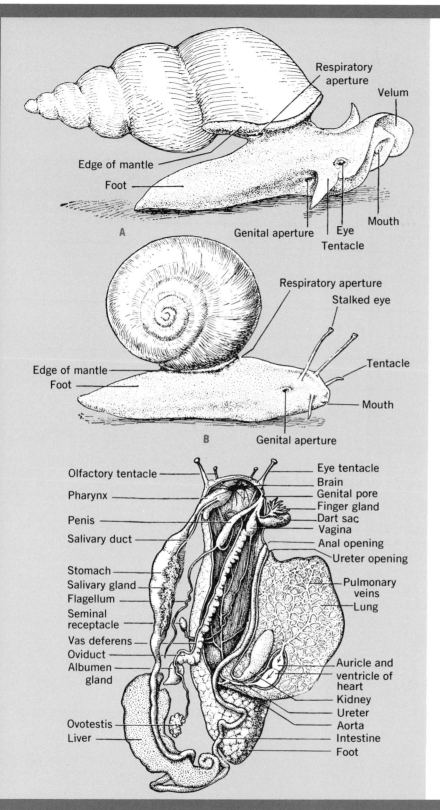

Respiratory
aperture
Velum

Edge of mantle

Foot

A

Genital aperture
Eye
Tentacle
Mouth

Respiratory aperture
Stalked eye

Edge of mantle
Foot

Tentacle

Mouth

B

Genital aperture

Olfactory tentacle
Pharynx
Penis
Salivary duct

Stomach
Salivary gland
Flagellum
Seminal
receptacle
Vas deferens
Oviduct
Albumen
gland

Ovotestis
Liver

Eye tentacle
Brain
Genital pore
Finger gland
Dart sac
Vagina
Anal opening
Ureter opening

Pulmonary
veins
Lung

Auricle and
ventricle of
heart
Kidney
Ureter
Aorta
Intestine
Foot

Fig. 20-7
A, Freshwater snail (*Lymnaea*); **B**, land
snail (*Humboldtiana*). Bodies are expanded
from the shell.
(*From Potter, G. E.: Textbook of zoology, ed. 2,
St. Louis, 1947, The C. V. Mosby Co.*)

Fig. 20-8
Snail (*Helix pomatia*). The roof of the
pulmonary sac is cut and turned to the
right; the pericardium and visceral sac are
opened, and the internal organs (viscera)
are somewhat separated. The finger gland
is also called the accessory, oviducal, or
mucous gland.

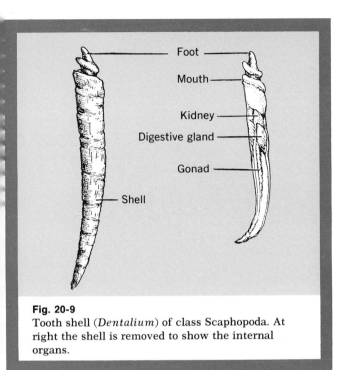

Fig. 20-9
Tooth shell (*Dentalium*) of class Scaphopoda. At right the shell is removed to show the internal organs.

Labels: Foot, Mouth, Kidney, Digestive gland, Gonad, Shell

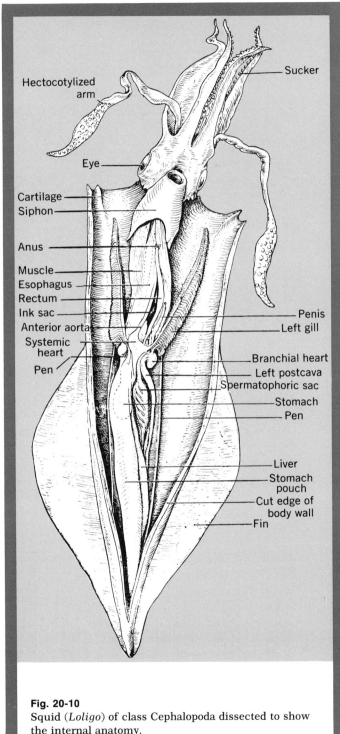

Fig. 20-10
Squid (*Loligo*) of class Cephalopoda dissected to show the internal anatomy.

Labels: Hectocotylized arm, Sucker, Eye, Cartilage, Siphon, Anus, Muscle, Esophagus, Rectum, Ink sac, Anterior aorta, Systemic heart, Pen, Penis, Left gill, Branchial heart, Left postcava, Spermatophoric sac, Stomach, Pen, Liver, Stomach pouch, Cut edge of body wall, Fin

pigment cells of blue, purple, yellow, and red, so that color changes for protective concealment can be produced quickly by changing the amount of the various pigments to blend with the surroundings. The head bears a mouth surrounded by ten sucker-bearing arms for capturing prey. One pair of arms (retractile tentacles) is longer than the others. A pair of well-developed eyes is present. A mantle encloses the internal organs within a mantle cavity. The mantle ends just back of the head, and its free margin is called a collar. Beneath the collar is a cone-shaped siphon, which propels water outward by a contraction of the mantle. By directing the siphon backward and forcing water through it the animal is jet-propelled forward; by directing the siphon forward the animal is propelled backward. A pair of posterior, triangular fins assists in locomotion. A feather-shaped plate (shell), called the pen, is just beneath the skin of the back near the anterior end and gives some support.

Squids capture fishes, other mollusks, and crustaceans for food.

Egg   Young

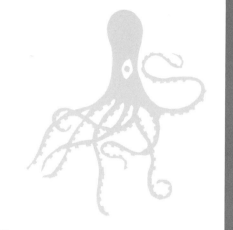

Fig. 20-11
Octopus or devilfish of class Cephalopoda.

Octopuses

The octopus (Gr. *okto,* eight; *pous,* foot), or devilfish (Fig. 20-11), has a body with the foot divided into eight sucker-bearing tentacles. Two large eyes are also evident. There is no internal shell. In general the characteristics of octopuses somewhat resemble those of other cephalopods. Some types are rather small and harmless, but the giant octopus of the Pacific may have a length of 30 feet and become unsociable and dangerous. Octopus eggs are protected by a tough case from which the young, immature octopus emerges.

ECONOMIC IMPORTANCE OF MOLLUSKS

Great quantities of clams and oysters are used for human food. More than 120 million pounds of oysters are collected in the waters of the United States annually. Extensive oyster beds have been cultivated along our Atlantic and Pacific coasts, and those in Chesapeake Bay and Long Island Sound are noteworthy. It takes about 4 years for an oyster to reach commercial size, and since oysters have natural enemies, their cultivation has become a great commercial endeavor. Many clams are used for chowder and other dishes. Valuable pearls are formed within oysters and clams when concentric layers of nacre are built up until a usable size is reached. Many types of irritating materials can initiate pearl formation. The Japanese have artificially cultured pearls by introducing foreign particles within the mantle, placing the oyster in an enclosure in the sea, and removing the pearl a few years later.

The shipworm *(Teredo)* (Fig. 20-5) burrows into the wood of wharves and ships by means of two movable valves on its anterior end. These burrows weaken the wood and permit it to decompose more rapidly. Snails are widely used as food in Europe, particularly in France. Garden snails and slugs (without shells) are great destroyers of plants. Certain species of snails may serve as intermediate hosts for such parasites as sheep liver flukes, blood flukes, and many others. Certain parts of squids may be used as food. A gastropod, the abalone, is also used as food. A type of squid, the cuttlefish *(Sepia),* produces sepia-colored ink, which is used commercially. This cuttlefish has a shell ("cuttlebone") that is used as a source of calcium for caged birds.

Important findings in neurophysiology and animal behavior have been made in recent years by using squids and octopuses as experimental animals.

1 What general characteristics describe the mollusks as a group?

2 What are the characteristics that distinguish members of the different classes of Mollusca?

3 Describe the structure and functions of the mantle.

4 Is it unusual for such high types of animals as mollusks not to be segmented?

5 Discuss the construction and function of the foot in different types of mollusks.

6 Explain the significance of the head and the "head-foot" in certain kinds of mollusks.

7 Describe the structure and function of the radula.

8 Explain the structures and functions of gills and lungs, giving examples of each.

9 Discuss the status of the various sensory organs in different mollusks.

10 Describe a glochidium, including its function.

11 Discuss the circulatory systems found in different mollusks, including the significance of the so-called open type, giving examples.

Selected references

Abbott, R. T.: American sea shells, Princeton, New Jersey, 1954, D. Van Nostrand Co., Inc.

Boycott, B. B.: Learning in the octopus, Sci. Amer. 212:56-66, 1965.

Keynes, R. D.: The nerve impulse and the squid, Sci. Amer. December, 1958.

Mead, A. R.: The giant African snail; a problem in economic malacology, Chicago, 1961, University of Chicago Press.

Meglitsch, P. A.: Invertebrate zoology, New York, 1967, Oxford University Press, Inc.

Morris, P. A.: A field guide to the shells of our Atlantic coast, Boston, 1947, Houghton Mifflin Company.

Morris, P. A.: A field guide to shells of the Pacific coast and Hawaii, Boston, 1952, Houghton Mifflin Company.

Morton, J. E.: Molluscs, London, 1958, Hutchinson & Co. (Publishers) Ltd.

Raven, C. P.: Morphogenesis; the analysis of molluscan development, New York, 1958, Pergamon Press, Inc.

Tressler, D. K., and Lemon, J. M.: Marine products of commerce, ed. 2, New York, 1951, Reinhold Publishing Corp.

Yonge, C. M.: Oysters, London, 1960, William Collins Sons & Co., Ltd.

chapter twenty-one

Phylum Arthropoda

Arthropods are members of the largest phylum of animals. The number of different species is well over one million. Arthropods occupy every type of habitat, consume every type of food, exhibit every type of relationship, and show every type of motility. An entire course in biology could be structured using only arthopods as examples. Table 21-1 gives some of the types and characteristics of the animals included in the phylum Arthropoda.

CLASS CRUSTACEA

Crayfishes

The two most common genera of freshwater crayfishes in North America are *Cambarus* (Figs. 21-1 and 21-2), found east of the Rocky Mountains, and *Astacus*, found principally west of the Rocky Mountains. In general the various species are structurally similar and resemble the large lobster, *Homarus*.

Members of the genus *Cambarus* are about 4 inches long and possess a calcareous exoskeleton. The body is divided into a cephalothorax (fused head-thorax) and an abdomen. The cephalothorax consists of twelve somites (segments); the abdomen consists of six. Each somite bears a pair of jointed appendages. The paired anterior antennules are not considered to be true, serially metameric appendages, because they are developmentally different from the eighteen pairs posterior to them. The

antennules are actually paired, prostomial sense organs, arising from a structure that seems to be a homologue of the prostomium of annelids. The stalked, compound eyes are also considered to be prostomial sense organs. The cephalothorax is covered dorsally and laterally by a shell-like carapace (Sp. *carapacho*, covering).

The eighteen pairs of appendages show considerable variations, depending on the functions they perform, but they all have fundamentally the same plan of structure, as can be noted from Table 21-2. With a few exceptions each appendage consists of a basal protopodite that bears a median endopodite and a lateral exopodite.

Any appendage composed of two branches is called biramous. All the metameric appendages of the crayfish arise embryologically from biramous structures. When structures are constructed along basically similar plans and have similar embryologic origins, they are said to be homologous structures, even though their functions may differ. When such structures appear in a serial sequence on a body, they illustrate serial homology (Fig. 21-1). Some distinguishing characteristics of crustaceans and other arthropods are given in Table 21-1.

The crayfish coelom is reduced in size and divided into several separate compartments, such as those enclosing the gonads and those enclosing the excretory green glands. The hemocoels contain blood and surround the digestive tract (Figs. 21-2 and 21-3).

The digestive system consists of a mouth (behind the mandibles); a short esophagus; a large stomach, composed of an anterior cardiac portion, which contains the gastric mill, and a posterior pyloric portion; this portion connects with a short midgut followed by a long intestine that exits through an anus beneath the telson. In the gastric mill chitinous teeth grind the food. A strainer of hairlike setae permits only fine particles to pass into the pyloric end. A pair of digestive glands ("liver") discharge secretions with enzymes into the midgut. All of the digestive system except the midgut is lined with chitin. Digested foods are absorbed by certain parts of the intestine and the digestive glands. Nearly anything that is edible seems to serve as food for a crayfish.

Table 21-1

Distinguishing characteristics of arthropods

Class	Antennae	Legs	Wings	Respiration	Habitat	Miscellaneous	Examples
Crustacea (L. *crusta*, shell)	2 pairs	Numerous	None	Gill breathing	Aquatic for most species (few para- sites)	Body composed of head, thorax, and abdomen; head and thorax may be fused into cephalothorax	Crayfish Lobster Shrimp Crab Barnacle Water flea Sow bug
Onychophora (Gr. *onyx*, claw; *phorein*, to bear)	1 pair	Numerous (with 2 claws on each)	None	Tracheal (air breathing)	Terrestrial (moist places)	Primitive, tropical or semi-tropical worm-like arthro- pods; possess numerous paired nephridia (an- nelidlike); limited in dis- tribution	*Peripatus*
Chilopoda (Gr. *cheilos*, lip; *pous*, foot)	1 pair (long)	Numerous (1 pair on most seg- ments)	None	Tracheal (air breathing)	Terrestrial (moist places)	Long, slender bodies, flat- tened dorso- ventrally, with 15 to 173 seg- ments; swift moving	Centipedes
Diplopoda (Gr. *diploos*, double; *pous*, foot)	1 pair (short)	Numerous (2 pairs on most seg- ments)	None	Tracheal (air breathing)	Terrestrial (moist places)	Long, slender bodies, subcylin- drical, with 25 to more than 100 segments; slow moving	Millipedes
Insecta (L. *insectus*, cut into)	1 pair	3 pairs (on thorax)	2 pairs, 1 pair, or none (depends on species)	Tracheal (air breathing)	Terrestrial (some live in water)	Body composed of head, thorax, and abdomen; wings, if pres- ent, attached to thorax; some species have wings at certain stages in life but wingless at others; many species harm- ful; few bene- ficial; most species neither	Bees Wasps Butterflies Moths True bugs Grasshoppers Flies Cicadas Beetles

Continued

Table 21-1
Distinguishing characteristics of arthropods—cont'd

Class	Antennae	Legs	Wings	Respiration	Habitat	Miscellaneous	Examples
Arachnoidea (Gr. *arachne*, spider; *eidos*, like)	None	4 pairs	None	Tracheal and book lungs (air breathing)	Terrestrial	Head and thorax fused into cephalothorax; abdomen present; no true jaws, but 1 pair of "nippers"; some species harmful; few beneficial; most neither	Spiders Scorpions Horseshoe crab Ticks Daddy longlegs

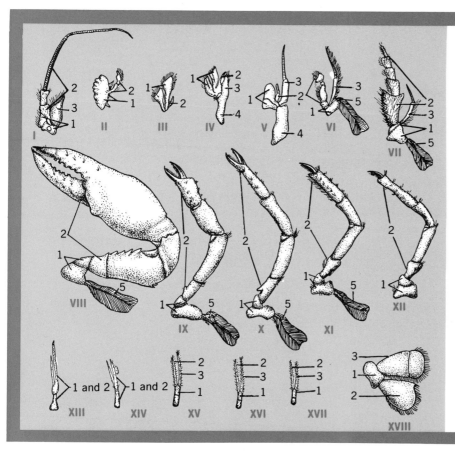

Fig. 21-1
Segmental appendages of the male crayfish (*Cambarus*) of class Crustacea, showing homology. The segmental appendages are removed from the left side, numbered from I to XVIII, and drawn somewhat to scale. 1, Protopodite; 2, endopodite; 3, exopodite; 4, epipodite; 5, gills. Appendage I is the antenna; II, the mandible; III and IV, first and second maxillae; V to VII, first, second, and third maxillipeds; VIII to XII, walking legs; XIII to XVII, swimmerets (pleopods); XVIII, uropod (sixth swimmeret). Antennules (first antennae), stalked eyes, and telson are not considered to be segmental appendages.

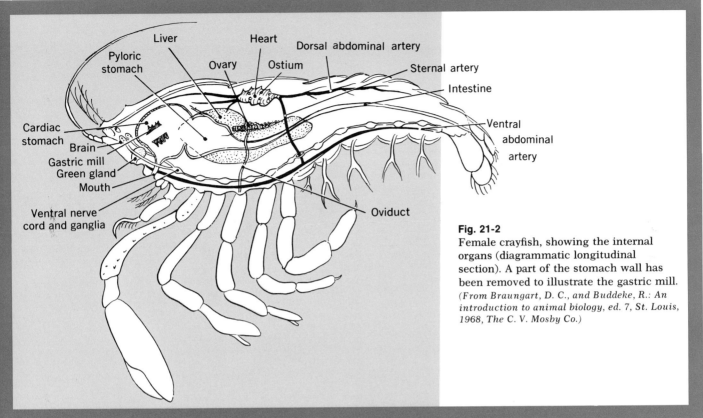

Fig. 21-2
Female crayfish, showing the internal organs (diagrammatic longitudinal section). A part of the stomach wall has been removed to illustrate the gastric mill.
(From Braungart, D. C., and Buddeke, R.: An introduction to animal biology, ed. 7, St. Louis, 1968, The C. V. Mosby Co.)

Respiration occurs in a series of delicate, featherlike gills attached to certain thoracic appendages. By moving a chitinous plate ("bailer"), located on the second maxilla, water currents are created over the gills.

Circulation is accomplished by an "open" system, consisting of a dorsal, muscular, rhythmically contracting heart, arteries, capillaries, and blood sinuses. The heart is located within a saclike pericardial sinus in the thorax. Blood from the heart enters seven main arteries (supplied with valves) and passes into fine capillaries that carry it to blood sinuses, which are actually spaces between tissues. These blood sinuses constitute the hemocoel ("blood cavity"). The blood sinuses return the blood to the pericardial sinus. Blood also passes from the sinuses to the gills, where oxygen and carbon dioxide are exchanged, and then returns to the pericardial sinus. Arthropod blood is commonly colorless and contains a number of ameboid cells and a respiratory, oxygen-carrying pigment called

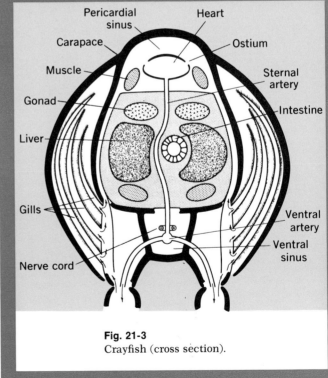

Fig. 21-3
Crayfish (cross section).

Table 21-2
Segmental appendages of the crayfish

Appendage	Protopodite	Endopodite	Exopodite	Function
Prostomium (No segment)				
Antennule (first antenna)	(Not considered to be segmental appendage)			Sensory
Eye (stalked)	(Not considered to be segmental appendage)			Sensory
Cephalothorax (12 segments)				
I. Antenna	2 segments; excretory pore in basal segment	Long, many-jointed "feeler"	Thin, broad, dagger-like	Touch; taste; balance
II. Mandible	2 segments; heavy jaw	Small; 2 distal segments of palp	None	Chew food
III. First maxilla	2 thin, medial, platelike lamellae	1 small, plate-like lamella	None	Handle food
IV. Second maxilla	2 bilobed, platelike lamellae; broad plate, epipodite	1 small, pointed segment	Dorsal plate, "bailer" (scaphognathite)	Handle food; create water currents in gill chamber
V. First maxilliped	2 thin segments extending inwardly; epipodite extending outwardly	2 small segments (reduced)	Long, basal segment with jointed filament	Handle food; touch; taste
VI. Second maxilliped	2 segments; gills	5 segments	Similar to V	Similar to V
VII. Third maxilliped	Similar to VI	5 larger segments	Similar to V	Similar to V
VIII. First walking leg (pincer, chela, or cheliped)	2 segments; gills	5 segments; distal 2 form large pincer	None	Offense; defense; walking; touch

Table 21-2
Segmental appendages of the crayfish—cont'd

Appendage	Protopodite	Endopodite	Exopodite	Function
IX. Second walking leg	Similar to VIII	Similar to VIII, but pincer smaller	None	Walking; grasping
X. Third walking leg	Similar to VIII; female bears genital pore	Similar to IX	None	Similar to IX
XI. Fourth walking leg	Similar to VIII	Similar to IX, but no pincer	None	Walking
XII. Fifth walking leg	Similar to VIII; male bears genital pore	Similar to XI	None	Walking; cleaning abdomen and eggs
Abdomen (6 segments)				
XIII. First abdominal (first swimmeret, or pleopod)	Fused with endopodite to form tube (in male); reduced or absent (in female)	Reduced or absent (in female)	None (in male); reduced or absent (in female)	Transfers sperms to female
XIV. Second abdominal (swimmeret, or pleopod)	Fused with endopodite (in male); 2 segments (in female)	Filament (in female)	None (in male); filament (in female)	As in XIII (in male); as in XV (in female)
XV. Third abdominal	2 segments	Filament	Filament	Creates water currents; used for attaching eggs and young (in female)
XVI. Fourth abdominal	2 segments	Similar to XV	Similar to XV	Similar to XV
XVII. Fifth abdominal	2 segments	Similar to XV	Similar to XV	Similar to XV
XVIII. Sixth abdominal (uropod)	1 short, broad segment	Flat, oval plate	Flat, oval plate divided crosswise into 2 parts	Swimming; egg protection (in female)
Telson (No segment)	(Not considered to be segmental appendage)			Swimming; egg protection

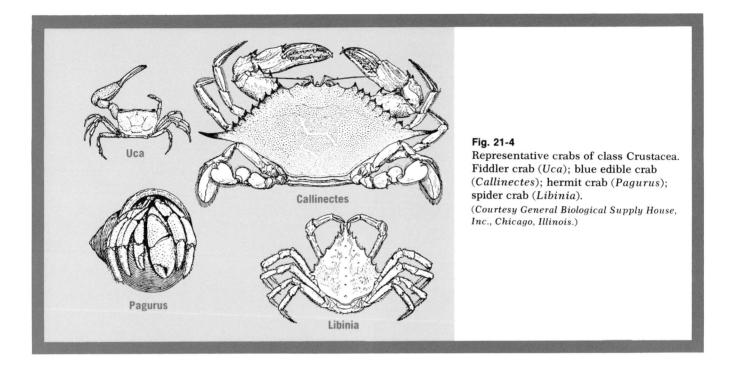

Fig. 21-4
Representative crabs of class Crustacea. Fiddler crab (*Uca*); blue edible crab (*Callinectes*); hermit crab (*Pagurus*); spider crab (*Libinia*).
(*Courtesy General Biological Supply House, Inc., Chicago, Illinois.*)

hemocyanin. The blood also has the ability to clot, which prevents some loss when an injury occurs.

Excretion is accomplished by green glands opening by a duct at the base of the antenna.

The well-developed central nervous system consists of a series of paired ganglia with a supraesophageal "brain." A circumesophageal connective connects the brain with the subesophageal ganglia. A double ventral nerve cord extends posteriorly from the subesophageal ganglia. Nerves pass from the various ganglia to different body parts.

The sinus gland at the base of the eye stalk produces endocrine hormones that control the spread of pigment in the epidermis and the compound eyes. They also regulate molting and the placing of limy salts in the exoskeleton.

The sensory organs are well developed. Two very efficient compound eyes at the ends of movable stalks are each composed of approximately 2,500 single eyes, called ommatidia (Gr. *ommation*, little eye). Each tubelike ommatidium does not take a complete picture but only a dark or light spot. The arrangement of these dark or light spots at the base of the eye gives the crayfish a flat, mosaic picture of the object. Moving objects can thus be detected.

The sense of equilibrium is regulated by two small, chitin-lined sacs (statocysts) in the base of the antennules. Grains of sand rolling within these sacs stimulate sensory hairs, thus helping the crayfish to ascertain its position. The crayfish is without a sense of hearing. The entire external body surface is sensitive to touch (by means of tactile hairs) and taste, but the pincers are most sensitive to touch and the outer (exopodite) antennules are most sensitive to taste.

Reproduction is diecious. Paired, hollow testes (or ovaries) are found in the center of the thorax. Sperms pass out of each testis through a vas deferens (sperm duct), which opens as a genital pore on the protopodite of the last (fifth) pair of walking legs. Eggs pass out of each ovary through an oviduct, which opens as a genital pore on the protopodite of the third pair of walking legs. As many as 300 eggs may be discharged in a stringlike manner at one time.

When copulation occurs, usually in the autumn, sperms are transferred to the seminal receptacle of the female (the cavity between the fourth and fifth pairs of walking legs) by using the first and second pairs of abdominal appendages of the male as a tube. Seminal receptacles are absent in certain kinds of crayfishes. The eggs are attached to the female swimmerets until the young larvae hatch in 5 to 6 weeks.

The larva, which looks like a miniature crayfish, sheds its external skeleton soon after hatching by molting. This occurs at intervals during the life of the animal, which averages about 4 years. Growth occurs in the short periods between molting, and a new exoskeleton is formed by the underlying tissues.

In general the eyes or any appendage may be regenerated when lost. Complete regeneration may require several moltings. The regenerated structure is not always the same as the lost one. If only part of an eye stalk is cut off, a normal eye will be regenerated; if an entire eye stalk is removed, an antennalike structure may be produced.

A crayfish may automatically break off walking legs at a specific predetermined joint. The power of self-mutilation (autotomy) is accomplished by a special muscle at the "breaking point." Regeneration occurs in the regular manner. In case such an appendage is caught, it can be severed, thus giving the animal a means of escape.

CLASS CHILOPODA (CENTIPEDES)
In addition to the information given in Table 21-1, the following may be added. All body segments bear one pair of jointed legs (Fig. 21-6), except the last two and the one just posterior to the head. The latter bears a pair of poison claws, maxillipeds, with which small animals may be killed. Centipedes live under stones and the bark of trees. Some tropical centipedes, which may become 1 foot long, are poisonous and dangerous to man. The common house centipede, with fifteen pairs of long legs, lives in damp places and destroys insects but is harmless to human beings.

CLASS DIPLOPODA (MILLIPEDES)
In addition to the information in Table 21-1 the following may be added. The two pairs of segmented appendages present on almost every body segment have probably come into being by a fusion

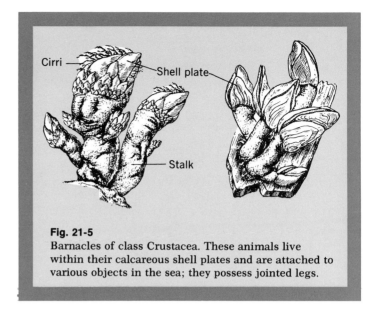

Fig. 21-5
Barnacles of class Crustacea. These animals live within their calcareous shell plates and are attached to various objects in the sea; they possess jointed legs.

of two segments. The mouth contains one pair of mandibles and one pair of maxillae. A series of scent glands may secrete an objectionable, odoriferous fluid for defensive purposes. Some millipedes may roll themselves into a ball. They live primarily on dead plant materials but may attack living plants. Common types may be found in gardens and grassy places.

CLASS INSECTA (INSECTS)
Insects are air-breathing arthropods (as are centipedes and millipedes) with a head, thorax, and abdomen. The head bears one pair of sensitive an-

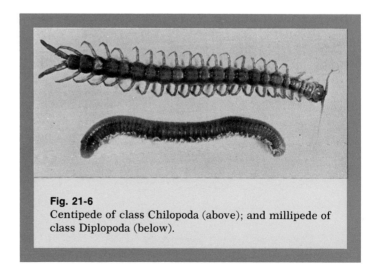

Fig. 21-6
Centipede of class Chilopoda (above); and millipede of class Diplopoda (below).

tennae. The mouthparts vary with different types of insects and include such kinds as chewing, as in grasshoppers; sucking, or siphoning, (with a tube for sucking foods), as in butterflies; or modifications of the two. Sometimes the sucking types are modified for piercing-sucking, as in the mosquitoes, flies, and true bugs.

The insect thorax has three segments, each with a pair of jointed legs, and it may have one or two pairs of wings or be wingless, depending upon the species, age, and sex. Sense organs may include simple eyes (ocelli), compound eyes, receptors for touch on various body parts, and receptors for sounds.

Grasshopper

The head of the grasshopper (Fig. 21-7) bears one pair of antennae, and the thorax bears three pairs of jointed legs. Although different species of grasshoppers vary in certain respects, the following descriptions apply to most species that are available for study. Grasshoppers have enlarged hind legs, sound-producing and sound-receiving structures, leathery forewings, and membranous hind wings. On either side of the head is a compound eye. On top of the head are simple eyes called ocelli. They have chewing mandibles (Fig. 21-8). Grasshoppers belong to the order Orthoptera (Gr. *orthos*, straight; *ptera*, wings).

Integument and skeleton

A flexible, noncellular, chitinous cuticle also serves as an exoskeleton to which organs, muscles, and tissues are attached on the inside. The cuticle is secreted by a cellular hypodermis beneath it. Beneath the hypodermis is a basement membrane.

Internally, the cavity in the body is not a true coelom but a hemocoel (Gr. *haima*, blood; *koilos*, hollow). It is filled with organs and a colorless blood. When the animal grows, it sheds its chitin at intervals by the process of molting, or ecdysis (ek' di sis) (Gr. *ekdyein*, to shed). The liquid secreted by the hypodermis hardens into new chitin.

Motion and locomotion

Grasshoppers may walk or jump by means of the three pairs of jointed legs or fly by means of two pairs of wings. Each segmented leg consists of (1)

a coxa (L. *coxa*, hip) attached to the thorax, (2) a trochanter (Gr. *trochanter*, runner), (3) a femur (L. *femur*, thigh), (4) a tibia (L. *tibia*, shin), and (5) a tarsus (Gr. *tarsos*, sole of foot). The tarsus is segmented, the proximal segment bearing three pads and the distal one a pair of claws. Between the claws is a fleshy pulvillus. The forewings are leathery and unfolded and cover the folded membranous hind wings. Chitinized, tubular veins in the wings give strength. The fine, strong, striated muscles attached to the inside of the chitinous skeleton help to move the wings, legs, and mouthparts.

Ingestion and digestion

The food of grasshoppers consists of vegetation. The principal parts of the digestive system include (1) a mouth, with a pair of salivary glands that secrete digestive juices, and the various mouthparts; (2) a tubular esophagus; (3) and enlarged crop for storage; (4) a gizzard (proventriculus) for grinding; (5) a stomach, with glandular, cone-shaped gastric ceca that secrete digestive juices; and (6) an intestine, which expands into a rectum opening through the anus.

Circulation

A single, tubular heart in the dorsal side of the abdomen is divided by valves into a series of chambers, each with a pair of ostia for the entrance of blood from the surrounding pericardial sinus. Valves close the ostia when the heart contracts. A tubular aorta (Gr. *aorta*, the great artery) extends anteriorly from the heart and opens into the hemocoel in the head region. The hemocoel contains the internal organs and circulates the blood. This so-called open system of circulation causes the blood to flow in vessels only part of the time. Most of the time it flows in tissue spaces or sinuses in the body and appendages. Eventually, the blood returns to the pericardial sinus. The liquid plasma of the blood contains colorless cells.

Respiration

Several pairs of external openings or spiracles (L. *spiraculum*, air hole) open into the tracheal (respiratory) system on either side of the thorax and abdomen. The spiracles permit the entrance of

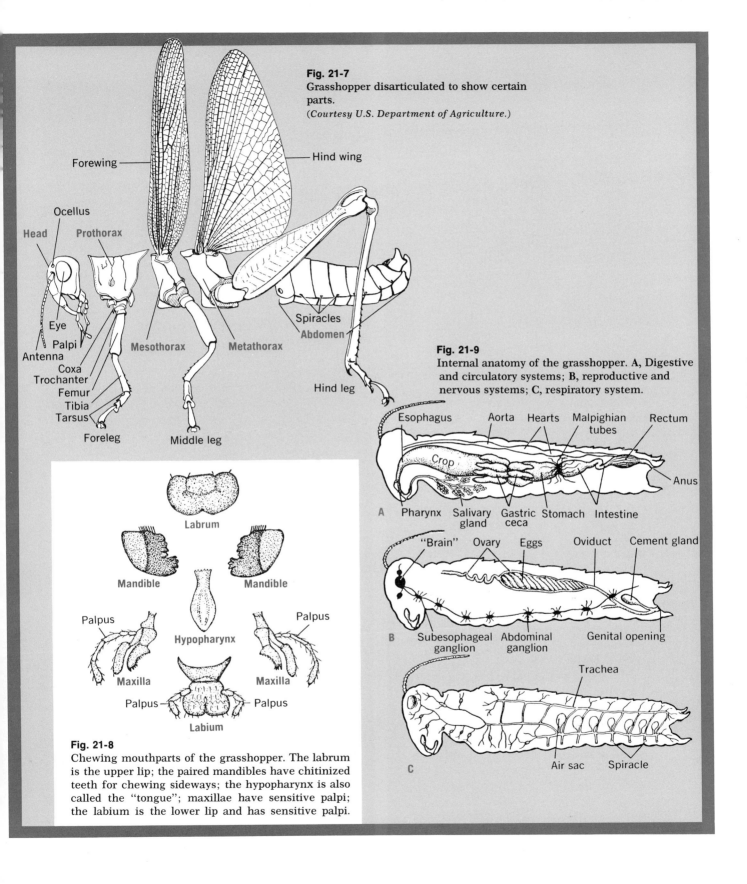

Fig. 21-7
Grasshopper disarticulated to show certain parts.
(*Courtesy U.S. Department of Agriculture.*)

Forewing — Hind wing

Ocellus
Head **Prothorax**
Eye
Palpi
Antenna
Coxa
Trochanter
Femur
Tibia
Tarsus
Foreleg
Mesothorax **Metathorax**
Middle leg
Spiracles
Abdomen
Hind leg

Fig. 21-8
Chewing mouthparts of the grasshopper. The labrum is the upper lip; the paired mandibles have chitinized teeth for chewing sideways; the hypopharynx is also called the "tongue"; maxillae have sensitive palpi; the labium is the lower lip and has sensitive palpi.

Labrum
Mandible Mandible
Palpus Palpus
Hypopharynx
Maxilla Maxilla
Palpus Palpus
Labium

Fig. 21-9
Internal anatomy of the grasshopper. **A,** Digestive and circulatory systems; **B,** reproductive and nervous systems; **C,** respiratory system.

Esophagus Aorta Hearts Malpighian tubes Rectum
Crop
Anus
A Pharynx Salivary gland Gastric ceca Stomach Intestine

"Brain" Ovary Eggs Oviduct Cement gland
B Subesophageal ganglion Abdominal ganglion Genital opening

Trachea
C Air sac Spiracle

Fig. 21-10
Portion of the grasshopper trachea with its branching tubules, both large and small (photomicrograph). The tracheal rings and nuclei are quite distinct.
(*Courtesy General Biological Supply House, Inc., Chicago, Illinois.*)

oxygen and the exit of carbon dioxide. The tubular tracheae ramify to all parts of the body and may have enlargements called air sacs (Figs. 21-9 and 21-10). The blood does not play an important role in respiration.

Excretion and egestion

The coiled malpighian tubules in the hemocoel collect wastes and empty them into the large intestine. Solid materials are eliminated through the anus.

Coordination and sensory equipment

A dorsal brain (three pairs of ganglia) is connected by a pair of circumesophageal connectives with a subesophageal ganglion. The ventral nerve cord continues posteriorly, with a pair of large ganglia in each thoracic segment and five pairs of ganglia in the abdomen. A sympathetic nervous system supplies the spiracles and muscles of the digestive system.

The compound eyes are covered with a cuticular cornea divided into numerous hexagonal facets. Each facet is the external surface of a unit called an ommatidium, composed of a long visual rod, and

the various units are separated from each other by a layer of dark pigment cells. Such an arrangement gives mosaic vision, in which each receptor receives a portion of the image.

Each simple eye, or ocellus, consists of a group of cells, the retinulae, a central optic rod, or rhabdom, and a transparent, cuticular "lens." The ocelli probably function as light-perception organs.

The pair of jointed, threadlike antennae bears sensory bristles, probably for olfactory purposes. Organs of taste are located on the mouthparts. Hairlike organs of touch are present on various body parts but particularly on the antennae. The pair of sound-receiving auditory organs located on the sides of the first abdominal segment consists of a membranous tympanum (Gr. *tympanon*, drum), which covers an auditory sac.

Reproduction

The sexes are separate. The female possesses a conspicuous ovipositor (L. *ovum*, egg; *ponere*, to place) at the tip of the abdomen for depositing eggs. In the female a pair of ovaries produces eggs that are discharged into a pair of oviducts. The oviducts unite to form a vagina, connected with the genital pore between the parts of the ovipositor. A seminal receptacle (spermatheca), connected with the vagina, receives sperms from the male during copulation and releases them to fertilize eggs. A secretion of the cement gland may stick eggs together as they are deposited.

In the male a pair of testes discharges sperm into a pair of vasa deferentia (sperm ducts), which unite to form the ejaculatory duct that opens through a penis. Accessory glands secrete a fluid into the ejaculatory duct to aid in the transfer of sperm to the female. Eggs are fertilized by sperm when they are deposited. A young grasshopper that hatches from an egg is called a nymph and resembles an adult without wings. As the grasshopper grows, it must shed its chitinous exoskeleton at certain intervals by the process of ecdysis (molting). Adult wings are eventually formed from wing buds (Fig. 21-11).

Honeybee

The honeybee, *Apis mellifica* (L. *apis*, bee; L. *mellificus*, honey), is placed in the order Hymenoptera (Gr. *hymen*, membrane; *ptera*, wings) because it has two pairs of membranous wings. Honeybees are more highly specialized in life habits and structure than grasshoppers. Colonies of honeybees consist of (1) workers, which are females with undeveloped reproductive organs, (2) male drones, and (3) female queens. A typical colony may contain 50,000 workers, a few hundred drones, and one adult queen (Fig. 21-12). The drones and queen are for reproductive purposes. Hymenoptera are characterized by having their mouthparts modified for both sucking and biting (chewing). The body is divided into head, thorax, and abdomen. The thorax is divided into an anterior prothorax, a middle mesothorax, and a posterior metathorax. The abdomen is also segmented.

Integument and skeleton

A tough, flexible cuticle covers the body and serves as an exoskeleton to which muscles and organs are attached internally. The cavity is a hemocoel, which carries blood.

Motion and locomotion

Locomotion is accomplished by two pairs of wings and three pairs of jointed legs. The wings

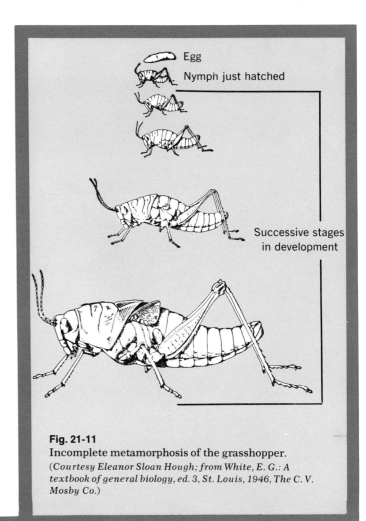

Fig. 21-11
Incomplete metamorphosis of the grasshopper.
(*Courtesy Eleanor Sloan Hough; from White, E. G.: A textbook of general biology, ed. 3, St. Louis, 1946, The C. V. Mosby Co.*)

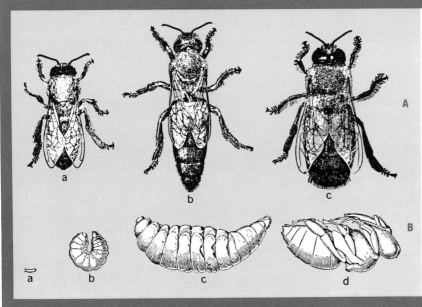

Fig. 21-12
Castes and development of the honeybee (*Apis*) of order Hymenoptera. In **A**, adults: a, worker; b, queen; c, drone. In **B**, immature stages: a, egg; b, young larva; c, old larva; d, pupa.
(*Courtesy U.S. Department of Agriculture.*)

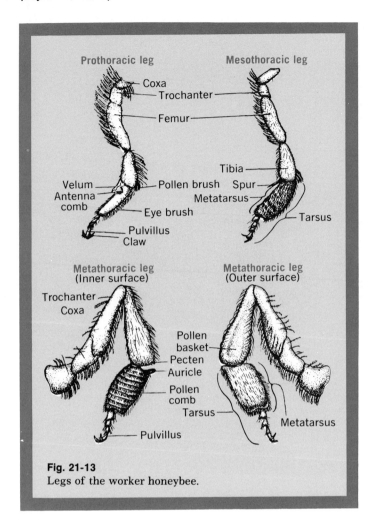

Fig. 21-13
Legs of the worker honeybee.

ing hairs and a flat, movable, spinelike velum; and (5) a segmented tarsus, the proximal segment of which may be called the metatarsus and which bears a semicircular antenna comb. This comb, together with the velum, constitutes the antenna cleaner, through which the antenna may be drawn to remove materials. On the opposite margin of the tibia from the velum is the pollen brush, composed of curved bristles. The last (distal) tarsal segment has claws with a padlike pulvillus between them. The pulvillus secretes a sticky substance for adherence.

2. Mesothoracic leg (second). The segments are the same as on the first pair of legs. A long pollen spur on the distal end of the tibia is used to remove pollen from the pollen basket and to clean wings.

3. Metathoracic leg (third). The pollen basket is on the outer, concave surface of the tibia and the long hairs curve over its depression to cover it somewhat. A pincerlike structure between the tibia and metatarsus is composed of rows of spines, the pecten (L. *pecten*, comb), and a liplike auricle (L. *auricula*, small ear). The pecten and auricle convey pollen to and pack pollen into the pollen basket. On the inner surface of the metatarsus are numerous transverse rows of stiff, bristlelike pollen combs to comb out pollen from various body parts and to handle wax. The wax is secreted in flat scales by a glandular area on the underside of the abdomen. The wax is masticated by the mandibles before it is used in building the "cells" of the honeycomb.

Ingestion and digestion

The mouthparts may be studied from Fig. 21-14. One pair of smooth mandibles lies beneath the upper lip (labrum) and is used in masticating wax. The sucking mouthparts assist the suction of the pharynx to convey fluids into the digestive tract. A long esophagus extends from the pharynx to the large honey sac (crop) in the abdomen. A large cylindrical stomach leads into the intestine, which joins the rectum, ending in the anus.

The nectar of flowers is sucked up and stored in the honey sac where it chemically changes into honey. The honey is regurgitated into the "cells" of the honeycomb. Here, the honey is still further dehydrated by currents of air, caused by the rapid

consist of a double layer of transparent membranes between which is a network of veins to strengthen them. When at rest, the wings are folded. During flight they are extended, and the forewings and hind wings are locked together by a row of tiny hooks on the front margin of the hind wings. Wings may vibrate over 400 times per second during flight.

The three pairs of legs are much more specialized than those of the grasshopper. Because of their complexity and differences, each leg will be considered separately (Fig. 21-13).

1. Prothoracic leg (first). The segments consist of (1) an oblong coxa, next to the thorax; (2) a short trochanter; (3) a long femur with branched, pollen-carrying hairs; (4) a tibia with pollen-carry-

vibrations of the wings. A minute drop of poison from the sting helps to preserve the honey. An average colony of bees in an average season collects about 40 pounds of honey.

Pollen is rich in the protein that honey lacks, so pollen ("bee bread") is essential in the diet of bees. The pollen collected in the pollen basket is placed in certain "cells" of the honeycomb. "Bee-glue" or propolis (Gr. *pro*, for; *polis*, city) is a resin collected from plants, and it is used in filling cracks and cementing loose parts. Various types of cells constitute the honeycomb.

Circulation

A long, delicate, tubular, muscular heart (Fig. 21-15, *B*) in the middorsal region of the body discharges the colorless blood toward the head region. Blood enters the heart through five pairs of ostia, each pair leading into a chamber of the heart. Valves prevent the backflow into the body during contraction. Blood discharges at the head region and passes into the hemocoel, from which it re-enters the heart chambers. The blood plasma contains white blood corpuscles.

Respiration

Respiration (Fig. 21-15, *C*) occurs through pairs of very small spiracles located along the sides of the thorax and abdomen and leading into a branched system of tracheae to convey air to all body parts. Certain tracheae may possess enlarged air sacs.

Excretion

Numerous, hollow, glandular, threadlike malpighian tubules excrete wastes into the intestine, much in the same manner as in grasshoppers.

Coordination and sensory equipment

A large "brain" (supraesophageal ganglion) in the dorsal part of the head supplies nerves to the eyes and antennae. The brain is connected by a ring (nerves) to the subesophageal ganglion, which supplies nerves to the mouthparts. A ventral nerve chain extends posteriorly from the subesophageal ganglion along the midventral side of the body. The chain is a double nerve strand and connects with two thoracic ganglia and five abdominal ganglia (Fig. 21-15, D).

The hairlike end organs of the sense of touch are

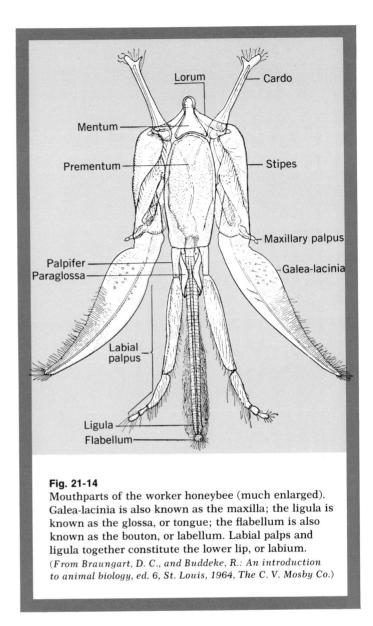

Fig. 21-14
Mouthparts of the worker honeybee (much enlarged). Galea-lacinia is also known as the maxilla; the ligula is known as the glossa, or tongue; the flabellum is also known as the bouton, or labellum. Labial palps and ligula together constitute the lower lip, or labium.
(*From Braungart, D. C., and Buddeke, R.: An introduction to animal biology, ed. 6, St. Louis, 1964, The C. V. Mosby Co.*)

present on various body parts but are particularly numerous on the tip of the antennae. The pair of jointed, hairy antennae has numerous, sound-sensitive pits that are thought to be for auditory purposes. Other pits on the antennae are thought to be for olfactory purposes. The so-called tongue bears numerous, bristlelike taste setae. Bees can be trained to estimate time intervals because some have been trained to come to a source of food at regular intervals. The pair of large compound eyes on the top and side of the head is constructed and functions similarly to that of the grasshopper. The

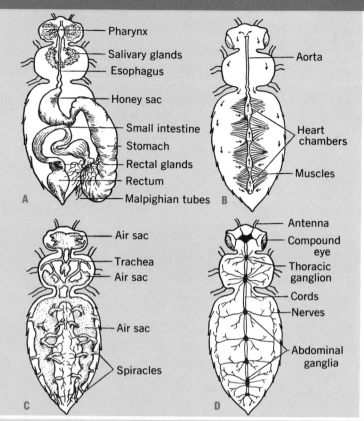

Fig. 21-15
Internal anatomy of the worker honeybee. **A,** Digestive system; **B,** circulatory system; **C,** respiratory system; **D,** nervous system.

A
- Pharynx
- Salivary glands
- Esophagus
- Honey sac
- Small intestine
- Stomach
- Rectal glands
- Rectum
- Malpighian tubes

B
- Aorta
- Heart chambers
- Muscles

C
- Air sac
- Trachea
- Air sac
- Air sac
- Spiracles

D
- Antenna
- Compound eye
- Thoracic ganglion
- Cords
- Nerves
- Abdominal ganglia

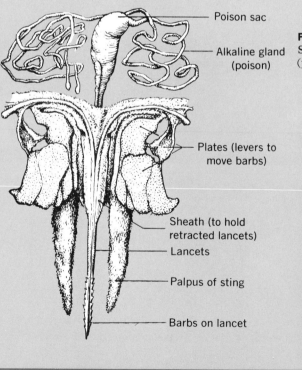

Fig. 21-16
Sting of the worker honeybee (parts somewhat separated and enlarged).

- Poison sac
- Alkaline gland (poison)
- Plates (levers to move barbs)
- Sheath (to hold retracted lancets)
- Lancets
- Palpus of sting
- Barbs on lancet

color sense of bees is better adjusted to the shorter wavelengths of the light spectrum; that is, toward the blue end of the spectrum. Three small simple eyes (ocelli) are present on the dorsal side of the head.

The sting is a modified ovipositor that is used for protection (Fig. 21-16). It is composed of two straight, grooved lancets (darts) with barbs at the tips and with muscles for their operation. A large, storage poison sac is connected with the base of the sting. Two acid glands and an alkaline gland mix their secretions to form the poisonous material that is injected when the bee stings. After stinging the worker bee leaves the sting, poison sac, and glands and dies. Males do not have a sting.

Language of bees

August Krogh, the Danish physiologist, described the amazing experiments of the Austrian zoologist, Karl von Frisch, on the ways in which bees convey information to other bees. Von Frisch began his investigations about 40 years ago and showed that bees are not totally color-blind. They have a definite color sense and can be trained to seek food on the background of a specific color that they distinguish from other colors. They are blind to the red end of the color spectrum.

Von Frisch constructed special, glass-covered hives in which he could make his observations, and he discovered that bees returning from a rich source of food perform special, dancing movements on the vertical surface of the honeycomb. He recognized two types of dances, (1) circling dance and (2) wagging dance. In the wagging dance the bee runs a certain distance in a straight line, swiftly wagging its abdomen from side to side and makes a turn. The dances excite the bees, and they follow the dancer closely, imitate the movements, and then proceed to search for the food as indicated. They know the type of food to seek from the precise information given by the odor of the flower nectar or pollen that adheres to the body of the bee.

The vigor of the dance that guides the bees is determined by the ease with which nectar is secured. When the supply of nectar in a certain type of flower is giving out, the bees visiting it slow down or stop their dance.

Von Frisch trained two groups of bees from the same hive to feed at different places. One group marked with a blue stain was trained to visit a feeding place a few meters away, whereas a group marked red was fed at a distance of 300 meters. All of the red bees performed wagging dances; all the blue ones performed circling dances. By a series of steps he then moved the nearer feeding place farther and farther from the hive. At a distance between 50 and 100 meters, the blue bees changed from a circling dance to a wagging dance. Conversely, the red bees, when brought gradually closer to the hive, changed from wagging to circling in the 50- to 100-meter interval. He also found that the frequency of the turns gave a fairly good indication of the distance. When the feeding site was 100 meters away, the bee made about 10 short turns in 15 seconds. To indicate 3,000 meters, only 3 long turns in 15 seconds were made.

Reproduction

The worker honeybee contains only vestigial (L. *vestigium*, trace) reproductive organs, since it is an undeveloped female. The reproductive organs of the male (drone) include a pair of bean-shaped testes, which produce sperms that are carried away by a pair of slender vasa deferentia. These expand to form the seminal vesicles for sperm storage. The two seminal vesicles combine to form one ejaculatory duct, which leads to the copulatory mechanism. One pair of large accessory glands secretes and empties nourishment into the ejaculatory duct.

In the female (queen) a pair of large ovaries produces eggs, which are carried by a pair of oviducts. These unite to form a tubular vagina leading to the exterior. A spermatheca attached to the vagina stores sperm received from the male during copulation. The queen is fertilized once in a lifetime, during a nuptial flight when swarming, and the sperms remain alive for years in the spermatheca. As an egg passes down the ovary toward the oviduct, it receives a shell with a small opening, the micropyle, through which a sperm may enter. A queen may lay an unfertilized egg to develop a drone or fertilized eggs to develop females, either queens or workers. A queen may lay 1,500 eggs per day for weeks at a time, and she may live several years. The eggs are small, oblong, and bluish white. Fertilized eggs are placed in worker or queen cells of the honeycomb; unfertilized eggs, in the drone cells. A wormlike, whitish larva ("grub") hatches from the egg in 4 days. All larvae are fed on a specially prepared and predigested mixture of honey and pollen ("royal jelly") for a few days, after which the drone and worker larvae are fed on plain honey and pollen, whereas the queen larva is kept on the "royal jelly" diet. This special food causes the larva to develop into a queen instead of a worker. After 6 days a larva develops into a pupa (L. *pupa*, baby) enclosed in a silken cocoon. A worker pupa changes into an adult bee in about 13 days, a queen in about 7 days, and a drone in about 15 days (Figs. 21-12 and 21-17).

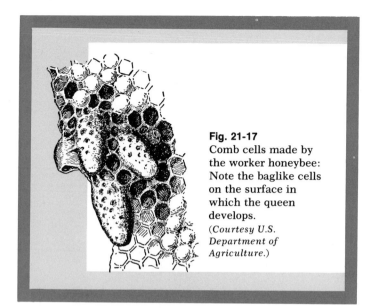

Fig. 21-17
Comb cells made by
the worker honeybee:
Note the baglike cells
on the surface in
which the queen
develops.
(*Courtesy U.S.
Department of
Agriculture.*)

Insect development (metamorphosis)

Many insects undergo marvelous changes as they develop from the egg to the adult stage. These changes are called metamorphosis.

Ametamorphosis

In ametamorphosis (or no metamorphosis) the egg develops into a young form that resembles the adult in all respects except size. After several molts the adult stage is reached.

Incomplete metamorphosis

In incomplete metamorphosis the egg develops into a larva that may display some resemblances to the adult and gradually transforms into an adult after a series of molts. Such larval forms are usually called nymphs, and the changes in body form occur gradually and are not very great between successive stages (Fig. 21-11).

If such larvae are aquatic, they are called naiads and often differ from the adult in having a respiratory system developed for aquatic breathing. The changes in body form are quite extensive (Fig. 21-18). In some classifications those possessing nymph stages are said to undergo gradual metamorphosis, whereas those with naiad stages are said to undergo incomplete metamorphosis.

Complete metamorphosis

In complete metamorphosis the egg develops into a segmented larval stage, to be followed by a quiescent pupal stage, which in turn develops into an adult (imago) (L. *imago*, image) stage (Fig. 21-19). In the larval stage the process of feeding and growth are outstanding; in the pupal stage the animal is enclosed in a case (such as a fibrous cocoon or a membranous chrysalis), and the pupa is completely transformed into an emerging adult; in the adult stage the propagation of the species is a chief function, although feeding and other activities may take place.

Hormones

Experiments have demonstrated that crustaceans and insects produce chemical substances called hormones which regulate such activities as metamorphosis, molting, growth, and pigmentation. Work in the field of genetics also has shown that minute quantities of gene-controlled chemical substances are released by various tissues and transported to different parts of the animal to influence metabolic activities in various regions. The production and release of these hormones are controlled by a number of internal and external environmental factors, possibly even several tissues may be involved in hormone secretion. Much information concerning insect hormones has been obtained from studies of the embryonic development of certain forms.

V. B. Wigglesworth in England experimented with *Rhodnius*, which has an elongated head and neck that can be cut at various levels or entirely removed. The larval nymphs suck blood and pass through a series of five molts before becoming adults. The experiments showed that two hormones are involved in the metamorphosis: a juvenile hormone, which causes the retention of nymphal characteristics, and a growth-and-differentiation (GD) hormone, which stimulates growth and differentiation of tissues and organs to the adult status. The juvenile hormone is secreted by a special cerebral gland, the corpus allatum (L. *corpus*, body; *allatus*, brought to, changed), which lies behind the brain. If this gland is properly removed from the larva of this insect, molting is prevented. It was shown that the brain cells secrete the GD hormone. In this insect it is possible to cut the head in such a way that the brain may be removed, leaving the corpus allatum intact, or to remove both areas. When the brain alone is removed, the nymph remains juvenile and undergoes extra molts. If both brain and corpus allatum are removed, the nymph changes to a miniature adult. When nymphs that were secreting large quantities of the juvenile hormone were grafted to headless adults, the latter molted and even developed certain nymphlike traits.

Experiments on the giant silkworm moth (*Cecropia*) by C. M. Williams of Harvard University showed that the juvenile hormone and the GD hormone were both present. These moths undergo complete metamorphosis, which makes it possible to analyze the actions of the hormones in more detail. The juvenile hormone is also secreted by a corpus allatum in moths, and the removal of the corpus allatum is quickly followed by transformation into the pupa stage. If a corpus allatum is surgically transplanted into a caterpillar that is ready to form a pupa, this stage of metamorphosis will be prevented. If extra glands that produce juvenile hormones are transplanted into mature larvae, the larvae can be induced to undergo additional larval molts and grow to giant size before pupating and eventually forming giant adults.

The pupal stage of the giant silkworm moth undergoes a period of dormancy of about 5 months at room temperature, which is called the diapause (Gr. *diapauvein*, to make to cease). If pupae are placed in a temperature of 3 to 5° C. for about 6 weeks, then at room temperature the pupae will become adults in about 1 month.

When brains from chilled pupae are surgically implanted into other pupae during their diapause, the latter metamorphose into adults. Even chilled pupae that are surgically grafted to other pupae in their diapause stage cause the latter to metamorphose into adults. If a pupa is cut into a head-thorax section and an abdominal portion and each of these sections is fastened to a plastic cover glass, the head-thorax piece transforms into an adult anterior end, but the abdominal section does not develop. However, the introduction of glandular tissue from the anterior end causes the abdominal section to develop into an adult posterior end.

It was observed that chilled brains alone did not

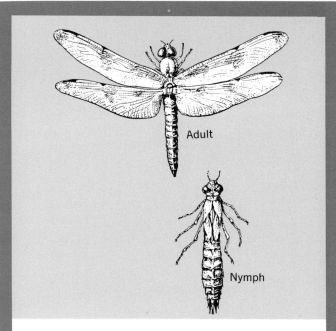

Fig. 21-18
Dragonfly of order Odonata, class Insecta, illustrating incomplete metamorphosis. The nymph (naiad) has large eyes and developing wings; the naiad stage is aquatic. The adult has large compound eyes and a characteristic jointlike nodus at the front margin of the wing.

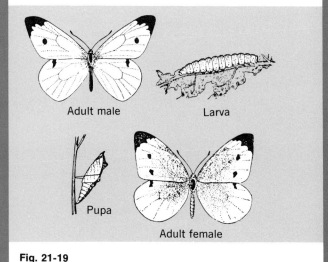

Fig. 21-19
Complete metamorphosis of the cabbage butterfly (*Pieris*). Note the differences in the adult male and female. The egg develops through successive stages (instars) of the larva into the pupa (chrysalis); the adult emerges from the pupa.
(*Courtesy General Biological Supply House, Inc., Chicago, Illinois.*)

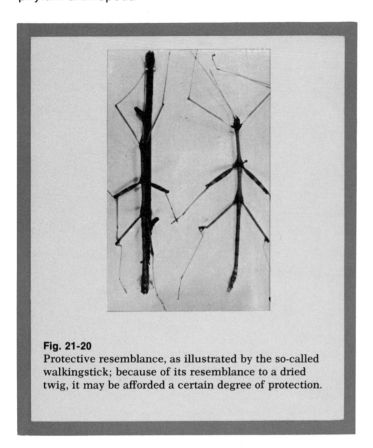

Fig. 21-20
Protective resemblance, as illustrated by the so-called walkingstick; because of its resemblance to a dried twig, it may be afforded a certain degree of protection.

INSECT MIMICRY

Fig. 21-21
Insect mimicry, as illustrated by the viceroy butterfly (*Basilarchia*), which resembles (mimics) the monarch butterfly (*Danaus*), shown at left.

cause the change but were effective only if administered together with tissue from the so-called prothoracic gland. Williams proved that chilling stimulates certain gland cells in the insect brain to secrete a hormone that stimulates the prothoracic glands to release the GD hormone. The absence of either the brain hormone or the prothoracic hormone prevents metamorphosis into an adult.

Coloration, color patterns, and protective resemblance

Many insects possess colors of various kinds, especially the butterflies, moths, and beetles. Colors may be classed as (1) chemical colors and (2) physical colors. Chemical colors are caused by pigments that in insects are mostly black or brown, although yellow, orange, red, and white may be present. Physical colors are caused by the structure of the surface that may diffract (scatter light into its different colored rays) or refract (deflect rays of light from a straight path to the oblique, passing from one medium to another). Common physical colors in insects include violet, blue-green, silver, and gold, as represented in the metallic colors of certain beetles, the iridescence of fly wings, and the iridescence of many butterflies. In some insects the colors and color patterns may vary with the seasons and with the sexes.

Certain insects have colors that blend in with their natural surroundings, thus giving them some protection. This may be a result of color, of the peculiar structures they possess, or of a combination of the two. Green katydids living among green plants are less conspicuous. The underwing moth has a mottled gray color resembling the bark of trees. The color of other insects resembles that of flowers that they customarily visit. The dead-leaf butterfly (*Kallima*) of India has the upper side of its wings brightly colored. However, the underside is colored like dead leaves, and when at rest with wings held together above the back, the butterfly is difficult to detect. It must rest on a twig in a particular way or it does not resemble a dead leaf, hence it may not be protected. The common walkingsticks are difficult to detect because their peculiar structure and coloration resemble a stick (Fig. 21-20).

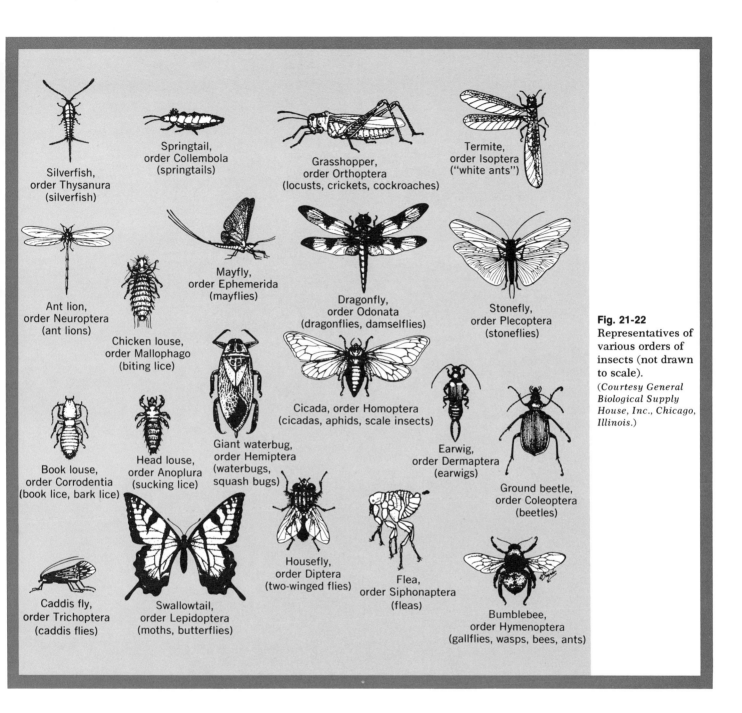

Fig. 21-22 Representatives of various orders of insects (not drawn to scale). (*Courtesy General Biological Supply House, Inc., Chicago, Illinois.*)

Silverfish, order Thysanura (silverfish)

Springtail, order Collembola (springtails)

Grasshopper, order Orthoptera (locusts, crickets, cockroaches)

Termite, order Isoptera ("white ants")

Ant lion, order Neuroptera (ant lions)

Chicken louse, order Mallophago (biting lice)

Mayfly, order Ephemerida (mayflies)

Dragonfly, order Odonata (dragonflies, damselflies)

Stonefly, order Plecoptera (stoneflies)

Book louse, order Corrodentia (book lice, bark lice)

Head louse, order Anoplura (sucking lice)

Giant waterbug, order Hemiptera (waterbugs, squash bugs)

Cicada, order Homoptera (cicadas, aphids, scale insects)

Earwig, order Dermaptera (earwigs)

Ground beetle, order Coleoptera (beetles)

Caddis fly, order Trichoptera (caddis flies)

Swallowtail, order Lepidoptera (moths, butterflies)

Housefly, order Diptera (two-winged flies)

Flea, order Siphonaptera (fleas)

Bumblebee, order Hymenoptera (gallflies, wasps, bees, ants)

It has been argued that certain insects possess what is called protective mimicry. In this kind of protective coloration one species with certain colors and disagreeable qualities is mimicked by another species with similar colors but without the disagreeable qualities; for example, the viceroy butterfly (*Basilarchia*) resembles (mimics) the disagreeable monarch butterfly (*Danaus*) to such an extent that it is not attacked (Fig. 21-21).

A brief study of typical representatives of some of the important orders (Fig. 21-22) is given in Table 21-3.

293

Table 21-3
Orders of the class Insecta (insects)

Name of order	Examples	Wings	Mouthparts	Metamorphosis	Miscellaneous
Thysanura (Gr. *thysanos*, tassel; *oura*, tail)	Bristletails, silverfish	Wingless	Chewing	None (larva resembles adult)	Primitive insects; 11 abdominal segments; 2 to 3 long, segmented caudal appendages; long antennae; common in dark, moist places
Collembola (Gr. *kolla,* glue; *ballo,* to put)	Springtails, snow fleas	Wingless	Chewing or sucking	None (larva resembles adult)	Primitive insects; 6 abdominal segments; springing organ on ventral side of abdomen (most species); sticky tube-like projection on ventral abdomen; 4 segments in antennae; common under decaying leaves or bark
Orthoptera (Gr. *orthos,* straight; *ptera,* wings)	Grasshoppers, katydids, cockroaches, crickets, walking-sticks, praying mantises	2 pairs (usually); forewings straight, leathery (wing covers); hind wings folded like fan when at rest; in some, wings absent or vestigial	Chewing	Incomplete	Most species live on vegetation; mantises feed on other insects
Isoptera (Gr. *isos,* equal; *ptera,* wings)	Termites	2 pairs; long, narrow, held flat on back; or wingless	Chewing	Incomplete	Social insects living in colonies with several castes; abdomen broadly attached to thorax; build earthen tubes for passageways
Neuroptera (Gr. *neuron,* nerve; *ptera,* wings)	Dobson flies (hellgrammite), ant lions, aphis lions (lacewing flies)	2 pairs; similar, usually with many cross veins	Chewing	Complete	Some species predacious; larvae may suck blood from their prey
Ephemerida (Gr. *ephemeros,* living but a day) or Ephemeroptera	Mayflies	2 pairs; membranous, triangular; forewings large; hind wings small, or absent; wings upright when at rest	Vestigial (in adults); chewing (in aquatic naiads)	Incomplete	Adults take no food and live for 1 day; 2 to 3 long abdominal "tails"; serve as fish food; aquatic naiads have lateral, tracheal gills

Table 21-3
Orders of the class Insecta (insects)—cont'd

Name of order	Examples	Wings	Mouthparts	Metamorphosis	Miscellaneous
Odonata (Gr. *odon*, tooth)	Dragonflies, damselflies	2 pairs; membranous; hind wings as large or larger than forewings; jointlike nodus near middle of front margin	Chewing	Incomplete	Adults and aquatic naiads both predacious; body long, tusklike; large head; large, compound eyes; small antennae
Plecoptera (Gr. *plekos*, folded; *ptera*, wings)	Stoneflies	2 pairs; membranous; hind wings usually larger and folded under forewings when at rest	Chewing	Incomplete	Larval naiads live in water under rocks; often with tufts of tracheal gills
Corrodentia (L. *corrodens*, gnawing)	Book lice, bark lice	Book lice wingless; bark lice have 2 pairs membranous wings that have few distinct veins; forewings larger than hind wings; held rooflike when at rest	Chewing	Incomplete	Book lice minute, grayish yellow; soft bodied; found in old papers; bark lice have oval bodies and feed on plants
Mallophaga (Gr. *plekos*, folded; *phagein*, to eat)	Biting lice, such as chicken lice, cattle lice	Wingless	Chewing	Incomplete	Flat, broad body and broad head; sharp claws, degenerate eyes; on skin, hair, feathers of mammals and birds
Thysanoptera (Gr. *thysanos*, fringe; *ptera*, wings)	Thrips, such as grass thrips, wheat thrips, fruit thrips, onion thrips	2 pairs (usually); long, similar, narrow, membranous, fringed with long hairs; some species wingless	Piercing-sucking	Incomplete	Minute, slender body; feet have bladderlike structures for adhering; suck plant juices
Anoplura (Gr. *anoplos*, unarmed; *oura*, tail)	Sucking (true) lice, such as human head lice, human body lice, or "cootie," dog lice; rat lice	Wingless	Piercing-sucking	None	Flat broad body with free horizontal head; claws for clinging to hair; degenerate eyes, or absent; parasites on mammals; eggs called "nits"

Continued

Table 21-3

Orders of the class Insecta (insects)—cont'd

Name of order	Examples	Wings	Mouthparts	Metamorphosis	Miscellaneous
Hemiptera (Gr. *hemi*, half; *ptera*, wings)	True bugs, such as stink bugs, squash bugs, chinch bugs, assassin bugs, bedbugs, water striders, back swimmers	2 pairs; forewings thickened at base with thinner extremities, which overlap on back; membranous hind wings folded; some wingless	Piercing-sucking (jointed beak)	Incomplete	Beak arises from front of head; many species on vegetation
Homoptera (Gr. *homos*, same; *ptera*, wings)	Cicadas, leafhoppers, aphids (plant lice), scale insects	2 pairs (usually); membranous and of uniform thickness, held rooflike over back; some species wingless	Piercing-sucking (beak)	Incomplete	Beak arises from hind part of lower side of head; many species on vegetation
Dermaptera (Gr. *derma*, skin; *ptera*, wings)	Earwigs	2 pairs; forewings small, leathery, meeting in middle of back; hind wings large, membranous and folded lengthwise and crosswise under forewings; some wingless	Chewing	Incomplete	One pair pincerlike appendages at tip of abdomen; elongated body; not common in United States
Coleoptera (Gr. *koleos*, sheath; *ptera*, wings)	Beetles, such as tiger beetles, ladybird beetles, June beetles, click beetles, ground beetles, snout beetles, dermestid beetles, etc.; some species wingless	2 pairs; forewings (elytra) hard, sheathlike; hind wings membranous and folded under elytra	Chewing	Complete	Chitinous covering usually heavy; sizes vary from small to very large; common in many places
Mecoptera (Gr. *mekos*, long; *ptera*, wings)	Scorpion flies	2 pairs; long, narrow, membranous, frequently dark spotted; some species wingless	Chewing	Complete	Clasping organ at tip of male abdomen resembles sting of scorpion; long, slender antennae; head prolonged into beak with chewing mouthparts

Table 21-3
Orders of the class Insecta (insects)—cont'd

Name of order	Examples	Wings	Mouthparts	Metamorphosis	Miscellaneous
Trichoptera (Gr. *trichos*, hair; *ptera*, wings)	Caddis flies	2 pairs; membranous, covered with long, silky hairs; hind wings usually shorter and broader; wings folded rooflike when at rest	Vestigial (in adults)	Complete	Soft, mothlike body; live near water and attracted by light; larva usually lives in caddis case formed of pebbles and sand
Lepidoptera (Gr. *lepis*, scale; *ptera*, wings)	Butterflies, skippers, moths	2 pairs; membranous and covered with overlapping scales; some wingless	Sucking (coiled under head)	Complete	Scales cover body; many with colors; moth antennae usually featherlike; butterfly and skipper antennae usually filamentous; larvae called caterpillars; some form cocoons of thread; pupa often called chrysalis
Diptera (Gr. *dis*, two; *ptera*, wings)	True flies, such as houseflies, fruitflies, horseflies, sandflies, crane flies, mosquitoes, gnats, midges	1 pair only; forewings membranous; hind wings represented by pair of knobbed threads, called halteres (for balancing); few species wingless	Piercing-sucking (proboscis)	Complete	Head attached to thorax by slender neck; compound eyes usually large; larvae called maggots
Siphonaptera (Gr. *siphon*, tube; *a-*, without; *ptera*, wings)	Fleas, such as human fleas, dog and cat fleas, rat fleas	Wingless	Piercing-sucking	Complete	Body small, oval, laterally compressed (sideways), covered with bristles; head small; no compound eyes; legs for leaping; parasitic on mammals and birds; larvae slender, legless, with chewing mouthparts
Hymenoptera (Gr. *hymen*, membrane; *ptera*, wings)	Honeybees, bumblebees, wasps (social and solitary), true ants, ichneumon wasps, gall flies, saw flies	2 pairs; similar membranous; forewings larger and with reduced venation; wings on each side held together by hooks during flight; some species wingless	Sucking or chewing	Complete	Abdomen of female usually has sting, piercer, or saw; many species live in colonies; some parasitic on other insects

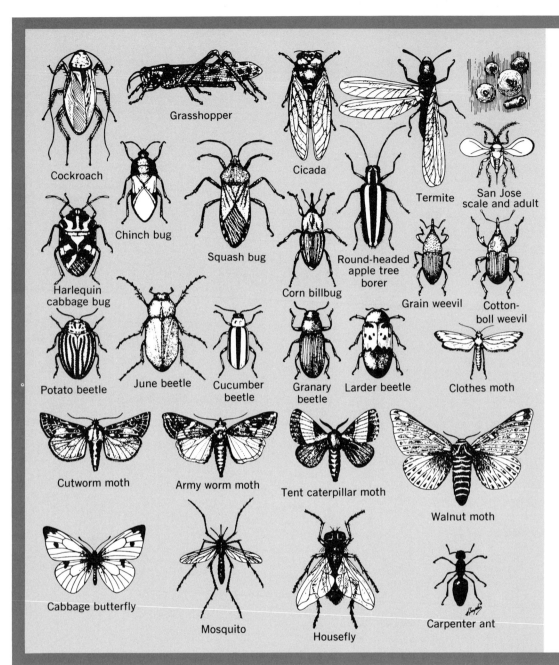

Fig. 21-23
Harmful or detrimental insects from various orders (not drawn to scale).
(*Courtesy General Biological Supply House, Inc., Chicago, Illinois.*)

Labels within figure: Cockroach, Grasshopper, Cicada, Termite, San Jose scale and adult, Chinch bug, Squash bug, Round-headed apple tree borer, Grain weevil, Cotton-boll weevil, Harlequin cabbage bug, Corn billbug, Potato beetle, June beetle, Cucumber beetle, Granary beetle, Larder beetle, Clothes moth, Cutworm moth, Army worm moth, Tent caterpillar moth, Walnut moth, Cabbage butterfly, Mosquito, Housefly, Carpenter ant

HARMFUL INSECTS

Space will not permit a detailed consideration of all the ways in which insects harm or benefit man, but there are probably more harmful than beneficial types. Harmful types (Fig. 21-23) may be grouped as follows.

1. Many insects destroy or damage plants, lumber and fruit trees, plant products, stored grains, fruits, and vegetables. The great numbers of insects involved are of many varieties. In recent estimates by the United States Department of Agriculture, insects cause $4,000,000,000 damage annually to farm crops, stored foodstuffs, forests, and domestic animals. Common examples include chinch bugs, Hessian flies, corn borers, cotton-boll

298

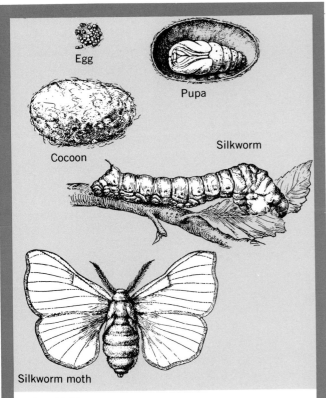

Egg

Pupa

Cocoon

Silkworm

Silkworm moth

Fig. 21-24
Silkworm (*Bombyx*) of order Lepidoptera. The larva, or silkworm, is shown feeding on a leaf. The pupa is shown with a part of the cocoon removed: Note the silk threads on the cocoon.

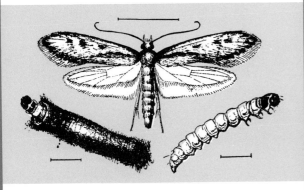

Fig. 21-25
Case-bearing clothes moth (*Tinea*) of order Lepidoptera (enlarged): adult moth (above); larva (lower right); larva partially concealed in a portable case (lower left). The indistinct dark spots on the buff-colored forewings distinguish the adult from the adult webbing clothes moth, the wings of which are uniformly buff colored.
(*Courtesy U.S. Department of Agriculture.*)

weevils, wireworms, grain weevils, army worms, grasshoppers, potato beetles, plant aphids (plant lice), and scale insects.

2. Insects that affect domestic animals and animal products include fleas, biting lice, sucking lice, mosquitoes, and various types of flies, such as horse botflies, hornflies of cattle, and warble flies, whose larvae burrow into the skin of cattle.

3. Insects are very destructive in houses. Termites may destroy wooden buildings and wooden contents. Furs, feathers, clothing, carpets, and fabrics may be damaged by clothes moths (Fig. 21-25) and carpet beetles. Papers may be destroyed by the simple silverfish, common in dark, damp places. Foods may be destroyed or damaged by ants, cockroaches, and weevils. Household pests also include mosquitoes, flies, and bedbugs.

4. Various diseases may be transmitted by many types of insects. Examples include malaria and yellow fevers, carried by mosquitoes; typhoid, dysentery, and diarrhea, carried by flies; African sleeping sickness, carried by tsetse flies; bubonic plague, carried by fleas; and typhus fever, carried by body lice.

BENEFICIAL INSECTS

It might seem that very few insects are beneficial. However, those activities of insects that are of benefit to man (Fig. 21-26) might be grouped as follows.

1. Several insects make products of importance to man, such as honey, beeswax, shellac (lac) wax, silk, and cochineal. Honeybees produce 500,000,000 pounds of honey in the United States annually, and about 10,000,000 pounds of beeswax is used annually. Lac insects secrete a wax from which shellac is made. Each cocoon of the silkworm moth contains about 1,000 feet of spun thread (Fig. 21-24); about 25,000 such cocoons are required to produce 1 pound of silk. About 50,000,000 pounds of silk are produced in the world annually. The dried bodies of a scale insect that lives on cactus may be used in making the dye cochineal, but aniline dyes have largely supplanted this source.

2. Several types of insects are necessary in the cross-fertilization of flowers of many of our important and useful plants. Bees and certain other

299

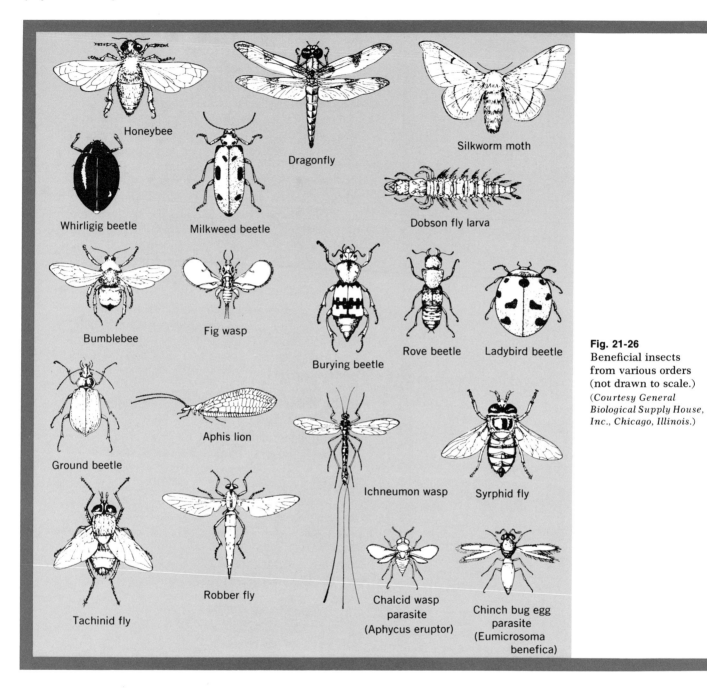

Honeybee

Dragonfly

Silkworm moth

Whirligig beetle

Milkweed beetle

Dobson fly larva

Bumblebee

Fig wasp

Burying beetle

Rove beetle

Ladybird beetle

Ground beetle

Aphis lion

Ichneumon wasp

Syrphid fly

Tachinid fly

Robber fly

Chalcid wasp parasite (Aphycus eruptor)

Chinch bug egg parasite (Eumicrosoma benefica)

Fig. 21-26
Beneficial insects from various orders (not drawn to scale.)
(Courtesy General Biological Supply House, Inc., Chicago, Illinois.)

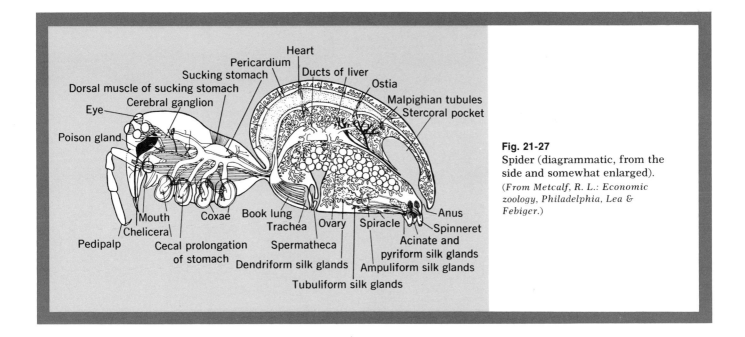

Fig. 21-27
Spider (diagrammatic, from the side and somewhat enlarged). (*From Metcalf, R. L.: Economic zoology, Philadelphia, Lea & Febiger.*)

insects pollenize the flowers of clovers, berries, apples, pears, and other fruits and crops. The Smyrna fig could not be grown in California until a small fig wasp (*Blastophaga*) was imported to pollenize the flowers.

3. Predacious insects are beneficial because they destroy great numbers of other insects, many of which are injurious. Common examples include ground beetles, ladybird beetles, aphis lions, praying mantises, and certain types of wasps. An Australian ladybird beetle was imported to California to destroy scale insects that parasitized orange and lemon trees. Many insects lay their eggs in the larvae of harmful types, and the young hatched from these eggs slowly destroy the parasitized host.

4. Several types of insects act as scavengers, the larvae eating great quantities of dead plant and animal materials, thus preventing their immediate decay and undesirable odors. Included in this group are flesheating flies, water-scavenger beetles, burying beetles, and tumble bugs. Insects also supply foods for animals, such as birds and fishes.

5. The maggots of certain common blowflies have been used in the healing of certain types of wounds. The maggots eat the dead tissues and secrete a substance called allantoin, which stimu-lates healing. It is an oxidation product of uric acid and is also present in the allantoic fluid of cows and in beet juice.

CLASS ARACHNOIDEA

Spiders, scorpions, ticks, mites, horseshoe crabs, and daddylonglegs are grouped together in the class Arachnoidea (Gr. *arachne*, spider; *eidos*, like) (Figs. 21-27 to 21-31). There are many differences between these arthropods with respect to structures, appendages, and habits. Many arachnids are predacious, sucking foods from their prey. Many types actually destroy injurious insects. The bite of the black widow spider (*Latrodectus*) may cause death in man. This is a black spider with an orange-red "hourglass" on the underside of the abdomen. The term widow is erroneously applied, because it was thought that she devoured her male immediately after mating. The sting of scorpions may be painful. Certain mites may transmit diseases, damage plants by sucking their juices, or cause annoying skin irritations.

The web-making behavior of spiders is now being used to advantage by pharmaceutical firms in the testing of drugs. The spiders make various types of abnormal webs under the influence of drugs.

Scorpions have elongated, segmented bodies and

301

Fig. 21-28
Scorpion of class Arachnoidea, showing the segmented appendages and sting at the tip of the abdomen.

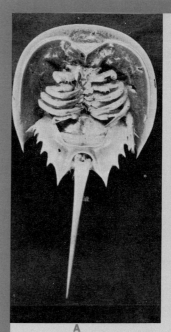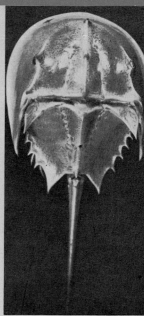

A   B

Fig. 21-30
Horseshoe crab (*Limulus*) of class Arachnoidea. A, ventral view; B, dorsal view.

Fig. 21-29
Wood ticks of class Arachnoidea. These external parasites on dogs and other animals suck blood and cause severe irritations.

Fig. 21-31
Peripatus of class Onychophora, sometimes considered as the connecting link between annelids and arthropods because it possesses certain traits of both.
(Courtesy Carolina Biological Supply Co., North Carolina.)

a poisonous sting at the tip of the abdomen. They are primarily inhabitants of the tropics and subtropics, where some forms may harm human beings. In courtship scorpions seize each other's claws and perform a mating dance.

Ticks and mites are arachnids with no external segments; even the cephalothorax and abdomen are fused. The mites are quite small and are found in many places. The female of the itch mite burrows into the skin to lay her eggs. Another species causes mange in dogs and other domestic animals. Chiggers (red "bugs") are the larvae of red mites and have only three pairs of legs. They are quite commonly found in vegetation at times. When a chigger attaches itself to the skin, it eats out a depression with the aid of a digestive secretion. This irritating fluid causes itching and red blotches. Ticks are usually larger than mites. They suck blood from vertebrate animals, and one large feeding may last for some time. The larvae are present in bushes and attach themselves to a host as it passes. Ticks may transmit such diseases as Texas cattle fever, relapsing fever, and Rocky Mountain spotted fever.

Horseshoe (king) crabs are common in shallow water along the Atlantic coast. These "living fossils" represent a very ancient group of animals. An unsegmented, horseshoe-shaped, dorsal cover and a broad, hexagonal abdomen with a long tail-like telson characterize them. Ventrally, the cephalothorax bears five pairs of walking legs and a pair of chelicerae, whereas the abdomen bears six pairs of thin, broad appendages fused in the median line. On certain abdominal appendages are exposed book gills for respiratory purposes. One pair of compound eyes and a pair of simple eyes are present.

Important studies on the physiology of vision are being made using the horseshoe crab as the subject.

Review questions and topics

1 What general characteristics describe the arthropods as a group?
2 What are the characteristics that distinguish members of the different classes of arthropods, including examples of each class?
3 Discuss the significance of jointed appendages and an exoskeleton of chitin.
4 Discuss the significance of the division of the body into head, thorax, and abdomen.
5 Discuss the importance of wings as found in certain insects.
6 Discuss the process of respiration by means of tracheae, book lungs, and gills, including examples of each type.
7 Describe the structure and functions of the "open" type of circulatory system and the presence of a hemocoel, including examples.
8 Explain the necessity of molting during the growth of arthropods.
9 Describe the structure and functions of typical arthropod nervous systems.
10 Discuss the status of sensory organ development in arthropods, including specific examples of each type of organ.
11 Describe the different types of metamorphosis, with examples of each type.
12 Describe the anatomy and functions of each part of the various systems in (1) the crayfish, (2) the grasshopper, and (3) the honeybee.
13 Discuss the biology of centipedes, millipedes, and spiders.
14 Discuss coloration, color patterns, and protective resemblance, giving examples of each.
15 Discuss hormones and insect activities, giving specific examples.
16 Discuss harmful and beneficial insects, giving examples of each.
17 Discuss the complex societies of bees.
18 List the distinguishing characteristics of each order of insects, giving examples of each order.

Selected references

Baker, E. W., and Wharton, G. W.: An introduction to acarology, New York, 1952, The Macmillan Company.

Borrer, D. J., and DeLong, D. W.: An introduction to the study of insects, New York, 1963, Holt, Rinehart & Winston, Inc.

Campbell, F. L. (editor): Physiology of insect development, Chicago, 1959, University of Chicago Press.

Carlisle, D. B., and Knowles, F.: Endocrine control in crustaceans, New York, 1959, Cambridge University Press.

Cheng, T. C.: The biology of animal parasites, Philadelphia, 1964, W. B. Saunders Company.

Esch, H.: The evolution of bee language, Sci. Amer. Apr., 1967.

Fox, R. M., and Fox, J. W.: Introduction to comparative entomology, New York, 1964, Reinhold Publishing Corp.

Gertsch, W. J.: American spiders, New York, 1949, D. Van Nostrand Co., Inc.

Green, J.: A biology of Crustacea, London, 1961, H. F. & G. Witherby, Ltd.

Horsfall, W. R.: Medical entomology, New York, 1962, The Ronald Press Company.

Jaques, H. E.: How to know the insects, ed. 2, Dubuque, Iowa, 1947, William C. Brown Company, Publishers.

Jones, J. C.: The sexual life of a mosquito, Sci. Amer. 218:108-16, 1968.

Meglitsch, P. A.: Invertebrate zoology, New York, 1967, Oxford University Press, Inc.

Metcalf, C. L., Flint, W. P., and Metcalf, R. L.: Destructive and useful insects, ed. 4, New York, 1962, McGraw-Hill Book Company.

Patton, R. L.: Introductory insect physiology, Philadelphia, 1963, W. B. Saunders Company.

Petrunkenitch, A.: The spider and the wasp, Sci. Amer. Aug., 1952.

Richards, A. G.: The integument of arthropods, Minneapolis, 1951, University of Minnesota Press.

Roeder, K. D.: Moths and ultrasound, Sci. Amer. 212:94-102, 1965.

Rothschild, M.: Fleas, Sci. Amer. 213:44-53, 1965.

Snodgrass, R. E.: A textbook of arthropod anatomy, Ithaca, New York, 1952, Comstock Publishing Associates.

Waterman, T. H. (editor): The physiology of Crustacea, 2 vol., New York, 1960, Academic Press Inc.

Williams, C. M.: The metamorphosis of insects, Sci. Amer. Apr., 1950.

Wilson, D. M.: The flight-control system of the locust, Sci. Amer. 218:83-90, 1968.

chapter twenty-two

Phylum Echinodermata

All echinoderms are marine animals. In contrast to most of the "higher" animals the "spiny-skinned" animals have radial symmetry. They share an endoskeleton and some embryologic developments with the chordates. A unique feature of the echinoderms is the presence of a water vascular system. Table 22-1 shows many of the characteristics of the animals included in the phylum Echinodermata.

CLASS ASTEROIDEA (STARFISHES)

The starfish, *Asterias,* (Fig. 22-1) has on its upper (aboral) surface numerous short spines attached to the calcareous skeletal plates (ossicles); a circular, sievelike madreporite plate, which is the entrance to its water-vascular system, and an anus. The calcareous plates are held together by muscles and connective tissues. Surrounding the spines are minute, pincerlike pedicellariae (Fig. 22-2), supplied with jaws that are operated by muscles. The pedicellariae are used to capture food, clean the body surface, and protect the soft, saclike dermal branchiae ("skin gills"), which carry on respiration.

On the under (oral) surface are a centrally located mouth and five ambulacral grooves (L. *ambulacrum,* covered), one in each arm, from which rows of muscular tube feet extend (Fig. 22-3). Other species of starfish may have more than forty arms. Starfish arms are moved by the action of muscles in the walls.

The water-vascular system is unique with the echinoderms (Fig. 22-4). In the starfish a porous, aboral madreporite leads to a short stone canal, which enters a circular ring canal around the mouth. From the ring canal five radial canals, one per arm, pass just above the ambulacral grooves and toward the tip of each arm. The radial canals give off connecting canals to which are attached the tube feet, with a bulb-shaped ampulla (L. *ampulla,* flask) at the inner end and a sucker at the outer end. The ampullae contain circular muscles, and the tube feet contain longitudinal muscles. When ampullae contract, valves in the connecting canals prevent the backflow of water into the radial canal, thus forcing water into the tube feet and causing the sucker to attach. The contraction of longitudinal muscles in the tube feet draws the animal forward. By regulating the internal water pressure the tube feet function in locomotion, in capturing food, and in adhering to rocks.

Nine small, spherical Tiedemann's bodies, located on the inner wall of the ring canal, are thought to produce the amebocytes of the water-vascular system.

Internally, a large coelom lined with cilia contains a fluid in which circulate the amebocytes, which function in excretion, respiration, and circulation. The digestive system contains a mouth and a short esophagus, which leads into a large, thin-walled stomach. The stomach consists of a large, oral cardiac portion and a small aboral pyloric portion. From the pyloric portion a tube passes into each arm and divides into two branches called digestive glands, or hepatic (pyloric) ceca. These are greenish in the living animal and together with the pyloric part of the stomach secrete digestive enzymes. Above the stomach is a slender intestine that exits at the anus. A pair of brownish, branched, pouchlike rectal ceca arise from the intestine and probably have an excretory function.

Foods of the starfish consist of oysters, clams, fishes, snails, barnacles, and worms. When a starfish opens an oyster or clam, it arches its body and attaches its tube feet to the two shells of the mollusk. A steady pull eventually draws the shells apart. After some time, because of fatigue, the mollusk relaxes, and the starfish everts its stomach through its mouth and between the shells. Diges-

305

Table 22-1
Distinguishing characteristics of echinoderms (phylum Echinodermata)

Class	Body	Ambulacral grooves	Tube feet	Pedicellaria (pincerlike organs)	Anus and madreporite plate	Examples
Asteroidea (Gr. *aster*, star; *eidos*, like)	Typically, 5 arms indistinctly marked off from central disk	Open on oral (under) side of animal	With suckers	Present	Anus aboral; madreporite aboral	Starfishes (*Asterias*)
Ophiuroidea (Gr. *ophis*, snake; *oura*, tail; *eidos*, like)	Typically, 5 long, slender, jointed arms sharply marked off from central disk	Absent or covered with ossicles (plates)	Without suckers	Absent	Anus absent; madreporite oral	Brittle stars (*Ophiura*)
Echinoidea (Gr. *echinos*, hedgehog; *eidos*, like)	Globe shaped (sea urchins), or disk shaped (sand dollars); compact skeleton (test); movable spines	Covered with ossicles	With suckers	Present	Anus aboral (sea urchins); anus marginal (sand dollars); madreporite aboral	Sea urchins (*Arbacia*); sand dollars
Holothuroidea (Gr. *holothurion*, sea cucumber; *eidos*, like)	Soft and cucumber shaped; no arms; no spines; 10 to 30 circumoral tentacles (modified tube feet); scattered ossicles (dermal plates) confined to body wall; cloaca usually with respiratory tree	Concealed	With suckers	Absent	Anus aboral; madreporite internal	Sea cucumbers (*Thyone*)
Crinoidea (Gr. *krinon*, lily; *eidos*, like)	Attached during part or all of life by aboral stalk of ossicles; some have 5 branched arms each bearing feather-like pinnules; no spines	Open, ciliated, and on oral surface	Tentacle-like in ambulacral grooves; without suckers	Absent	Anus oral; no madreporite	Sea lilies; feather stars (*Antedon*)

tive juices begin digestion of the mollusk within its shells. The partially digested food is placed into the stomach, which is drawn back through the mouth.

The body fluid in the coelom is circulated by cilia, thus carrying absorbed foods to various body parts. The reduced circulatory system consists of a circumoral vessel (around the mouth) that connects with one radial vessel in each arm located beneath the radial canal.

Excretion is performed by amebocytes of the coelomic fluid, which collect wastes and pass them to the outside through the ciliated walls of the dermal branchiae.

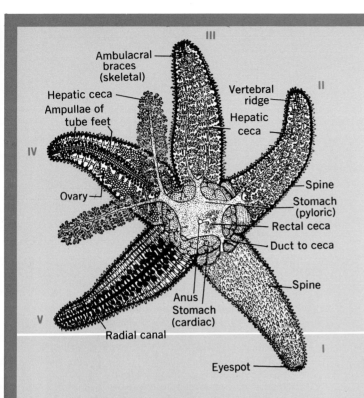

Ambulacral
braces
(skeletal)

III

Vertebral
ridge

II

Hepatic ceca

Hepatic
ceca

Ampullae of
tube feet

IV

Ovary

Spine

Stomach
(pyloric)

Rectal ceca

Duct to ceca

Anus
Stomach
(cardiac)

Spine

V

Radial canal

I

Eyespot

Fig. 22-1
Common starfish (*Asterias*) dissected from the aboral
side to show the digestive, locomotor, reproductive, and
skeletal systems. **I,** Arm (ray), showing the spines and
eyespot; **II** and **III,** arms with the aboral surface removed;
IV, arm with the aboral surface removed and the hepatic
ceca moved to show the bulblike ampullae of the tube
feet, etc.; **V,** arm with the internal organs and vertebral
ridge removed to show the rows of ampullae of the tube
feet, the connecting canals, and the radial canal (all parts
of the water-vascular system).

Fig. 22-2
Pedicellaria of a starfish.

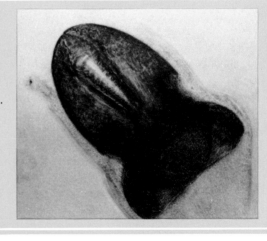

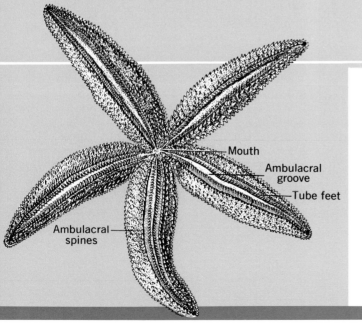

Mouth

Ambulacral
groove

Tube feet

Ambulacral
spines

Fig. 22-3
Starfish from the oral or underside, showing rows of tube
feet extending from the ambulacral grooves. Long,
movable ambulacral spines protect the tube feet when
they are retracted within the groove; similar oral spines
surround the mouth for protection.
(*From Braungart, D. C., and Buddeke, R.: An introduction
to animal biology, ed. 6, St. Louis, 1964, The C. V. Mosby Co.*)

Respiration is accomplished by soft, saclike dermal branchiae, thin protrusions of the coelom lining that pass through small openings in the skeleton. Oxygen and carbon dioxide are exchanged through the walls of these branchiae.

The nervous system consists of a single oral nerve ring around the mouth, which is connected with a radial nerve cord running beneath the radial canal of the water-vascular system in each arm. A double nerve ring above the other ring also passes nerves to various tissues; the tube feet are particularly sensitive. A light-sensitive eyespot and a small tactile tentacle are present at the end of each arm.

Reproduction is diecious. Five pairs of saclike, branched gonads (testes or ovaries) are present between the bases of the arms. Fertilization occurs in the water, and the zygote develops regularly through the blastula and gastrula stages. The larva formed is called a bipinnaria (L. *bis,* twice; *pinna,* feather), which is bilaterally symmetrical and swims by means of two bands of cilia (see Chapter 28 for detailed illustrations). One female may release over 200 million eggs per season, and a male produces many times that number of sperms.

An arm may be automatically broken off near its base by a process known as autotomy, and any arm with part of its central disk may regenerate

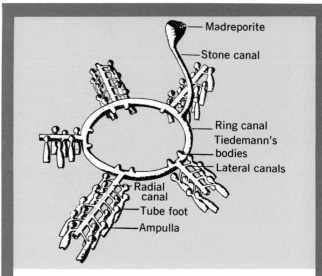

Fig. 22-4
Water-vascular system of the starfish (somewhat diagrammatic). Madreporite is a porous plate on the upper surface of the starfish, between the bivium rays, through which water enters. The ring canal is also known as the circular or circumoral canal. There are nine Tiedemann's bodies (the tenth is replaced by the stone canal), which produce amebocytes found in the fluid within the water-vascular system. The lateral canals are also known as connecting canals.
(*From Braungart, D. C., and Buddeke, R.: An introduction to animal biology, ed. 7, St. Louis, 1968, The C. V. Mosby Co.*)

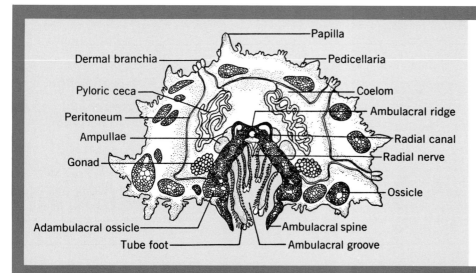

Fig. 22-5
Starfish arm or ray (diagrammatic cross section).
(*From Braungart, D. C., and Buddeke, R. An introduction to animal biology, ed. 6, St. Louis, 1964, The C. V. Mosby Co.*)

Fig. 22-6
Starfish, showing the
regeneration of one arm.

Fig. 22-7
Brittle star of class
Ophiuroidea (oral view).

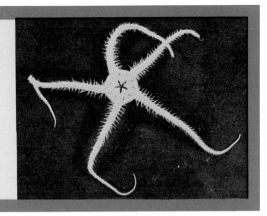

an entire body. A missing arm may be regenerated
to form a complete animal again (Fig. 22-6).

CLASS ECHINOIDEA

Sea urchins

Typical sea urchins are somewhat globular and
have a shell (test) composed of ten pairs of columns
of calcareous plates, bearing long, sharp, movable
spines (Fig. 22-8). Five pairs of these columns are
homologous to the five arms of a starfish and bear
pores for the long tube feet. Spines are moved
freely on knoblike elevations by muscle action.
The three-jawed pedicellariae capture food and
keep the body clean.

The water-vascular system consists of a madreporite; stone canal; ring canal; five radial canals;
five interradial, pouchlike Pollian vesicles, and
tube feet with ampullae (Fig. 22-4).

Plant and animal materials on the sea bottom are
ingested by a complex apparatus called Aristotle's
lantern, in which pairs of teeth are attached.
Respiration occurs in ten branched, pouchlike gills
located around the mouth. Many other structures
resemble comparable ones in the starfish. Some sea
urchins possess very long poisonous spines. Sea
urchins may be green, purple, black, gray, white,
and other colors.

Sand dollars

Sand dollars have flat, disk-shaped bodies with
small calcareous spines (Fig. 22-9). There are no
free arms, the spaces between them being more
or less filled in. The aboral surface shows a starlike
arrangement of areas, or "petals." The oral (under)

Fig. 22-8
Sea urchins of class Echinoidea. **A,** Oral view: Note
the whitish teeth (Aristotle's lantern) in center; **B,**
skeleton ("test") with the spines removed and showing
the plates of the ambulacral and interambulacral zones;
tubercles for the attachment of spines and pores for
the tube feet are visible on the skeleton; **C,** oral view
of another species.

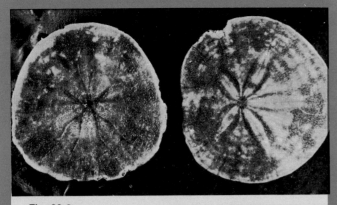

Fig. 22-9
Sand dollar of class Echinoidea. Furrowlike ambulacral areas shown radiating from the central mouth in the oral view at left; aboral areas or "petals" shown at right.

surface shows five furrowlike areas radiating from the central mouth region. Many other structures resemble comparable ones in other echinoderms. The anus is located in the marginal edge of the animal.

CLASS HOLOTHUROIDEA (SEA CUCUMBERS)

Sea cucumbers (Fig. 22-10) have leathery bodies elongated along an oral-aboral axis and a mouth surrounded with ten to thirty retractile tentacles at the oral end and an anus at the aboral end. The tentacles are modified tube feet that capture small marine organisms and transport them to the mouth. Small calcareous plates are embedded in the body. The dorsal (upper) side bears two longitudinal

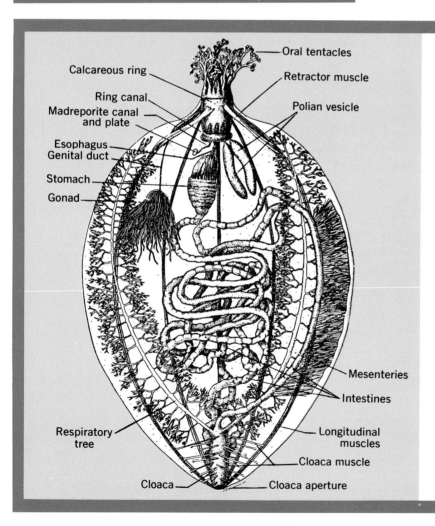

Calcareous ring
Ring canal
Madreporite canal and plate
Esophagus
Genital duct
Stomach
Gonad
Oral tentacles
Retractor muscle
Polian vesicle
Respiratory tree
Mesenteries
Intestines
Longitudinal muscles
Cloaca muscle
Cloaca
Cloaca aperture

Fig. 22-10
Sea cucumber (*Thyone*) of class Holothuroidea (in longitudinal section and somewhat diagrammatic).

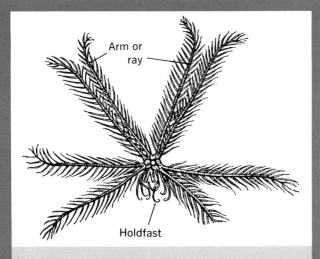

Fig. 22-11
Feather star (*Antedon*) of class Crinoidea, showing only four of the five branched arms, each with many small pinnules (oral view). Holdfasts attach it to the ground in the sea.

Labels: Arm or ray; Holdfast

zones of tube feet, and the ventral side has three similar zones. Sea cucumbers move slowly by means of the tube feet and the action of body wall muscles. The coelom is extensive. The water-vascular system consists of a small, internal madreporite, a ring canal with two saclike Polian vesicles, and five radial canals that are connected to the rows of tube feet.

The digestive system consists of a mouth; a short esophagus; an oval, muscular stomach; a long, looped intestine; an enlarged, muscular cloaca; and an anus. Two long, branched respiratory trees attached to the cloaca pump water in and out of the tree for respiration and excretion purposes. Respiration is also accomplished by the cloaca, tube feet, tentacles, and body wall.

The nervous system includes a nerve ring with five radial nerves. Sensory organs include those affected by touch and light.

Reproduction is diecious, with the gonads opening to the outside behind the tentacles. When strongly stimulated, these animals cast out much of their viscera by contracting the body wall. This phenomenon is known as autotomy. They also have great powers of regeneration of lost parts. Sea cucumbers have a special kind of larval stage, known as auricularia (L. *auricula*, small ear), which bears short processes with diffuse cilia. The adult bodies may have such colors as brown, yellow, red, black, pink, and purple.

ECONOMIC IMPORTANCE OF ECHINODERMS

Starfishes destroy large numbers of mollusks, particularly oysters and clams. One adult may eat nearly a dozen oysters or clams daily. Control measures in shellfish beds include catching them with nets or spreading quicklime, which destroys them. Echinoderms may be used as agricultural fertilizers because of their lime and nitrogen content. Because of their nature, adults are not good sources of food, but eggs may be used. Echinoderm eggs are used in biologic experiments, especially those of embryologic development, including artificial parthenogenesis. The latter phenomenon was first discovered in 1900 by Jacques Loeb, when it was found that eggs would develop without the use of sperms when the chemical composition of the sea water was changed. Dried sea cucumbers (trepang) are used for preparing soup in the Orient.

Review questions and topics

1 What general characteristics describe the echinoderms as a group?
2 What are the characteristics that distinguish members of the different classes of echinoderms?
3 Discuss the significance of calcareous skeletons and spines, as found in echinoderms, including examples of the various types.
4 Describe the structure and function of the various parts of the unique water-vascular system.
5 Discuss the phenomena of bilateral symmetry in larval stages and radial symmetry in adults. When the positions of certain structures in adults are carefully considered, might one apply the term biradial?
6 Explain the structure and function of the pincerlike pedicellaria.
7 Explain the structure and function of the dermal branchiae ("skin gills").
8 Explain the structure and function of the respiratory tree as found in sea cucumbers.
9 Is it unusual for echinoderms not to have complex organs of excretion?
10 Discuss such phenomena as parthenogenesis, autotomy, and regeneration as displayed by certain echinoderms.

311

Some contributors to the knowledge of echinoderms

Agassiz

Alexander Agassiz (1835-1910)
An American who made important studies of the structure and relationships of echinoids (1872-1874) and of North American starfishes (1877). (*The Bettmann Archive, Inc.*)

Oskar Hertwig (1849-1922)
He studied the fertilization of sea urchin eggs by sperms (1875). (*Historical Pictures Service, Chicago.*)

H. Ludwig
He wrote a classic monograph on the holothurians (1892).

H. L. Clark
He made systematic descriptions of most species of echinoderms in New England (1904).

E. W. MacBride
He contributed a most important account of echinoderms (1906).

Wesley R. Coe
Coe, of Yale University, made detailed studies of various echinoderms, especially those of Connecticut (1912).

Hertwig

Selected references

Feder, H. M.: On the methods used by the starfish *Pisaster ochraceus* in opening three types of bivalved mollusks, Ecology 36:764-767, 1955.

Galtsoff, P. S., and Loansanoff, V. L.: Natural history and methods of controlling the starfish (*Asterias forbesi* Desor.), U.S. Bureau of Fisheries Bulletin 49: 75-132, 1950.

Harvey, E. B.: The American Arbacia and other sea urchins, Princeton, New Jersey, 1956, Princeton University Press.

Hyman, L. H.: The invertebrates; volume IV, Echinodermata, New York, 1955, McGraw-Hill Book Company.

Meglitsch, P. A.: Invertebrate zoology, New York, 1967, Oxford University Press, Inc.

Nichols, D.: Echinoderms, London, 1962, Hutchinson & Co. (Publishers) Ltd.

chapter twenty-three

Phylum Chordata

Distinguishing characteristics of the Chordata (Gr. *chorde*, cord) include (1) a semirigid skeletal axis, the dorsal notochord, which is present at some time during the life cycle; (2) a hollow dorsal nervous system, which is dorsal to the digestive system; and (3) paired pharyngeal clefts (gill slits), which connect the pharynx with the exterior at some stage in the life cycle.

All lower chordates up to and including the fishes are marine animals and respire by means of gills throughout life. In higher vertebrates gills or gill slits or traces of them are usually present only in the larval or embryonic stages. In mammals the gill slits never open but are modified for other purposes.

The phylum Chordata includes the subphyla Urochordata, Cephalochordata, and Vertebrata. Most chordates have a segmented vertebral column, which places them in the latter phylum. Since the following chapters deal with chordates (vertebrates) in detail, only the general characteristics of the subphyla will be considered in this chapter.

SUBPHYLUM UROCHORDATA

Urochordata are widely distributed marine animals; examples include ascidians, sea squirts, and tunicates (Fig. 23-1).

A notochord is present in the tadpolelike, free-swimming larva, but is absent in the adult animal. In the larval stage there is a dorsal neural tube in the tail, which enlarges in the trunk and ends in the vesicle ("brain"). In the adult there is a single nerve ganglion with nerves.

Respiration takes place through numerous pharyngeal gill slits, which are ciliated.

The body of the adult Urochordata is usually barrel shaped or saclike. In some species the body is brightly colored. The body is enclosed by a covering called a tunic (L. *tunica*, mantle). The animals are generally attached to the marine substratum.

SUBPHYLUM CEPHALOCHORDATA

Cephalochordata are marine, sand-dwelling animals that swim with swift body movements. Amphioxis (*Branchiostoma*) is an example (Fig. 23-2).

A notochord is present along the entire length of the body throughout all stages of the life cycle. The nerve cord is tubular and extends along the entire dorsal side of the body. Like the notochord, it is present throughout the life of the animal. A small anterior cerebral vesicle is also present.

Respiration occurs through numerous pairs of pharyngeal gill slits.

The body of the Urochordata is transparent and fishlike and is pointed at both ends. The body is compressed laterally, and there is no distinct head present. There are no paired lateral appendages, although there is a long dorsal fin along the entire length of the body. There are numerous V shaped muscles (myotomes) along the body.

SUBPHYLUM VERTEBRATA

Vertebrates include cyclostomes (jawless fishes), sharks (cartilaginous fishes), bony fishes, amphibians, reptiles, birds, and mammals.

All vertebrates have an axial, dorsal rod (notochord) present at some time during the life cycle. This notochord is present throughout the life in some of the lower vertebrates but disappears before adulthood in others. In the higher vertebrates the notochord becomes modified into a segmented vertebral column. The vertebrae may be either cartilaginous or bony. A dorsal nerve tube, which is connected anteriorly with the brain, is also present.

Pharyngeal clefts (gill slits) are present during some stage of the life cycle. These slits are present only during the larval or embryonic stage in the higher vertebrates.

313

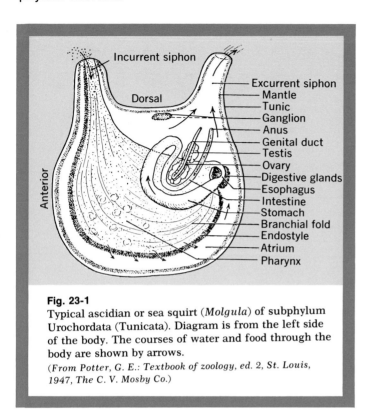

Fig. 23-1
Typical ascidian or sea squirt (*Molgula*) of subphylum
Urochordata (Tunicata). Diagram is from the left side
of the body. The courses of water and food through the
body are shown by arrows.
(*From Potter, G. E.: Textbook of zoology, ed. 2, St. Louis,
1947, The C. V. Mosby Co.*)

The vertebrate body is divided into a head, neck
(generally), and trunk. In some species a tail is
present. Vertebrates usually have two pairs of
lateral appendages and possess an internal jointed
skeleton.

The heart is situated ventrally and has from two
to four chambers. Red blood corpuscles are present.
These animals are either cold or warm blooded,
depending upon the class.

Class Agnatha (Cyclostomata) (lampreys, hagfishes)

The notochord of animals in the class Agnatha
is present throughout life. A brain and dorsal
nerve cord are also present, and there are from
eight to ten pairs of cranial nerves.

The skeleton is cartilaginous and fibrous and the
body is round and slender. There are no scales present and the skin is soft. Mucous glands are present.
These animals possess unpaired fins with rays of
cartilage (Fig. 23-3).

The respiratory system consists of from six to
fourteen pairs of gills in gill pouches.

The digestive system consists of a sucking mouth
with no jaws and a rasping tongue. An intestine
with a fold (typhlosole) is present, but there is no
stomach.

The heart consists of one auricle and one ventricle; there are aortic arches in the gill region. The
blood has both red and white corpucles; these animals are cold blooded.

The excretory system consists of a pair of kidneys
(pronephros) and paired pronephric ducts that
empty into the cloaca.

Lampreys are diecious; hagfishes are monecious.
Fertilization in all animals of this class is external.

These animals possess the senses of taste, sight,

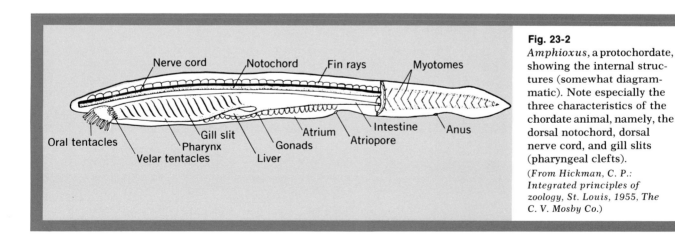

Fig. 23-2
Amphioxus, a protochordate,
showing the internal structures (somewhat diagram-
matic). Note especially the
three characteristics of the
chordate animal, namely, the
dorsal notochord, dorsal
nerve cord, and gill slits
(pharyngeal clefts).
(*From Hickman, C. P.:
Integrated principles of
zoology, St. Louis, 1955, The
C. V. Mosby Co.*)

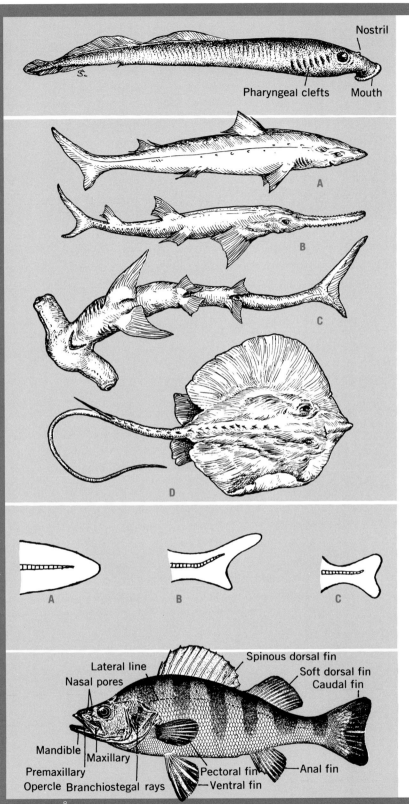

Fig. 23-3
Lamprey of class Cyclostomata. Note the circular sucking mouth, median unpaired nostril, and seven pharyngeal clefts (gill slits). Lampreys frequently attack and kill fishes.

Nostril
Pharyngeal clefts   Mouth

Fig. 23-4
Representatives of class Elasmobranchii (not drawn to scale). **A,** Spiny dogfish shark (*Squalus*); **B,** sawfish (*Pristis*); **C,** hammerhead shark (*Sphyrna*); **D,** southern stingray (*Dasyatis*), which is common in the Gulf of Mexico.

Fig. 23-5
Symmetries as shown by the tails of fishes. **A,** Diphycercal (Gr. *diphyes,* of double nature; *kerkos,* tail), as in the lungfish, in which the vertebral column extends to the posterior part of the body, and the tail develops symmetrically above and below the column; **B,** heterocercal (Gr. *heteros,* different), as in sharks, in which the vertebral column extends into the larger dorsal lobe; **C,** homocercal, as in most modern fish, in which the vertebral column extends slightly into the dorsal lobe but the tail is symmetrical.

Fig. 23-6
External features of the perch.
(From Hegner, R. W.: College zoology, New York, The Macmillan Company.)

Lateral line
Nasal pores
Mandible
Maxillary
Premaxillary
Opercle   Branchiostegal rays
Spinous dorsal fin
Soft dorsal fin
Caudal fin
Pectoral fin
Ventral fin
Anal fin

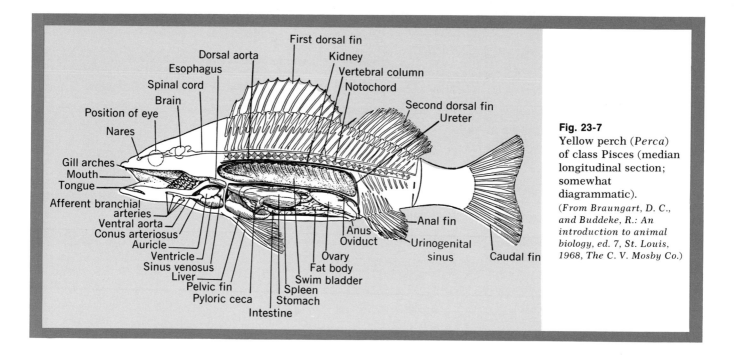

Fig. 23-7
Yellow perch (*Perca*) of class Pisces (median longitudinal section; somewhat diagrammatic). (*From Braungart, D. C., and Buddeke, R.: An introduction to animal biology, ed. 7, St. Louis, 1968, The C. V. Mosby Co.*)

smell, and hearing. There is one nasal sac present for smell; the auditory organs consist of two semi-circular canals.

Class Chondrichthyes (Elasmobranchii) (sharks, rays)

Animals of the class Chondrichthyes possess a notochord throughout life. A brain and dorsal nerve cord are present and there are ten pairs of cranial nerves.

The endoskeleton is composed entirely of cartilage and separate vertebrae are present. The body is spindle shaped and there are platelike (placoid) scales on the skin. There are mucous glands present on the skin also. These animals have a caudal (tail) fin that is asymmetrical and paired pectoral and pelvic fins. Two unpaired dorsal and median fins and fin rays are also present (Fig. 23-4).

Class Osteichthyes (Pisces) (bony fishes)

Bony fishes possess a notochord that may persist in part throughout life. A brain and a dorsal nerve cord are present, and there are ten pairs of cranial nerves.

The skeleton is more or less bony and replaces the original cartilaginous skeleton of earlier classes.

The respiratory system consists of paired gills that are supported by bony arches. There may be a gill cover (operculum) and an air bladder in some species.

Bony fishes have various body shapes (Fig. 23-5). The caudal fin is usually symmetrical (homocercal), and there are paired and unpaired fins with fin rays of bone or cartilage. The skin may possess different types of dermal scales, although some fishes are scaleless. Mucous glands are present in the skin. Bony fishes have a terminal mouth with jaws and teeth (Fig. 23-6).

The circulatory system consists of a heart with one auricle and one ventricle. There are four pairs of aortic arches, and both arterial and venous systems are present. The red blood cells are nucleated. These animals are cold blooded.

The excretory system consists of a pair of kidneys (mesonephros) and paired ureters that empty into a cloaca.

Fishes are diecious and have paired gonads. Fertilization is external; the eggs are laid in the water by the female.

Class Amphibia (frogs, toads, salamanders)

Amphibians have a notochord that does not persist throughout life. A brain is present with a dorsal

Fig. 23-8
Representatives of class Amphibia (not to scale). A, Redspotted newt (*Diemictylus*); B, tiger salamander (*Ambystoma*).
(*Courtesy General Biological Supply House, Inc., Chicago, Illinois.*)

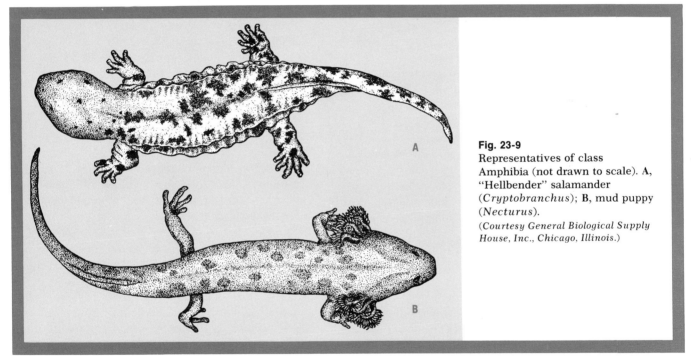

Fig. 23-9
Representatives of class Amphibia (not drawn to scale). A, "Hellbender" salamander (*Cryptobranchus*); B, mud puppy (*Necturus*).
(*Courtesy General Biological Supply House, Inc., Chicago, Illinois.*)

nerve cord, and there are ten pairs of cranial nerves. The skull has two occipital condyles.

The endoskeleton is made up mostly of bone. Ribs may be present or absent, depending on the species. There are usually two pairs of appendages and the feet are often webbed. Some amphibians, however, have small limbs or are legless.

Amphibians may have various body shapes. A head and trunk is present, and a tail is present or absent, depending upon the species. The skin is smooth and moist and is usually scaleless. Many glands are present and pigment cells are common. A terminal mouth is present with small teeth in the jaws.

The respiratory system consists of paired gills or lungs. The external gills of larval stages persist throughout the life of some species. The skin may assist in respiration.

317

The circulatory system is made up of a heart with two auricles and one ventricle. There is double circulation through the heart. Numerous blood vessels are present in the skin. Amphibians are cold blooded.

The excretory system consists of a pair of kidneys (mesonephros) and paired ureters which empty into a cloaca.

Amphibians are diecious and possess paired gonads. Fertilization may be external or internal. Eggs with a jellylike cover are laid in water.

Amphibians possess the senses of taste, sight, hearing, smell, and touch. There are paired nostrils connected with the mouth.

Class Reptilia

Reptilia include turtles, lizards, chameleons, snakes, crocodiles, and alligators.

These animals possess a brain and a dorsal nerve cord. There are twelve pairs of cranial nerves present. The skull has one occipital condyle.

The endoskeleton of reptiles is bony and ribs and a sternum are present. They may have various body shapes. The skin is dry and rough, and usually has horny scales or bony plates. There are few glands in the skin. There are usually two pairs of limbs present, although snakes have no appendages. Most reptiles have five toes on each appendage to aid in running, climbing, or paddling.

Paired lungs are present throughout life for respiration.

The heart consists of two auricles and one ventricle, except in crocodiles, which have two auricles and two ventricles. There is usually one pair of aortic arches. All reptiles are cold blooded.

The excretory system consists of a pair of kidneys (mesonephros) and paired ureters, which empty into the cloaca.

Reptiles are diecious and possess copulatory organs. Fertilization is internal and the eggs have a leathery, limy shell. The embryo develops within the amnion.

Reptiles possess the senses of taste, sight, hearing, touch, and smell (paired nostrils are present).

Class Aves (birds)

The endoskeleton of birds is bony and some cartilage is present. Birds have small ribs, and a sternum with a keel is present. The skull has one occipital condyle.

The nervous system is made up of a brain, a dorsal nerve cord, and twelve pairs of cranial nerve. A well-developed system of endocrine glands assists in coordination.

Paired lungs are present, and there are air sacs in the skeleton and among the viscera. A voice box (syrinx) is present at the junction of the trachea and the bronchi.

Birds may have various body shapes, but they are frequently spindle shaped. Feathers cover the body, except the legs, which are covered by scales. There is an oil gland at the base of the tail for preening. Paired wings are present for flying, although some birds possess only vestigial wings. There are paired hindlimbs for walking, perching, and swimming. Birds typically have four toes.

Birds possess a terminal mouth with paired jaws and a horny beak; there are no teeth present. Many birds have a crop for storage and a muscular gizzard for grinding.

The circulatory system consists of a heart and an aortic arch on the right side. The heart has two auricles and two ventricles. There are separate systemic and pulmonary systems. The red blood cells are nucleated. Birds are warm blooded (homoiothermic).

The excretory system consists of a pair of kidneys (mesonephros) and paired ureters, which empty into a cloaca. There is no urinary bladder and urine and feces are eliminated together.

Birds are diecious. The male possess paired testes; the female has one ovary (left). There are no copulatory organs in most birds. Fertilization is internal. The eggs have a limy shell and much yolk is present within them. The embryos develop within the amnion.

Class Mammalia

Mammals include man, monkeys, horses, dogs, cats, pigs, whales, bats, and so on. The mammalian endoskeleton is bony with some cartilage present. Elongated tails may or may not be present. There are ribs and a sternum present. The skull has two occipital condyles.

The brain of mammals is well developed, as is the dorsal nerve cord. There are twelve pairs of cranial

Fig. 23-10
Representative reptiles of class Reptilia (not to scale).
A, Horned "toad" (*Phrynosoma*); B, giant-collared
lizard (*Crotaphytus*).
(*Courtesy General Biological Supply House, Inc., Chicago,
Illinois.*)

Fig. 23-11
Representative reptiles of
class Reptilia (not to scale).
A, Black snake (*Coluber*); B,
box turtle (*Terrapene*).
(*Courtesy General
Biological Supply House,
Inc., Chicago, Illinois.*)

Fig. 23-12
Representative reptiles
of class Reptilia (not to
scale). A, True chameleon
(*Chamaeleo*); B,
American alligator
(*Alligator*).
(*Courtesy General
Biological Supply House,
Inc., Chicago, Illinois.*)

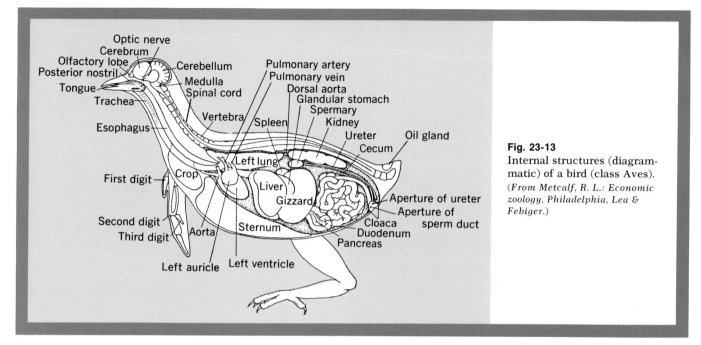

Fig. 23-13
Internal structures (diagrammatic) of a bird (class Aves). (*From Metcalf, R. L.: Economic zoology, Philadelphia, Lea & Febiger.*)

nerves. There is also a well-developed endocrine system for coordination and other purposes.

The respiratory system consists of paired lungs, a voice box (larynx), and an epiglottis.

Mammals have various body shapes. There is a unique muscular diaphragm that separates the thorax from the abdomen. The skin possesses hair at some time in life. In some types of mammals there are scales present on the tail. Glands include sweat, tear, mammary, scent, and sebaceous (hair) glands.

Mammals possess two pairs of appendages, although they are reduced in some animals. The appendages are adapted for walking, digging, climbing, swimming, or flying, depending upon the species. There are typically five digits on each appendage. All mammals (except certain whales and a few others) possess a mouth with teeth on both jaws.

The circulatory system consists of a heart with two auricles and two ventricles, and one aortic arch on the left side. There are separate systemic and pulmonary systems. The red blood cells are nonnucleated. Mammals are warm blooded.

The excretory system consists of a pair of kidneys (metanephros). Paired ureters empty into a urinary bladder and out through the urethra.

Mammals are diecious and possess paired testes or ovaries, and copulatory organs are present. The unique mammary glands provide nourishment for the young. The young develop in the amnion.

Mammals possess the senses of sight, smell, taste, touch, and hearing. There are external ears and paired movable eyelids (the third eyelid is reduced).

ECONOMIC IMPORTANCE OF VERTEBRATES

The economic importance of the various members of such a large and varied group is so extensive that limited space will not permit a complete and detailed consideration.

Hagfishes live in the mud of the sea and destroy fishes, especially if they have been caught in nets or on lines. Lampreys are serious enemies of game and food fishes. The Atlantic sea lamprey (*Petromyzon*) has reduced the commercial catch of fishes in the Great Lakes. Sea lampreys suck blood and injure fishes in other ways.

Dogfish sharks destroy large numbers of food

Fig. 23-14
Representative mammals of class Mammalia (not to scale). **A**, Orangutan (*Pongo*); **B**, llama (*Lama*); **C**, American bison or buffalo (*Bison*); **D**, red fox (*Vulpes*); **E**, Bactrian camel (shedding) (*Camelus*); **F**, Russian bear (*Ursus*); **G**, beaver (*Castor*); **H**, armadillo (*Dasypus*). *(Courtesy General Biological Supply House, Inc., Chicago, Illinois.)*

fishes, lobsters, and crabs, besides destroying the nets of fishermen. In some places the meat of sharks is eaten by man. Sharkskin leather is used in manufacturing bags, shoes, bookbindings, and jewel cases. Shark livers are good sources of vitamin A.

Many fishes serve as sources of food and provide human recreation. Over 2 billion pounds of codfishes are caught annually, and cod-liver oil is a good source of vitamins A and D. The salmon of the Pacific coast are caught and canned in great quantities. The eggs of the sturgeon and other fishes are made into caviar. Certain types of fishes may help to keep waters clean by destroying mosquitoes and other insect larvae.

Frogs and toads are great enemies of insect pests. Frog skins may be used in making glue and bookbindings. Frog legs and the meat of mud puppies are eaten by man. Frogs are widely used in laboratories for dissection and experimental purposes. They are used in a test for human pregnancy.

Reptiles kill many destructive rodents and harmful insects. Turtles and tortoises are widely used as human food. Canned rattlesnake meat is eaten by man. The skins of crocodiles and alligators are used in making leather goods. The tortoise shell obtained from the horny covering of the hawkbill turtle is used in making combs and ornaments.

Birds are of great importance because they eat large quantities of harmful insects and the seeds of weeds. However, they may also eat great quantities of fruits and seeds of beneficial plants. Some birds supply food and feathers for man. In some regions the excrement, called guano, of certain sea birds accumulates in such quantities as to be a valuable source of fertilizer. Certain game birds afford much recreation and some food for hunters.

The economic importance of mammals is great and varied. Domestic mammals are of enormous value to man, serving as foods, beasts of burden, pets, sources of clothing, and sources of recreation. In the latter category might be listed polo, horse racing, and dog racing. Ivory is obtained from the tusks of elephants and walruses; oil from the fat of whales; ambergris, used in manufacturing perfumes, from the intestine of whales; furs are obtained from the muskrat, mink, fox, otter, weasel, marten, wolverine, badger, beaver, skunk, raccoon,

bear, and rabbit. The worst mammalian pest is the rat, which causes damage amounting to hundreds of millions of dollars annually. Rats also carry the flea that harbors the cause of the dreaded bubonic plague ("black death"). When certain mammals are first imported into a country, they may be beneficial but later may become serious pests. After rabbits were introduced into Australia they became such pests that large numbers had to be destroyed. The mongoose destroys lizards, snakes, and rats in India. After it was imported into Jamaica it was very beneficial, but later it became a serious pest by destroying fruits, birds, poultry, and domestic animals.

Review questions and topics

1 Discuss the characteristics that apply to the vertebrates as a group.

2 List the characteristics by which the classes of the subphylum Vertebrata may be distinguished from each other, giving examples of each class.

3 Which systems seem to be developed best in the vertebrates? Which the least?

4 In the series of vertebrates from the lowest to the highest, discuss the status of (1) notochord and skeleton, (2) nervous system, (3) gills and lungs, (4) body and integument, (5) appendages, (6) digestive system, (7) circulatory system, (8) excretory system, (9) sensory organs, and (10) reproductive system.

5 Discuss the relative importance of two-, three-, and four-chambered hearts, giving examples of each.

6 Discuss the importance of systemic and pulmonary circulatory systems. What important consequence results from such an arrangement?

7 Discuss the importance of the division of the coelom into such cavities as pericardial, visceral, and thoracic, giving examples of each.

8 List the characteristics that distinguish mammals, giving examples.

9 Describe the structure and functions of a muscular diaphragm in mammals.

10 Discuss how the paired jointed appendages may be modified and adapted in various vertebrates to perform certain functions.

11 Discuss what happens to the pharyngeal (gill) slits, especially in higher vertebrates.

12 Describe the development of the nervous system, especially in higher mammals.

13 Discuss the well-developed endocrine systems of higher vertebrates, including some functions performed.

Some contributors to the knowledge of certain vertebrates

Francis Willughby (1635-1672)
A British naturalist who laid the foundation for modern ornithology by writing *Ornithology*, which was completed (1676) by **John Ray (162?-1705)**, Willughby's teacher.

Georges Cuvier (1769-1832)
A Frenchman who was largely responsible for establishing the science of comparative anatomy. He emphasized the relation between structure and function. He studied not only living animals but fossil remains as well.

John J. Audubon (1785-1851)
An American who published one of the great works on birds, *The Birds of America* (1828-1838), with unsurpassed artistic plates. The Audubon Society was named after him. (*Historical Pictures Service, Chicago.*)

Audubon

Alphonse Toussenel (1803-1885)
He was the author of a number of valuable works on birds.

Richard Owen (1804-1892)
An English anatomist who made comparative studies of fishes, birds, apes, etc. He proposed the concepts of homology and analogy. (*The Bettman Archive, Inc.*)

Owen

Louis Agassiz (1807-1873)
An American, was a great investigator and professor of zoology and geology at Harvard University, where he founded the Museum of Comparative Zoology.

Ernest Haeckel (1834-1919)
He gave us a theory of embryonic germ layers and a principle of recapitulation, or "biogenetic law" (embryos in their development repeat in an abbreviated manner the stages of the ancestral history of the race). Both theories have required modification in recent times. (*Historical Pictures Service, Chicago.*)

Haeckel

Thomas H. Huxley (1825-1895)
An English biologist, made important contributions in the fields of comparative anatomy and paleontology.

John Burroughs (1837-1921)
An American naturalist who wrote many books on the lives and habits of living organisms.

Edward D. Cope (1840-1897)
An American naturalist and comparative anatomist who studied many vertebrates, including fossil reptiles and mammals.

David Starr Jordan (1851-1931)
An American biologist who made a lifelong study of fishes, and his publications and teachings assisted in the establishment of this phase of biology in the United States.

Henry Fairfield Osborn (1857-1935)
An American paleontologist who studied many types of living and fossil vertebrates.

14 Are most vertebrates diecious? Does fertilization occur externally or internally? Include examples of each.

15 What characteristics do all protochordates as a group have in common?

16 Discuss the characteristics that distinguish the members of the three subphyla from each other. Include examples of each subphylum.

17 Discuss each characteristic animals must have in order to qualify as chordates.

18 Which systems in the chordates seem to be developed to the greatest extent? Which the least?

19 What is the status of the muscular system in such a type as *Amphioxus*?

20 What is the status of the development of the head and sensory equipment in protochordates?

21 What might be some disadvantages to an animal of being attached (sessile)?

22 Is there any significance to the fact that all protochordates are marine?

23 System by system, do the chordates as a group show greater complexity of structure and function when compared with some of the higher nonchordates?

Selected references

Applegate, V. C., and Moffett, J. W.: The sea lamprey, Sci. Amer. **192:**36-41, 1955.

Barrington, E. J. W.: The biology of Hemichordata and Protochordata, London, 1965, Oliver & Boyd, Ltd.

Berrill, N. J.: The origin of vertebrates, New York, 1955, Oxford University Press, Inc.

Blair, W. F., and others: Vertebrates of the United States, New York, 1957, McGraw-Hill Book Company.

Braun, M. E. (editor): The physiology of fishes, New York, 1957, Academic Press Inc.

DeBeer, G. R.: Vertebrate zoology, London, 1951, Sidgwick & Jackson, Ltd., Publishers.

Orr, R. T.: Vertebrate biology, Philadelphia, 1961, W. B. Saunders Company.

Parker, T. J., and Haswell, W. A.: A textbook of zoology, ed. 7, New York, 1962, The Macmillan Company.

Ruibal, R. (editor): The adaptations of organisms, Belmont, Calif., 1967, Dickenson Pub. Co., Inc.

Torrey, T. W.: Morphogenesis of the vertebrates, New York, 1962, John Wiley & Sons, Inc.

Young, J. Z.: The life of vertebrates, ed. 2, New York, 1962, Oxford University Press, Inc.

chapter twenty-four

The frog

The common leopard or grass frog is known as *Rana pipiens* (L. *rana*, frog; *pipiens*, piping) (Fig. 24-1). Its body is smooth and covered with mucus, which is secreted by glands in the skin. The frog has the ability to change color because of changes in the black and yellow pigment cells in the skin. Because of its coloration, the frog is afforded a certain degree of protection from enemies. When in water, it need keep only the tip of its nose above the surface because of the location of the nostrils (external nares). Two large eyes are located on the top of the head. The tympanum (eardrum) is external and just posterior to each eye. The body may be divided into head and trunk, and the latter bears two pairs of appendages.

INTEGUMENT AND SKELETON

The loose-fitting skin is composed of (1) a rather thin outer layer, called the epidermis and (2) a thicker, inner layer, the dermis (Fig. 24-2). The epidermis consists of several layers of cells. The outer cells, composing the stratum corneum (L. *stratum*, layer; *corneus*, horny), are flat, compact, and horny (they are shed several times during the active season when the frog molts) and the inner ones located next to the dermis and composing the malpighian layer are columnar and give origin to the outer layer. The dermis consists of connective tissues that contain glands, blood vessels, pigments (Fig. 24-3), nerves, muscle fibers, and lymph spaces. The dermis is made of two layers. The outer layer, called the stratum spongiosum, consists of loose connective tissue and contains (1) pigment bodies that give the frog its spotted pattern (pigments may also be present in the epidermis), (2) small, spherical mucous glands that pour out a slimy secretion upon the surface of the skin, (3) larger spherical poison glands that secrete a whitish, acrid fluid for protection, and (4) numerous sensory and tactile papillae (just below the epidermis) for sensory purposes. The inner layer, called the stratum compactum, consists of dense connective tissues in which the fibers run somewhat parallel to the surface of the skin and among which are blood vessels. The smooth, scaleless, hairless skin functions as an organ of respiration, gives protection, and is used for sensory purposes. The pigment bodies are responsible for some protective coloration.

The bony endoskeleton (Fig. 24-4) consists of an axial skeleton (skull and vertebral column) and an appendicular skeleton (the pectoral girdle with its forelimbs and the pelvic girdle with its hind limbs). The frog has no ribs. Most of the bones of the skull, except those of the upper and lower jaws and the hyoid bone to which the tongue is attached, form the brain case (cranium). The brain and spinal cord connect through a large opening (foramen magnum) at the base of the cranium. The cranium articulates with the first vertebra (atlas) by means of a pair of rounded prominences, the occipital condyles. The pair of prootid bones, one bone on either side of the posterior part of the

Fig. 24-1
Grass frog
(*Rana pipiens*).
(*Courtesy, Carolina Biological Supply Company.*)

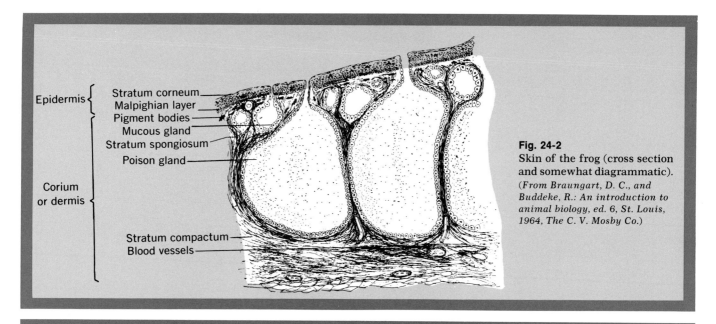

Epidermis {
 Stratum corneum
 Malpighian layer
 Pigment bodies
Corium
or dermis {
 Mucous gland
 Stratum spongiosum
 Poison gland

 Stratum compactum
 Blood vessels

Fig. 24-2
Skin of the frog (cross section and somewhat diagrammatic). *(From Braungart, D. C., and Buddeke, R.: An introduction to animal biology, ed. 6, St. Louis, 1964, The C. V. Mosby Co.)*

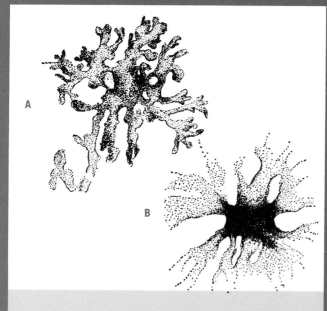

A

B

Fig. 24-3
Pigment melanophore from the frog (*Rana temporia*). **A**, Pigment distributed in response to light; **B**, pigment contracted.
(Redrawn and modified from Noble, G. K.: Amphibia of North America, New York, 1931, McGraw-Hill Book Company; from Potter, G. E.: Textbook of zoology, ed. 2, St. Louis, 1947, The C. V. Mosby Co.)

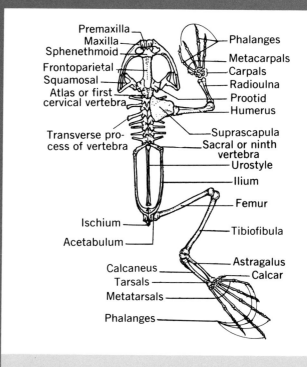

Premaxilla
Maxilla
Sphenethmoid
Frontoparietal
Squamosal
Atlas or first cervical vertebra
Transverse process of vertebra
Ischium
Acetabulum

Phalanges
Metacarpals
Carpals
Radioulna
Prootid
Humerus
Suprascapula
Sacral or ninth vertebra
Urostyle
Ilium
Femur
Tibiofibula
Astragalus
Calcar

Calcaneus
Tarsals
Metatarsals
Phalanges

Fig. 24-4
Skeleton of the frog (dorsal view; appendages of left side are not shown). The acetabulum is not a bone but is the joint at the proximal end of the femur.

cranium, forms the rounded auditory capsule that encloses the inner ear. The dorsal roof of the cranial cavity consists of two bones, the frontoparietals, each formed by the fusion of a frontal and a parietal bone. At the anterior end of the brain case is the tubular sphenethmoid bone. A pair of vomer bones helps to form the ventral wall of the olfactory sacs and also helps to form the roof of the mouth. The vomer bones bear vomerine teeth on the ventral surface. The upper jaw (maxilla) consists of a pair of premaxillae, a pair of maxillae, and a pair of quadratojugal bones. The first two pairs of bones bear teeth. The lower jaw (mandible) is the only part of the two jaws that moves. The hyoid apparatus consists of a large flat plate of cartilage in the floor of the mouth cavity. Rods of cartilage and bone extend anteriorly and posteriorly from its central part.

The vertebral column consists of nine vertebrae and a bladelike, posterior urostyle. A typical vertebra consists of (1) an oval, basal centrum (for articulation), (2) a neural arch, through which the spinal cord passes, (3) a single dorsal spine (neural spine), attached to the neural arch, and (4) a pair of transverse processes (except on the atlas), which extend laterally for the attachment of muscles. The articulating processes at each end of the neural arch are called zygapophyses. Ligaments hold the vertebrae together but allow a certain amount of movement.

The pectoral girdle (L. *pectus*, breast), to which the forelimbs are attached, is attached to the vertebral column by muscles. Compare this attachment with that in man (see Chapter 25). The sternum (breastbone) is located on the ventral median line and is composed of a number of bones and cartilages. The ventral part of the pectoral girdle consists of an anterior clavicle and a posterior coracoid. Other smaller bones help make up this part of the girdle. The dorsal part of the girdle is composed of the bony scapula, dorsal to which is the cartilaginous suprascapula. The glenoid fossa is the cavity with which the humerus of the forelimb articulates. The radioulna of the forearm is a fusion of the radius and ulna bones. Contrast this with the radius and ulna in man. The wrist consists of six carpal bones, and the hand is supported by five metacarpals. Distal

to the hand are the bones (phalanges) of the digits.

The pelvic girdle (L. *pelvis*, cavity), with which the hindlimbs articulate, is attached to the transverse processes of the ninth, or sacral, vertebra. The girdle is composed of a pair of long ilium bones (pl. ilia), a pair of ischium bones, and a pair of pubis bones. These three pairs of bones articulate so that a cavity is formed (acetabulum) with which the femur of the hind limb joins. The anterior part of the acetabulum is formed by the ilium and the posterior part by the ischium, whereas the ventral part is formed by the cartilaginous pubis. The tibiofibula is a fusion of the tibia and fibula bones. Contrast this with the tibia and fibula in man. The tarsals (ankle bones) are arranged in two rows, the proximal one consisting of long bones, the astragalus and calcaneus. The distal row contains a series of smaller bones. Distal to this are the five elongated metatarsals (foot). The five toes (digits) are composed of a total of fourteen phalanges. On the tibial side of the first toe there is a rudimentary digit called the calcar or prehallux. There are no claws.

MOTION AND LOCOMOTION

Well-developed muscles are present in the body, appendages, and head (Fig. 24-5). Other muscles move the lower jaw, aid in breathing, obtain foods, and produce sounds by means of the vocal apparatus. The muscles attached to the skeleton are called skeletal muscles, each of which has an origin that is the more fixed end and an insertion that is the more movable end. Pulsating lymph "hearts" (two near the third vertebra and two near the end of the vertebral column) force lymph into a branch of the renal portal and internal jugular veins.

INGESTION AND DIGESTION

Living insects, worms, and similar organisms are captured by a rather sticky, extensile tongue, attached at its front end. The tongue is thrown forcibly forward by the rapid filling of a lymph space beneath it. The large mouth cavity bears cone-shaped maxillary teeth on the upper jaw (Fig. 24-6). The two vomer bones in the roof of the mouth bear vomerine teeth. A constricted, hori-

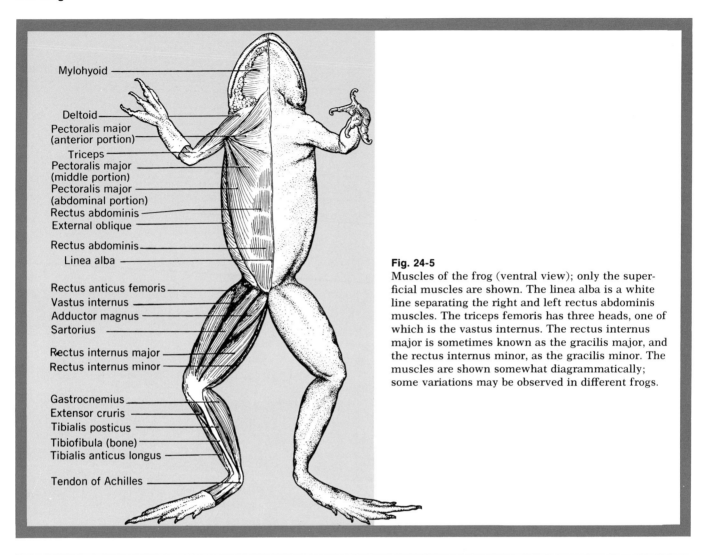

Fig. 24-5
Muscles of the frog (ventral view); only the superficial muscles are shown. The linea alba is a white line separating the right and left rectus abdominis muscles. The triceps femoris has three heads, one of which is the vastus internus. The rectus internus major is sometimes known as the gracilis major, and the rectus internus minor, as the gracilis minor. The muscles are shown somewhat diagrammatically; some variations may be observed in different frogs.

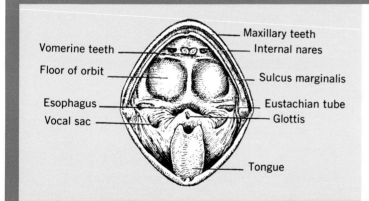

Fig. 24-6
Mouth of the bullfrog opened to show the internal structures.
(*From Potter, G. E.: Textbook of zoology, ed. 2, St. Louis, 1947, The C. V. Mosby Co.*)

zontal slit separates the mouth cavity from the esophagus (Gr. *oisophagos*, gullet). The stomach is crescent-shaped and is composed of a larger, anterior cardiac part and a smaller, posterior pyloric part (Gr. *pylorus*, gatekeeper), which connects with the coiled small intestine. The small intestine consists of an anterior duodenum and a much-coiled ileum (L. *ileum*, groin), which widens into the large intestine. The large intestine connects with the saclike cloaca, which also receives tubes from the kidneys and reproductive system. The cloaca empties to the exterior through the anus.

The pancreas (Gr. *pan*, all; *kreas*, flesh) is a much-branched, tubular organ that lies between the stomach and the duodenum. It passes its alkaline digestive juices into the common bile duct (Fig. 24-7). The large, reddish, trilobed liver secretes an alkaline bile, which is carried to the gallbladder. From the gallbladder the bile enters the duodenum through the common bile duct.

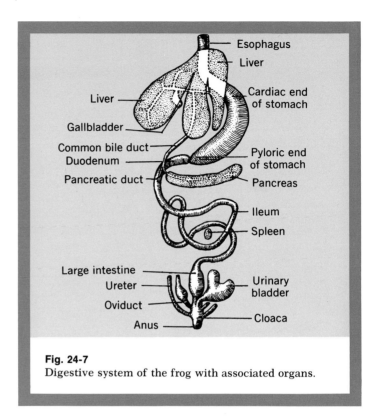

Fig. 24-7
Digestive system of the frog with associated organs.

Table 24-1
Uses of foods in the body of the frog

Food types	Uses	How used	Where stored	By-products
Carbohydrates	Serve as fuel and furnish energy; may help build certain tissues	Unite with oxygen through process of oxidation	As glycogen (animal starch) in liver, muscles, ovaries, nerves, and skeleton	Carbon dioxide, water
Fats	Same as above	Same as above	As adipose tissue in body, liver, and fat bodies	Carbon dioxide, water
Proteins	Build living tissues; repair destroyed tissues	Broken down for release of energy and their constituent elements	Probably all parts of body	Urea, carbon dioxide, water

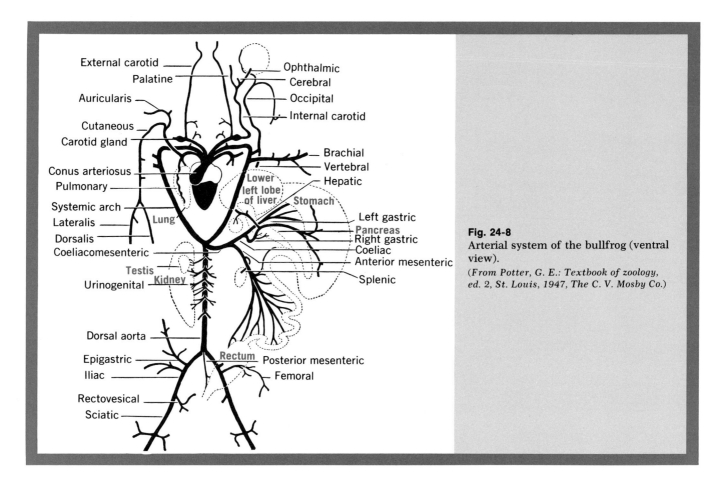

Fig. 24-8
Arterial system of the bullfrog (ventral view).
(*From Potter, G. E.: Textbook of zoology, ed. 2, St. Louis, 1947, The C. V. Mosby Co.*)

Physiology of digestion

Digestion breaks down complex, insoluble foods, such as proteins, fats, and carbohydrates, into simple, soluble compounds capable of being absorbed by the cells and assimilated into living protoplasm (Table 24-1).

The various foods are acted upon by specific enzymes that hasten the conversion processes without being used up themselves.

In the mouth there is no mastication, digestion, or enzymatic action. In the esophagus certain glands produce an alkaline mucous secretion that is mixed with the acid gastric juice secreted by the glands in the walls of the stomach. The cardiac end of the stomach has long, tubular, branched, deeply set glands for the secretion of mucus. The pyloric end of the stomach has short, tubular, shallow glands for secreting gastric juice, which contains the enzyme pepsin and about 0.4% hydrochloric acid (HCl). In other words the reaction is as follows:

$$\text{Proteins} + H_2O \xrightarrow[\text{Pepsin}]{\text{HCl}} \text{Soluble peptones}$$

After the partially digested foods pass through the pyloric valve from the stomach into the duodenum, they are mixed with the alkaline pancreatic juice. The alkalinity of the pancreatic juice is due to sodium carbonate (Na_2CO_3). The three specific enzymes of the pancreatic juice are amylase, trypsin, and lipase.

The hepatic cells (Gr. *hepar,* liver) of the tubular glands of the liver secrete a greenish bile that is mixed with the pancreatic juice in the common bile duct before they enter the duodenum. Certain

330

Fig. 24-9
Arterial system of the bullfrog that has been injected with colored latex (x-ray view). Observe the flexures of arteries at the elbow and knee joints to permit free movements.
(Courtesy Carolina Biological Supply Co., North Carolina.)

bile enzymes convert fats, when in an alkaline environment, into a soapy emulsion capable of passing through the intestinal walls into the blood and lymph systems. The liver also stores glycogen, which is changed by certain liver enzymes into usable sugar when needed.

The production and roles of intestinal juices and their enzymes in the frog are probably similar to those in higher animals. Possibly starches may be converted into sugars in the intestine. The various types of foods acted upon by specific enzymes in their proper environments are eventually absorbed by the cells of the intestine, passed into the lymph and blood vessels, and transported to body tissues to be utilized.

CIRCULATION

The heart, located within the thin, saclike pericardium, is three-chambered, consisting of two thin-walled auricles (right and left) and one muscular, cone-shaped ventricle. A thick-walled, tubular conus arteriosus (truncus arterious) arises from the base of the ventricle. A thin-walled, triangular sinus venosus, located on the dorsal side of the heart, is connected with the right auricle. In the adult frog the blood is pumped from the ventricle into the conus arteriosus, which has many branches (Fig. 24-8).

After passing from the arteries into thin-walled capillaries the blood is returned from the various

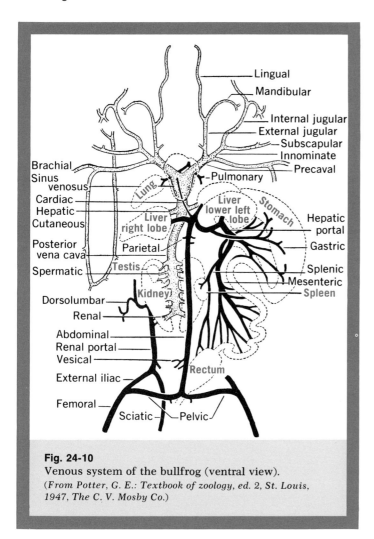

Fig. 24-10
Venous system of the bullfrog (ventral view).
(*From Potter, G. E.: Textbook of zoology, ed. 2, St. Louis, 1947, The C. V. Mosby Co.*)

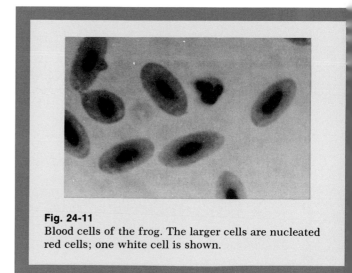

Fig. 24-11
Blood cells of the frog. The larger cells are nucleated red cells; one white cell is shown.

tissues and organs of the body by a system of veins (Fig. 24-10). The right and left pulmonary veins return the oxygenated (aerated) blood from the right and left lungs to the left auricle. The blood from all other parts of the body is returned to the sinus venosus through three large veins known as (1) the posterior vena cava (postcaval) and its branches, (2) the right anterior vena cava (right precaval) and its branches, and (3) the left anterior vena cava (left precaval) and its branches. The blood from the sinus venosus enters the right auricle. The right and left auricles send their blood into the one ventricle, which forces its mixture of

oxygenated blood (from the left auricle) and non-oxygenated blood (from the right auricle) into the conus arteriosus through pocket-shaped semilunar valves.

Frog blood (Fig. 24-11) consists of the following: (1) Oval, biconvex, nucleated red blood corpuscles (erythrocytes) contain hemoglobin. Hemoglobin unites temporarily with oxygen in the lungs and skin to form oxyhemoglobin, which in turn gives up its oxygen to cells and tissues when or where it is needed. (2) Ameboid white blood corpuscles (leukocytes), which are able to move independently and are of different sizes, pass through the walls of blood vessels and tissues. They destroy bacteria and other organisms by ingesting them, thus serving to prevent infections. (3) The spindle cells assist in the clotting of blood upon their disintegration. Blood corpuscles originate principally in the marrow of the bones and in the spleen. (4) The plasma or liquid part of the blood carries foods, carbon dioxide, wastes, proteins, and mineral salts. Blood coagulates, especially after injuries, to form a clot, which includes fibrin, red and white corpuscles, and tissue cells.

RESPIRATION

In the earlier tadpole stages external gills are present, but these are later covered to form internal gills, which communicate with the exterior through a small opening. The internal gills are eventually absorbed, and typical lungs develop in the air-breathing adult frog. In the adult frog respiration takes place through the skin and lungs, but probably more through the skin. During hibernation the lungs are inactive, yet skin respiration continues, even though the rate may be reduced. In lung respiration the air is admitted into the mouth cavity from the outside through the external nares (nostrils) and then through the slitlike glottis (Gr. *glotta*, tongue) into the short, tubular larynx (Gr. *larynx*, voice box). From the larynx the air passes into the trachea (windpipe) and finally into the thin-walled, saclike, paired lungs. The lungs are ovoid, distensible, and internally divided by folds (septae) into a number of compartments known as alveoli (L. *alveolus*, small cavity) to increase the surface exposed to the air. Thin-walled capillaries line the inner surfaces of the alveoli and permit the exchange of oxygen and carbon dioxide between the air in the lungs and the blood in the circulatory system. The amount of exchange of these gases depends upon the concentration of each on either side of the lung and blood vessel membranes. Air is forced into the lungs through the slitlike glottis by closing the nares and contracting the floor of the mouth. Air is expelled from the lungs through the glottis into the mouth cavity by the contraction of the muscles of the body walls. It may be expelled or drawn into the mouth through the nares by closing the glottis and alternately raising or lowering the floor of the mouth. Sounds may be produced by forcing air back and forth through the glottis.

EXCRETION

Some wastes are excreted by the frog skin and intestine, but many are removed from the blood by a pair of elongated kidneys in the dorsal abdominal cavity (Figs. 24-14 and 24-15). Internally, a kidney contains a number of malpighian bodies, each consisting of an enclosing membrane known as Bowman's capsule, which surrounds a coiled mass of thin-walled capillaries known as a glomerulus (L. *glomus*, ball). Wastes are collected from the blood in the glomeruli and carried by uriniferous tubules to collecting tubules, to the tubular ureter, and finally to the saclike cloaca. From the cloaca

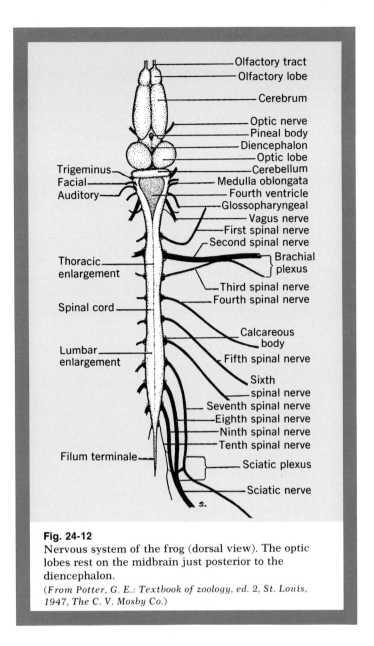

Fig. 24-12
Nervous system of the frog (dorsal view). The optic lobes rest on the midbrain just posterior to the diencephalon.
(*From Potter, G. E.: Textbook of zoology, ed. 2, St. Louis, 1947, The C. V. Mosby Co.*)

the urine may be stored in the thin-walled, distensible urinary bladder, which voids only at certain intervals. Ciliated, funnel-shaped nephrostomes in the ventral part of the kidney open into the coelom, from which wastes may be obtained and later eliminated.

COORDINATION AND SENSORY EQUIPMENT

The nervous system may be divided into (1) a central nervous system, consisting of the brain and spinal cord, (2) a peripheral nervous system, con-

333

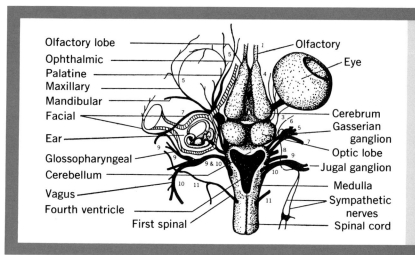

Olfactory lobe
Ophthalmic
Palatine
Maxillary
Mandibular
Facial
Ear
Glossopharyngeal
Cerebellum
Vagus
Fourth ventricle
First spinal

Olfactory
Eye
Cerebrum
Gasserian ganglion
Optic lobe
Jugal ganglion
Medulla
Sympathetic nerves
Spinal cord

Fig. 24-13
Brain and cranial nerves of the bullfrog (*Rana catesbeiana*) (somewhat diagrammatic dorsal view). 1 to 10, cranial nerves; 11, first pair of spinal nerves. Certain cranial nerves show some of their branches.
(*From Atwood, W. H.: Comparative anatomy, ed. 2, St. Louis, 1955, The C. V. Mosby Co.*)

sisting of ten pairs of cranial nerves (Table 24-2) and ten pairs of spinal nerves, and (3) a sympathetic nervous system, consisting of nerves and ganglia that supply the internal (visceral) organs (Figs. 24-12 and 24-13).

The brain has the following structures: (1) two small, fused olfactory lobes for the sense of smell, (2) two large, elongated cerebral hemispheres (cerebrum), (3) two large optic lobes for the sense of sight, (4) the midbrain, (5) the small, narrow cerebellum, and (6) the wide medulla oblongata, which connects with the enlarged portion of the spinal cord. When the brain (except the medulla oblongata) is removed, the frog is still able to breathe, jump, swim, swallow food, and use its sense of equilibrium.

On the ventral side of the brain the following structures are distinguishable: (1) the optic chiasma, or the crossing of the optic nerves, (2) the pituitary body (hypophysis), and (3) the infundibulum, a bilobed extension of the diencephalon.

The spinal cord is composed of a central mass of gray matter (principally nerve cells) in the shape of the letter H and an outer mass of white matter made up of nerve fibers. The hollow central canal extends throughout the entire cord and may be seen in the middle of the crossbar of the H in a cross section of the spinal cord. The central canal connects anteriorly with the cavities (ventricles) of the brain. The spinal cord has two surrounding membranous meninges, the outer dura mater and the inner pia mater.

There are ten pairs of spinal nerves, each arising from the gray matter of the spinal cord by a dorsal and ventral root. The union of these two roots at the side of the cord forms a spinal nerve. Each spinal nerve passes out between the bony arches of adjacent vertebrae.

The sympathetic nervous system consists of two main trunks that parallel the spinal cord, one on either side of it. Each trunk has ganglia or enlargements where the ten pairs of spinal nerves unite with it.

The skin, by way of its sensory nerve endings, receives tactile, chemical, heat, and light stimuli. The eyes resemble the eyes of other vertebrates. There are three eyelids: the rather motionless upper lid, the lower lid, which is fused with the third eyelid, or nictitating membrane (L. *nictare*, to beckon). The lens permits objects, especially moving objects, to be seen at definite distances. The pupil contracts and regulates the amount of light that enters the eye. The sensitive retina (L. *rete*, net) within the eye is stimulated by light and transfers the impulses to the optic nerve, which carries them to the brain to give the sensation of sight. The eyes lie in orbits (sockets) at the side of the skull and are moved by six eye muscles known as the external and internal recti, the superior and inferior recti, and the superior and inferior oblique.

Table 24-2
Cranial nerves of vertebrates

Number	Name	Origin	Distribution	Function
I	Olfactory	Olfactory lobe	Mucous membrane lining nose	Sensory (smell)
II	Optic	Second vesicle of forebrain (dienceph-alon)	Cells of retina of eye	Sensory (sight)
III	Oculomotor	Ventral part of midbrain	Superior, inferior, inter-nal recti; and inferior oblique muscles of eye	Motor (eye movement)
IV	Trochlear	Dorsal part of midbrain	Superior oblique muscle of eye	Motor (eye movement)
V	Trigeminal	Laterally from medulla (hindbrain)	Face, tongue, and mouth, and to muscles of jaws or mandibles	Sensory and motor
VI	Abducens	Ventral part of medulla	External rectus muscle of eye	Motor
VII	Facial	Laterally from medulla	Muscles of face, roof of mouth, hyoid	Motor (principally)
VIII	Auditory	Laterally from medulla	Cells of semicircular canal and other parts of inner ear	Sensory (hearing and equilibrium?)
IX	Glosso-pharyngeal	Laterally from medulla	Membranes and muscles of tongue and pharynx	Sensory and motor
X	Vagus	Laterally from medulla	Heart, lungs, pharynx, stomach, intestine, visceral arches	Sensory and motor
XI*	Spinal accessory	Laterally from medulla	Muscles of shoulder	Sensory and motor
XII*	Hypo-glossal	Ventral part of medulla	Tongue and neck muscles	Motor

*The XI and XII pairs are absent in fishes and amphibia.

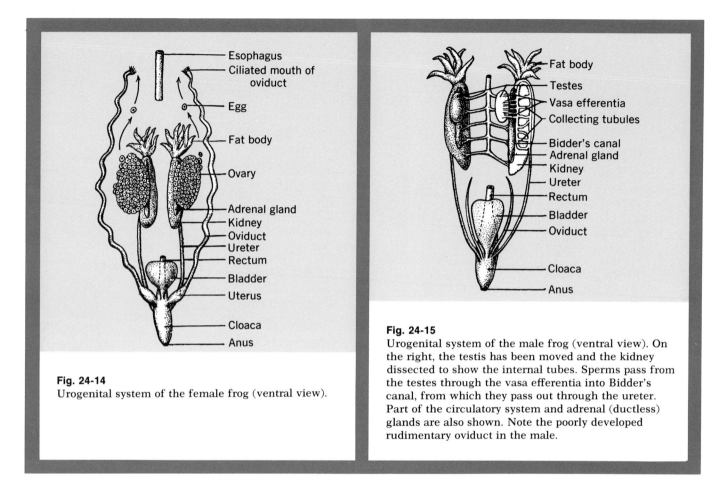

Fig. 24-14
Urogenital system of the female frog (ventral view).

Fig. 24-15
Urogenital system of the male frog (ventral view). On the right, the testis has been moved and the kidney dissected to show the internal tubes. Sperms pass from the testes through the vasa efferentia into Bidder's canal, from which they pass out through the ureter. Part of the circulatory system and adrenal (ductless) glands are also shown. Note the poorly developed rudimentary oviduct in the male.

The tympanic membrane of the outer ear communicates with the inner ear by a bony columella that vibrates with the sound stimuli received. The auditory nerve carries the impulses to the brain, where the sensation of hearing is really produced. There are no external ears. The middle ear communicates with the mouth cavity by means of the eustachian tube, which aids in equalizing air pressures on the eardrums. The inner ear also contains organs of equilibrium.

The olfactory sense is located in a pair of nasal cavities lined with folds of sensitive, epithelial, nasal membranes. The external nares (anterior nares) connect with the nasal cavity. The internal nares (posterior nares) connect the nasal cavity with the mouth cavity. The nares in amphibia and other vertebrates (above the fishes) are used for both respiratory and olfactory purposes. The olfactory nerves connect the epithelial nasal membranes of the nasal cavities with the olfactory lobes of the brain. The elevated papillae of the mouth and tongue contain organs of taste, especially if foods and chemicals are in solution.

REPRODUCTION

The sexes are separate (diecious) (Figs. 24-14 and 24-15). The sperms of the male originate in paired, small, oval testes. The sperms pass through the vasa efferentia into the kidneys, into the ureter by means of Bidder's canal, into the cloaca, and out through the anus. The eggs originate in the large paired ovaries and later break out through the ovary walls into the coelom. From here the eggs find their way into the much-coiled, paired oviducts, the

336

funnel-shaped openings of which are located near the anterior edge of the abdominal cavity. Each oviduct leads into a thin-walled uterus that empties into the cloaca. The cloaca leads to the anus. The eggs are supplied with a layer of gelatinous food, which is produced by glandular cells of the oviduct. There are no copulatory organs. A yellowish hand-shaped fat body is located anterior to each reproductive organ and serves for food storage. The embryology of the frog is considered in Chapter 28.

Some contributors to the knowledge of the frog

Carolus Linnaeus (1707-1778)
He first used the word "amphibia" for a general group of more or less aquatic vertebrates.

A. M. Marshall
He authored one of the first texts, *The Frog* (1882).

A. Ecker
He wrote the *Anatomy of the Frog* (1889).

H. Gadow
He described the biology and natural history of amphibia in a volume in the *Cambridge Natural History* (1901).

G. B. Howes
He discussed frogs in some detail in his *Atlas of Elementary Zootomy* (1902).

Joseph Leidy (1823-1891)
He studied the worm parasites of frogs (1904).

M. Dickerson
He authored a widely used text called *The Frog Book* (1906).

Samuel J. Holmes
He wrote a fine book called *The Biology of the Frog* (1930).

G. Kingsley Noble
Noble, of the American Museum of Natural History, authored an excellent volume called *Biology of the Amphibia* (1931).

Franz Werner
He gave an extensive account of amphibia in Kükenthal's *Handbuch der Zoologie* (1931).

Review questions and topics

1 List the characteristics that place the frog in the phylum Chordata, subphylum Vertebrata, and class Amphibia.
2 Explain the significance of (1) protective coloration, (2) internal bony skeleton, (3) closed system of arteries, capillaries, and veins, (4) three-chambered heart, (5) lymph hearts and lymph, (6) paired appendages with digits, (7) well-developed skeletal muscles that function in opposition, and (8) red blood corpuscles that carry oxygen.
3 Describe the structure and functions of the following for the frog: (1) integument, (2) locomotion, (3) ingestion and digestion, (4) circulation, (5) respiration, (6) excretion, (7) coordination and sensory organs, and (8) reproduction. In what specific ways do these show improvements over comparable structures and phenomena in lower animals?
4 Explain: (1) origin and insertion of a muscle, (2) pectoral and pelvic girdles, (3) auricle and ventricle, (4) vasa efferentia and ureter, (5) oviduct and ureter, and (6) aorta and vena cava.
5 Describe the structures and functions of the various parts of the frog digestive system, including the physiology of digestion in detail.
6 Explain how frog blood is carried from the heart to the lungs and skin.
7 Explain the structure and functions of the blood cells of the frog.
8 When excretory functions are localized in a pair of kidneys, must there be a comparable development of the circulatory system to take wastes to them?
9 Explain the role of the circulatory system in the respiratory process in the lungs and skin.
10 Discuss the improvements of the structure and functions of the nervous system and sensory organs over those of lower animals.
11 List the number, names, origin, distribution, and functions of the ten pairs of cranial nerves in amphibia. In what ways do reptiles, birds, and mammals differ from amphibia and fishes in regard to cranial nerves?

Selected references

Cochran, D. N.: Living amphibians of the world, Garden City, New York, 1961, Doubleday & Company, Inc.

Holmes, S. J.: The biology of the frog, New York, 1928, The Macmillan Company.

Moore, J. A. (editor): Physiology of the Amphibia, New York, 1964, Academic Press Inc.

Muntz, W. R. A.: Vision in frogs, Sci. Amer. **210:**110-119, 1964.

Noble, G. K.: Biology of the Amphibia, New York, 1931, McGraw-Hill Book Company.

Oliver, J. A.: The natural history of North American amphibians and reptiles, Princeton, New Jersey, 1955, D. Van Nostrand Co., Inc.

Stuart, R. R.: The anatomy of the bullfrog, Chicago, 1940, Denoyer-Geppert Co.

part four

THE BIOLOGY OF MAN

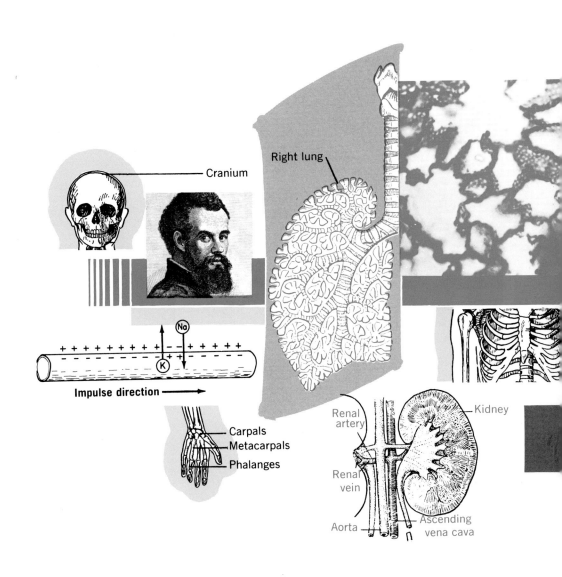

Cranium

Right lung

Impulse direction ⟶

Carpals
Metacarpals
Phalanges

Renal artery

Kidney

Renal vein

Aorta

Ascending vena cava

Biology of man—the human form
Biology of man—metabolism
Biology of man—control mechanisms and reproduction

chapter twenty-five

Biology of man —
the human form

The form and function of the body is a result of the repeated division and specialization of a fertilized egg. The resulting embryo contains trillions of cells, each part of a particular tissue and organ. These function as part of the following systems.

Integumentary (skin) system—protection, support, heat regulation, absorption, excretion, stimuli reception

Skeletal system—support, protection, posture, motion, locomotion, manufacture of blood corpuscles (by bone marrow), transmission of sound waves (ear bones)

Muscular system—locomotion, movement of body parts, such as the stomach, intestines, heart

Digestive system—ingestion, digestion, absorption of foods, elimination of wastes

Circulatory system—transportation of foods, wastes, oxygen, heat, carbon dioxide, endocrine hormones

Respiratory system—supply of oxygen and elimination of carbon dioxide and other wastes

Excretory system—excretion and elimination of metabolic waste materials

Nervous and sensory system—reception of stimuli, transmission and interpretation of impulses for purposes of correlation, movement, locomotion, behavior, secretion; centers of sight, hearing, taste, smell, equilibrium

Endocrine (ductless) gland system—production of hormones for correlating and regulating body processes

Reproductive system—production of sex cells for the continuation of the species

INTEGUMENT

The functions of the human skin and its accessory structures may be summarized briefly as follows: (1) regulation of heat; (2) excretion of wastes; (3) protection against injury, harmful light rays, loss of water, and disease-producing organisms; (4) prevention of the absorption of various deleterious materials; (5) assistance in normal respiration; (6) supplying of information about our surroundings through the various types of sensory organs; and (7) production of hairs, nails, glands, and teeth.

The two layers of the human skin are (1) the external epidermis (cuticle) and (2) the deeper dermis (corium) (Fig. 25-1). The epidermis is composed of stratified squamous and columnar epithelium and contains no blood vessels but has fine nerve fibrils. The hairs, nails, and numerous glands are modified epidermis. When people "peel" after a sunburn, the epidermis is shed.

Hair is formed when the pigmented malpighian layer of the epidermis extends downward into the dermis to form tubelike hair follicles. The cells at the base of a follicle produce a hair, which is fused epidermal cells supplied with a horny, protein material called keratin (Gr. *keras,* horny).

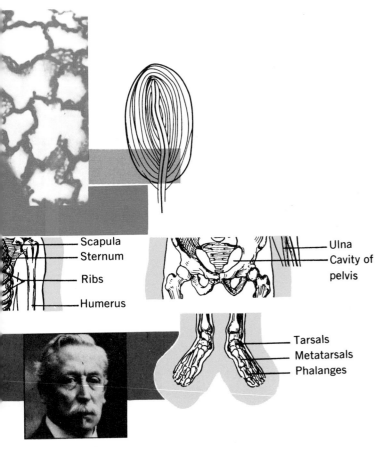

Scapula
Sternum
Ribs
Humerus

Ulna
Cavity of pelvis

Tarsals
Metatarsals
Phalanges

When a hair is being formed, it first appears as a tiny elevation below the skin surface. As more is formed, the hair is pushed from the skin surface. The part of the hair within the follicle is called the root, and the remainder is called the shaft. All skin is provided with follicles, except the palms of the hands, the soles of the feet, and the last portions of the fingers and toes. The consistency of hairs depends upon the structure of the follicle: a round follicle (in cross section) gives rise to straight hair, an oval follicle to curly hair, and a rather flat, ribbon-shaped follicle to wavy (kinky) hair. Hair color is determined by the quantity and quality of the pigments present and their relation to the transparent air within the hair. The base of each hair is supplied with a nerve and with blood vessels for nourishment. Muscle fibers in the dermis are attached to the hair follicle for hair movement.

Sebaceous (oil) glands are formed by the invagination of the malpighian layer of the epidermis. They are usually associated with hair follicles, being especially numerous on the face and scalp. The oils pass from the gland into the hair follicle and then to the skin surface, thereby keeping the hair and skin from becoming dry and preventing undue evaporation or absorption of water and other liquids by the skin.

Nails are formed from closely packed epithelial tissues along a furrow at the base of the nail. The nail is formed by a fusion of clear, dead, horny, keratinized cells.

The dermis of the skin is thicker than the epidermis and has (1) an upper papillary layer, containing numerous papillae (elevations) that increase the surface for nerves and tactile organs, blood vessels, hair follicles, and sebaceous glands, and (2) a lower reticular layer, containing yellow elastic and white fibrous connective tissues, which contain adipose tissue ("fat") and sweat glands.

Sweat glands are present in all human skin but are most numerous under the arms and on the forehead, soles of the feet, and palms of the hands. These coiled, tubular glands are located in the dermis and empty their excretions through pores on the skin surface. Over two million sweat glands in the entire skin eliminate more than a quart of liquid per day under normal conditions. Perspiration eliminates body wastes and regulates body

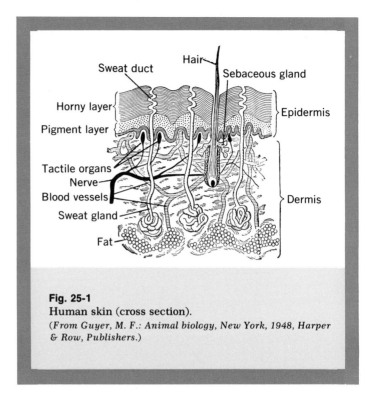

Fig. 25-1
Human skin (cross section).
(*From Guyer, M. F.: Animal biology, New York, 1948, Harper & Row, Publishers.*)

heat through the evaporation of water. Heat that is produced in various tissues, especially muscles, is distributed by the blood throughout all parts of the body, thus producing an average, normal body temperature of 98° to 99° F. Since the skin is well supplied with blood, it can act efficiently as a heat-regulating mechanism. Enlargement (dilation) of blood vessels and relaxation of the muscle fibers allow more blood to lose more heat, whereas a contraction of blood vessels and muscles has the opposite effect.

The dermis is attached to the deeper tissues by connective tissue, called subcutaneous tissue. In some parts of the body the skin is tightly attached to the deeper tissues, but in other parts it is loosely attached to permit more freedom of movement. The dermis contains many sensory nerve endings for the reception of such stimuli as heat, cold, pain, pressure, and touch. Receptors are present in most areas of the skin but are much more concentrated in some regions than in others; for example, tactile (touch) receptors are more numerous on fingertips than on the back of the hand. On the inner surface of the hands and fingers and on the soles of the feet there are many minute ridges that in-

341

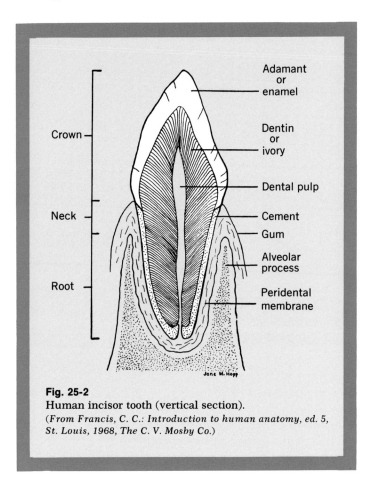

Fig. 25-2
Human incisor tooth (vertical section).
(From Francis, C. C.: Introduction to human anatomy, ed. 5, St. Louis, 1968, The C. V. Mosby Co.)

Fig. 25-3
Cast showing the thirty-two permanent human teeth.
(Courtesy American Dental Association.)

crease friction and form distinctive fingerprint and footprint patterns.

Teeth are derived embryonically from epithelial tissues and are embedded in the upper and lower jaws. The part of the tooth (Fig. 25-2) above the gum is called the crown and is covered with hard enamel. The remainder of the tooth is composed of softer dentin with its central canal (pulp cavity), which in turn contains blood vessels and nerves. The teeth are anchored in their sockets by a cement that also protects the dentin. Man, like other mammals, has a temporary (baby) set of teeth, twenty in number, which appears between the ages of 6 months and 2½ years. The permanent set of thirty-two (Fig. 25-3) is composed (on each side of each jaw) of two flat, sharp incisors (front teeth) for cutting foods as they overlap, one pointed canine (eye tooth or cuspid) for tearing foods, two broad-surfaced premolars (bicuspids) for grinding, and three large molars for grinding. The premolars have two surface elevations, and the larger molars have four or more elevations for grinding. The last pair of molars (wisdom teeth) may not erupt until later in life, or not at all. The normal dental formula for man is as follows:

$$I \frac{2}{2}; \ C \frac{1}{1}; \ P \frac{2}{2}; \ M \frac{3}{3} \ \text{(for each jaw)}$$

SKELETON

The skeletal system consists of 206 named bones, cartilage, and ligaments. The bones are illustrated in Fig. 25-4, and the names and numbers are given in table form in Fig. 25-5. Bones are classified concerning shape as follows: (1) long (arms and legs), (2) short (wrist), (3) flat (shoulder blade, patella), and (4) irregular (vertebrae). It will be noted that the teeth are not listed as part of the skeleton but are included with the integument because of their epithelial origin.

Study of a human skeleton will show that there are several types of joints, each with its specific functions. Joints are classed as (1) immovable (irregular, dovetail connections [sutures] of the bones of the cranium) and (2) movable, with movements of various types for specific purposes. Movable joints may be further classed as (1) ball-and-socket (femur and pelvic girdle, humerus and pectoral girdle), (2) hinge (femur and tibia, humerus and ulna), (3) sliding (most vertebrae), and (4) rotating (radius and ulna).

Most bones originate in the embryo as cartilage, hence they are known as cartilage bones in contrast to the membrane bones formed by the gradual ossification of soft, fibrous, membranous tissues (skull bones). Certain parts of the skeleton remain as cartilage, such as the external ear, tip of the nose, tip of the breastbone, areas between the vertebrae, and articulatory surfaces of movable joints. The great strength, elasticity, and reduced friction of cartilage make it an efficient part of the skeleton.

Ligaments are bands of tissue that connect the bones and support the interior organs. The ligaments bind the body parts together into an efficient structure.

Briefly stated, the functions of the human skeleton are (1) to form a framework to support other organs and give posture to the body; (2) to give protection to vital organs, such as the brain, spinal cord, heart, and lungs; (3) to form solid attachments for muscles so that they may act as a system of levers in motion and locomotion; (4) to store certain mineral reserves; (5) to form blood corpuscles by the bone marrow; and (6) to transmit sound waves as accomplished by the hammer, anvil, and stirrup bones of the ears.

MOTION AND LOCOMOTION

The bones of the skeleton are passive structures to which the active skeletal muscles are attached. By means of these muscles, body parts are moved,

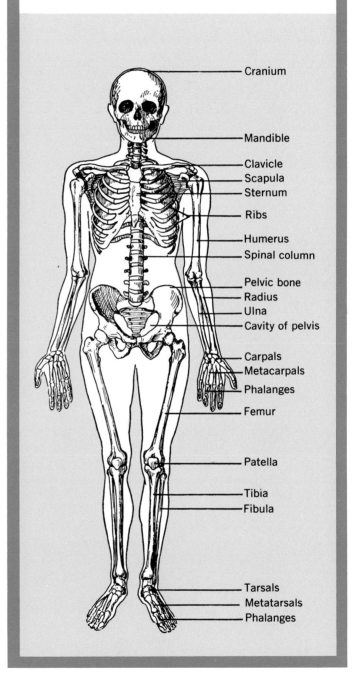

Fig. 25-4
Human skeleton. Clavicle is commonly called the collarbone; scapula, the shoulder blade; patella, the kneecap.
(*From Braungart, D. C., and Buddeke, R.: An introduction to animal biology, ed. 7, St. Louis, 1968, The C. V. Mosby Co.*)

Cranium

Mandible

Clavicle
Scapula
Sternum

Ribs

Humerus
Spinal column

Pelvic bone
Radius
Ulna
Cavity of pelvis

Carpals
Metacarpals
Phalanges

Femur

Patella

Tibia
Fibula

Tarsals
Metatarsals
Phalanges

Human skeleton*

Axial (80)

- Skull
  - Cranium (brain case) (8)
    - **Occipital** (base of skull) (1)
    - **Parietal** (top of head) (2)
    - **Frontal** (forehead) (1)
    - **Temporal** (above ears) (2)
    - **Ethmoid** (back of nose) (1)
    - **Sphenoid** (back of eye) (1)
  - Face (14)
    - **Mandible** (lower jaw) (1)
    - **Maxilla** (upper jaw) (2)
    - **Palate** (2)
    - **Malar or zygomatic** (cheek) (2)
    - **Lacrimal** (inner orbit) (2)
    - **Inferior turbinated** (nose) (2)
    - **Vomer** (nasal septum) (1)
    - **Nasal** (bridge of nose) (2)
- Ear bones (6)
  - **Malleus** (hammer) (2)
  - **Incus** (anvil) (2)
  - **Stapes** (stirrup) (2)
- Vertebral column (26)
  - **Cervical** (neck) (7)
  - **Thoracic** (chest) (12) with ribs
  - **Lumbar** (lower trunk) (5)
  - **Sacral** (sacrum) (1)†
  - **Coccygeal** (caudal or tail) (1)‡
- **Hyoid** (base of tongue) (1)
- **Sternum** (breastbone) (1)
- **Ribs** (24)

Appendicular (126)

- Pectoral girdle (shoulder)
  - **Clavicle** (collarbone) (2)
  - **Scapula** (shoulder blade) (2)
- Arms
  - **Humerus** (2)
  - **Radius** (2)
  - **Ulna** (2)
  - **Carpals** (wrist) (16)
  - **Metacarpals** (hand) (10)
  - **Phalanges** (fingers) (28)
- Pelvic girdle (hip) (2)§
- Legs
  - **Femur** (thigh) (2)
  - **Tibia** (shin) (2)
  - **Fibula** (2)
  - **Tarsals** (ankle and heel) (14)
  - **Metatarsals** (foot) (10)
  - **Phalanges** (toes) (28)
- Kneecap (patella) (2)

Fig. 25-5
Outline of bones in human skeleton.

*This does not include the variable number of **sesamoid bones** (ses′ a moid) (Gr. *sesamon*, sesame seed; *eidos*, like), embedded in the tendons of the hand, knee, and foot, or the **wormian bones** (wur′ mi an) (after Worm, a Danish anatomist), which are isolated bones in the sutures or joints, especially those of the skull.

The figures in parentheses give the number of bones of each type.

†Five bones fused.

‡Four bones fused.

§Three bones (ilium, ischium, and pubis) fused.

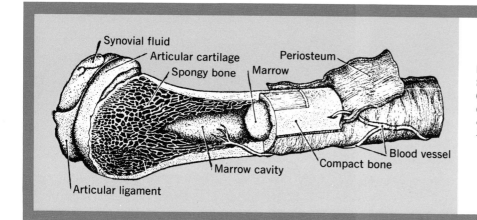

Fig. 25-6
Structure of a long bone and joint (diagrammatic). (*From Neal and Rand: Chordate anatomy, New York, McGraw-Hill Book Company.*)

Labels: Synovial fluid; Articular cartilage; Spongy bone; Marrow; Periosteum; Compact bone; Blood vessel; Marrow cavity; Articular ligament

or the body as a whole is moved from place to place.

A muscle functions by contracting. This is initiated by the transmission of a nerve impulse, which may be voluntary or involuntary. The muscles of the skeleton are under control of the will, and those of the viscera (stomach, intestines) and heart are not.

The fine structure of the muscles differs and thus, skeletal muscle is also known as striated, visceral muscle as smooth, and heart muscle as cardiac. Under the microscope striated muscle shows a series of fine lines or striae, and it lacks a distinct cellular structure, having nuclei in a scattered pattern. By contrast, smooth muscle is composed of a series of distinct, spindlelike cells, each with its own nucleus. Cardiac muscle appears to have the fine striation and also individual nuclei and cellular structure. A series of branchings and joinings is also characteristic of cardiac muscle.

Skeletal muscle

Muscles functioning in movement are attached to two bones across a joint; contraction then pulls one bone toward (or away from) the other. The muscle end attached to the immovable bone is called the origin, while the muscle end attached to the movable bone is called the insertion. Muscles that move a part away from the median line are called abductor muscles; those that move a part toward the median line are called adductor muscles. The terms flexor and extensor indicate the types of movements muscles effect. The biceps of the forearm is a flexor (flexes or bends the forearm), and the triceps of the forearm is an extensor

(extends or straightens the forearm). These muscles act as opposing pairs (Fig. 25-7).

Man has more than 600 skeletal muscles, to which scientific names are applied in various ways: (1) in relation to the structure, or bone, with which they are associated (*triceps brachii* muscle on the back of the upper arm, or brachium) (Fig. 25-8); (2) in relation to the number of "heads" with which they originate (*triceps brachii*, meaning three "heads," or *biceps brachii*, with two "heads" and located on the front of the upper

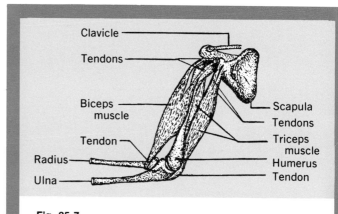

Labels: Clavicle; Tendons; Biceps muscle; Tendon; Radius; Ulna; Scapula; Tendons; Triceps muscle; Humerus; Tendon

Fig. 25-7
Human biceps and triceps muscles (diagrammatic). The origin of the biceps (two "heads") is on the scapula (shoulder blade), and the insertion is on the radius bone; the origin of the triceps (three "heads") is on the scapula; and the insertion is on the ulnar bone. The two muscles work in opposition to each other.

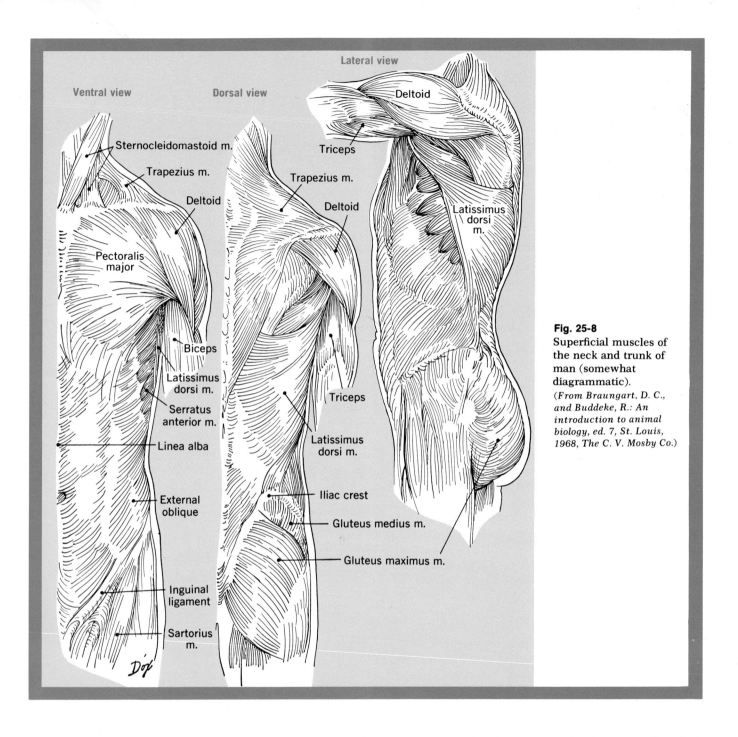

Ventral view

- Sternocleidomastoid m.
- Trapezius m.
- Deltoid
- Pectoralis major
- Biceps
- Latissimus dorsi m.
- Serratus anterior m.
- Linea alba
- External oblique
- Inguinal ligament
- Sartorius m.

Dorsal view

- Trapezius m.
- Deltoid
- Triceps
- Latissimus dorsi m.
- Iliac crest
- Gluteus medius m.
- Gluteus maximus m.

Lateral view

- Deltoid
- Triceps
- Latissimus dorsi m.

Dú

Fig. 25-8
Superficial muscles of the neck and trunk of man (somewhat diagrammatic).
(*From Braungart, D. C., and Buddeke, R.: An introduction to animal biology, ed. 7, St. Louis, 1968, The C. V. Mosby Co.*)

arm); (3) in reference to the shape of the muscle (*deltoid*, the delta-shaped muscle of the top of the shoulder); (4) in reference to the direction in which they run (*external oblique*, a strong muscle of the abdominal wall that lies obliquely); (5) in reference to their location (*external intercostals*, superficial muscles between the ribs, and *internal*

intercostals, deeper muscles between the ribs); (6) in reference to their function (*adductor longus*, adducts the thigh toward the median line); and (7) in reference to the length and size of the muscle (*peroneus longus*, the large muscle attached to the fibula bone and *peroneus brevis*, the smaller muscle attached to the fibula bone).

346

Fig. 25-9
Striated muscle from the rabbit (enlarged 24,000 diameters by the electron micrograph). Each diagonal ribbon is a thin section of myofibril (muscle fibril) with its dense A-bands bisected by less dense H-zones, and lighter I-bands bisected by more dense, narrow Z-lines, or membranes.
(*From Huxley: Scientific American, November, 1958.*)

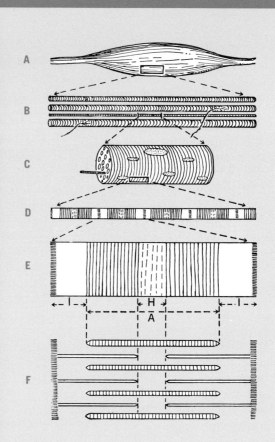

Fig. 25-10
Striated muscle dissected and represented by schematic drawings. Muscle, **A**, is made up of muscle fibers, **B**, which appear striated (banded) in the light microscope. Small branching structures at the surface are "end-plates" of motor nerves, which signal the fibers to contract. Single muscle fiber, **C**, is made up of myofibrils, beside which lie cell nuclei and mitochondria. In the single myofibril, **D**, striations are resolved into a repeating pattern of dark and light bands. Single unit of this pattern, **E**, consists of Z-line, then I-band, then A-band which is interrupted by H-zone, then next I-band, and finally next Z-line. Electron micrographs show that the repeating band pattern is a result of the overlapping of thick and thin filaments, **F**.
(*From Huxley: Scientific American, November, 1958*).

Muscle structure

A medium-sized muscle contains approximately ten million cells (fibers), so that over six billion muscle cells are contained in the total musculature. Each striated, skeletal muscle is composed of numerous bundles, and each bundle consists of hundreds of delicate fibrils, each having a diameter of 10 to 100μ.

Fibrils are composed of a series of alternate dark and light areas, thus giving a banded (striated) appearance to skeletal muscles. Each skeletal muscle cell contains several nuclei. Skeletal muscles are voluntary (under the control of the will), distinctly striated, and have a rather rapid rate of action and a relatively high rate of fatigue compared with other types of muscles.

The contractile structure of a muscle fiber is made up of long, thin elements called myofibrils, each about 1μ in diameter. A myofibril has cross striations like the fiber of which it is a part. These striations arise from a repeating variation in the density, or concentration, of protein along the myofibrils (Figs. 25-9 and 25-10). The striations can be seen in isolated myofibrils. Under high magnification of a light microscope there is a regular alternation of dense bands (A-bands) and lighter bands (I-bands). The central region of an A-band is frequently less dense than the rest of the band and is known as the H-zone. The I-band is bisected by a narrow, dense line known as the Z-line (or Z-membrane). From one Z-line to the next, the repeating units of the myofibril may be summarized as Z-line, I-band, A-band (interrupted by the H-zone), I-band, and Z-line.

When examined with an electron microscope, the myofibril is found to consist of still smaller filaments, each of which is 50 to 100 angstrom units in diameter (1 angstrom unit is 1 ten-thousandth of a micron). When properly prepared thin sections of muscle are observed with an electron microscope, the myofibrils are seen to consist of two kinds of filaments, one of which is twice as thick as the other. Each filament is arranged in register with other filaments of the same kind, and the two arrays overlap for part of their length. This overlapping gives rise to the crossbands of the myofibril, the dense A-band consisting of overlapping thick and thin filaments, the lighter I-band consisting of thin filaments alone, and the H-zone consisting of thick filaments alone. Halfway along their length the thin filaments pass through a narrow zone of dense material that comprises the Z-line.

The two kinds of filaments are linked together by an intricate and well-ordered system of cross bridges, which may play an important role in muscle contraction. The bridges project outward from a thick filament at rather regular intervals of 60 to 70 angstroms, and each bridge is 60 degrees around the axis of the filament with respect to the adjacent bridge. The bridges thus form a helical pattern that repeats every six bridges along the filament. This pattern joins the thick filament to each one of its six adjacent, thin filaments once every 400 angstroms.

Muscle contraction

Knowledge of the band structure of striated muscle made it apparent that changes in the pattern during contraction should give insight into the molecular nature of the process. It has been found that over a wide range of muscle lengths, during both contraction and expansion,

the length of the A-bands remains constant. On the other hand, the length of the I-bands changes in accord with the length of the muscle. The length of the A-band is equal to the length of the thick filament. But the length of the H-zone increases and decreases with the length of the I-band, so that the distance from the end of one H-zone through the Z-line to the beginning of the next H-zone remains approximately the same. This distance is equal to the length of the thin filaments, so they, too, do not alter their length appreciably. From all of this it is concluded that when a muscle changes length, the two sets of filaments slide past each other (sliding-filament phenomenon) (Fig. 25-10).

When a muscle is treated with an appropriate salt solution and examined with a light microscope, it is observed that the A-bands have disappeared. Such a salt solution will remove myosin (a muscle protein), which demonstrates that the thick filaments of the A-band are composed of myosin. If the myofibril is treated to extract its actin (another protein), a large part of the material in these segments is removed, which indicates that the thin filaments of the I-bands are composed of actin. It is likely that the physical expression of the combination of actin and myosin is found in the bridges between the two kinds of filaments; thus it seems that the sliding movement is mediated by the bridges.

It is suggested that the cross bridges seem to form a permanent part of the myosin filaments; probably they are those parts of the myosin molecules that are involved directly in the combination with actin. Since the number of myosin molecules in a given volume of muscle is surprisingly close to the number of bridges in the same volume, it is suggested that each bridge may be part of a single myosin molecule.

How can the bridges cause contraction? One theory is that the bridges may oscillate back and forth and hook up with specific sites on the actin filament. The bridges then could pull the filament a short distance and return to their original location, ready for another pull. One would expect that each time a bridge went through such a cycle a phosphate group would be split from a single molecule of adenosine triphosphate (ATP), which would supply the energy for the cycle. When the muscle has relaxed, it may be supposed that the removal of phosphate groups from the ATP has ceased, and that myosin bridges can no longer combine with actin filaments, thus permitting the muscle to return to an uncontracted condition.

ATP is one of the most important substances in muscle. The terminal phosphate group of this compound is linked to the structure by an "energy-rich" bond. Rupture of this bond produces inorganic phosphate and adenosine diphosphate (ADP) with the release of a large amount of energy that can be used in performing work.

$$ATP \rightarrow ADP + H_3PO_4 + Energy$$

Myosin is the actual enzyme or is at least associated with the catalytic breakdown of ATP to ADP. The reaction is activated by calcium ions and inhibited by magnesium ions. In resting muscle the enzymatic activity is inhibited, and ATP maintains the extensibility of the fiber. When stimulated, the enzymatic activity breaks down ATP to ADP, with the sudden release of energy for contraction. The chemical interaction of ATP, myosin, and actin results in the closer interdigitation of the thick and thin filaments, with consequent muscle shortening. The ATP is subsequently restored, to be used later.

Muscle stimulation

Around each muscle fiber is an electrical, polarized membrane, the inside of which is about 1/10 of a volt negative with respect to the outside. If the membrane is depolarized temporarily, the muscle fiber contracts; it is by this means that muscle activity is controlled by the nervous system. An impulse traveling over a motor nerve is transmitted to the muscle membrane at the motor end plate or neuromuscular junction. It then passes deep into the myofibril by way of a system of fine T-tubules, which cross the area where actin and myosin overlap. At this point the tubules come in contact with the sarcoplasmic reticulum, which runs parallel to the myofibril. An impulse travels by way of the T-tubules to the reticulum, which then releases calcium ions that in turn initiate the muscle contraction. Then a wave of depolarization, the "action potential," passes down the muscle fiber causing it to twitch. When nerve impulses arrive in rapid succession through the motor nerve, the twitches run

together, and the muscle maintains its contraction as long as the stimulation continues (or until the muscle becomes exhausted). The muscle relaxes automatically when nerve stimulation ceases.

As long as a man is conscious and his muscles are not actively contracting, the muscles are not completely relaxed but are constantly partially contracted. This phenomenon is known as tonus ("tone") (Gr. *tonos*, tension). Tonus, by a series of nerve impulses, is responsible for body posture and keeps muscles in readiness for actual contraction. Severing of the nerve to a skeletal muscle eliminates tonus. During tonus only a small fraction of the muscle fibers are involved, and it is thought that individual fibers work in relays, thus giving them a chance to recover between contractions.

An entire muscle cannot contract maximally, but a single fiber can respond only maximally or not at all. This is explained by the all-or-none law and can be demonstrated by dissecting out a single muscle fiber and subjecting it to repeated stimuli of increasing intensity, starting with those too weak to cause contraction. Response occurs only when a certain level of stimulus strength is reached, and at that time the fiber will contract completely. Stimuli of even greater intensity do not cause any greater contraction. The nature and amount of contraction of an entire muscle depend upon the number of fibers that are contracting and upon their contracting simultaneously or alternately.

When a muscle is given a single stimulus, such as a single electric shock, it responds with a single, quick twitch that lasts about 0.05 second in a human muscle and about 0.1 second in a frog muscle. A single twitch of a frog muscle consists of three periods: (1) a latent period (0.01 second), or the interval between stimulus application and the start of visible contraction; (2) the contraction period (0.04 second), during which the muscle contracts and does work; and (3) the relaxation period (0.05 second), during which the muscle relaxes (returns to its original length). After a twitch the muscle uses oxygen and gives off carbon dioxide and heat at a rate greater than that during rest, suggesting a recovery period (several seconds) in which the muscle is restored to its original condition. If stimuli are repeated so that successive contractions occur before the muscle has recovered from the previous stimuli, the muscle becomes fatigued, and the twitches grow fewer and finally stop. Sufficient rest will allow a fatigued muscle to regain its ability to contract. However, normal contractions of muscles do not occur as single twitches but as sustained contractions caused by a series of nerve impulses (stimuli) reaching them in rapid succession. Such sustained contraction is called tetanus (Gr. *tetanos*, stretched), and the stimuli occur so rapidly that relaxations cannot occur between successive contractions. In most normal contractions the various muscle fibers are stimulated in rotation, so that although individual fibers contract and relax, the muscle as a whole remains contracted partly. In the weak contraction of a muscle only a small percentage of the contained fibers receives nervous stimuli, whereas in a stronger contraction, more fibers contract at the same time. If a muscle has contracted many times, it will exhaust its supplies of glycogen and organic phosphates and produce lactic acid, so that it can contract no longer.

Visceral muscles

Many unnamed, mononucleated, smooth muscles are part of the internal organs, such as the esophagus, stomach, and intestines. These muscles are involuntary, have a rather slow, rhythmic rate of action, and do not fatigue easily. The churning and mixing movements of the digestive system, called peristalsis, and the rhythmic contractions of the uterus during childbirth are examples of their activity.

Cardiac muscle

Numerous indistinctly striated, branched, cardiac muscles compose the walls of the heart and blood vessels. Because of the branches or connections between some adjacent cells, there may be more than one nucleus per cell. Cardiac muscles are involuntary, have a variable rate of action, and under normal conditions do not fatigue easily.

Review questions and topics	
	1 What are the systems of the human body and their functions?
	2 Describe the integument of man and its various functions.
	3 Describe the human skeleton and its functions.
	4 Define the following bones: sesamoid, wormian, patella, phalanges, hyoid, pelvic girdle.
	5 Discuss the band structure of striated muscles, including changes during contraction.
	6 Explain how energy is released when muscles perform.
	7 Explain the all-or-none law and its significance.
	8 Discuss the periods when a muscle responds to a single stimulus.
	9 Describe the following types of muscles: skeletal, visceral, and cardiac.
	10 Define: myofibril, fatigue, myosin, actin, adenosine triphosphate (ATP), motor end-plate, "action potential," tonus, tetanus, and sliding-filament phenomenon.

Biology of man — metabolism

The material presented in this chapter deals with the "logistics" of living. While it must be emphasized that no living process takes place in a vacuum—that the interaction of all systems is what constitutes life—for convenience we will consider only certain systems here.

These are, in general, the various systems that deal with the obtaining of food, its processing, distribution and use, and related activities that contribute to or result from this energy cycle. The following areas will be discussed: (1) foods and nutrition, (2) respiration, including breathing and cell respiration, (3) excretion, and (4) circulation and related areas.

FOODS AND NUTRITION

In the strict sense a food is a substance that can be broken down by the organism to release energy. This definition would apply primarily to fats, carbohydrates, and proteins, and their related substances. A broader definition would include other materials not synthesized by the cells but still necessary for metabolism. Water, vitamins, and inorganic chemicals would be part of an expanded list of basic nutrients.

DIGESTION

Because of their size and chemical make-up, most large organic molecules cannot be absorbed until they have been digested. They are then absorbed in the form of smaller molecules. Often these same small food molecules are introduced through the veins of a person too weak to eat solid food. It is important to note that the actual molecules available to the cells are the same after digestion as those introduced in intravenous feeding.

Digestion, as such, is primarily a process of enzymatic hydrolysis, resulting in a change of the size of large organic food molecules. This results, then, in the splitting of food molecules by water under the influence of the appropriate enzyme.

Digestive system

The digestive system in man is basically a long tube, of varying diameter, that provides the proper environment for the chemical change of food molecules. It consists of the mouth, esophagus, stomach, duodenum, small intestine, colon or large intestine, rectum, and anus. The liver, gallbladder, and pancreas may be considered as associated digestive structures (Fig. 26-1).

The ducts of the three pairs of salivary glands open into the mouth and provide saliva, which contains enzymes important in carbohydrate digestion. After food has been swallowed, it passes through the esophagus into the stomach. The rate of movement depends on the fluidity of the food and the action of peristalsis, the squeezing activity of smooth muscle. Anyone who has taken a cold drink on a hot summer day is aware of the speed with which liquids pass through the esophagus and into the stomach.

The stomach is an expanded, sacklike structure with its larger area at the anterior end. It tapers down to form a constriction at the pylorus, the site of the pyloric sphincter. The glands of the stomach secrete gastric juices containing hydrochloric acid (HCl) in ion form, and enzymes, in an inactive state. The primary action in the stomach is that of protein digestion. It is interesting to note that the enzymes are active only at the low pH provided by the H^+ ions. Contraction of the smooth muscles in the stomach wall provides a vigorous churning or mixing of the food and gastric juices. Waves of contraction push the food masses forward toward the pylorus, which periodically opens, and the semiliquid food, called chyme, squirts through the valvelike sphincter and into the duodenum. The

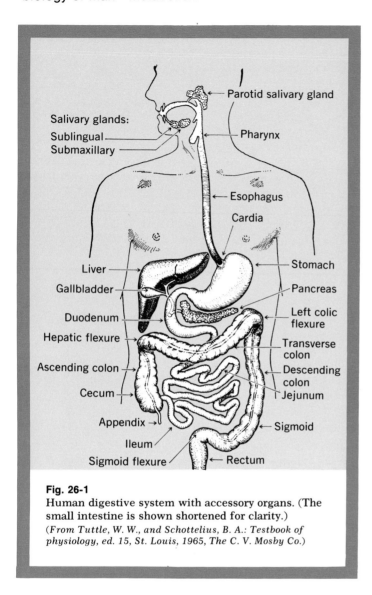

Fig. 26-1
Human digestive system with accessory organs. (The small intestine is shown shortened for clarity.)
(*From Tuttle, W. W., and Schottelius, B. A.: Testbook of physiology, ed. 15, St. Louis, 1965, The C. V. Mosby Co.*)

area; these are then digested to produce glycerin and fatty acids. All of these are absorbed by cells lining the small intestine. The rate of absorption varies with the size and shape of molecule and heat of food.

Unabsorbed materials in a fluid state pass into the large intestine where water is absorbed and the residue concentrated as the feces. Bacterial activity may produce some vitamin precursors at this time. The entire residue then moves into the rectum and is later eliminated by defecation.

Control of digestion

Before discussing details of the digestive process, we shall consider another important factor—the control of the entire process.

The common sensation we have on entering a bakery—the pleasurable smell of cakes and pastry—results in a flow of saliva into the mouth. We are, in a sense, ready to begin eating. This same flow of saliva may be produced by such things as the sight of food, the odor of food, and even the memory of food. We might add to this, perhaps, the sound of the "sizzle" of a steak frying, and of course, the contact of food in the mouth.

These examples emphasize the variety of factors that may initiate the flow of saliva. This secretion is controlled by nerve impulses from the brain. As food is being chewed, other nerve impulses stimulate the flow of gastric and pancreatic juices.

As food is swallowed and actually enters the stomach, the stomach lining is stimulated to release into the blood a hormone, gastrin, which then influences further gastric secretion. Contact by the food entering the duodenum helps stimulate the flow of intestinal juices. The low pH of food entering the duodenum activates a hormone, secretin, which passes through the blood to the pancreas, which then releases pancreatic juices. The same hormone may stimulate the liver to release bile. Previously secreted bile has been stored in the gallbladder. The presence of fats in the food acts as the stimulus for the release of another hormone, cholecystokinin, which is carried by the blood to the gallbladder, which contracts to release stored bile. Finally, the passage of foods further along the intestine initiates the release of still another hormone, enterogastrone, whose

chyme is mixed with fluids from the intestinal mucosa, liver, gallbladder, and pancreas. Fluids from the pancreas and gallbladder enter through the common bile duct.

In the duodenum and small intestine the process of digestion is finally completed. Carbohydrate digestion, started in the mouth, proceeds to the point of producing simple sugars. Protein digestion, started in the stomach, is completed and results in amino acids. Fat is emulsified by bile, producing smaller particles with a larger surface

Table 26-1

Summary of digestive processes in man

Region	Secretion and source	Enzyme	pH	Substances acted upon	Products formed
Mouth	Saliva (from salivary glands)	Amylase	Neutral	Starches	Maltose
Stomach	Gastric juice (from gastric glands)	Pepsin*	Acid	Proteins	Polypeptides
		Rennin (in infants)	Acid	Milk proteins (caseins)	Curdled milk
Small intestine	Pancreatic juice (from pancreas)	Amylase	Alkaline	Starches	Double sugars
		Lipase	Alkaline	Fats	Fatty acids and glycerol
		Trypsin	Alkaline	Proteins and polypeptides	Peptides and amino acids
	Intestinal juice (from glands of small intestine)	Peptidase	Alkaline	Peptides	Amino acids
		Maltase	Alkaline	Maltose (malt sugar)	Glucose
		Lactase	Alkaline	Lactose (milk sugar)	Glucose and galactose
		Sucrase	Alkaline	Sucrose (cane sugar)	Glucose and fructose
		Enterokinase (not a true digestive enzyme)†	Alkaline	†See footnote	†See footnote

*When acting with hydrochloric acid (HCl).
†Activates the inactive trypsinogen of the pancreatic juice, changing it to active trypsin. (See information under Substances acted upon and Products formed.)

action is the inhibition or turning off of the gastric phase of activity.

The process of digestion is controlled by a variety of psychologic, nervous, and chemical factors, all of which provide the proper digestive fluids in the order of their activity, finally stopping the activity when no more food is present.

As we are aware, the malfunction of any of these processes may lead to complications, such as gastric ulcers.

Digestive process

Digestion is primarily the result of enzymatic hydrolysis. This may be shown by the following scheme:

$$\text{Foods} + \text{H}_2\text{O} \xrightarrow[\text{pH, ions}]{\text{Enzyme}} \text{Absorbable products}$$

Table 26-1 summarizes the digestive processes. The following scheme shows the general outlines of digestion of the primary foods.

353

$$\text{Complex carbohydrates} + H_2O \xrightarrow{\text{Amylase}} \text{Disaccharides}$$

$$\text{Disaccharides} + H_2O \xrightarrow{\text{Disaccharases}} \text{Monosaccharides}$$

$$\text{Proteins} + H_2O \xrightarrow{\text{Pepsin}} \text{Polypeptides}$$

$$\text{Polypeptides} + H_2O \xrightarrow{\text{Peptidases}} \text{Amino acids}$$

$$\text{Fats} + H_2O \xrightarrow{\text{Lipase}} \text{Fatty acids} + \text{Glycerol}$$

Fate of absorbed foods

With the exception of fats, which pass into the lacteals and lymphatic circulation, most absorbed foods move into the capillaries of the intestinal villi and then pass by way of the hepatic portal system to the liver. In this, the largest organ of the body, hundreds of metabolic activities take place.

Among these processes are: (1) glucose-glycogen activities; (2) amino acid breakdown with the formation of urea; (3) storage of proteins, minerals, and vitamins; (4) formation of plasma proteins, such as albumin and fibrinogen; and (5) some fatty acid metabolism.

Carbohydrate metabolism

The liver is the primary storage site of carbohydrate in the body. The formation of glycogen serves both to store glucose and to regulate its content in the blood. The amount of glucose normally present in the circulating blood is approximately 0.1%. A higher concentration is present in the blood leaving the intestine immediately following a meal. As this quantity of glucose reaches the liver, much is taken out of circulation and converted to glycogen. The process is controlled by insulin, a hormone secreted by the pancreas. When the supply of circulating glucose is low, glycogen breakdown occurs, and the level is raised. In the stress reaction the hormone adrenalin promotes the breakdown of glycogen and leads to an increased supply of glucose in the circulating blood.

Diabetic patients have abnormal quantities of insulin in the blood. If too little is present, glycogen is not formed, and glucose accumulates in the blood and is excreted through the kidneys. When too much insulin is present, the supply of circulating glucose drops as it is converted to glycogen.

This produces a type of cell "starvation" and may result in the person going into a coma.

In general, carbohydrates (1) form the principal energy source available to cells, (2) form a storage product, and (3) supply a basis for synthesis of such substances as nucleotides. When carbohydrates are not available, fats and proteins may break down to form glucose.

Amino acid metabolism

Amino acids are not stored as such in the body, but they are required in the daily diet where their chief function is the formation of proteins. These include enzymes, antibodies, tissue proteins, blood proteins, and cell membranes.

As amino acids are used in metabolism, the amino group, $-NH_2$, is removed and forms ammonia (NH_3), a potential poison. Detoxification occurs as the $-NH_2$ is converted to urea, which is then excreted by the kidneys.

Amino acids may also be utilized as food and converted to fat or carbohydrates.

Fat metabolism

Fats form an important component of all cell membranes, being part of a lipoprotein complex. They may be readily converted to glucose and undergo respiration, hence, they are a major storage food. When other kinds of foods are present, they may be converted to fats and stored. Thus, any type of excess food may be stored as fat.

■
Mineral salts

Mineral salts are inorganic substances that, in combination with other food constituents, promote the formation and maintenance of various body structures and functions. Minerals not only aid in metabolic processes but also actually help to form certain body tissues and fluids. The daily excretion in feces, urine, and perspiration of about 30 gm. of mineral salts must be replaced. The absence of the proper amount of minerals in a diet is more serious than lack of the proper amounts of carbohydrates, fats, or proteins and when prolonged, causes the use of the body reserve of salts, with possibly serious consequences. Fortunately, many common foods are rich in minerals, but actual deficiencies of calcium, iron, and iodine may occur at times.

About fifteen elements are known to be essential as mineral salts in the human diet, some being required only in traces.

In addition to the formation of muscles and bones, minerals are important as the ions that are involved in various activities, such as respiration, heartbeat, and nerve impulse transmission.

Water

About two-thirds of the human body is water. It is the medium in which chemicals are dissolved and in which chemical reactions occur. Water is essential for the separation of the molecules of carbohydrates, fats, and proteins during digestion. It helps to form the blood plasma and lymph. Water equalizes body heat and removes metabolic wastes. Over 2 quarts of water are lost daily, the specific amount changing with such variables as climate and activities. This loss must be replaced by liquid foods and water-containing foods, such as fruits and green vegetables.

Vitamins

Vitamins are organic compounds that are essential for a number of living processes. They differ chemically and cannot be formed by the body cells, although bacteria in the intestine may synthesize certain types. It was found in 1912 that animals could not survive on a diet of purified carbohydrates, fats, and proteins alone, but that "accessory growth factors" were essential. Vitamins were originally named because they were considered to be amines ($-NH_2$) that were essential to life (L. *vita*, life). Some vitamins, originally considered as a single unit, are now known to be complexes of vitamins, for example, original vitamin A (now divided into A, D, and E) and vitamin B (divided into about a dozen kinds).

Lunin (1888), in Switzerland, fed mice a diet of carbohydrates, fats, proteins, and inorganic salts that he had isolated or synthetically prepared in the laboratory. The quantities of these ingredients were thought to be in the same proportions as those in milk. When mice were fed this diet, they died, but when milk was added to the prepared diet, the mice lived. He concluded that milk contained an unknown substance essential to life.

In 1913, McCollum and Davis, and Osborne and Mendel experimentally proved the presence of a fat-soluble dietary substance in butterfat that promoted growth and well-being of rats. The substance was first called "fat-soluble A" factor.

Probably all plants and animals require vitamins of some types, but their specific requirements vary. Vitamins serve as integral parts of coenzymes for the fundamental enzyme reactions in all protoplasm. Plants normally synthesize vitamins, but animals do not ordinarily. Following is a brief account of important facts concerning vitamins.

Vitamin A

Important sources and characteristics

Vitamin A is found primarily in fatty, oily foods of animal origin, hence is fat-soluble. Examples are fish-liver oils, cream, butter, milk, and egg yolk. We derive much of our vitamin A from a pigment, carotene, found in green-, red-, or orange-colored plants. The beta type of carotene ($C_{40}H_{56}$) is a water-soluble precursor, or provitamin, which is transformed into vitamin A chiefly by the liver. Vitamin A is not affected by light or heat but is destroyed by ultraviolet light or oxidation.

Functions and effects of deficiencies

Vitamin A maintains normal functioning of epithelial tissues (skin, eyes, eyelids, and digestive and respiratory systems). The insoluble proteins of the most superficial layer of the skin become transformed into a special hornlike, protective form known as keratin. Abnormal keratinization impairs the function of the involved tissues. When sweat and oil glands are affected, the skin becomes dry, scaly, and pimpled (acne).

Night vision (dim light vision) depends upon the presence in the rods (of the eye retina) of a pigment called visual purple. When stimulated by light, the rods and cones generate nerve impulses, which create a sensation of light in the brain. Bright light entering the eye quickly bleaches the visual purple, which accounts for the temporary blindness when we go from bright light into dim light; but in reduced light the visual purple normally regains its color quickly. This is known as dark adaptation and permits seeing dimly lighted objects that were temporarily invisible. Dark adaptation depends on

Fig. 26-2
Vitamin A deficiency in white rats.
The animals were litter mates, 21 days
old at the start of the experiment. For
33 days the animal at left received a
diet containing all nutritive substances
except vitamin A, whereas the animal
at right received all nutritive
substances plus a sufficient amount
of vitamin A: Note the eye symptoms
(xerophthalmia) in the vitamin
A-deficient rat.
*(Courtesy Upjohn Co., Kalamazoo,
Michigan; from Tuttle, W. W., and
Schottelius, B. A.: Textbook of physiology,
ed. 15, St. Louis, 1965, The C. V. Mosby
Co.)*

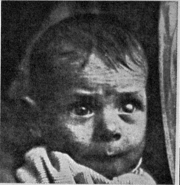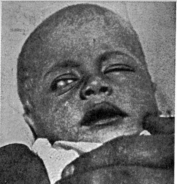

Fig. 26-3
Xerophthalmia, an eye disease caused by a dietary
deficiency of vitamin A; the eye becomes dry and
a layer of horny tissue forms upon the cornea.
*(From Harris: Vitamins in theory and practice; courtesy
Cambridge University Press and The Macmillan Company.)*

vitamin A so insufficient vitamin A may result in night blindness. Vitamin A is one of many factors associated with growth and a deficiency may lead to increased susceptibility to infections, especially of the respiratory tract.

Vitamin B complex

Of the numerous water-soluble vitamins grouped together as the vitamin B complex, only a few are known to be of importance to man. The B complex vitamins are of fundamental importance to life, being found in all living organisms. They are as-

sociated with metabolic processes throughout the body, as is shown by the fact that the amount of these vitamins required is largely determined by the amount of energy expended in work and the maintenance of body temperature.

Generally speaking, the absence of these vitamins may lead to lowered general well-being, increased fatigue, and emotional disturbances, such as depression and abnormal irritability. The B complex vitamins play important roles as coenzymes in the many successive chemical changes in the glucose and glycogen molecules. Since these

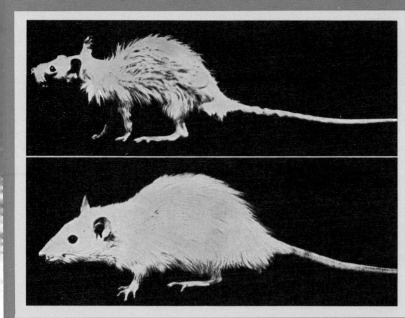

Fig. 26-4
Results of riboflavin (vitamin B_2) deficiency. **A,** Rat, 28 weeks old, received no riboflavin and weighed 63 gm.; **B,** after receiving food rich in riboflavin for 6 weeks the same rat weighed 169 gm.
(Courtesy Bureau of Human Nutrition and Home Economics, U. S. Department of Agriculture.)

chemical changes occur in all types of cells and are necessary for life, their absence is of great importance. The following vitamins of the B complex will be considered: thiamine, riboflavin, niacin, pyridoxine, pantothenic acid, biotin, folic acid, choline, and vitamin B_{12}.

Thiamine (B_1)
Important sources and characteristics

Good food sources of thiamine, or vitamin B_1, include cereal grains, beans, peas, nuts, wheat germ, and pork. Many additional foods are now enriched to supply vitamins and necessary minerals. Vitamin B_1 has an odor somewhat like yeast and a salty, nutlike taste. It is soluble in water and is not destroyed by cooking, but the cooking water may contain much vitamin B_1 that was originally present in the food.

Functions and effects of deficiencies

Vitamin B_1 acts within living cells as an essential part of carbohydrate metabolism. As the body converts foods into energy through chemical means, it employs enzymes, which are essential to the process. Vitamin B_1 is converted to cocarboxylase and acts as a part of one of these essential enzymes. In vitamin B_1 deficiency not all of the carbohydrate

is used, and pyruvic acid builds up in the body.

Vitamin B_1 is not stored in quantity in the body, hence a daily intake is necessary. The body takes in only the amount needed at the time, and any excess is destroyed or excreted. Serious, prolonged deficiency of vitamin B_1 leads inevitably to beriberi (multiple peripheral neuritis). Symptoms include inflamed peripheral nerves, which become painful and degenerate, heart changes, muscle weakness, and swellings caused by excess fluids. The disease can be fatal.

Partial deficiency may cause loss of appetite, fatigue, poor sleep, irritability, and digestive system disorders.

Riboflavin (B_2)
Important sources and characteristics

Vitamin B_2, or riboflavin, is slightly water soluble and has a bitter taste. Good food sources include milk and its products, eggs, meats, legumes, green leaves, and whole-grain cereals. Riboflavin is essential to life. We have no special storage tissues in our bodies for riboflavin, yet a certain level is maintained in various places, especially in the liver and kidneys. It is not destroyed by heat, but exposure to light for 1 hour destroys about 40% of riboflavin in milk.

357

Functions and effects of deficiencies

Riboflavin is a vital link in nature's chain of reactions for using carbohydrate energy. It is a constituent of many enzyme systems and thus is intimately connected with essential life processes. It is required by the metabolic processes of mammals, birds, fishes, insects, and lower forms of life. In certain animals the requirement of riboflavin may be met by synthesizing it through the activities of bacteria within the intestine.

Within cells riboflavin is attached to a phosphate group, where it is known as riboflavin-5-phosphate, or flavin mononucleotide. This in turn may be attached to still another essential substance, adenylic acid, forming flavin-adenine dinucleotide (FAD). Either nucleotide then is attached to a protein, forming an enzyme, and takes part in important oxidation-reduction reactions.

Deficiencies of riboflavin occur in man in several ways. Symptoms include inflammation of eyes and skin, and nerves and blood are also affected. In riboflavin-deficient rats there is stunted growth, hair loss, inflamed and cataractous eyes, and even death (Fig. 26-4).

Nicotinic acid (niacin)

Important sources and characteristics

Nicotinic acid is another B complex vitamin. It is not destroyed by boiling, but it may dissolve in cooking water. It is said to be formed by bacterial activity within the intestine. Good food sources include yeast, roasted chicken, beef liver, tuna fish, halibut, peanuts, lean meats, poultry, and enriched flour.

Functions and effects of deficiencies

Nicotinic acid deficiency gives rise to pellagra ("skin seizure"), a disease characterized by lack of vitality and strength, loss of appetite, indigestion, diarrhea, pain, skin eruptions, and sometimes severe mental disturbances. Formerly, the cause of these symptoms was thought to be nicotinic acid deficiency alone, but now it is known that a lack of thiamine, riboflavin, and possibly other B vitamins are also involved, thus making pellagra a multiple-deficiency disease.

A diet lacking the essential amino acid tryptophan produces effects similar to those caused by nicotinic acid deficiency. Tryptophan can also be substituted for nicotinic acid in abolishing most of the pellagra symptoms. Although milk and eggs are poor sources of nicotinic acid, they are excellent pellagra preventives because of their supply of tryptophan.

Combined with a highly complex phosphorus compound, nicotinic acid forms nicotinamide-adenine-dinucleotide (NAD) and nicotinamide-adenine-dinucleotide-phosphate (NADP), often called coenzymes I and II, which play important roles in carbohydrate catabolism.

Pyridoxine (B$_6$)

Important sources and characteristics

Pyridoxine is another B complex substance that seems to be essential for animals. Good food sources include liver, eggs, fish, lettuce, celery, lemon, whole wheat, and milk.

Functions and effects of deficiencies

Many ill effects have been attributed to the lack of vitamin B$_6$, such as growth impairment in young experimental animals and body weight loss in adults. Skin lesions are common, with anemia in some instances. This vitamin plays an important part in the chemical reactions involving amino acids, apparently acting as a coenzyme. There is much controversy concerning the value of this vitamin in treating illness.

Pantothenic acid

Important sources and characteristics

Pantothenic acid is present in all living cells. Common sources include eggs, meat, liver, fresh vegetables, sweet potatoes, cane molasses, and milk. If diets contain sufficient eggs, meat, and milk, the requirements of this substance apparently are met.

Functions and effects of deficiencies

Pantothenic acid functions as one of the important coenzymes of metabolism (coenzyme A of the Krebs cycle; see Fig. 26-12). It seems to be necessary for normal skin and nerves, and its presence is probably needed for many fundamental chemical reactions in cells. Even though many disturbances in lower animals are attributed to its lack, none have been definitely established for man.

Biotin

Important sources and characteristics

Biotin is a substance that plays important roles in living processes. Good sources include liver, kidney, eggs, milk, and most fresh vegetables. Balanced human diets supply sufficient amounts of biotin.

Functions and effects of deficiencies

Only a most unbalanced diet would evoke such symptoms of biotin deficiency as skin and tongue lesions. Its lack causes dermatitis in chicks and rats. Biotin, acting as a coenzyme, takes part in many enzyme reactions, including carbon dioxide utilization, deamination (splitting of the amine group [NH$_2$] from the amino acid), and oxidation of intermediate products in protein catabolism. It is involved also in carbohydrate and fat metabolism.

Folic acid

Important sources and characteristics

Folic acid occurs in plant and animal tissues. Good sources include liver, asparagus, broccoli, spinach, yeast, soybeans, and egg yolk.

Functions and effects of deficiencies

Folic acid is usually formed in sufficient amount in the human intestine to meet daily needs. It is necessary for the growth and formation of blood cells and for certain metabolic processes. Lack of folic acid causes anemia in man and, in chickens, such conditions as kidney hemorrhages and bone deformities.

Choline

Important sources and characteristics

Choline is a complex nitrogenous substance that performs a number of vital functions.

Functions and effects of deficiencies

Choline aids in fat storage and usage. When absent, the liver becomes filled with fat. Choline also aids in forming phospholipids (phosphorus-containing fat). Lecithin, a common phospholipid in the body, is formed when one of three fatty acids is replaced by choline and phosphoric acid in a molecule of a neutral fat (such as palmitin). Lecithin is essential in cells, where it decreases surface tension, causing it to adhere to the protoplasm surface. It forms a constituent of the plasma membrane and contributes to its selective permeability—a highly important phenomenon in cell physiology.

Choline is used in the construction of acetylcholine, a compound formed by the union of choline and acetic acid. Acetylcholine plays roles in nerve impulse transmission, skeletal muscle contraction, and cardiac inhibition.

Cobalamin (B$_{12}$ group)

Important sources and characteristics

Cobalamin is sometimes referred to as the "erythrocyte maturation factor." It is unique in that it contains the element cobalt. It is manufactured by certain fungi within the intestine. Common sources include beef liver, milk, egg yolk, oysters, and yeast.

Functions and effects of deficiencies

Cobalamin serves as a factor in forming and maturing red blood corpuscles. Its lack causes pernicious anemia, poor growth, and wasting of tissues.

Ascorbic acid (vitamin C)

Important sources and characteristics

Ascorbic acid is a water-soluble substance that has a slightly acidic taste. Since it is able to "attach" itself to oxygen, it is very important as an antioxidant, being used for certain foods that have a tendency to discolor and lose their natural flavor (for example, frozen peaches, apples, bananas, pears, and pineapples). Common sources include citrus fruits, tomatoes, potatoes, cabbage, and spinach.

Functions and effects of deficiencies

We cannot live without vitamin C. It is essential for maintaining healthy bones and for forming collagen, an important protein of the skin, tendons, cartilage, and connective tissues, where it synthesizes "cell cement." It is required for the normal functioning of blood vessels and also promotes healing of wounds and fractures. It is essential for tissue respiration, in which foods are burned to provide energy and heat for the body.

359

Lack of vitamin C causes scurvy, or scorbutus, in man and certain other animals. Scurvy symptoms include great fragility of capillary walls, fracture of bones, swollen, spongy, bleeding gums, loose teeth, painful joints, and slow healing wounds.

Vitamin D (calciferol)

Important sources and characteristics

Vitamin D is a fat-soluble vitamin that is very limited in its distribution. It is never found in natural foods of plant origin. Normal diets probably contain very little, if any, of it. It is quite stable, withstanding boiling, and is not oxidized readily. The most abundant sources are liver oil and body oils of such fishes as salmon, halibut, mackerel, cod, and swordfish. Egg yolk may contain some vitamin D, but the amount depends on the irradiation of the hen and the amount of vitamin in her food.

It is doubtful that man needs much vitamin D in foods because the irradiation of his skin (by ultraviolet light) changes the precursor (ergosterol) in the skin into this vitamin.

Functions and effects of deficiencies

Vitamin D regulates calcium and phosphorus metabolism. Its lack causes rickets (Fig. 26-5), characterized by soft bones, bowed legs, beaded ribs, defective teeth, and enlarged wrists, knees, and ankles.

Vitamin E (fertility vitamin)

Important sources and characteristics

Vitamin E is a fat-soluble substance that is abundant in vegetable oils. Common sources include green leaves, wheat germ, yeast, vegetable oils (corn, cottonseed, soybean, and peanut), meats, and eggs.

Functions and effects of deficiencies

Vitamin E is considered necessary for cell maturation and differentiation in vertebrates and for nuclear growth and activity. It is doubtful if this vitamin normally influences the reproductive functions in man. Male rats that have been experimentally fed with sufficient supplies of other known vitamins but deprived of vitamin E become sterile, their germ-producing seminiferous tubules

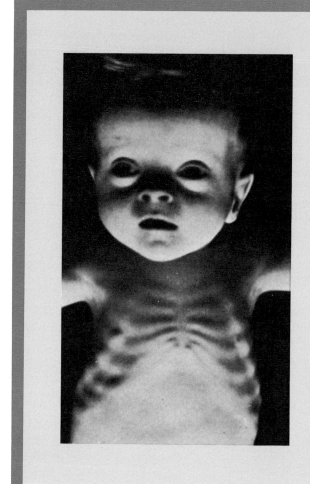

Fig. 26-5
Rickets disease caused by vitamin D deficiency, showing the square head and enlarged junctions between the ribs and rib cartilages.
(*From Jeans, P. C., and Marriott, W. McK.: Infant nutrition, ed. 4, St. Louis, 1947, The C. V. Mosby Co.*)

are destroyed, and their sperms perish. However, in female rats the ovaries remain normal, but embryos die a few days after fertilization. These effects are avoided by feeding yeast or wheat germ, the beneficial effects being attributed to vitamin E.

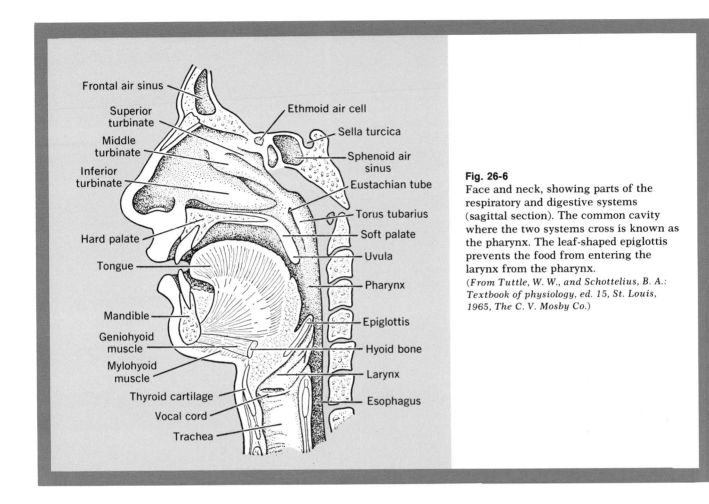

Fig. 26-6
Face and neck, showing parts of the respiratory and digestive systems (sagittal section). The common cavity where the two systems cross is known as the pharynx. The leaf-shaped epiglottis prevents the food from entering the larynx from the pharynx.
(*From Tuttle, W. W., and Schottelius, B. A.: Textbook of physiology, ed. 15, St. Louis, 1965, The C. V. Mosby Co.*)

Vitamin K (antihemorrhagic vitamin)

Important sources and characteristics

Vitamin K is a fat-soluble compound, and its absorption into the blood is dependent upon bile, as is true of fatlike substances in general. Common sources include green leaves, spinach, soybean oil, kale, liver, and egg yolk.

Functions and effects of deficiencies

Vitamin K assists in prothrombin formation in the liver, necessary for blood coagulation. In obstructive jaundice, if bile fails to enter the intestine, vitamin K cannot be absorbed. Man and certain animals do not need the vitamin in foods, but it is manufactured by bacterial action within the intestine. A lack of vitamin K results in an increased tendency for hemorrhages. Vitamin K is sometimes given before surgical operations and to pregnant women shortly before delivery.

RESPIRATION

The term "respiration" has been used in several ways to indicate some feature of the overall process of oxygen activity in the organism. We will modify its use and describe the entire process under the following headings: (1) breathing, the movement of air from the atmosphere into the lungs; (2) gas exchange, the movements of oxygen and carbon dioxide between the lungs and the blood, and between the blood and the tissues; (3) gas transport, the various processes involved in carrying these gases; and (4) cellular respiration, the immediate activity of individual cells in what

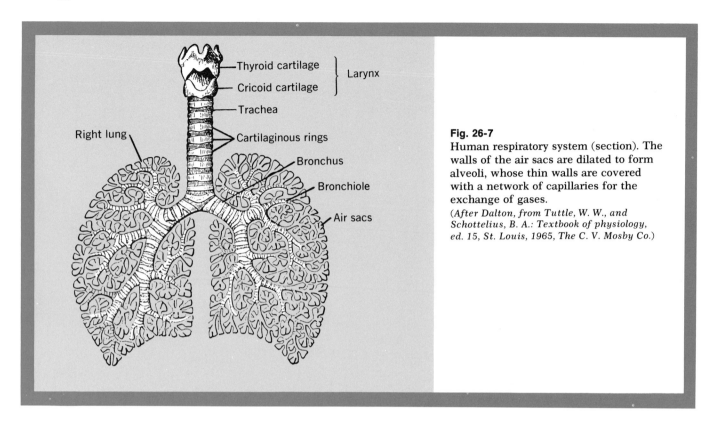

Fig. 26-7
Human respiratory system (section). The walls of the air sacs are dilated to form alveoli, whose thin walls are covered with a network of capillaries for the exchange of gases.
(After Dalton, from Tuttle, W. W., and Schottelius, B. A.: Textbook of physiology, ed. 15, St. Louis, 1965, The C. V. Mosby Co.)

is called aerobic respiration. This last process will be considered later in some detail.

Breathing system

The breathing system is composed of the nose, pharynx, larynx (voice box or Adam's apple), trachea (windpipe), bronchi, and lungs (Figs. 26-6 and 26-7).

The nose is divided by a partition (septum) to form two wedge-shaped cavities, lined by a highly vascular, mucous membrane, the upper layer of which is ciliated. Sinuses in the bones are associated with the nasal cavities, so that inflammations may spread to the sinuses easily. The lateral surface of each nasal cavity has three light, spongy, bony projections called conchae, which make the upper part of the nasal passages very narrow. The nose has the following functions: (1) to remove dust and other foreign materials by hair, cilia, and mucus (secreted by goblet cells), (2) to give warmth and moisture to the inhaled air, (3) to detect odors

by means of the olfactory nerve endings in the upper passages, and (4) to act as a sounding board for the voice.

The pharynx is a common cavity that connects the nasal cavities with the larynx and the mouth with the esophagus. This is the site of the sore throat associated with a cold that is often caused by breathing unfiltered air through the mouth.

The larynx is a cartilaginous box that forms the prominence in the midline of the front part of the neck. Within the laryngeal cavity are two folds of mucous membrane, extending from front to back but not quite meeting in the middle. Embedded in the edges of these folds are fibrous and elastic ligaments that constitute the vocal cords, because they function in voice production as air passes between them. Above the vocal folds are two smaller folds that do not aid in voice production but protect the larynx during swallowing, help keep the true vocal folds moist, and assist in holding the breath. The opening between the true vocal folds is the glottis. The size of the glottis and the tension of the vocal folds regulate the tone produced. The glottis is protected above by a leaf-shaped fibrocartilage, the epiglottis, which prevents food from entering the trachea. The trachea is a membranous tube about 4 inches long, located in front of the esophagus. The walls are strengthened by sixteen to twenty cartilaginous, C-shaped structures. It extends from the lower end of the larynx to the two branches, or bronchi. Each bronchus divides and subdivides, the smallest branches being the bronchioles. Each bronchiole terminates in a series of saclike alveoli. The thin-walled alveoli are surrounded by thin-walled capillaries through which the exchange of gases occurs (Fig. 26-8).

The lungs are cone shaped and lie in the thorax (Fig. 26-9), separated by the thick mediastinum, which contains the heart, larger blood vessels, and trachea. The left lung is smaller and longer than the right because the heart occupies part of this space. Each lung is enclosed in a serous sac called the pleura, consisting of an outer layer, or parietal pleura, which adheres closely to the diaphragm and the walls of the thorax, and a visceral pleura, which covers the lungs. The two pleurae are separated by a thin layer of serum to reduce

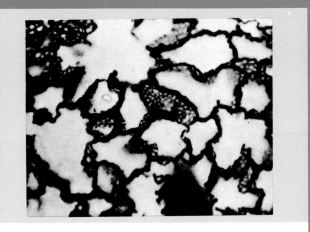

Fig. 26-8
Capillary network from the alveoli of the human lung.

friction. Inflammation of the pleura is called pleurisy.

Breathing process

Breathing may be defined as the rhythmic inhalation of air into the lungs and the exhalation of carbon dioxide and other gases from them. During inhalation the size of the thorax is increased by the

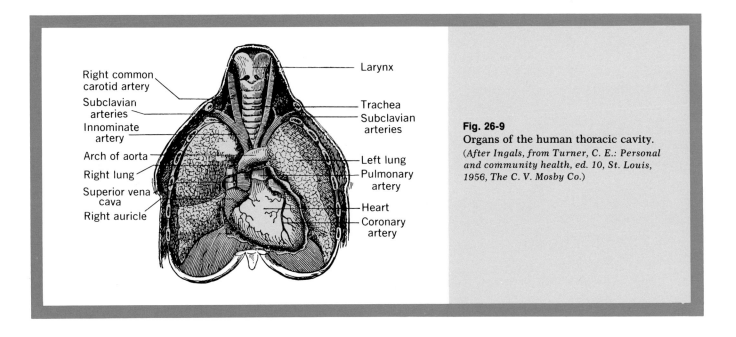

Fig. 26-9
Organs of the human thoracic cavity.
(After Ingals, from Turner, C. E.: Personal and community health, ed. 10, St. Louis, 1956, The C. V. Mosby Co.)

contraction of the breathing muscles, the diaphragm and rib muscles, thereby decreasing the pressure within the lungs and allowing the greater pressure of the external air to force air into the lungs until the pressures are equalized. The decrease in pressure actively enlarges the size of individual alveoli, much as a balloon is enlarged by air rushing into it. During exhalation the size of the thorax is decreased by the relaxation of these muscles, thus forcing out a quantity of the air from the lungs. About 500 ml. of air is moved with each breath. With a breathing rate of approximately sixteen times a minute, this means that an average of 8 liters of new air enters the lungs each minute.

An increase in the force of breathing results in moving more air per breath. The maximum amount of air involved is about 4,000 ml. and is called the vital capacity of the person. As with other physiologic activities, many factors influence the vital capacity.

The air eventually comes in contact with the alveolar lining, where the following take place: (1) loss of an estimated 5% of the oxygen from the inhaled air, (2) gain of about 4% of carbon dioxide, (3) gain of approximately 1% of nitrogen, (4) saturation of the expired air with moisture (about 1 pint daily), (5) warming the expired air to nearly that of the blood (98.6° F.), thereby losing body heat, and (6) transfer of oxygen and carbon dioxide through the thin-walled air sacs of the lungs (Table 26-2).

Table 26-2
Composition of inspired and expired air

	Inhaled air percent	Exhaled air percent
Oxygen	20.96	15.8
Carbon dioxide	0.04	4
Nitrogen	79	80.2

Breathing control

The coordination of both chemical and nervous stimuli regulates the breathing rate. In general, as the tissues are more active, they produce more carbon dioxide, which is carried by blood to the brain where certain cells of the medulla oblongata are stimulated. This causes an impulse to be transmitted via the phrenic nerve to the diaphragm and rib muscles, which then contract, enlarging the thoracic cavity. The resulting enlargement of the alveoli stimulates stretch receptors, and the impulse is transmitted by the vagus nerve back to the medulla, stopping the impulses of the phrenic nerve. The muscles relax, reducing the cavity, and we exhale. Seconds later the process is repeated.

Gas exchange

The result of the preceding activity is a change in gas concentration in the alveoli. Very simply, it is here that oxygen moves into the blood and carbon dioxide into the lungs. A similar diffusion takes place wherever there are differing gas concentrations. Thus every cell is a site of gas exchange. The following transfers are the result of diffusion involving high and low concentration of gases.

Blood entering the lungs through the pulmonary artery has previously passed over the tissues. As the blood passes through the capillaries, a system of differing concentrations of gases, separated by a membrane, is set up. The blood loses carbon dioxide to the alveoli and picks up oxygen from them. Thus there is a higher oxygen and lower carbon dioxide content in the blood leaving through the pulmonary veins and passing to the various tissues. As a result of respiration, there is a buildup of carbon dioxide and a utilization of oxygen in the various cells. Therefore the blood and cells provide another system of diffusion, with carbon dioxide moving out of the cells and oxygen moving into the cells. The blood then moves back to the lungs.

Gas transport

Oxygen is carried in the blood by hemoglobin in the red blood cells. The combination is known as oxyhemoglobin (HbO_2). Briefly, this combination is stable in places where there is a high concentration of oxygen, as in the capillaries of the lungs. It breaks down, releasing the oxygen in places where there is a low concentration of oxygen, as in the tissues. The rate of release of oxygen to the cells is therefore controlled by the rate of usage by the cells.

Red blood cells are also important in the transport of carbon dioxide from tissues to the lungs. Approximately one third of the carbon dioxide combines with hemoglobin, forming carboxyhemoglobin ($HbCO_2$) in much the same way that oxyhemoglobin is formed. The other two thirds of the carbon dioxide combines with water in the red blood cells, forming carbonic acid and then a bicarbonate ion. In the lung capillaries these compounds break down releasing carbon dioxide, which diffuses into the alveoli.

Cellular respiration

Respiration is the release of energy by the chemical breakdown of foods. This process takes place in the mitochondria, and the energy is trapped in molecules of adenosine triphosphate (ATP). The energy in ATP is then used by the cell in such activities as synthesis, secretion, contraction, and active transport. Molecules of the energy-rich ATP are generally formed and used within the same cell.

The source of the energy in the original foods is photosynthesis. The energy is passed through a series of processes involving plants, herbivores, and carnivores, finally becoming concentrated, in the high-energy phosphate bonds of ATP, by a series of biologic oxidations involving electron transfers.

From various investigations it is evident that the fundamental reactions in respiration in all cells are quite similar. The steps by which glucose is converted to pyruvic acid are the same in man, plants, and animals, and the remainder of breakdown reactions are similar if oxygen is available.

For convenience, the entire process of respiration will be considered under the three headings of glycolysis, pyruvic acid breakdown, and citric acid cycle.

Glycolysis

The breakdown of glucose, with the release of energy and formation of ATP, fits the preceding description of cell respiration.

The process begins with the phosphorylation, or addition of a phosphate group —Ⓟ (actually, —PO_3H_2), of glucose. It ends with the formation of pyruvic acid. During the glycolytic cycle (Fig. 26-10), the 6-carbon glucose has split, producing two 3-carbon triose phosphates (PGAL). Through a series of enzymatic reactions these 3-carbon molecules are oxidized by splitting off hydorgen, producing an energy-rich phosphate bond. The high-energy phosphate, ~Ⓟ, is now transferred to a molecule of adenosine diphosphate (ADP), forming the energy-rich ATP. The process of oxidation, high-energy ~Ⓟ formation, and transfer to ADP is repeated and ends in the formation of pyruvic acid.

It should be noted that in this cycle the following three things are outstanding: (1) phosphorylation, providing the —Ⓟ that will later become ATP; (2) oxidation and the removal of hydrogen from the triose, resulting in a high-energy bond; and (3) the formation of ATP, by transfer of this energy-rich ~Ⓟ to ADP.

Phosphorylation

By consulting Fig. 26-10 it can be seen that four separate phosphorylations take place. In the two at the 6-carbon level the source of the —Ⓟ is ATP. At the 3-carbon level the source is the inorganic —Ⓟ of the cell. Although all four of these phosphates will eventually be raised to the high-energy phosphate, ~Ⓟ, and form ATP, the net gain will be 2 ATP. The other two ATP molecules merely replace the starting molecules.

Oxidation

Of the several types of energy-rearranging reactions during respiration, two of these—dehydrogenation and dehydration—take place during glycolysis. This removal of hydrogen and water can be seen in Fig. 26-10. These and a third, decarboxylation (removal of carbon dioxide), also take place in the later stages of respiration.

The immediate result of oxidation is the removal of a food fraction (hydrogen, water, carbon dioxide) and the resulting formation of a high-energy bond.

ATP formation

As mentioned earlier, ATP is formed by the transfer of the high-energy ~Ⓟ from the oxidized

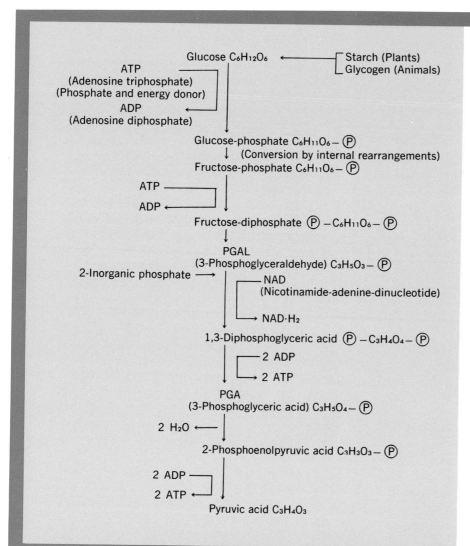

Fig. 26-10
Glycolytic (Myerhof-Embden) cycle, in which glucose (sugar) is converted anaerobically by enzymatic actions into pyruvic acid in living organisms. ⓟ represents $-PO_3H_2$ (phosphorylation).

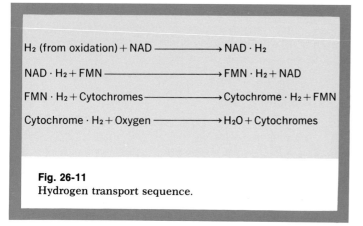

Fig. 26-11
Hydrogen transport sequence.

triose phosphate to ADP. A second, and more plentiful, source of ATP is available when the type of oxidation involves the removal of hydrogen. The hydrogen does not simply dissipate into the cell contents but combines temporarily with a series of hydrogen acceptors, which are then said to be "reduced."

The first of these reducing agents is a nucleotide derivative of the B vitamin nicotinic acid, which has been discussed previously in connection with photosynthesis. This carrier is nicotinamide-adenine-dinucleotide or NAD (formerly called TPN). As an acceptor in its reduced form it is $NAD \cdot H_2$.

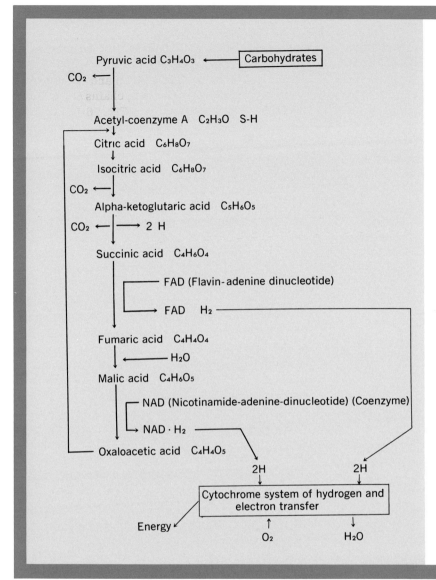

Pyruvic acid $C_3H_4O_3$ ← Carbohydrates

CO_2 ←

Acetyl-coenzyme A C_2H_3O S-H

Citric acid $C_6H_8O_7$

Isocitric acid $C_6H_8O_7$

CO_2 ←

Alpha-ketoglutaric acid $C_5H_6O_5$

CO_2 ← → 2 H

Succinic acid $C_4H_6O_4$

— FAD (Flavin-adenine dinucleotide)

→ FAD H_2 —

Fumaric acid $C_4H_4O_4$

← — H_2O

Malic acid $C_4H_6O_5$

— NAD (Nicotinamide-adenine-dinucleotide) (Coenzyme)

NAD · H_2 —

Oxaloacetic acid $C_4H_4O_5$

2H   2H

Cytochrome system of hydrogen and electron transfer

Energy ←   ↑   ↓
                  O_2 H_2O

Fig. 26-12
Krebs citric acid cycle and cytochrome system in which pyruvic acid is converted by enzymatic actions into CO_2 and H_2O, with liberation of energy (much simplified). ATP and pantothenic acid are used in making acetylcoenzyme A. Each cycle (from pyruvic to oxaloacetic acid) makes a total of 15 high-energy phosphate ATP molecules available.

In a cell that uses oxygen this hydrogen may be transferred further through a "bucket brigade" of acceptors. The final acceptor is oxygen, and the result of the final transfer is water.

These acceptors include a second nucleotide, flavin mononucleotide (FMN) or its dinucleotide (FAD), and the cytochrome series mentioned with photosynthesis. The entire sequence may be summarized as in Fig. 26-11.

It should be noted that after each transfer the previous acceptor is freed and may again function as an acceptor or reducing agent for further oxidation. Also, and of more importance at this time, for every hydrogen molecule transferred three molecules of ATP are formed. Each transfer has served as an additional oxidation and produced more energy that is used to form more high-energy ~Ⓟ.

The net energy available as a result of glycolysis is thus seen to be 8 ATP, two from the oxidation of the triose and six from the hydrogen-transfer sequences.

Pyruvic acid breakdown

The result of the sequence of pyruvic acid breakdown is the formation of carbon dioxide and a 2-carbon fraction, combined with a coenzyme,

367

called acetyl-coenzyme A. The process involves many steps and, as in glycolysis, depends on oxidation, hydrogen acceptors, nucleotides, and vitamin derivatives, as well as other cell components. Coenzyme A is a derivative of the B vitamin, pantothenic acid.

Carbon dioxide is formed by the breakdown of the 3-carbon pyruvic acid ($C_3H_4O_3$) to form a 2-carbon molecule. This then combines with coenzyme A, splitting off hydrogen and forming acetyl-coenzyme A. The hydrogen then enters the transfer cycle, forming an additional 6 ATP (three from each of the pyruvic acid molecules resulting from the original glucose breakdown).

Citric acid cycle

The series of reactions in the citric acid cycle was postulated by the English biochemist, H. A. Krebs, and is called the Krebs citric acid cycle (tricarboxylic acid cycle), because citric acid ($C_6H_8O_7$) is the first substance formed in the series (Fig. 26-12).

The 2-carbon acetyl-coenzyme A unites with 4-carbon oxaloacetic acid to form 6-carbon citric acid. Then successive enzymes break down the citric acid, step by step, through eight different intermediate compounds to oxaloacetic acid, which then combines with another molecule of acetyl-coenzyme A, to continue the cycle.

In the cycle, carbon dioxide is given off, hydrogen atoms are removed, and the electrons of the hydrogen atoms are transferred by cytochromes to the oxygen, which then unites with the hydrogen ions to form water.

As the two molecules of acetyl-coenzyme A, derived from the two molecules of pyruvic acid are metabolized in the Krebs cycle and cytochromes, about twenty-four energy-rich phosphate compounds are formed.

The carbon and oxygen atoms are removed together from a substrate molecule by decarboxylation, with thiamine pyrophosphate as a necessary cofactor. One such decarboxylation process occurs as 3-carbon pyruvic acid is converted to 2-carbon acetyl-coenzyme A. Two decarboxylation processes occur in the Krebs cycle as 6-carbon citric acid is converted into 4-carbon oxaloacetic acid.

The Krebs cycle is the final common pathway for the oxidation of amino acids and fatty acids, as well as for carbohydrates. The fatty acids usually found in tissues contain sixteen and eighteen carbons in a long chain. The long chains are split into 2-carbon acetyl-coenzyme A (Fig. 26-12), and the latter enter the Krebs cycle by uniting with oxaloacetic acid.

Certain amino acids, through enzyme actions, can be converted to pyruvic acid, and others are converted to other members of the Krebs cycle. By various means the amino groups are removed, and the carbon chains of the amino acids eventually enter the Krebs cycle to be oxidized to yield energy, carbon dioxide, and water.

Anaerobic respiration

If the "bucket brigade" of the hydrogen transfer system is blocked by a lack of any of the factors, pyruvic acid may serve as an acceptor. While this is a temporary activity in oxygen-using organisms, it should be noted that a group of anaerobic, or non-air users, survives on the small amounts of energy available by glucose breakdown (Fig. 26-13). The ATP formed by glucose breakdown alone is only two molecules, in contrast to the thirty-eight formed when the six carbons of glucose are finally converted to 6-carbon dioxide.

EXCRETION

Excretion may be considered as the elimination from the body of the waste products of metabolism. Organs involved in this process include the kidneys, lungs, and skin. Although it is sometimes included in the list, the intestine strictly speaking is not an excretory organ, since the products of defecation have not been previously absorbed. (Table 26-3).

The main products excreted by the lungs are carbon dioxide and water. The condensing of the breath on a cold surface is a good example of this water loss (up to 1 pint per day).

The many sweat glands in the skin account for further loss of water, carbon dioxide, and mineral salts.

The kidneys

Two bean-shaped kidneys, located at the back of the abdominal cavity, one on each side of the ver-

Aerobic   Net yield

1. Glucose ———————————→ Pyruvic acid
   C$_6$ 2 C$_3$ 8 ATP

2. Pyruvic acid ———————————→ Acetyl-coenzyme A
   2 C$_3$ 2 C$_2$ + 2 CO$_2$ 6 ATP

3. Acetyl-coenzyme A ———————————→ Carbon dioxide
   2 C$_2$ [Krebs cycle]   4 CO$_2$ 24 ATP
                                                                          ———————
                                                        Total   38 ATP

Anaerobic

1a. Glucose ———————————→ Triose phosphate

1b. Triose —(P)+ NAD $\xrightarrow[\text{Dehydrogenation}]{\text{Oxidation}}$ Pyruvic acid + NAD · H$_2$ 2 ATP

1c. Pyruvic acid + NAD · H$_2$ ———————→ Lactic acid (in animals)   0 ATP

                                        Alcohol + CO$_2$ (in plants)   ———————
                                                        Total   2 ATP

Fig. 26-13
Summary of cellular respiration.

Table 26-3
Excretion of human wastes

	Primary	Secondary
Kidneys	Water and soluble salts	Carbon dioxide and heat
Lungs	Carbon dioxide (12 cubic feet daily)	Water (250 ml. daily) and heat
Skin	Water, salts, carbon dioxide, and heat	Dead skin, nails, etc.
Alimentary canal	Solids and secretions	Water, carbon dioxide and other gases, salts, and heat

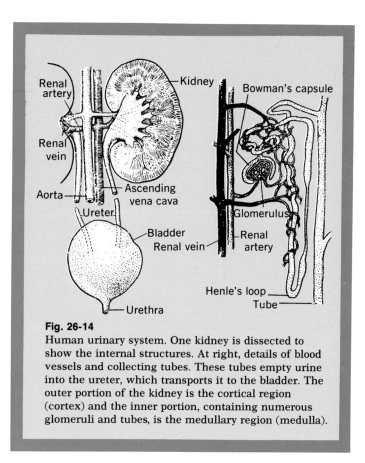

Fig. 26-14
Human urinary system. One kidney is dissected to show the internal structures. At right, details of blood vessels and collecting tubes. These tubes empty urine into the ureter, which transports it to the bladder. The outer portion of the kidney is the cortical region (cortex) and the inner portion, containing numerous glomeruli and tubes, is the medullary region (medulla).

tebral column, select wastes from the blood and pass them through the ureters to the urinary bladder where wastes are stored (Fig. 26-14). Each kidney consists of an outer cortex and an inner medulla. When examined microscopically, the cortex contains approximately one million globelike, renal corpuscles, or nephrons, each of which is composed of (1) a coiled mass, or glomerulus, of thin-walled capillaries arising from the renal arteries, and (2) a thin, double-walled, enclosing glomerular capsule (Bowman's capsule), which is the beginning of a renal tubule. The convoluted renal tubules travel irregularly and empty into the straighter collecting tubes, which in turn pass the urine into the basinlike pelvis of the kidney then out through the ureter. The muscle contractions in the walls of the ureters cause the urine to pass toward the muscular bladder, located in the pelvic cavity. The bladder normally holds about 1 pint,

and the contraction of its three layers of muscles forces the urine to the exterior through the tubular urethra.

The structure of the individual renal corpuscle is such that the walls of the tubules and the surface of the blood vessels are in contact at two sites. The first site is in the capsule itself, where the blood vessels form the highly branched glomerulus. The second is in the capillary network that surrounds the renal tubule. The dual kidney function of filtration, followed by reabsorption, takes place at these areas of contact.

Blood pressure in the glomerulus is such that most of the fluid and molecules smaller than proteins are forced out of the vessels and into the capsule itself. This capsular fluid then passes through the renal tubules and comes in contact with the area surrounded by capillaries. Here, most of the water is reabsorbed, and the ions and molecules are selectively removed.

While approximately 170 quarts of fluid pass through the kidneys each day, only about 1 quart leaves the body as urine. This urine is primarily water, with varying amounts of urea, carbon dioxide, and ions, depending on the type of diet. Thus the kidneys, formerly thought to function only in excretion, also control the water content, ion and mineral balance, and pH of the blood.

CIRCULATION

The function of the circulatory system is to carry blood through all tissues of the body. The system itself is composed of (1) the heart, a biologic pump, (2) blood vessels, consisting of arteries, capillaries, and veins, and (3) blood, the circulating fluid.

While the interaction of heart, vessels, and blood results directly in circulation, the system may also be considered to function in all of those activities influenced by the materials it carries.

The heart

The heart is functionally a pump, forcing blood through arteries in a double path—the pulmonary circulation to the lungs and the systemic circulation to the rest of the body. Its main structure, the muscular myocardium, is organized into four chambers, the right and left atria and the right and left ventricles. The chambers on each side are con-

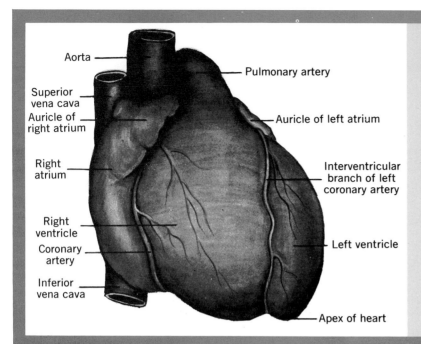

Aorta

Pulmonary artery

Superior vena cava

Auricle of right atrium

Auricle of left atrium

Right atrium

Interventricular branch of left coronary artery

Right ventricle

Coronary artery

Left ventricle

Inferior vena cava

Apex of heart

Fig. 26-15
Human heart, showing some blood vessels (front view).
(From Francis, C. C.: Introduction to human anatomy, ed. 5, St. Louis, 1968, The C. V. Mosby Co.)

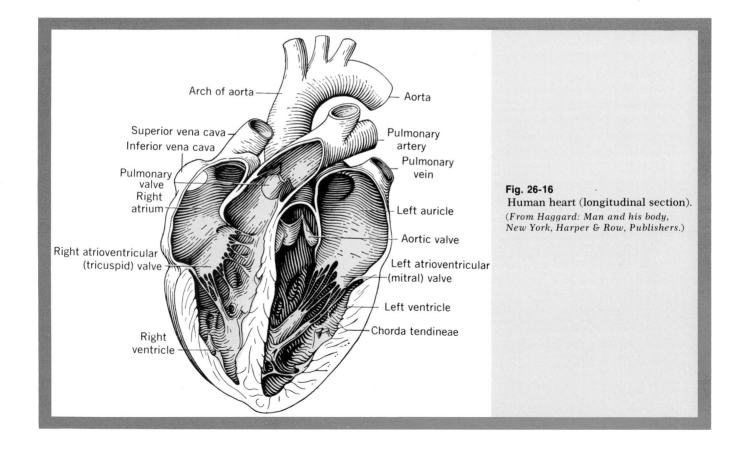

Arch of aorta

Aorta

Superior vena cava

Inferior vena cava

Pulmonary artery

Pulmonary vein

Pulmonary valve

Right atrium

Left auricle

Aortic valve

Right atrioventricular (tricuspid) valve

Left atrioventricular (mitral) valve

Left ventricle

Chorda tendineae

Right ventricle

Fig. 26-16
Human heart (longitudinal section).
(From Haggard: Man and his body, New York, Harper & Row, Publishers.)

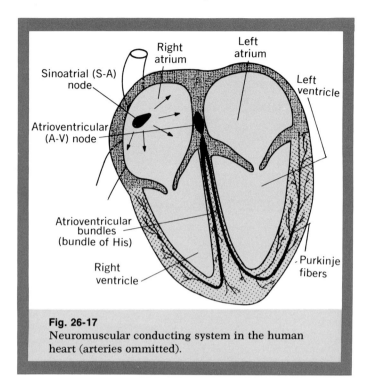

Fig. 26-17
Neuromuscular conducting system in the human heart (arteries ommitted).

nected through atrioventricular valves. Each atrium and each ventricle has an epithelial lining, the endocardium. The entire heart is enclosed in a saclike pericardium.

Blood from the systemic circulation flows through the venae cavae into the right atrium, which contracts, forcing it through the right atrioventricular (tricuspid) valve into the right ventricle. Contraction of the ventricle forces blood through the pulmonary semilunar valve into the pulmonary artery and then to the lungs. Blood from the lungs reenters by way of the pulmonary veins into the left atrium, then through the left atrioventricular (mitral) valve, to the left ventricle. Contraction of the ventricle forces blood through the aortic semilunar valve and into the systemic circulation.

It should be understood that blood enters both sides of the heart at the same time and leaves through the pulmonary artery and aorta at the same time.

Heartbeat

The simultaneous contraction of the atria, followed by the simultaneous contraction of the ventricles, is controlled by a series of unique structures. The heartbeat begins in the sinoatrial node or "pacemaker," located in the right atrium between the venae caval openings. A wave of contraction, similar to the ripples in a pond, spreads out through both atria, stimulating an atrioventricular (A-V) node located in the wall separating these chambers. The contracting impulse then spreads rapidly through the right and left atrioventricular bundles (bundles of His), located in the wall between the ventricles. Each bundle terminates in a series of branching fibers (Purkinje fibers) in the outer walls of the ventricles.

The speed of transmission through these fibers initiates a uniform contraction of the ventricles, forcing blood out of the heart at the rate of approximately 72 times per minute.

The rate of contraction is influenced by such factors as activity, emotion, and diet. Impulses from a "heart-rate center" in the brain inhibit the pacemaker through the vagus nerve or accelerate it through the spinal nerves. The interaction of these nerves controls the beating of the heart and therefore the supply of blood throughout the body.

The blood

Blood is a liquid tissue that consists of clear, faint-yellow plasma in which are suspended the red blood corpuscles (erythrocytes), the various types of white blood corpuscles (leukocytes), and the blood platelets. Blood forms about one-thirteenth of the total body weight and in an average man totals about 6 liters (over 6 quarts). It is somewhat viscous, slightly heavier than water, and slightly alkaline (pH of 7.35) (Table 26-4).

Erythrocytes (e -rith′ ro site) (Gr. *erythros*, red; *kytos*, cell) constitute about 50 percent of the blood volume. When mature, they are without a nucleus and consist of an elastic framework known as the stroma (Gr. *stroma*, bedding) and hemoglobin (he mo -glo′ bin) (Gr. *haima*, blood; *globos*, sphere). Hemoglobin consists of a protein and an iron-containing compound, the latter being responsible for the chemical affinity for oxygen. When hemoglobin carries oxygen, it is known as oxyhemoglobin and liberates its contained oxygen where needed. Anemia (Gr. *an-*, not; *haima*, blood) is a

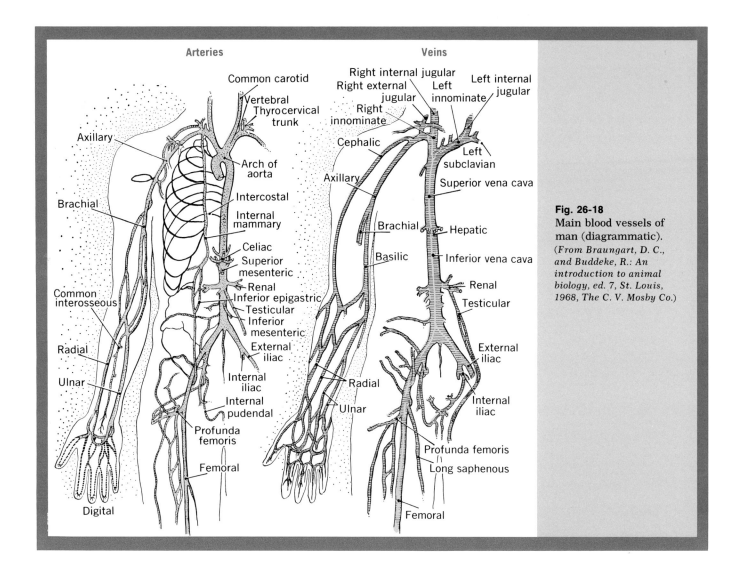

Arteries

Common carotid
Vertebral
Thyrocervical trunk
Axillary
Arch of aorta
Brachial
Intercostal
Internal mammary
Celiac
Superior mesenteric
Renal
Inferior epigastric
Testicular
Inferior mesenteric
Common interosseous
External iliac
Radial
Internal iliac
Ulnar
Internal pudendal
Profunda femoris
Femoral
Digital

Veins

Right internal jugular
Right external jugular
Left internal jugular
Left innominate
Right innominate
Cephalic
Left subclavian
Axillary
Superior vena cava
Brachial
Hepatic
Basilic
Inferior vena cava
Renal
Testicular
Radial
External iliac
Ulnar
Internal iliac
Profunda femoris
Long saphenous
Femoral

Fig. 26-18
Main blood vessels of man (diagrammatic). *(From Braungart, D. C., and Buddeke, R.: An introduction to animal biology, ed. 7, St. Louis, 1968, The C. V. Mosby Co.)*

condition in which there is a decrease in the number of erythrocytes, in the amount of hemoglobin, or in both. These conditions may result from impaired blood formation, increased destruction of erythrocytes, or both.

An abnormal increase in the number of erythrocytes is called polycythemia (pol e si -the′ me ah) (Gr. *polys*, many; *kytos*, cell; *haima*, blood), which often results from a loss of body fluids, hence of blood plasma. Often there is an overproduction of erythrocytes in the red bone marrow.

Leukocytes (Gr. *leukos*, white; *kytos*, cell), because of their ameboid movements, are able to escape from the blood vessels and penetrate into

the body tissues. Leukocytes may be classified as (1) granulocytes, with distinguishing granules in the cytoplasm, and (2) agranulocytes, without granules in the cytoplasm. Because of the variations in the lobes of the nuclei of the granulocytes, they are sometimes referred to as polymorphonuclear leukocytes (Gr. *polys*, many; *morphe*, form). They act as phagocytes (Gr. *phagein*, to eat; *kytos*, cell) by engulfing bacteria, cell fragments, and foreign materials. The granulocytes are classified in three groups according to the type of granules in their cytoplasm: (1) eosinophils, which stain readily by eosin (acid) stains; (2) neutrophils, which stain by neutral dyes; and (3) basophils,

Table 26-4
Human blood

	Red blood corpuscles (erythrocytes)	White blood corpuscles (leukocytes)	Blood platelets
Nucleus	No nucleus when mature	Always nucleated	None
Shape	Flat, biconcave disks	Variable	Spherical
Motility	None (may bend)	Ameboid movement	None
Diameter	7.7μ	8 to 20μ	3μ (average)
Number	5,000,000 per cubic millimeter	5,000 to 9,000 per cubic millimeter	250,000 per cubic millimeter
Where formed	Red bone marrow	Red bone marrow, lymphoid tissue, and spleen	Bone marrow
Where lost	Liver and spleen	Liver and lumen of intestine	Disintegrate rapidly, which may explain variations in number
Length of life	125 days	Variable, few hours to months	4 days
Functions	Carry oxygen and aid in transportation of carbon dioxide	Protect against bacteria, cell fragments, and foreign particles; repair tissues	Clotting of blood

which stain well with basic stains (Table 26-5).

The agranulocytes lack cytoplasmic granules and are classified in two groups: (1) lymphocytes (L. *lympha*, lymph or water; Gr. *kytos*, cell) and (2) monocytes (Gr. *monos*, alone; *kytos*, cell), both being formed in the lymphoid tissue. Lymphocytes contribute to the repair of wounds by connective tissue formation. They are important in the process of immunity, being antibody carriers. Monocytes are active in phagocytosis, engulfing bacteria and cell debris. A decrease in the number of leukocytes is called leukopenia (Gr. *leukos*, white; *penes*, poor) and an increase is leukemia (Gr. *leukos*, white; *haima*, blood).

Platelets (Gr. *platys*, flat) of mammals are small and colorless, and their various sizes average about 3μ. They are roundish or oval biconvex disks that look spindle shaped when seen in profile. In man they usually average approximately 250,000 per cubic millimeter. When they leave the blood vessels, they tend to adhere to one another and to surfaces that they contact. They originate in the

Table 26-5

Summary of leukocytes

Type	Percent of total leukocytes	Diameter in microns	Characteristics (with Wright's stain)
Granulocytes			
Neutrophils	60 to 70	10 to 12	Fine, light blue cytoplasmic granules; 3- to 5-lobed nucleus
Eosinophils	1 to 4	12	Bright red cytoplasmic granules; 2-lobed nucleus
Basophils	0.5	10	Large, dark purplish blue cytoplasmic granules; irregular nucleus; often S-shaped
Agranulocytes			
Lymphocytes	25	8	Thin layer of nongranular, robin's-egg blue cytoplasm; large, bright purple nucleus
Monocytes	5 to 10	12 to 20	Thick layer of nongranular cytoplasm; large horseshoe- or kidney-shaped, purple nucleus

bone marrow and are associated with the process of blood clotting.

Functions of blood

The functions of blood are as follows: (1) respiratory, transporting oxygen to the tissues and carbon dioxide from them; (2) excretory, carrying waste materials from the tissues to the organs of excretion; (3) nutritive, transporting sugars, amino acids, fats, and ions from the intestine to the tissues; (4) regulatory, providing a fluid environment for cells and osmotic balance, maintaining pH by buffer action, distributing hormones to all cells, and controlling body temperature by water movement and sweating; and (5) protective, by the phagocyte action of white blood cells, by antibody action, and by the clotting mechanism.

Blood vessels

Arteries are vessels whose walls are rather thick, contractile, and elastic. They consist of (1) an inner layer of endothelial cells and elastic tissue, (2) a middle or intermediate layer of muscle and elastic tissue, and (3) an external layer of elastic tissues. Arteries carry blood away from the heart, and veins carry blood back toward the heart.

Veins are vessels whose structure is similar to that of arteries, being composed of three layers, but whose walls are thinner and less elastic because of a poorly developed middle layer and the lack of very much muscle and elastic tissue. Certain veins, especially those of the legs, have a series of semilunar valves to prevent the backflow of blood. In general, the systemic veins parallel the systemic arteries, frequently having the same names as the arteries.

Capillaries are thin-walled vessels that form a network connecting the arteries and veins. The walls are a single layer of flat endothelial cells. The capillaries are so numerous that there is hardly any body part that does not contain them. The exchange of materials takes place through them because of their thin walls and the slow movement of the blood within. An average capillary is about 8μ in diameter. Compare this with the diameter of a red blood corpuscle.

Blood plasma

Blood plasma is the fluid part of the blood and is approximately 90% water. It serves as the basic carrier for the cells and other materials. Plasma proteins form about 8% of the remaining fraction, and inorganic ions, foods, wastes, and hormones account for the rest.

Most of the protein is albumin, which is important in maintaining the osmotic pressure of the blood, globulins, which are essential to antibody formation and immunity, and fibrinogen, which functions in clotting of the blood.

The ion concentration is important in maintaining pH and all metabolism as indicated earlier.

Clotting of blood

Loss of blood is known as hemorrhage; this initiates the following protective activities in the body: (1) clotting of blood at the site of the injury; (2) decrease in the general blood pressure; (3) contraction of the small vessels of the skin, muscles, and intestines in order to supply the vital parts of the body; (4) increase in the blood volume by the contraction of the spleen, which normally contains a large quantity of blood; and (5) passage of water and salts from the tissues into the capillaries because of increased osmotic pressure.

Shortly after blood leaves blood vessels it loses its normal fluidity and becomes jellylike through a process known as coagulation, or clotting. Maintaining the fluidity of the blood while it flows through normal blood vessels and yet preserving its inherent coagulability is a marvelous phenomenon of vital importance.

When a quantity of freshly drawn blood stands for a few minutes, a meshwork of microscopic, threadlike fibrils is formed throughout the blood mass. The fibrils entangle blood corpuscles and transform the original liquid into a gel. Then the fibrils shrink and press out of the gel a slightly yellowish fluid, the blood serum. This shrinkage continues with the formation of more fluid, and eventually the gel grows firmer to form a clot, composed of fibrils and entangled blood corpuscles.

When the stringy material is washed properly, it is found to be composed of a white, fibrous, elastic, protein substance. Since there is no insoluble protein in the blood plasma, what is its origin and what causes its formation? The red and white corpuscles do not contribute to this, because after their removal, the remaining plasma still has the ability to coagulate. Thus, the insoluble fibrin of the clot must originate from some soluble protein, such as serum albumin, serum globulin, or from a substance in the blood called fibrinogen (Fig. 26-19). By proper techniques these three can be separated. The serum albumin and globulin have been found not to be associated with blood coagulation. Fibrinogen, however, after being obtained free from the other proteins and when dissolved in a salt solution, will coagulate in a way similar to that of plasma. Thus in blood coagulation soluble fibrinogen is changed into insoluble fibrin. As long as blood remains in normal blood vessels, no fibrin is formed, and the fluidity of the blood is maintained.

When it is realized that approximately thirty-five compounds take part in clot formation, it is understandable that this phenomenon is one of the most complicated chemical processes in our body. Many theories have been proposed concerning the roles of the various chemicals involved, but we shall limit our consideration to a summary of a few salient facts that have been accepted generally.

The conversion of the soluble protein, fibrinogen, into the insoluble protein, fibrin, is caused by an enzyme called thrombin (Gr. *thrombos*, clot). In blood, thrombin is present in an inactive form, called prothrombin. For the formation of prothrombin an adequate supply of vitamin K is essential. Since bile is required for absorption of vitamin K from the intestine into the blood, a bile deficiency will reduce the amount of prothrombin and hence cause deficient clotting, even if sufficient amounts of the vitamin are present in the intestine. Certain types of hemorrhages can be prevented by the administration of vitamin K before operations.

Injury to tissues or blood vessels, or the disintegration of blood platelets, gives rise to the formation of a substance called thromboplastin. In the

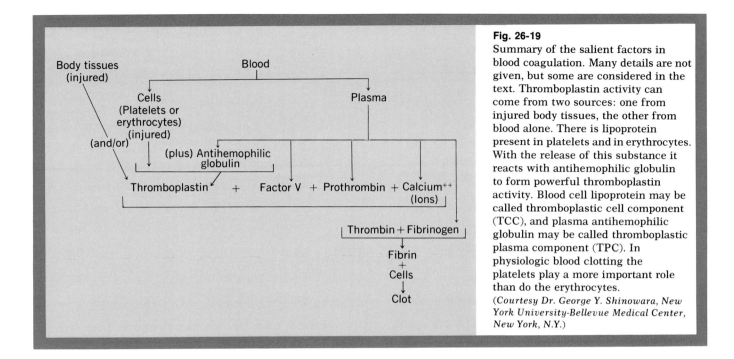

Fig. 26-19

Summary of the salient factors in blood coagulation. Many details are not given, but some are considered in the text. Thromboplastin activity can come from two sources: one from injured body tissues, the other from blood alone. There is lipoprotein present in platelets and in erythrocytes. With the release of this substance it reacts with antihemophilic globulin to form powerful thromboplastin activity. Blood cell lipoprotein may be called thromboplastic cell component (TCC), and plasma antihemophilic globulin may be called thromboplastic plasma component (TPC). In physiologic blood clotting the platelets play a more important role than do the erythrocytes. (*Courtesy Dr. George Y. Shinowara, New York University-Bellevue Medical Center, New York, N.Y.*)

presence of calcium ions this transforms the inactive prothrombin into active thrombin.

It is highly important that shed blood should coagulate, but it is equally desirable that the fluidity of blood be maintained so long as it remains in blood vessels normally. Two important factors in this connection are intact blood platelets and intact blood vessels, especially the endothelial lining of the latter. Injury to blood vessels not only liberates thromboplastin but also increases the destruction of platelets. Since platelets live only about 4 days, they constantly supply thromboplastin.

The formation of a clot within a blood vessel that is not severed is called a thrombus. This may be caused by an injury of the vessel wall from a blow or from toxins of bacteria that injure the blood platelets. If a part of a thrombus circulates in the vessels, it is called an embolus, which, if it should block circulation to a vital part, has serious consequences.

Under normal conditions clotting is inhibited by the action of two enzyme-inactivating substances, antithrombin and heparin.

Functions of human lymph

The composition of the lymph is similar to that of the blood plasma. It ranges from colorless to a yellowish color, has an alkaline reaction, contains no blood platelets, clots slowly and not firmly, may contain a few red blood corpuscles, and contains lymphocytes.

The lymph is derived from (1) the blood plasma by filtration through the thin walls of the capillaries and (2) secretions of the endothelial cells that line the numerous capillaries.

The functions of human lymph are summarized as follows: (1) It bathes all parts of the body not reached directly by the blood, thus supplying foods and oxygen and receiving carbon dioxide and wastes. There is a continuous interchange between the blood plasma and the lymph through the processes of osmosis and diffusion. (2) It aids in the fight against foreign materials, such as bacteria and protozoans. (3) It helps to equalize body temperature. (4) It helps to regulate the acid-alkaline balance of the various body parts. (5) It helps to collect and transport fatigue products that are the result of cellular activity. (6) It aids in transporting enzymes and other secretions to various body parts (Fig. 26-20).

Human lymph is found in various places and consequently has a variety of functions. The principal locations are (1) in the lymph ducts and their enlargements, the lymph nodes, (2) in tissue spaces

377

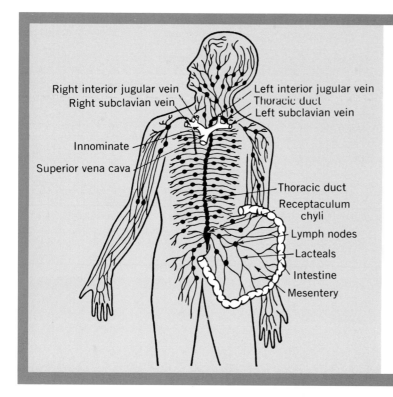

Fig. 26-20

Lymph system and parts of certain veins of the upper part of the body. Lymph from the abdominal organs and lower limbs flows into the thoracic duct, which empties into the left subclavian vein. Lymphatics from the left arm and left sides of the thorax, neck, and head also empty into the thoracic duct. Lymph from the right arm and right sides of the thorax, neck, and head flows into the right subclavian vein.

(From Zoethout, W. D., and Tuttle, W. W.: Textbook of physiology, ed. 11, St. Louis, 1952, The C. V. Mosby Co.)

(tissue sinuses) or cavities in various tissues, (3) in the pleural cavity (around the lungs), (4) in the pericardial cavity around the heart, (5) in the peritoneal cavity (abdominal cavity), (6) in the perineural cavities (spaces between the various linings of the brain and spinal cord), and (7) in the lacteals or lymphatics, which originate in the small, finger-like villi of the intestine. Fats are absorbed from the intestine by the lacteals and eventually placed in the bloodstream.

Human blood groups

Blood types

The typing of human blood is based on the presence or absence of certain substances in the red blood cells or plasma. The substance in the cells is a type of antigen called an agglutinogen. The plasma substance is a type of antibody called an agglutinin. When the proper kinds of each are mixed, clumping, or agglutination, of the red blood cells results.

The name of the blood type is derived from the kind of agglutinogen (antigen) present in the red blood cells. In the A-B-O system, type A has agglutinogen A, type B has agglutinogen B, type AB has both A and B, and type O has neither A nor B.

Since a mixture of plasma from a person of type A and red blood cells from a type B will result in a clumping of cells, it is apparent that type A plasma contains the agglutinins against type B cells. Table 26-6 shows the scheme in detail. Note that type O may be given by transfusion to all blood types, and type AB may receive from all types. This, of course, assumes that all precautions have been taken. Under normal circumstances the same type would be transfused.

Since type O blood contains agglutinins a and b and type AB contains agglutinogens A and B, an apparent contradiction exists in this "universal donor—universal recipient" scheme. This is solved when it is demonstrated that the agglutinins lose their effect when diluted, as they are during transfusion.

Other blood groups, such as the Rh and MN systems, depend on similar sets of agglutinogens and agglutinins.

Table 26-6

Human blood groups

Blood group	Agglutinogen (antigen) in red blood corpuscles	Agglutinin (antibody) in blood plasma	May receive blood from group	May give blood to group	Percent of the white population of the United States represented by each blood group*
O	None	a and b	O	O, A, B, AB	41
A	A	b	O, A	A, AB	45
B	B	a	O, B	B, AB	10
AB	A and B	None	O, A, B, AB	AB	4

*Different populations will show other percentages.

BODY DEFENSES

The mere presence of microorganisms in air, food, milk, and water does not produce a disease in all instances. Among the most important deterrents to disease production in man are the various body defenses.

Defenses (first line)

The main defenses of the body are: skin, which acts as a mechanical barrier; nose, in which mucus collects the organisms that cilia move toward the exterior, after which the enzyme lysozyme destroys the bacteria, and sneezing and coughing occur; eyes, from which organisms are washed away mechanically by tears, which also contain the enzyme lysozyme; mouth, a mucous membrane acting as a mechanical barrier; stomach, which contains the acid of the gastric juices; intestine, which contains a mucous membrane acting as a mechanical barrier and in which the antagonistic action of other organisms occurs; urethra, in which the action of urine takes place.

Defenses (second line)

The body's second line of defenses consists of (1) the production of inflammations, in which many organisms that have penetrated the deeper tissues are trapped and destroyed (inflammations are usually characterized by redness and swelling because of increased supplies of blood; temperature, because of increased metabolic activity in the area; pain, because of abnormal activities; pus formation in later stages because of the destruction of microorganisms and tissues); and (2) phagocytic action of certain white blood corpuscles, in which the phagocytic cells engulf and destroy microorganisms in the various body tissues, in the bloodstream, and lymph nodes.

Defenses (third line)

Other defensive reactions of the body that depend upon present or previous contact with the infectious organisms, or their products, are known as immunologic reactions. Pathogenic organisms may injure the hosts, which they attack in two ways: (1) they consume or destroy vital tissues directly, or (2) they produce substances that poison host tissues in the immediate vicinity of the organisms or are carried to other parts of the body. These poisonous substances, called toxins, are either produced by living organisms as they grow and reproduce, or are liberated when organisms die and disintegrate. Some toxins are general in their action, affecting a variety of host tissues, while others are more selective in their action, affecting heart muscle, nervous tissue, or other specific body tissues. Some toxins are proteins, while others are of unknown chemical composition.

Any substance that stimulates the body of an animal to produce a specific antibody is called an antigen. When animals are attacked by certain pathogenic bacteria or their toxins, they produce substances that neutralize the poisonous effects

379

of the toxins or that inactivate or destroy the bacteria. These protective substances, called antibodies, are of several types, among which are antitoxins, agglutinins, bacteriolysins, and opsonins. Antitoxins are substances that neutralize specific toxins and are formed in animal bodies in direct response to the toxins. Antitoxins are specific in their action; for example, diphtheria antitoxin reacts only with diphtheria toxin and not with the toxin of the tetanus organism. Agglutinins are antibodies that agglutinate (clump) the specific organisms that acted as antigens in their formation in a body. The agglutination may occur in the body or outside, where it can be used for identifying specific organisms for diagnosis of certain diseases. For example, if the blood serum of an animal immunized against typhoid organisms is mixed with a suspension of typhoid organisms, the latter are agglutinated. Bacteriolysins are antibodies that dissolve the specific organisms that have stimulated the body cells to produce them. Opsonins, meaning "to prepare food for," are antibodies that weaken bacteria so that they are more readily destroyed by the phagocytes.

If the production of antibodies by the host is sufficiently extensive and rapid, pathogenic bacteria are destroyed and their toxins neutralized, and the animal may recover from the disease. An animal that has formed sufficient antibodies is therefore immunized against the specific disease caused by the type of bacterium that led to the production of the antibodies. This immunity often persists for the life of an individual animal that has experienced and has recovered from a particular disease. Thus, one attack of measles or typhoid fever usually renders the victim immune to these diseases indefinitely.

It is not always necessary for an animal to have a disease in order to be immunized against it. One method of immunization involves the administration of vaccines, which are preparations of dead or weakened bacteria or viruses. When such vaccines enter the body, they stimulate the formation of antibodies, which in turn may immunize the animal for certain periods of time. Such antibodies are specific for the organisms from which the vaccines were prepared. Vaccines are used in immunizing against such diseases as smallpox, poliomyelitis, typhoid, cholera, and bubonic plague. Since vaccines stimulate the animal body to produce its own antibodies, this type of immunity is similar to that acquired as a consequence of actually having that disease.

The condition of active immunity after either vaccination or having the disease is the result of antibody production by the individual himself. A varying period of time must elapse before immunization is effective.

On occasion, and especially prior to the development of newer antibiotics, the severity of some particular diseases necessitated the development of passive immunity. This is the result of receiving a serum containing antibodies from a person or animal who had active immunity previously. Passive immunity has the advantage of being effective immediately; however, it does not last very long.

ALLERGIES

Tissue cells that have responded to stimulation by an antigen are immunologically different from tissue cells that have not so responded. The antigenically stimulated cells often retain antibodies for some period of time. Later, when a specific antigen again contacts such cells, they react in a distinctive manner and are said to be hypersensitive, or in an allergic state (Gr. *allos,* changed; *ergon,* activity). When cells have produced antibodies, the allergic reaction may be vigorous, especially if some time has elapsed between the first contact with the sensitizing antigen and that which causes the later allergic reaction. Sometimes normal cells that have not been stimulated actively by an antigen may passively acquire the antibodies necessary for certain types of allergic reactions by merely absorbing them from the blood.

In some types of allergies the antibodies remain associated closely with the cells that produce them and are called allergic reagins. The antigens that stimulate their production often differ from substances commonly considered as antigens and are called allergens. Common allergens include certain cosmetics, some plastics, sulfonamides and other drugs, certain dyes on fabrics, and certain antibiotics (Fig. 26-21). Hypersensitivity to any of them may occur. There are hundreds of allergic reactions, and their specific characteristics depend

Fig. 26-21
Forty-eight causes of allergic reactions.
(Courtesy Lederle Laboratories, Pearl River, N. Y.)

on the kind and location of tissues involved; the nature and dosage of the allergen; the nature of the reagin; and the degree of hypersensitiveness.

While the term allergy may include all reactions resulting from tissue hypersensitiveness, some classifications exclude those reactions that do not occur naturally and have unique properties. For example, if a guinea pig is injected with 1 ml. of a foreign protein, such as horse serum, in about 3 weeks the guinea pig is hypersensitive to a second injection of horse serum and will exhibit a striking response, called an anaphylactic reaction (Gr. *ana*, against; *phylaxis*, protection). Shortly after the injection of even a small second dose of the horse serum the guinea pig may display such symptoms as uneasiness, coughing, scratching of the nose, gasping for breath, urination, defecation, finally cessation of breathing, and death. If the injection is not fatal, the animal may recover within 2 hours. Most symptoms in anaphylaxis are caused by con-

381

tractions of smooth muscles, such as are found in lungs, bladder, and intestines. Common allergic reactions include the so-called hay fevers (caused by pollens), the early stage of the common cold, asthma, food allergies (caused by such things as eggs and strawberries), cosmetic rashes, and certain industrial dust reactions.

To test for an allergy a small quantity of sterile solution of a particular protein is injected into (not under) the patient's skin. If the patient is allergic to this protein, a large, localized inflamed area results because of the allergen-reagin (antigen-antibody) reaction. An allergic tendency is inheritable.

Review questions and topics

1 Describe the various parts of the human digestive system, including the specific functions performed by each.
2 Discuss the digestive process and the control of digestion.
3 Discuss the metabolism of carbohydrates, amino acids, and fats.
4 Discuss the roles of the following in nutrition: mineral salts, water, and vitamins.
5 List the characteristics and functions of each of the important vitamins.
6 Explain the terms breathing and respiration, including gas transport and cellular respiration.
7 Discuss the various parts of the breathing system and the function of each part.
8 Explain in detail the process of glycolysis, including phosphorylation, oxidation, and ATP formation.
9 Explain pyruvic acid breakdown and Krebs citric acid cycle (tricarboxylic acid cycle).
10 Explain anaerobic respiration.
11 Discuss the 2-carbon acetyl-coenzyme A and its function in respiration.
12 Explain how the various wastes are eliminated by the human body.
13 Describe the structure of the human circulatory system and the functions of the various parts.
14 Discuss the functions of the human blood, including the characteristics and specific functions of its various parts.
15 Explain in detail the functions of each correlated phenomenon in the clotting of blood.
16 Describe the human lymph system and the functions of lymph.
17 Discuss each of the human blood groups and their significance.
18 Define agglutinogen (antigen) and agglutinin (antibody) and describe where each is found.
19 Discuss in detail the body defenses against microorganisms.
20 Discuss allergies and their significance.

chapter twenty-seven

Biology of man—control mechanisms and reproduction

All living protoplasm is necessarily irritable or subject to stimulation. A stimulus is any external or internal substance, material, or condition that affects a cell or group of cells, thereby setting up a change known as a response. General types of stimuli can be chemical, electrical, thermal, or mechanical; general types of responses can be moving, secretory, thermal, or chemical. The responsive mechanisms of man are complex and varied. The three steps involved are as follows: (1) a special structure called a receptor must be stimulated; (2) there must be some method of conduction of the effects of stimulation to (3) a specialized structure called an effector, which must respond in some way.

While the actual coordinating substances in man are quite varied, the more obvious mechanisms are found in the nervous and endocrine systems.

NERVOUS SYSTEM

Organization

The human nervous system will be considered under the headings of (1) the central nervous system, which consists of the brain and spinal cord and their related nerves; (2) the autonomic nervous system, which is associated with involuntary control; and (3) the sense receptors, which are in contact with the environment.

Neurons

The basic unit of the nervous system is the individual neuron, or nerve cell. This is a living, metabolizing cell that is specialized to receive and transmit nervous impulses. While the structure may vary, an individual cell has an irregular shape with several protoplasmic extensions. According to the direction of impulse transmission, these extensions are referred to as dendrites, leading toward the cell body, and axons, leading away from it. The length of these extended fibers varies tremendously.

An individual fiber such as an axon has an outer fatty covering called the myelin sheath, surrounded by a cellular neurilemma (Fig. 27-1). These apparently serve as nutritive, protective, and insulating sheaths. Nerve fibers usually terminate in a series of branching fibers.

Based on the direction of impulse transmission, neurons are called (1) sensory, leading from sense organs to the brain or spinal cord, (2) motor, leading from the brain or spinal cord to a muscle or gland acting as an effector, and (3) interneurons, connecting the other two in the brain or spinal cord.

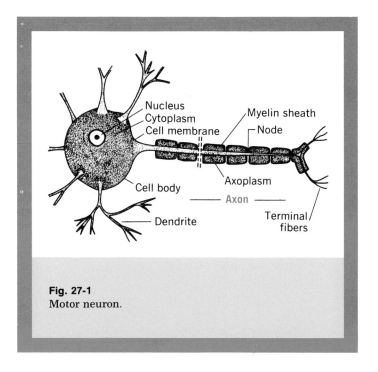

Fig. 27-1
Motor neuron.

Nerve impulses

How do nerves carry a message by what is commonly called a "nerve impulse"? Luigi Galvani, professor of anatomy at the University of Bologna, accidentally discovered that if a frog's leg touched a piece of iron, an electric current was propagated (1786). Sixty years later it was proved that nerve and muscle cells actually possess electrical charges and are capable of generating an electric current.

To measure a nerve impulse two pairs of electrodes are placed on a dissected nerve. One pair (the "transmitter") stimulates the nerve, and the other pair (the "receiver") is applied at various points to ascertain the strength and velocity of the signal. The signal, which travels at a constant velocity, arrives at the receiver after a delay that represents the time required to travel along the nerve between the two electrodes. In the fastest nerve fibers of warm-blooded animals this velocity is about 300 feet per second.

A nerve fiber, or axon, is a long, thin tube that is attached to a microscopic nerve cell, or neuron. These fibers vary in diameter from less than thirty-thousandths to one-thirtieth of an inch. In limbs of large animals the fibers may be several feet long.

The explanation of the electrical activity of the fiber lies in its chemistry and that of the tissue fluid around it.

There are two types of nerve fibers: one is enclosed by a myelin sheath and the other is almost bare (nonmyelinated) and freely exposed to the surrounding tissue fluid. However, in both there seems to be a type of membrane that acts as a resistant barrier to the movement of ions.

The nerve membrane is extremely delicate and is intimately associated with signal transmission along the fiber. This membrane is probably a very thin surface layer of fatty material, only one or two molecules thick, thus making it invisible with high-powered magnifications. The surface membrane displays selective permeability, allowing certain chemical substances to pass through much more readily than others.

Under normal conditions the ion content of the fluids outside and inside the membrane would be the same. However, the selective permeability of the membrane appears to prevent this equilibrium, especially with regard to positive sodium (Na^+)

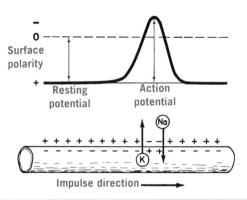

Fig. 27-2
Action potential and "sodium theory" of nerve impulse transmission. The action potential wave (top) spreads along the surface of the nerve fiber (bottom). During the rise of action potential, sodium ions (Na) enter the fiber and make it positive; during the resting state of the nerve, outward pressure of potassium ions (K) keeps the interior of the fiber negative. It is believed that the so-called resting cell may operate a "sodium pump" that constantly drives out sodium ions as fast as they enter. This may explain how a nerve cell manages to hold down its concentration of sodium ions against the combined forces of diffusion and electrical pressure.
(*From Katz, B.: The nerve impulse, Sci. Amer.* 187:55-64, 1952.)

ions, which are "pumped" out of the fiber as fast as they enter. This results in relatively more positive (+) ions outside the fiber than inside it—a condition known as polarization (Fig. 27-2).

Stimuli such as electricity, pressure, or chemicals upset the selective permeability of the fiber membrane. This in turn permits the inrush of positive sodium ions (Na^+) and causes depolarization, with more positive ions inside than outside. This spreads to the neighboring areas, resulting in more depolarization, and generates the nerve impulse. In other words the nerve impulse is a spreading wave of depolarization caused by the temporary upset of the selective permeability of the fiber membrane.

Synapses and impulse transmissions

Individual neurons are placed end to end, forming long pathways. The axon tip of one neuron "connects" functionally with a dendrite of the next.

This is not a structural connection because the fiber terminals are separated by a microscopic gap, called a synapse (Gr. *synapsis*, union).

The tip of an axon usually terminates in a series of brush-like fibers. Recent studies with electron microscopy have shown that the ends of the fibers are swollen into synaptic knobs. The knobs contain many small sacs called synaptic vesicles that hold the transmitter substances. Apparently the action of the nerve impulse is to force the emptying of the synaptic vesicles into the gap between the axon and dendrite. After discharging the transmitter substances, the vesicles move back into the synaptic knob where they are refilled.

Two hormones, epinephrine (an adrenalin-like substance) and acetylcholine, play important roles in synaptic transmission. Epinephrine is secreted by the axon terminals of some fibers of the sympathetic nervous system. Such epinephrine-producing fibers are often said to be adrenergic. Other fibers of the sympathetic system, and of the so-called parasympathetic nervous system and possibly the nerve fibers of the central nervous system, secrete a second hormone, acetylcholine, a rather simple chemical that is normally present in many parts of the body and has important physiologic functions. Acetylcholine brings about impulse transmission across many neural synapses. Acetylcholine-secreting fibers are called cholinergic.

In these synapses a potent enzyme, cholinesterase, is present that splits acetylcholine into acetyl and choline fractions, thus making the hormone ineffective. If cholinesterase were absent, acetylcholine would remain in the synapse and would stimulate dendrites constantly. If two impulses entered a synapse in quick succession, acetylcholine produced by the first would still be effective when the second produced its own hormone. Hence, the impulses would merge, and the result would be one long, extended response and not two distinct responses. Thus cholinesterase keeps the impulses separated. There is just sufficient time for acetylcholine to stimulate dendrites once, and then the hormone is immediately destroyed.

The consequences of impulse transmission in synapses due to chemicals include the following: (1) A diffusion through a synapse takes a relatively longer time than does impulse transmission within a nerve fiber. Thus a reflex through many synapses requires more time than is expected on the basis of impulse speeds through fibers alone. (2) Different types of impulses within fibers probably release different concentrations of hormones from axon terminals. If a certain terminal secretes an insufficient amount of hormone, it may be unable to activate an adjacent dendrite. However, if each of several axons at a synapse secretes subthreshold amounts of hormone at the same time, the total may suffice to initiate an impulse in the adjacent dendrite. This type of synaptic summation probably plays an important role in normal nervous activity. (3) Synapses direct the impulses in only one direction, from axon toward dendrites. Only axon terminals are sensitive to these hormones. (4) Synapses fatigue fairly easily, whereas nerve fibers rarely become tired. When axon terminals function intensely, they may exhaust their hormone-secreting ability temporarily, and synaptic transmission may slow down even more, or stop completely for some time.

Central nervous system
The brain

The human brain (Figs. 27-3 and 27-4) consists of (1) cerebrum, (2) cerebellum, (3) midbrain, (4) medulla oblongata, and (5) pons varolii. The cerebrum, which is the largest and most prominent part of the brain, consists of the right and left cerebral hemispheres. Each hemisphere is divided by sulci into five distinct areas known as lobes (frontal, parietal, temporal, and occipital lobes, and the insula, which is invisible from the surface). The outer layer of the cerebrum, known as the cortex, has numerous, foldlike convolutions that greatly increase the surface area. Specific regions of the cerebral cortex include the following: motor area, sensory areas (heat, cold, pain, touch, light pressure, and muscle sense), auditory area, visual area, olfactory area (taste and smell), and speech area. Beneath the gray cortex of the cerebrum is a mass of nervous tissue known as white matter.

The brain contains cavities (ventricles) as follows: (1) two lateral ventricles, one in each cerebral hemisphere; (2) the third ventricle behind the lateral ventricles and connected with each by an opening called the foramen of Monro; and (3) the

385

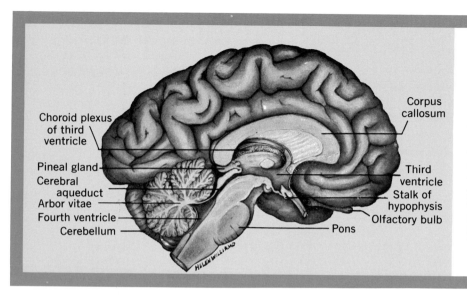

Fig. 27-3
Human brain, showing the medial aspect of the left half (sagittal section). The convoluted cerebrum (cerebral hemisphere) is shown above the corpus callosum, and the spinal cord is shown below the pons. The hypophysis is also called the pituitary gland.
(From Francis, C. C.: Introduction to human anatomy, ed. 5, St. Louis, 1968, The C. V. Mosby Co.)

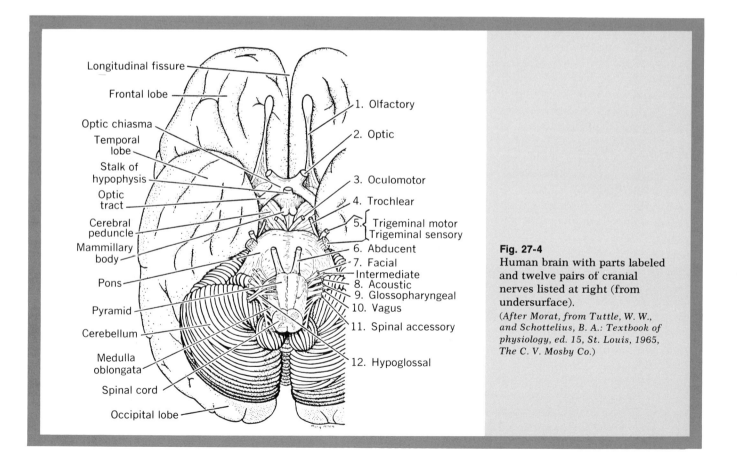

Fig. 27-4
Human brain with parts labeled and twelve pairs of cranial nerves listed at right (from undersurface).
(After Morat, from Tuttle, W. W., and Schottelius, B. A.: Textbook of physiology, ed. 15, St. Louis, 1965, The C. V. Mosby Co.)

Fig. 27-5
Summary of the human nervous system.

Central nervous system	Brain	**Cerebrum,** which is large, ovoidal, convoluted, and made of 2 hemispheres with 5 lobes **Cerebellum,** which is smaller, oval, nonconvoluted but with smaller furrows (sulci) **Midbrain,** which is short and connects the cerebellum with the pons varolii **Medulla oblongata,** which is pyramid shaped and continues with the spinal cord **Pons varolii,** which is in front of the cerebellum between the midbrain and the medulla oblongata and which connects the parts of the brain
		Cranial nerves (12 pairs) and their end organs **Spinal cord** for reflexes and pathways to and from the higher nervous centers **Spinal nerves** (31 pairs) and their end organs
Autonomic nervous system		**Sympathetic,** which has centers, ganglia, and plexuses in the cervical, thoracic, and lumbar regions of the spinal cord **Parasympathetic,** which consists of the centers and ganglia of the cranial and sacral parts of the autonomic system

fourth ventricle in front of the cerebellum and behind the pons and medulla, connected with the third ventricle by a small canal called the aqueduct of Sylvius. The coverings of the brain are called meninges and are the same as for the spinal cord (dura mater, outer layer; arachnoid, middle layer; pia mater, inner layer). Thin layers of fluid separate the various coverings.

Functions of the cerebrum, in addition to those already mentioned, are as follows: It governs all man's mental activities (reason, will, memory, intelligence, higher feelings, and emotions); it is the seat of consciousness, the interpreter of sensations, and the originator of voluntary acts; it acts as a control on many reflex acts that originate involuntarily (weeping, laughing, defecation, urination).

The cerebellum lies at the base or posterior part of the brain. The outer cerebellar cortex is made of gray matter, which is not convoluted but is traversed by numerous furrows (sulci). All functions of the cerebellum are below the level of consciousness, the main function being the reflex control of skeletal muscle activities.

The midbrain connects the cerebral hemispheres with the cerebellum and pons. Two pairs of round elevations, the corpora quadrigemina, act as centers for auditory and visual reflexes. Important pathways to and from other parts of the brain pass through the midbrain.

The pons lies in front of the cerebellum and above the medulla. Its fibers connect the two halves of the cerebellum and join the medulla with the midbrain.

The medulla oblongata lies between the pons and the spinal cord, being much like the spinal cord structurally. The fourth ventricle of the brain is located within the medulla and connects with the central canal of the cord. The medulla contains such vital centers as cardiac, respiratory, and vasoconstrictor centers, the latter for the control of arterial pressure.

Cranial nerves

Nerve impulses pass to and from the brain by way of the twelve pairs of cranial nerves. In some of these, such as the optic nerve, impulses pass only in the direction of the brain and are called sensory nerves. In others, called motor nerves (such as the facial nerve), impulses pass from the brain. Still others carry impulses both to and from the brain; these are mixed nerves, for example, the vagus nerve. The following is a listing of the cranial nerves:

I. **Olfactory**—sense of smell
II. **Optic**—sense of sight
III. **Oculomotor**—control of the following eye muscles: ciliary; inferior oblique; superior, inferior, and internal (medial) recti; sphincter of the iris of the eye
IV. **Trochlear (pathetic)**—superior oblique muscle of the eye
V. **Trigeminal**—sensory to the head; motor for the muscles of mastication

387

VI. Abducens (abducent) — external (lateral) rectus muscle of the eye

VII. Facial — motor to the face and scalp; sensory to the tongue; secretory to the submaxillary and sublingual (salivary) glands of the mouth

VIII. Auditory (acoustic) — to the cochlear part of the ear for hearing; to the semicircular canals of the ear for equilibrium

IX. Glossopharyngeal — motor to the pharynx; sensory to the tongue, mucous membranes of the pharynx, tonsils, eustachian tube, and tympanic cavity of the ear; secretory to the parotid gland (salivary) of mouth

X. Vagus (pneumogastric) — sensory to the larynx, trachea, lungs, esophagus, stomach, small intestine, and part of the large intestine; motor for respiration, heart action, and digestion (inhibits heart action); secretory for gastric and pancreatic glands

XI. Accessory — to muscles of the shoulder

XII. Hypoglossal — to the tongue

Spinal cord

The spinal cord consists of a central canal surrounded by a core of gray matter, which is surrounded by white matter. The gray matter in cross section resembles the letter H, the two forward projections being called anterior columns and the two backward projections, the posterior columns. The spinal cord serves as a center for spinal reflexes and as a pathway to and from the brain. The white matter of the cord has (1) long ascending tracts to transmit afferent impulses from the spinal nerves to the brain and (2) long descending tracts to transmit efferent impulses from the motor centers of the brain to the anterior columns of the cord to control muscular movements. A summary of the spinal nerves is given in Table 27-1.

Table 27-1
Spinal nerves

Spinal nerve	Number of pairs
Cervical (neck)	8
Thoracic (thorax)	12
Lumbar (back)	5
Sacral (pelvis)	5
Coccygeal (tail)	1
Total	31 pairs

Reference to Fig. 4-14 will show the general relationship of neurons, spinal nerves, and spinal cord. Dendritic fibers of a sensory neuron lead back to the spinal cord by way of a spinal nerve, which branches just before entering the spinal cord. The dorsal root has an enlargement, called a ganglion, which contains the cell bodies of the sensory neurons. The sensory axon then passes into the cord where it forms a synapse with an interneuron. The entire interneuron is located in the gray matter of the spinal cord. A second synapse contacts a motor neuron whose cell body is also in the cord. The motor axon leaves by way of the ventral root, which rejoins the dorsal root forming the intact spinal nerve. The spinal nerve branches at its terminus, supplying the various organs of that area.

Reflex pathways

A nervous reflex involves the transmission of an impulse, initiated in a receptor, causing a response in an effector. This system involves receptors, sensory neurons, the spinal cord or brain, motor neurons, and an effector. The familiar knee jerk reflex is one of the simplest examples.

Usually, a small hammer is used to strike the patellar tendon below the kneecap. This stretches the quadriceps muscle (in front of the thigh) and results in an impulse being carried by way of a sensory neuron to the spinal cord. Here, a synapse is made with a motor neuron, sending an impulse back to the quadriceps, which contracts, "jerking" the lower leg.

Other reflex patterns may be more complex than this, involving one or more interneurons. In some cases a particular stimulus may result in a variety of responses, the impulse being carried over several interneurons in the spinal cord. At the other extreme several sensory neurons have to be involved before a single response will take place. If the knee jerk may be considered an example of a simple reflex pathway, consider the type of pathway involved in an action such as balancing on one foot.

Autonomic nervous system

The autonomic nervous system consists of the sympathetic system (thoracolumbar), which has centers, ganglia, and plexuses in the cervical, thoracic, and lumbar regions of the spinal cord, and the parasympathetic (craniosacral) system, which consists of centers and ganglia of the cranial and sacral parts of the autonomic system. The involuntary control of the viscera by this system is essential to normal metabolism.

The two divisions are usually considered to be antagonistic to each other, since they control opposite effects in various organs. See Table 27-2 for some examples.

Table 27-2
Some functions of autonomic nerves

Organ	Sympathetic	Para-sympathetic
Iris muscles (pupil of eye)	Dilates	Constricts
Heart rate	Increases	Slows
Bronchi	Dilates	Constricts
Intestine	Decreases peristalsis	Increases peristalsis
Bladder	Fills	Empties
Sphincters	Contracted	Relaxed

The neuron units of the autonomic or visceral systems differ from the other parts of the central nervous system in that there are usually two motor units involved. Thus a second cell body is located outside the brain or spinal cord in a ganglion. Fibers from this innervate the visceral organs.

A reflex arc in the autonomic system would involve: (1) a sensory neuron, with a cell body in the dorsal root ganglion of the spinal nerve, (2) a preganglionic neuron, with a cell body in the spinal cord and axon, leaving to terminate in a ganglion outside the cord, and (3) postganglionic fibers, from a cell body in the ganglion, leading to the particular parts of the heart, lungs, and intestine.

Perhaps the major difference between the sympathetic and parasympathetic neurons is that the postganglionic fibers usually secrete different neurohormones. As in the fibers of the central system, preganglionic neurons secrete acetylcholine and are called cholinergic fibers. Postganglionic fibers of the parasympathetic system are primarily cholinergic also. Sympathetic postganglionic fibers, however, secrete adrenalin and are called adrenergic.

■

Sense receptors

As indicated previously, a neuron may be depolarized by a variety of stimuli. In many cases a particular type of stimulus results in an impulse sooner than another type. Accordingly, then, this efficient stimulus is the basis of the designation of certain sensory neurons as receptors for a particular type of stimulus.

Simple receptors

The sensory nerves of the skin transmit sensations of pressure, pain, heat, and cold from the specific sense organs to the proper parts of the central nervous system to be interpreted (Fig. 27-6). Pressure receptors of the skin, sex organs, and internal organs give us a sense of touch and automatic regulation of muscular coordination. The skin contains about 500,000 pressure receptors, which are especially abundant on the fingertips and the tip of the tongue.

The endings of the afferent nerve fibers in muscles are branched terminal arborizations (internal-pressure receptors), which frequently extend over the ends of the muscle fibers. Such receptors are stimulated by pressure caused by muscular contractions. However, afferent nerve fibers have other types of endings in muscles, such a spirals wound around muscles, tiny circular end plates, or club-shaped terminal organs. These receptors also have an effect on sensation. This conscious reaction is known as muscle sense, or kinesthetic sense (Gr. *kinein*, to move; *aisthesis*, perception), and through it we know whether muscles are contracted or moving, or whether a joint is bent.

Thermoreceptors in man are heat receptors,

389

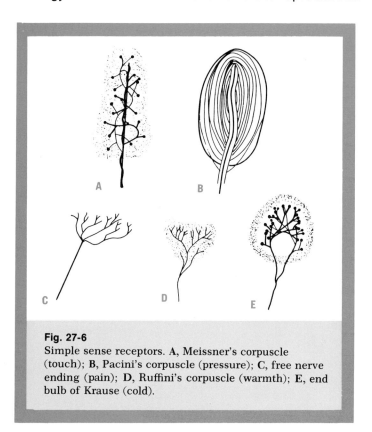

Fig. 27-6
Simple sense receptors. **A,** Meissner's corpuscle
(touch); **B,** Pacini's corpuscle (pressure); **C,** free nerve
ending (pain); **D,** Ruffini's corpuscle (warmth); **E,** end
bulb of Krause (cold).

which give the primary taste qualities of sweet,
sour, salty, and bitter.

The human taste buds are the end organs of
nerve filaments arising from the trigeminal, facial,
and glossopharyngeal nerves (cranial nerves).

The importance of smell to the taste of things
is well known, as demonstrated by foods losing
their taste when we have a cold with a stuffy nose.

Smell

Smell (olfactory) receptors are specialized epithelial cells in the mucous membranes of the nose
for the detection of several types of odors.

Although there is still much doubt as to the correct explanation for the sense of smell, recent investigations have suggested that the actual stimulus for an impulse in an olfactory fiber is caused
by the shape of the particular molecule. The terminal fibers in the nasal epithelium have a series of
pits, or depressions, into which a particular molecule may fit. Thus, a variety of substances having
the same general molecular shape would produce
the sensation of similar odors. A list of suggested
primary odors appears in Table 27-3. Complex
odors would result from a combination of these,
stimulating a variety of receptors.

stimulated by a temperature higher than that of
the skin, and cold receptors, stimulated by temperatures lower than that of the skin. The thermoreceptors of the tongue are particularly sensitive,
those of the face less so, and those of the trunk and
limbs least sensitive.

Complex receptors

Complex receptors are similar to the preceding
simple receptors in that they are based on single
sensory neurons. They differ in the number of
cells associated in the particular sense organ and
in their somewhat complex supporting tissues.

Chemoreceptors

Chemoreceptors are stimulated by suitable concentrations of definite chemical substances.

Taste

Taste (gustatory) receptors are groups of specialized epithelial cells, located chiefly on the tongue,

Table 27-3
Primary odors

Primary odor	Example
Camphoraceous	Moth balls
Musky	Skunk
Floral	Roses
Pepperminty	Mints
Ethereal	Ether
Pungent	Vinegar
Putrid	Rotten eggs

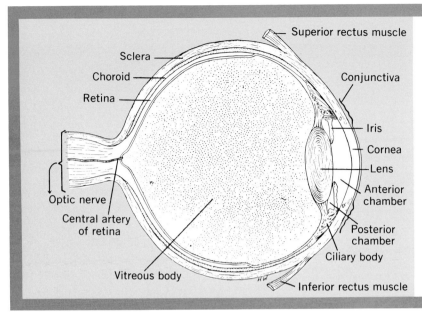

Fig. 27-7
Human eyeball (diagrammatic vertical section).
(*From Francis, C. C.: Introduction to human anatomy, ed. 5, St. Louis, 1968, The C. V. Mosby Co.*)

Vision

The visual apparatus consists of the eyes, their muscles that control movement, the tear glands, and the eyelids (Fig. 27-7). The eye itself is a spherical body formed of three coats. The outermost covering, the sclera, is a fibrous, protective layer, which is transparent in front, forming the slightly bulging cornea. In the back it continues as the covering of the optic nerve. The second, or middle layer, is pigmented and contains many blood vessels, giving it a dark color. This choroid coat continues in the front of the eye as the iris. The adjustable opening in the iris is called the pupil. A lymphlike fluid, the aqueous humor, fills the space on either side of the iris. Just behind this is the lens, held in place by ligaments. The innermost layer is the retina, the actual light receptor area. This does not continue in the front of the eye. The interior of the eyeball is filled with a clear, jellylike vitreous body.

The amount of light entering the eye is determined by the iris muscles, which in turn control the diameter of the pupil. These muscles are under autonomic control and dilate or constrict with varying light intensity. The actual fine focusing of the light waves on the retina is determined by the shape of the lens, which is controlled by muscles and is alternately bulging or slender, depending upon whether the object being viewed is near or far.

The retina is a multilayered structure that is specialized for light reception and impulse conduction. Most light receptor cells are sensitive to black and white light and are called rod cells. A smaller number of cells, called cones, function in color vision. There is evidence that separate cones exist that are sensitive to red, blue, and green light. For both rods and cones the action of a light wave appears to split a particular pigment molecule, resulting in a stimulus to a sensory neuron that leads to the brain by way of the optic nerve. In the rods this substance is called rhodopsin, or visual purple. Vitamin A is necessary for its formation.

Sound

The human auditory apparatus (Fig. 27-8) consists of (1) an external ear, with its auditory canal, which has a membranous tympanum (eardrum) at its inner end; (2) the middle ear, with its eustachian tube connecting it with the pharynx to equalize air pressure; the middle ear bones—hammer or malleus (L. *malleus*, hammer), anvil or incus (L. *incus*, anvil), and the stirrup or stapes (L. *stapes*, stirrup); and the two openings of the middle ear into the inner ear, which are known as the fenestra vestibuli (ovalis) and the fenestra cochleae (rotunda); (3) the internal ear, with its vestibule, its cochlea (shaped like a snail shell), and the three semicircular canals, which serve the purpose of equilibrium; and (4) the auditory,

391

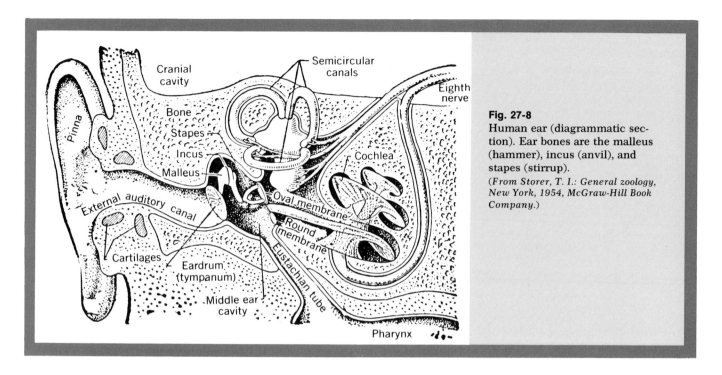

Fig. 27-8
Human ear (diagrammatic section). Ear bones are the malleus (hammer), incus (anvil), and stapes (stirrup).
(From Storer, T. I.: General zoology, New York, 1954, McGraw-Hill Book Company.)

or acoustic, nerve leading from the internal ear to the central nervous system.

Sound receptors are located in the coiled cochlea of the inner ear and are composed of vibrating "hair cells" that are stimulated by vibratory pressure waves brought to them from the ossicles (ear bones) of the middle ear. Sound vibration frequencies, which range from about 16,000 to 20,000 per second, are detectable. Gravity receptors consist of three pairs of membranous semicircular canals (labyrinth) within the inner ear. The canals are so arranged that one lies in a plane at right angles to both planes of the other two canals in its labyrinth and is in a plane parallel to that of one canal in the labyrinth of the opposite ear. An enlarged ampulla near the end of each canal contains most of the receptor organs. Their receptor cells bear flexible hairs projecting into the internal lymph. Movements create currents in the lymph of the canals, thus stimulating the hairs, which send impulses to the brain.

ENDOCRINE SYSTEM

The structure and functions of various organs in the human body are also affected by substances produced in other organs and transmitted primarily by the blood. This chemical coordination is brought about by specific chemical substances known as hormones (Gr. *hormaein,* to excite). These hormones are manufactured in certain organs from ingredients brought to them by the blood and are carried away by the blood instead of through ducts or tubes (Figs. 27-9 to 27-10).

Endocrine glands and their secretions that contain the specific hormones are influenced by such factors as the quantity and quality of foods brought to them by the blood, the action of hormones from other endocrine glands, and the action of certain parts of the nervous system affecting the medulla of the adrenals. Endocrine glands were formerly studied separately, but recent work has shown the great interdependence of many of them, and this new approach to their study has helped in obtaining a more correct picture of them. Some organs, such as testes, ovaries, and the pancreas, may function as both ductless and duct glands. These organs produce some substances that move through special tubes, such as the pancreatic duct, and other substances that are carried away in the blood. Some glands, such as the pituitary, thyroid, and adre-

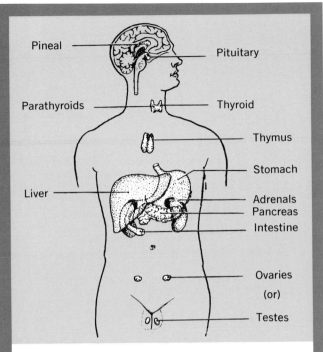

Fig. 27-9

Approximate locations of certain endocrine (ductless) glands. The liver shows the relative positions of other organs.

nals, function only as ductless glands. Several produce a number of hormones with more or less specific functions, which complicates the problem of their investigation.

Pituitary gland

The bilobed pituitary gland (Fig. 27-9) produces many hormones that influence many parts of the body, and consequently it is often called the master control gland of the body. In man it weighs about 0.5 gm. and is about the size of a pea. It lies between the roof of the mouth and the floor of the brain (hypothalamus) in a depression of the sphenoid bone. It is a double gland, consisting of (1) the anterior lobe, which together with the intermediate part (pars intermedia), arises embryologically from a pouch on the roof of the mouth and (2) the posterior lobe, which arises from the brain. This gland thus has a dual origin and quite different functions. It is also called the hypophysis. The various hormones and some of their functions are summarized in Table 27-4.

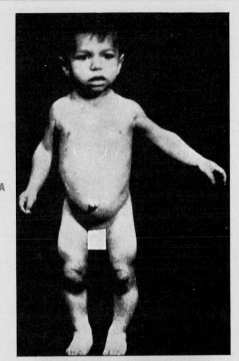

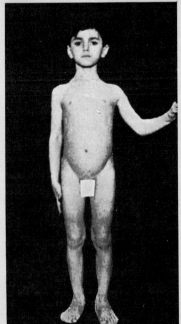

Fig. 27-10

A, Hypothyroidism in a 6-year-old child; B, same individual 2 years after the beginning of replacement treatment.

(Courtesy Dr. Jacob Rosenblum, from Wolf: Endocrinology in modern practice, Philadelphia, W. B. Saunders Co.)

Table 27-4
Summary of endocrine glands, hormones, and functions

Gland	Hormone	Function
1. Pituitary a. Anterior lobe	1. Somatotropic (growth-stimulating) hormone (STH) (somatotropin)	1. Regulates growth of somatic (body) cells
	2. Follicle-stimulating hormone (FSH)	2. Regulates growth of graafian follicles in ovaries; regulates spermatogenesis in seminiferous tubules of testes
	3. Interstitial cell-stimulating hormone (ICSH), or luteinizing hormone (LH)	3. Forms corpus luteum (ovaries); stimulates interstitial cells to produce male sex hormone
	4. Lactogenic hormone (LTH) (prolactin, luteotropin)	4. Stimulates lactation (milk production); stimulates secretion of corpus luteum hormone
	5. Thyroid-stimulating hormone (TSH) (thyrotropin)	5. Controls thyroid gland; controls iodine intake by thyroid and synthesis of thyroxine from diiodotyrosine
	6. Adrenocorticotropic hormone (ACTH)	6. Controls cortex of adrenal glands
b. Posterior lobe	1. Vasopressin; antidiuretic hormone (ADH)	1. Stimulates smooth muscles, urinary bladder, and gallbladder; elevates blood pressure (by constricting small arteries); decreases volume of urine secreted
	2. Oxytocin (Pitocin)	2. Contracts muscles of uterus; ejects milk from mammary glands
2. Adrenals (suprarenals) a. Cortex	1. Cortisone (hydrocortisone, glucocorticoids)	1. Controls food metabolism (glucose, proteins, and fats); suppresses inflammation
	2. Aldosterone (desoxycorticosterone [DOCA], mineralocorticoids)	2. Controls secretion of salt and water

Gland	Hormone	Function
	3. Androgens (androsterone and others similar to testosterone)	3. Similar to testosterone; androsterone produced by both men and women, and increased functioning in women causes such masculine traits as beard, deep voice, and regression of certain female reproductive organs
b. Medulla	1. Epinephrine	1. Increases blood pressure, blood sugar, and heart rate; inhibits gastrointestinal tract; hastens blood coagulation; decreases glycogen in liver
	2. Norepinephrine	2. Effects similar to adrenalin but with wider vasoconstrictor influences; resembles action of sympathetic nerves
3. Thyroid	1. Thyroxine	1. Controls rate of metabolism, growth, development, and maturation
	2. Triiodothyronine	2. Similar to thyroxine but acts more rapidly
4. Parathyroids	1. Parathormone (Parathyrin)	1. Regulates metabolism of calcium and phosphorus (necessary for bone growth, osmotic balance, and muscle contraction)
5. Islands of Langerhans of pancreas	1. Insulin	1. Controls metabolism of glucose (sugar) and fats; lack of insulin causes diabetes mellitus
	2. Glucagon	2. Controls breakdown of glycogen in liver; effect on carbohydrate metabolism opposite that of insulin
6. Thymus	"Thymic hormone"	Aids in forming lymphocytes and antibodies

Continued

Table 27-4
Summary of endocrine glands, hormones, and functions—cont'd

Gland	Hormone	Function
7. Pineal	Suspected of secreting a hormone influencing Na$^+$	
8. Testes (interstitial tissue)	1. Testosterone 2. Androsterone	Controls secondary sex traits, including larger muscles, body hair, beard, and deepened voice
9. Ovaries a. Graafian follicles	1. Estrone (estrogen, estradiol)	1. Controls secondary sex traits, menstrual cycle, mammary glands, and sex changes at puberty
b. Corpus luteum	1. Progesterone	1. Regulates menstruation, mammary glands, and pregnancy (by preparing uterus wall to receive fertilized egg)
10. Placenta	1. Chorionic gonadotropin	1. Regulates growth of corpus luteum during pregnancy
	2. Relaxin	2. Relaxes pelvic ligaments during childbirth
11. Gastrointestinal mucosa a. Stomach	1. Gastrin	1. Stimulates gastric juice secretion
b. Small intestine	1. Secretin	1. Stimulates pancreatic juice secretion when chyme from stomach enters upper intestine
	2. Enterogastrone	2. Decreases secretion and motility of stomach when hormone is carried by blood to stomach
	3. Cholecystokinin	3. Causes gallbladder contraction to force out stored bile; hormone is liberated from intestinal mucosa by fatty foods

When the pituitary is experimentally removed from young animals, growth stops and sexual maturity fails to develop. If removed from adults, male and female reproductive organs atrophy, along with the atrophy of the adrenal gland cortex and thyroid. If pituitary extracts are administered to normal young animals, they become giants who are sexually mature at an early age, and the sex glands, adrenal cortex, and thyroid enlarge because of increased secretions. An underactive pituitary during growth causes a small, properly proportioned midget. Oversecretion of the growth-stimulating hormone, after normal growth ceases, causes acromegaly (ak ro -meg' a ly) (Gr. *akron*, tip; *megalou*, large), with grossly enlarged hands, feet, facial bones, and broad long jaws.

Adrenal glands

One soft, cup-shaped adrenal gland is attached to the top of each kidney and is composed of (1) an outer cortex and (2) a darker, inner medulla. The two parts differ concerning embryologic origin, structure, and functions. Each adrenal gland weighs about 5 gm.

All of the hormones secreted by the cortex chemically are called steroids (Gr. *stereos*, solid; L. *oleum*, oil). The surgical removal of adrenal glands from animals results in decreased sodium and chlorine in the blood, a loss of water with decreased blood plasma (hence lowered blood pressure), and increased potassium in the blood. Insufficient cortical secretion results in Addison's disease (first described by the English physician, Addison, in 1855). Symptoms include muscular weakness, low blood pressure, digestive upsets, and bronzed skin caused by the depositing of a pigment called melanin (Gr. *melas*, black).

Thyroid gland

The paired, shield-shaped lobes of the thyroid gland, weighing about 25 gm., are connected by an isthmus and located in front of the trachea just below the voice box, or larynx. They are present in all vertebrates.

The thyroid contains a protein (thyroglobulin) that possesses a large amount of iodine. From the protein a hormone, thyroxine, can be obtained that contains about 65% iodine. A deficiency of thyroxine causes a reduction in the basal metabolism rate (BMR) of 30 to 50% of normal, whereas excess thyroxine causes increased oxygen usage, increased heat production, and increased metabolic wastes.

Hypothyroidism

Underactivity or degeneration of the thyroid, resulting in a deficiency of thyroxine, produces a disease in adults called myxedema (Gr. *myxa*, mucus; *eidema*, swelling). This condition is characterized by a low metabolic rate (decreased heat production); slow pulse; mental and physical lethargy; waxy, puffy skin caused by mucus in the subcutaneous tissues; usually a normal appetite, which because of the lowered metabolic rate results in food storage and a tendency toward obesity; and usually a falling of the hair. When hypothyroidism (Fig. 27-10) is present from birth, the disease is called cretinism, and the child has low intelligence and does not mature sexually. Successful treatments can occur if started early. Another type of hypothyroidism is produced by insufficient iodine in the diet, resulting in thyroxine deficiency. In an attempt to compensate the thyroid enlarges, producing simple goiter, in which the gland may be slightly or greatly enlarged with symptoms resembling those of myxedema but of a milder nature.

Hyperthyroidism

Hyperthyroidism is a condition that results either from the overactivity of a normal-sized gland or an increase in its size, with a highly increased metabolic rate (increased heat production) and excessive perspiration. Other symptoms include loss of weight (foods are quickly used), high blood pressure, irritability and nervous tension, weakness and involuntary trembling of muscles, protrusion of the eyeballs, known as exophthalmos (Gr. *ex*, out; *ophthalmos*, eye) and giving the name exophthalmic goiter to this condition.

Parathyroid glands

There are usually two pairs of parathyroid glands, about the size of a pea, which are embedded in the thyroid. When the parathyroids are removed surgically from an animal, death follows in a few days unless the hormone (parathormone) is adminis-

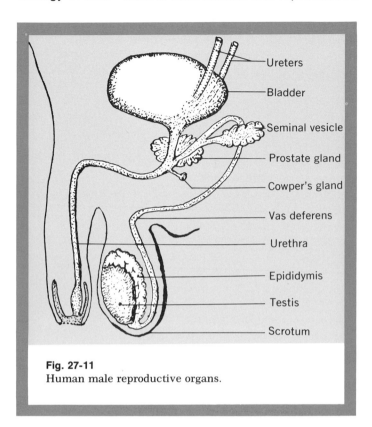

Fig. 27-11
Human male reproductive organs.

Labels: Ureters, Bladder, Seminal vesicle, Prostate gland, Cowper's gland, Vas deferens, Urethra, Epididymis, Testis, Scrotum

tive pancreatic juice (not an endocrine hormone).

Insulin was extracted from the pancreas by two Canadians, Banting and Best, in 1922, and has been used successfully in treating sugar diabetes (diabetes mellitus), which is a disease that has been known since the first century A.D. In sugar diabetes there is excess sugar in the blood and urine, causing excessive excretion of urine, and producing dehydration and thirst. The patient becomes progressively weaker and may die eventually unless treated. If too much insulin is administered, the blood-sugar level may be so lowered as to produce insulin shock, and unconsciousness may result unless the blood-sugar level is restored quickly.

Thymus gland

The thymus gland lies in the upper part of the chest near the lower part of the trachea. It is rather large during childhood but regresses at puberty (12 to 17 years of age).

Recent studies have indicated a possible endocrine function of the thymus in mice. Apparently, a substance is formed in the thymus that passes to the lymph glands and initiates the formation of a type of lymphocytes called plasma cells. These cells are antibody-producing cells and are the basis of immunity.

Pineal gland

The small, cone-shaped pineal gland between the two cerebral hemispheres is located at the base of the brain (thalamus) and posterior to the pituitary. Exact functions of this gland are not well established. It appears to be involved in light-controlled reactions, pigment functions, and aldosterone secretion.

Testes

Two ovoid bodies, the testes (L. *testis*, testicle), are suspended in the scrotum (Fig. 27-11), and their interstitial cells produce endocrine hormones; the seminiferous tubules of the testes produce sperms. Castration (removal of the testes) results in improperly developed secondary sex traits. A eunuch (Gr. *eune*, couch; *ehcein*, to hold) is a man who was castrated before maturity and

tered. Symptoms in animals after removal include cramps, involuntary muscle twitching, and convulsions. This condition, called tetany, results from an increased irritability of nerves and muscles caused by a decrease of calcium in body fluids and blood. If calcium solution is injected into a tetanized animal, the convulsion stops.

In case of increased parathyroid functioning caused by gland enlargement or a disease such as a tumor, the blood calcium increases. Part of this calcium may come from bones, rendering them soft and easily fractured. Muscles become painful, decrease in size, and are less irritable than normal. The excess blood calcium may be deposited in various organs.

Islands of Langerhans of the pancreas

The flat, elongated pancreas lies in the curvature between the stomach and small intestine. Groups of certain cells within the pancreas secrete the hormone, insulin, whereas other cells secrete a diges-

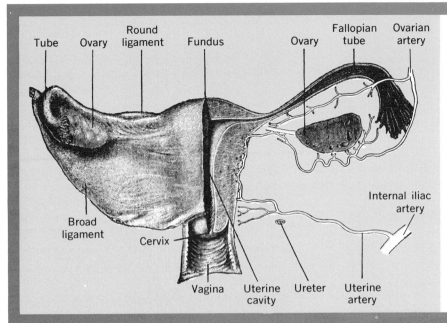

Fig. 27-12
Human female reproductive organs (the left half shows posterior [back] view and the right half shows diagrammatic section). The ovary shows several internal follicles in which the eggs (ova) are formed. The outer funnel-shaped end of the fallopian tubes (near ovaries) opens into the body cavity. The ureter is shown in cross section. Round and broad ligaments give support to ovaries and tubes.

(From Pitzman, M.: The fundamentals of human anatomy, St. Louis, The C. V. Mosby Co.)

as a result is beardless and has a high-pitched voice. Injection of testosterone into females causes some secondary male sex traits to develop. The male sex hormone also partially influences male sexual behavior.

Ovaries

Two bean-shaped ovaries, about 1½ inches long, near the female uterus (Fig. 27-12) contain an outer layer of germinal epithelium for forming eggs within the graafian follicles. Afer a fluid-filled follicle breaks and releases its egg it fills with yellowish cells, which constitute the corpus luteum (L. *corpus;* body; *luteum,* yellow). Between the onset of menstruation and the menopause one or more follicles mature and release an egg (ovulation), once each month. The ovarian hormones and their functions are summarized in Table 27-4.

Placenta

The placenta (L. *placenta,* flat cake) is a flattish organ that attaches the developing embryo to the uterus wall of the mother, supplying it with nourishment.

During the first eight weeks of pregnancy the placenta secretes a hormone, chorionic gonadotropin, which stimulates the corpus luteum to continue secretion of its hormones. In the third and fourth months, as the amount of chorionic gonadotropin decreases, the placenta begins to produce progesterone and estrogens.

Chorionic gonadotropin secreted in the urine has been used as an almost unfailing test for pregnancy (Aschheim-Zondek test). A small quantity of urine of a pregnant woman is injected into an immature female mouse, rat, or rabbit, where it stimulates the reproductive organs and results in estrus ("heat") and the characteristic changes in the genital tract.

Gastrointestinal mucosa

The mucous membranes lining the stomach and small intestine produce several hormones that are summarized in Table 27-4. One hormone, cholecystokinin (Gr. *chole,* bile; *kystis,* bladder; *kinesis,* to move), causes the gallbladder to empty its stored bile into the small intestine.

Interrelationships of hormones

Studies of several hormones and their functions reveal an involved interrelationship among the various endocrine glands, such as shown by the pituitary, adrenals, and reproductive organs. The hormones are exceedingly powerful agents, and in some instances their activities cover practically the entire body. What regulates the amount of each hormone produced? How do they interact so well and normally in most cases? In some instances the body requirements for a particular hormone will regulate its production automatically. Under normal conditions an increase in blood sugar stimulates the production of insulin, and the production is decreased when the need has been met.

The pituitary and its "target glands" serve as a good example of endocrine relationship (Fig. 27-13). The anterior pituitary secretes the following hormones: (1) Thyroid-stimulating hormone (TSH), which stimulates the thyroid gland to secrete thyroxine, which controls cell respiration and oxygen use as well as inhibiting further TSH release; (2) Somatotropic hormone (STH), which promotes protein formation and body growth; (3) Adrenocorticotropic hormone (ACTH), which stimulates the production of several hormones, including one that inhibits further production of ACTH; (4) Gonadotropic hormones, which are associated with sex and reproduction in females. The follicle-stimulating hormone (FSH), the luteinizing hormone (LH), and the lactogenic hormone (LTH) individually and collectively control the development of female sex characteristics, ovary maturation, egg production and release, milk production, and the release of hormones including estrogens and progesterone, which in turn influence further female development and inhibit the anterior pituitary.

These same FSH and LH, called the interstitial cell-stimulating hormone (ICSH), influence the development of male characteristics and especially the supporting interstitial cells of the testes and the seminiferous tubules that produce sperm. The testes in turn produce the androgens, such as testosterone, which in addition to supporting masculine characteristics, also inhibit the pituitary.

Antihormones

Another method of preventing an over-supply of hormones is the production of antihormones (Gr. *anti*, against). Several examples of these are known, including the antithyrotropic hormone. When an animal receives injections of the thyroid-stimulating hormone (TSH) of the anterior lobe of the pituitary, it shows an increase in the size of the thyroid and all the symptoms of increased thyroxine stimulation, such as increased basal metabolic rate. However, when these injections are continued for 30 days or longer, the animal no longer responds even to very large doses of TSH, the metabolic rate drops to normal, and the animal responds to thyroid hormone at this time. When the serum of an animal that has become resistant to TSH is injected into a normal animal, the latter also becomes refractory to the subsequent administration of TSH, because it shows no increase in its metabolism. This gradually acquired resistance is attributed to a substance (antihormone) that is antagonistic to the injected hormone. It is unknown whether the body relies upon this method of checking hormonal activities to any great degree.

REPRODUCTIVE SYSTEM
The formation of the reproductive system and its later functioning is intimately connected with the endocrine system.

Male reproductive system
The human male reproductive system (Fig. 27-11) consists of (1) a pair of testes, suspended in the scrotum; (2) numerous vasa efferentia, which lead into a single, highly convoluted collecting tubule, the two constituting the epididymis, which is attached to each testis; (3) the pair of vasa deferentia, or sperm ducts, which lead from the collecting tubules to the pair of saclike seminal vesicles, just behind the bladder; (4) the small prostate gland, surrounding the urethra and ejaculatory ducts; (5) the ejaculatory ducts, leading from the seminal vesicles to the single tubular urethra, which leads to the outside through the penis; and (6) the pair of small Cowper's glands posterior to the urethra and connected to it by a pair of small ducts.

The testes contain many seminiferous tubules that produce sperm (spermatozoa) by a proliferation of the spermatogonia cells that line the tubules. The number of sperm discharged at one time may be about 200 million, suspended in a small amount of seminal fluid (semen), which is secreted by the seminiferous tubules, epididymides, vasa deferentia, and primarily by the prostate gland. The sperm is extremely small and has a globular head with a nucleus, a neck, and a complex flagellar.

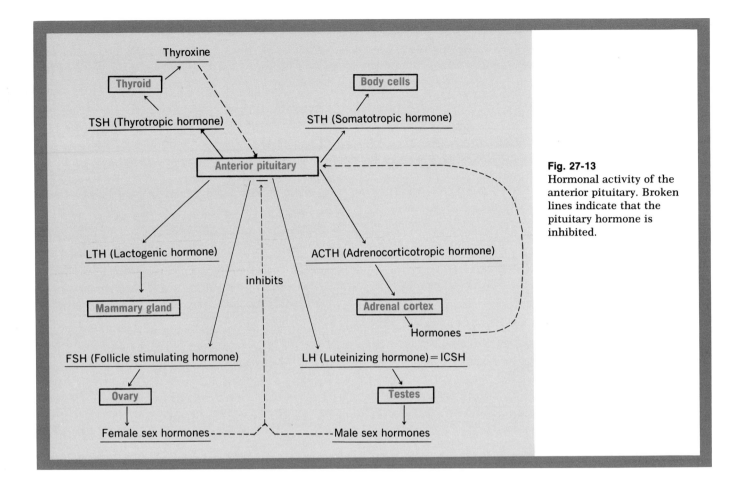

Fig. 27-13
Hormonal activity of the anterior pituitary. Broken lines indicate that the pituitary hormone is inhibited.

(Diagram labels:)

Thyroxine

Thyroid

Body cells

TSH (Thyrotropic hormone)

STH (Somatotropic hormone)

Anterior pituitary

LTH (Lactogenic hormone)

inhibits

ACTH (Adrenocorticotropic hormone)

Mammary gland

Adrenal cortex

FSH (Follicle stimulating hormone)

LH (Luteinizing hormone) = ICSH

Hormones

Ovary

Testes

Female sex hormones

Male sex hormones

Female reproductive system

The human female reproductive system (Fig. 27-12) consists of (1) the pair of oval ovaries in the lower abdominal cavity and (2) the pair of fallopian tubes (oviducts), the anterior ends of which are funnel shaped and lie near the ovary; the anterior opening of the tubes is the ostium (infundibulum), which picks up the ovum (egg) that has been produced and liberated by the ovary; the fallopian tubes carry the ovum to (3) the pouchlike uterus, in which the embryo develops; and (4) the vagina, which connects the uterus with the exterior. The walls of the uterus contain smooth muscles that contract vigorously under certain conditions, such as childbirth. The inner lining of the uterus, the endometrium, is a heavy, mucous, glandular layer to which the fertilized ovum may adhere. The uterus is well supplied with blood vessels for the nourishment of the future embryo.

With the onset of sexual maturity (puberty) the female begins to ovulate (produce and mature an ovum in the ovary). The ovum is ripened within the ovary and released into the fallopian tube where it may be fertilized by a male sperm or die if not fertilized. The sperm are deposited in the vagina during copulation and move up the fallopian tubes. Each developing ovum in the ovary is contained within a graafian follicle, which in later stages of its development occupies a position near the surface of the ovary, appearing there as a small bump. In fact, an ovary may possess several graafian follicles in various stages of development at the same time. When the ovum is mature, the follicle ruptures the wall of the ovary and deposits the ovum in the coelom (body cavity) from which it passes into the fallopian tube. Most of the graafian follicle cells remaining in the ovary organize themselves into a yellowish ductless gland called the corpus luteum, which is described in Table 27-4. (If the ovum is not fertilized, the corpus luteum

401

degenerates.) The human ovum is very small (0.15 mm. in diameter) because of a minimum of food (yolk). Consequently, the developing embryo must have nourishment from the mother. The ovum, while in the upper part of the fallopian tube, produces a small, nonfertilizable polar body (polocyte) and a second polar body after fertilization. Ovulation occurs at regular intervals with a series of interrelated phenomena, including the preparation of the uterus for the implantation of the fertilized ovum. The menstrual cycle in the human female occurs more or less within 28 days but may be altered by emotional shocks, psychic disturbances, worry, physical illness, or climatic changes. If fertilization does not occur, the superficial mucous layer of the uterus is shed, accompanied by a rupturing of blood vessels (hemorrhage). This produces menstruation, in which tissues and blood leave the uterus through the vagina.

Spermatogenesis

Testes contain thousands of cylindrical seminiferous tubules in which millions of sperms develop. The tubules contain a layer of germinal epithelium, which produces cells called spermatogonia that divide by mitosis to increase their number and provide for testis growth. After sexual maturity some spermatogonia undergo spermatogenesis, which consists of two meiotic cell divisions followed by the cell changes necessary to form mature sperms (Fig. 27-14). Other spermatogonia continue to divide by mitosis to produce more spermatogonia for later spermatogenesis.

Spermatogenesis starts by increasing the size of the spermatogonia to form primary spermatocytes. These divide by the first meiotic division into cells of equal size known as secondary spermatocytes. Each of these in turn divides by the second meiotic division to form four oval spermatids. Each spermatid contains a nucleus with the haploid number (N) of chromosomes. A spermatid must undergo changes (without further cell division) to become a mature sperm. Some of these changes include a decrease in the size of the nucleus, which becomes more dense as it is covered by membranes secreted by the Golgi complex and later forms the head of the sperm. Much of the cytoplasm is diminished, and the centrioles pro-

duce the flexible tail of the sperm to be used by it in locomotion. Mitochondria of the cell concentrate at the junction of the head and tail to form the middle piece of the sperm.

Oogenesis

Ovaries contain a layer of germinal epithelium, which by mitosis, produces numerous cells known as oogonia, and from which mature eggs will develop eventually. When the individual is sexually mature, an oogonium enlarges to form a primary oocyte, which contains yolk to serve as food in case the egg is fertilized. After it has completed its growth phase the primary oocyte divides by the first meiotic division (Fig. 27-14). The resulting cells are of unequal size, the larger, or secondary oocyte, receiving most of the cytoplasm and yolk, and the smaller, or polar body (polocyte), receiving mostly nuclear materials.

The secondary oocyte divides by the second meiotic division, again with an unequal distribution of cytoplasm, to form a large ootid, with most of the cytoplasm and yolk, and a second polar body. The first polar body may divide to form two additional polar bodies. The polar bodies disintegrate and disappear. The ootid undergoes further changes (without cell division) and becomes a mature egg (ovum). Hence, each primary oocyte produces one egg, in contrast to the four sperms formed by each primary spermatocyte in the male.

During fertilization the union of a haploid egg and a haploid sperm reestablishes the diploid chromosome number. The zygote (fertilized egg) and all body cells that develop from it by mitosis have this diploid number of chromosomes, one half of each pair (and one half of the genes) coming from the male and the other half of each pair (and half of the genes) coming from the female. Because of the interaction of genes, an offspring may resemble one parent more than the other, even though the two parents make equal contributions to its inheritance.

Endocrine control

In human males the gonadotropic secretions of the anterior pituitary, the interstitial cell-stimulating hormone (ICSH), control the development of the supporting cells. These interstitial cells

Fig. 27-14
Spermatogenesis and oogenesis. Diploid number of chromosomes in the body (somatic) cells is shown as two pairs, whereas sex cells have the haploid (N) number. **A,** Primordial germ cells of the male are formed in embryo and are inactive until the approach of sexual maturity; **B,** spermatogonia, which arise by mitosis at the approach of sexual maturity; **C,** primary spermatocyte in which meiosis begins; only one cell is drawn, showing the bundle of four components or chromatids (tetrad); during this time homologous pairs of chromosomes come together (synapsis) and crossing-over may take place between them; **D,** secondary spermatocytes, showing double chromosomes (dyads); **E,** spermatids, containing the haploid chromosome number; **F,** sperms in which much cytoplasm is lost and a tail is formed; **G,** primordial germ cells of the female are formed in embryo; **H,** oogonia, which arise by mitosis; **I,** primary oocyte in which meiosis begins (showing tetrad); **J,** secondary oocyte, which receives most of the cytoplasm; **K,** polar body (polocyte) is smaller but has half of the chromosomes; **L,** ootid is larger than its polar body; **M,** egg (ovum); polar bodies are nonfunctional and disintegrate; unequal distribution of cytoplasm gives the egg sufficient food for the development of the embryo after fertilization.

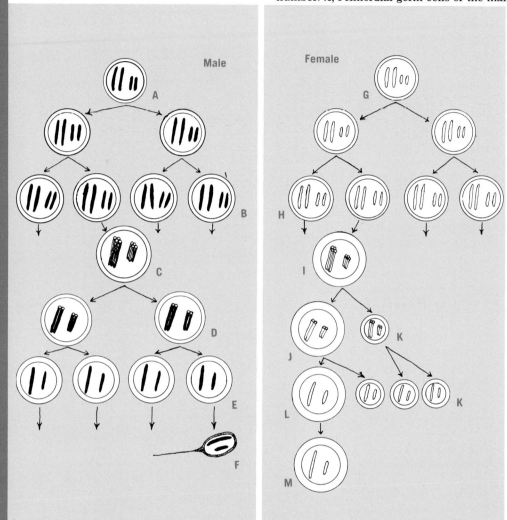

secrete testosterone, which controls the development of the male sex characteristics, including beard, muscles, and voice. Another anterior pituitary secretion, the follicle-stimulating hormone (FSH), influences the seminiferous tubules, which produce sperm.

In females the anterior pituitary hormone, FSH, controls the development of egg follicles and the secretion of estrogens from the follicle cells. The sex characteristics of females are maintained by these estrogens. In the mature female estrogens also inhibit further FSH secretion and stimulate the release of another anterior pituitary hormone, lutenizing hormone (LH). The decrease in concentration of FSH and the increase in LH stimulates ovulation, the release of the egg from the follicle and from the ovary.

The empty follicle now forms a yellowish corpus luteum. As more LH is secreted, the corpus luteum begins to release a new substance, progesterone. This hormone controls the preparation of the uterine lining for pregnancy. If a fertilized egg is present, the uterus is prepared to receive and nourish it; if not, menstruation takes place.

Menstruation

If fertilization does not take place, the corpus luteum continues to secrete progesterone. Increased concentrations of this inhibit LH from the pituitary; the corpus luteum disintegrates and progesterone secretion decreases. Without the influence of progesterone the uterine lining begins to break up, and the resulting blood and tissue are passed through the vagina as the menstrual flow (L. *mensis*, month). This takes place about 2 weeks after ovulation and indicates the end of one cycle of preparation for pregnancy. It is also the start of another one, because a tiny follicle is beginning to develop.

As has been indicated, the pituitary secretions, follicle developments, ovarian secretion, and uterine changes are all interrelated. The first 2 weeks of the cycle are referred to as the follicular phase, since it is during this time that the follicle develops, terminating in ovulation. The second 2 weeks are referred to as the luteal phase, since it is during this period that the corpus luteum is formed and is secreting. This phase terminates in menstruation. The following are the events of the follicular phase: (1) FSH secretion, (2) follicle development and maturation, (3) estrogen secretion and activity, (4) buildup of uterine lining, and (5) ovulation. The luteal phase both follows and depends on the follicular phase; it includes: (1) LH secretion, (2) ovulation, (3) corpus luteum formation, (4) progesterone secretion, (5) further development of uterine lining, and (6) menstruation.

■
Pregnancy and development

See Embryology of man in Chapter 28 for a discussion of pregnancy and development.

Some contributors to the knowledge of the biology of man

Hippocrates (460-370 B.C.)
A Greek who presented methods of medical treatments and the responsibilities of a physician to the patient and to the profession. His code, called the "Hippocratic Oath," is still followed by doctors. He made a science of medicine and is called the "Father of Medicine."

Herophilus (300 B.C.?)
A Greek who worked in Alexandria, Egypt, and was called the "Father of Anatomy." He is said to have dissected animals and human bodies. He wrote a book on anatomy, recognized the brain as the center of nervous activity, and distinguished between the kinds of nerves. At that time Alexandria was a great center of Greek learning in Egypt.

Galen (Claudius Galenus) (130-200 A.D.)
A Greek who was born in Asia Minor but received his medical training in Alexandria, Egypt, and then did much of his studying in Rome. In order to solve problems of anatomy and physiology he resorted to experiments and observations by dissections. For a period of 500 years after the time of Herophilus the dissection of human bodies was prohibited, so that Galen dissected animals primarily. His dissections showed in detail the brain, spinal cord, and nerves. His book on dissection, entitled *Anatomical Procedure*, originally consisted of sixteen volumes. His work was the authority in anatomy and physiology for 1,000 years, until the Renaissance.

Leonardo da Vinci (1452-1519)
The great Italian painter who believed that experience and direct observations, not classical authority, were the source of knowledge. He studied anatomy because of its relationship to painting, and he made comparisons between human anatomy and that of animals.

Andreas Vesalius (1514-1564)
A Belgian anatomist who published a large treatise, *On the Structure of the Human Body* (1543), based on dissections and direct observations rather than on the previous assumptions of so-called authorities. He went to Paris at the age of 16 years to study medicine, where in dissection he excelled over the barber dissectors who demonstrated the structures. He taught anatomy at the University of Padua, Italy, and was called the "Father of Modern Dissective Anatomy." He stressed the importance of direct observations and discovered that the right and left ventricles of the heart are completely separated from each other. (*Historical Pictures Service, Chicago.*)

Vesalius

Michael Servetus (1511-1553)
A Spanish anatomist, religious philosopher, and mystic who postulated the circulation of the blood, later proved by William Harvey. He was burned at the stake for his antireligious views.

Hieronymus Fabricius (153?-1619)
An Italian anatomist at Padua who discovered the valves in veins—knowledge that was later used by his student, William Harvey.

William Harvey (1578-1657)
An English physician who proved that blood circulates through the body in arteries and veins, although he did not see capillaries because he had no microscope. He proved that the heart was muscular and pumps actively and that veins controlled the direction of flow.

Giovanni Borelli (1608-1679)
An Italian physiologist who established physiology in what was later to become an effort to describe organic functions in mechanical and physicochemical terms. He taught Marcello Malpighi.

Richard Lower (1631-1691)
An English physician who first published an account of a successful blood transfusion, but it is probable that unsuccessful transfusions had been tried as early as 1492.

Stephen Hales (1677-1761)
An English physiologist and botanist who conceived experiments in animal and plant physiology, which served as a connection between physiologic studies of the 17th century and the 19th century.

Luigi Galvani (1737-1798)
An Italian naturalist who studied the roles of electricity in organisms, particularly those in muscular movements.

Edward Jenner (1749-1823)
An English physician who discovered that vaccination with cowpox would immunize man against the more serious smallpox (1798), thus ushering in a new era in preventive medicine.

William Beaumont (1785-1853)
An American surgeon and physiologist who performed famous experiments on digestion using the incompletely healed stomach of a wounded patient. His work accelerated the establishment of the modern physiology of digestion.

Thomas Addison (1793-1860)
An English physician who contributed greatly to establishing modern physiology of glandular secretions, especially those of the adrenal glands.

Johannes Müller (1801-1858)
A German who laid the broad outlines for present-day physiology, using chemistry and physics as tools. He stressed comparative physiology by studies of the functions in lower and higher animals.

Ivan P. Pavlov (1849-1936)
A Russian physiologist who performed many experiments with the digestive system and whose book, *Digestive Glands* (1897), was classic in digestive studies. He received a Nobel Prize in 1904. He proposed the concept that acquired reflexes play a role in the nervous reaction patterns in animals (1910).

Eijkman

Christian Eijkman (1858-1930)
A Dutch physiologist who produced experimental polyneuritis in fowls by feeding them polished rice. He called attention to rice hulls as being the source of an agent for preventing human beriberi (1897). This later led to the discovery of the antineuritic vitamin B_1. (*Historical Pictures Service, Chicago.*)

Frederick G. Hopkins (1861-1947)
An English biochemist and physiologist who made historic studies of muscular chemistry, cellular respiration, and biochemical dietary deficiencies.

Edward C. Kendall (1886-)
An American biochemist who first isolated the hormone thyroxine from the thyroid gland in crystalline form (1914) and isolated cortisone from the adrenal glands. ACTH (adrenocorticotropic hormone) was isolated from the pituitary gland (1943).

Frederick G. Banting (1891-1941) and Charles H. Best (1899-)
Banting and Best, Canadians, extracted the hormone insulin from the pancreas (1921), thus paving the way for the treatment of diabetes mellitus. This work was based on results of the removal of the pancreas from dogs by **Mering** and **Minkowski** (1889). (*Historical Pictures Service, Chicago.*)

Banting

Review questions and topics

1 Explain coordination in man by (1) nervous system and (2) chemical substances.
2 Describe the structure and function of (1) the various kinds of receptors, (2) the various kinds of effectors, and (3) the different types of conductors.
3 Explain in detail the theory regarding the initiation and conduction of nerve impulses.
4 List the more important structural and functional characteristics of the parts of the human nervous system.
5 Describe the structure and important functions of each endocrine gland.
6 Describe the anatomy and physiology of the male and female reproductive systems.

Selected references

Barrington, E. J. W.: The chemical basis of physiological regulation, Glenview, Ill., 1968, Scott, Foresman and Company.

Brown, J. H., and Barker, S. B.: Basic endocrinology, Philadelphia, 1962, F. A. Davis Co.

Carlson, A. J., Johnson, V., and Cavert, H. M.: The machinery of the body, ed. 5, Chicago, 1961, University of Chicago Press.

Case, J.: Sensory mechanisms, New York, 1966, The Macmillan Company.

Comroe, J. H.: The lung, Sci. Amer. 214: 56, 1966.

Davis, D. E.: Integral animal behavior, New York, 1966, The Macmillan Company.

Eccles, J.: The synapse, Sci. Amer. 212:56-66, 1965.

Francis, C. C.: Introduction to human anatomy, ed. 5, St. Louis, 1968, The C. V. Mosby Co.

Glassman, E., editor: Molecular approaches to psychobiology, Belmont, Calif., 1967, Dickenson Pub. Co., Inc.

Greisheimer, E. M.: Physiology and anatomy, ed. 8, Philadelphia, 1963, J. B. Lippincott Co.

Guyton, A. C.: Function of the human body, ed. 3, Philadelphia, 1969, W. B. Saunders Company.

Huxley, H. E.: The mechanism of muscular contraction, Sci. Amer. 213:18-27, 1965.

Montagna, W.: The skin, Sci. Amer. 212: 56-66, 1965.

Porter, K. R., and Franzini-Armstrong, C.: The sarcoplasmic reticulum, Sci. Amer. 212:72-80, 1965.

Prosser, C. L., and Brown, F. A.: Comparative animal physiology, ed. 2, Philadelphia, 1961, W. B. Saunders Company.

Ruch, T. C., and Fulton, J. F.: Medical physiology and biophysics, ed. 18, Philadelphia, 1960, W. B. Saunders Company.

Wood, J. E.: The venous system, Sci. Amer. 218:86-94, 1968.

Wurtman, R. J., and Axelrod, J.: The pineal gland, Sci. Amer. 213:50-60, 1965.

part five

THE CONTINUITY OF LIFE

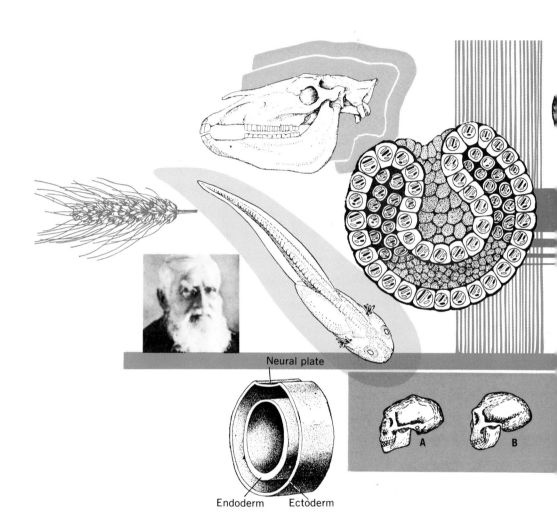

Neural plate

Endoderm   Ectoderm

Animal development
Genetics
Genes and gene action

chapter twenty-eight

Animal development

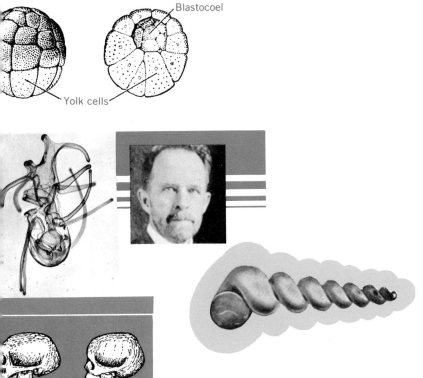

Blastocoel

Yolk cells

C   D

Changes in form during the growth and development of an organism result from several distinct processes. The total number of cells may increase by simple mitosis. The size of individual cells may increase while the number remains the same or also increases, and finally, the position of one cell with reference to another may change. The same number of cells may be present as a flat sheet, a solid mass, a hollow ball, or a mound. All types of increase in size and shape occur during the development of an animal.

Three distinct stages in the formation of a new individual are recognized; these are (1) cleavage, during which the fertilized egg divides repeatedly, resulting in an increase in the number of cells and a decrease in the size of cells; (2) gastrulation, which is the formation of the primitive tissues; and (3) morphogenesis, which is the organization of distinct organs and tissues characteristic of the particular animal.

MORPHOGENESIS

Morphogenesis deals with the study of the origin and development of form and structure. It includes the differentiation of tissues, organs, and systems. Both heredity and environmental factors (chemical, physical, and physiologic), external and internal, are influential in determining morphogenesis in an organism. The appearance and development of certain traits may be used to determine the age of the developing organism.

An egg, such as a frog egg, can be separated experimentally into two halves by tying a ligature through its middle, allowing development to be observed in each half. If the knot is tied rather tightly, the two halves begin to develop somewhat independently, resulting in a two-bodied embryo with two heads, two eyes in each, and complete duplication of limbs, gills, and hearts. When the ligature is made still tighter, two complete embryos result from the original egg. Evidently, the various structures of the embryo are not prefixed in the egg. It is even possible to interchange experimentally the different parts of an egg without unusual development. The factors that influence the ultimate use of cells in development probably are chemical, although other factors might well be considered.

408

Fig. 28-1

Starfish development. **A,** Unfertilized egg; **B,** zygote with polar body; **C,** 2 blastomeres; **D,** 4 blastomeres; **E,** 8 blastomeres; **F,** 16 blastomeres; **G,** 32 blastomeres; **H,** 64 blastomeres; **I,** nonmotile blastula; **J,** motile blastula; **K,** early gastrula; **L,** late gastrula (ventral view); **M,** early bipinnaria (lateral view); **N,** late bipinnaria (lateral view); **O,** brachiolaria; **P,** young starfish. (*Courtesy Carolina Biological Supply Company.*)

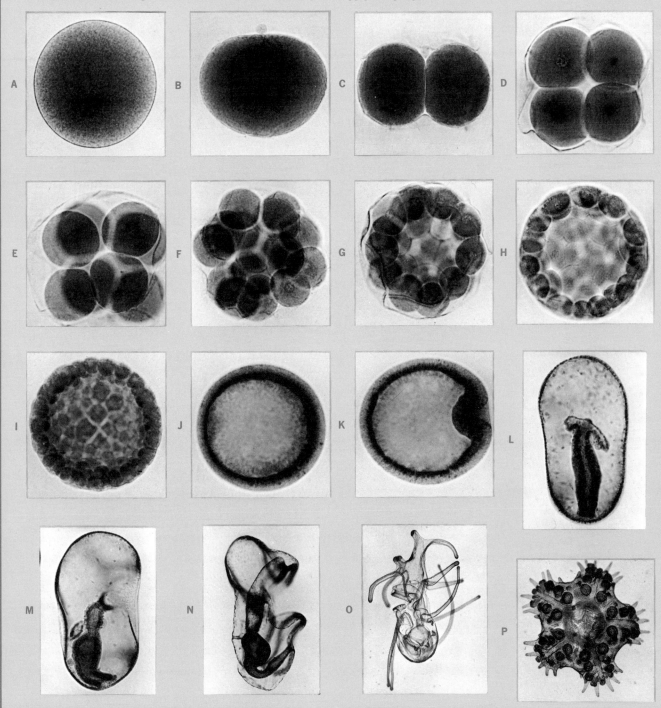

Fig. 28-2

Fertilization and early embryonic development in animals. The diploid number of chromosomes is considered to be two pairs in the body (somatic) cells, whereas the sex cells have the haploid number because of meiosis. **A,** Entrance of the sperm into the egg during fertilization; **B,** zygote or first cell of the new individual, in which the nuclear walls of the sperm and egg disappear, and their contents tend to fuse; note the tail of the sperm is left outside the egg; **C to H,** various stages in mitosis in which two cells are formed, each with equal chromosomes; **I,** four-cell stage arising from two cells by mitosis; **J,** many-cell (morula) stage; **K,** blastula or hollow-sphere stage (half section); **L,** early gastrula stage (half section); **M,** later gastrula stage, in which three primary germ layers (ectoderm, mesoderm, and endoderm) are shown.

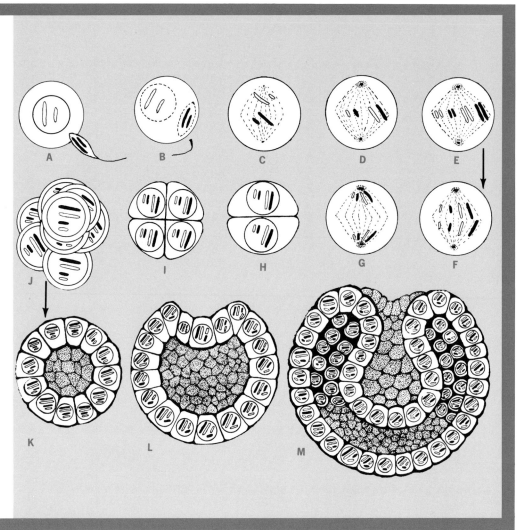

EMBRYOLOGY OF THE STARFISH

The stages in cleavage and gastrulation in the starfish zygote are quite easy to follow because of its small size. The entire series may be studied on one slide.

In general, after fertilization the zygote goes through a number of equal divisions resulting in 2, 4, 8, 16, 32 cells, and so on (Fig. 28-1). Although the number of cells increases, the total mass or bulk does not change. The repeated divisions result in the formation of a hollow ball of cells called the blastula (Gr. *blastos,* bud).

After this stage some of the cells divide faster than others, resulting in an invagination in one side of the blastula. This invagination represents the beginning of the gastrula stage (Gr. *gastros,* digestion).

Eventually, a distinct two-layered embryo is formed. The outer layer is the ectoderm, and the inner is the endoderm. The cells of the endoderm form a third layer, the mesoderm. These three primitive tissues form the distinct structures of the larval stages of the starfish.

EMBRYOLOGY OF THE FROG

The embryologic development of many animals is similar (Fig. 28-2), with only minor differences in certain stages. The frog has been chosen to illustrate the general principles because (1) the frog has

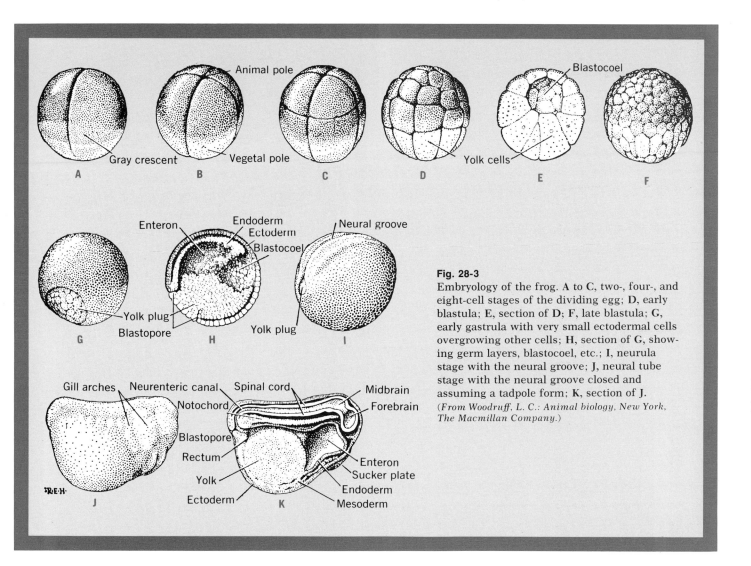

Animal pole

Gray crescent — Vegetal pole

A   B   C   D   Yolk cells   E   Blastocoel   F

Enteron   Endoderm   Neural groove
Ectoderm
Blastocoel

Yolk plug
Blastopore   Yolk plug
G   H   I

Fig. 28-3
Embryology of the frog. A to C, two-, four-, and eight-cell stages of the dividing egg; D, early blastula; E, section of D; F, late blastula; G, early gastrula with very small ectodermal cells overgrowing other cells; H, section of G, showing germ layers, blastocoel, etc.; I, neurula stage with the neural groove; J, neural tube stage with the neural groove closed and assuming a tadpole form; K, section of J.
(*From Woodruff, L. C.: Animal biology, New York, The Macmillan Company.*)

Gill arches   Neurenteric canal   Spinal cord   Midbrain
Notochord   Forebrain
Blastopore
Rectum   Enteron
Yolk   Sucker plate
Endoderm
Ectoderm   Mesoderm
J   K

a rather typical, representative method of development and (2) the materials used in the study are available and rather inexpensive.

Frog eggs (and also eggs of fish, reptiles, and birds) have a large amount of yolk concentrated toward the lower or vegetal pole, while the cytoplasm is at the upper or animal pole.

Early cell stages

In 2 or 3 hours after fertilization the zygote divides by mitosis to form the 2-cell stage. A second division occurs by mitosis in about 1 hour and at right angles to the first plane of division, thus forming the 4-cell stage. These four, more or less equal cells are called blastomeres (Gr. *meros*, part).

The next plane of cleavage is horizontal and slightly above the middle or equator, thus dividing each of the four previous cells to form the 8-cell stage (Fig. 28-3). Of these eight cells, four are pigmented, smaller, located at the animal pole, and known as micromeres (Gr. *mikros*, small); the other four are unpigmented, larger, located at the vegetal pole, and known as macromeres (Gr. *makros*, large).

Morula and blastula stages

These cells continue to divide until there is a large number of cells, packed together in a somewhat solid mass known as the morula stage (L. *morum*, berry). The micromeres continue to divide

411

more rapidly than the macromeres at this time. It is evident that growth cannot continue indefinitely in this manner, or the animal would be solid, without cavities in which tissues and organs could be placed. Consequently, at a certain stage in the cleavage process, the cells of the morula stage line up to form a one-layered, hollow sphere known as the blastula stage. This sphere has a central, fluid-filled cavity known as the blastocoel (Gr. *koilos*, hollow). The blastula stage consists of (1) an outer, transparent, jellylike capsule; (2) a dark, pigmented animal hemisphere, composed of smaller and more numerous cells; and (3) a light-colored, unpigmented vegetal hemisphere, which is composed of larger and fewer cells. The vegetal cells are quite large and contain yolk or food that is supplied to the cells of the animal hemisphere. This arrangement makes the blastula wall on the vegetative side much thicker than that on the upper or animal side. The active growth in this stage occurs primarily in the animal hemisphere region.

Gastrula stage

The gastrula (yolk plug) stage is formed as follows. At a certain point between the animal and vegetal hemispheres the vegetal cells turn in toward the blastocoel. The pigmented animal cells (ectoderm) grow over the lighter, unpigmented vegetal cells (endoderm) and fold in with them to some extent at that point. Thus, an inner layer of cells is continuous with the outer layer. Because of more rapid mitosis, the animal hemisphere continues to grow almost entirely over the vegetal hemisphere, leaving a small, light yolk plug exposed. The space between the boundaries of the infolded layers of cells, which surround the yolk plug, is the blastopore, or primitive mouth. The ingrowth of the latter is shown on the surface by a thin, crescent-shaped fold or groove. The outer layer of cells, the ectoderm, is continuous with the inner inturned layer or endoderm. The point at which the endoderm cells of the vegetal region turn in is one side of the yolk plug and is known as the dorsal lip of the blastopore. This inturned endoderm forms a cavity, known as the archenteron (Gr. *arche*, beginning; *enteron*, gut), or primitive intestine (primitive gut). The blastocoel now appears as a reduced cavity at the opposite side and

is gradually being crowded out by the developing archenteron and endoderm. The indentation on the side of the yolk plug opposite the dorsal lip of the blastopore is known as the ventral lip of the blastopore.

Neurula stage

The neurula (Gr. *neuron*, nerve) stage follows the gastrula stage. The neural groove, which is the forerunner of the future nervous system, begins as a small depression on the dorsal side of the blastopore and grows anteriorly along the dorsal side of the embryo as a thickened neural plate (medullary plate) in the ectoderm (Fig. 28-4). During this time the embryo grows longer and develops definite anterior and posterior ends. A thickened fold at each margin of the original neural plate forms a neural fold (medullary fold). These folds at first are flat and far apart; later they arch toward the median dorsal line and unite to form the future neural tube. At this time the neural plate sinks to form a definite neural groove along the middorsal side of the embryo. The neural groove is composed of ectoderm cells.

An elongated mass of cells dorsal to the archenteron forms the long, rodlike notochord (Gr. *noton*, back; *chorde*, rod), which still may be connected with the endoderm from which it originates. The mass of cells at either side of the neural groove is known as the mesoderm (middle germ layer).

Neural tube stage

The neural tube stage closely follows the neurula stage. The two neural folds on the dorsal surface of the embryo at this time have met and fused into an elongated neural tube. At this stage the neural tube probably will be free from the outer ectoderm from which it originated. The anterior part of the neural tube constricts and enlarges by well-regulated mitosis to form the future fore-, mid-, and hindbrains. The notochord is now free from the archenteron and is just below the neural tube.

The mesoderm completely surrounds the archenteron ventrally, and near the middorsal part there appears a small split or break that is the forerunner of the coelom (body cavity). This break continues ventrally, thus forming the body cavity between the two layers of the mesoderm. The inner

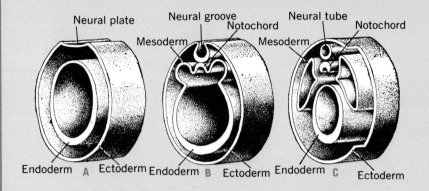

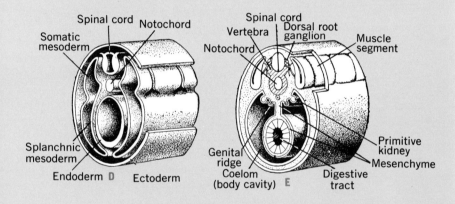

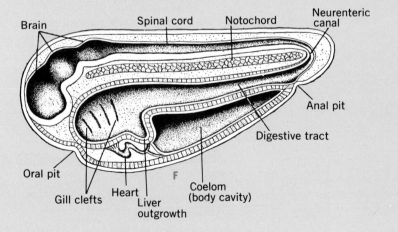

Fig. 28-4
Development of a vertebrate (somewhat diagrammatic). Stages **A** to **E** are a cross section of the mid-body region. **A,** Neural plate stage; **B,** neural groove stage; **C,** neural tube stage; **D,** spinal cord stage; **E,** spinal cord stage still later; **F,** embryo cut lengthwise to show the internal structures.

(From Guyer, M. F.: Being well born, Indianapolis, The Bobbs-Merrill Co., Inc.)

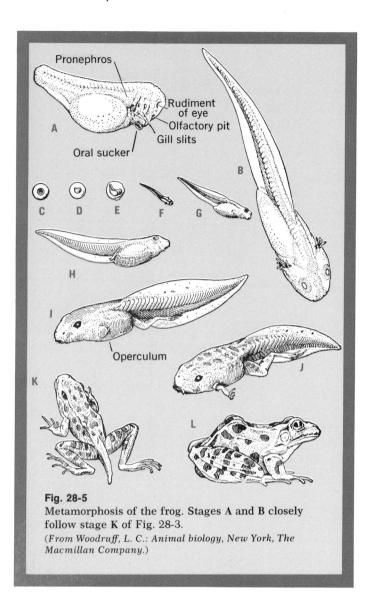

Fig. 28-5
Metamorphosis of the frog. Stages A and B closely
follow stage K of Fig. 28-3.
(*From Woodruff, L. C.: Animal biology, New York, The
Macmillan Company.*)

suckers on the ventral side of the tadpole serve
for attachment purposes. The stomodeum (primitive mouth) appears as an oval pit in front of the
suckers. The olfactory pits are a pair of small
depressions above and anterior to the stomodeum.
The three pairs of external gills are fingerlike
processes on either side of the head that act as
specialized organs of respiration. The proctodeum
(primitive anus) is located on the dorsoposterior
part of the tadpole. A tail and a pair of eyes are also
present.

The larval stage with internal gills (Fig. 28-5)
follows the stage with external gills. The external
gills are covered by a fold of skin known as the
operculum (gill cover), which has a single opening,
the spiracle. As the three pairs of external gills
are resorbed, four pairs of internal, fishlike gills
are formed. In this stage the suckers are small
projections just behind the mouth. The mouth is
surrounded by a number of small projections
known as the circumoral papillae and has a pair of
horny jaws. The intestine shows through the
transparent ventral body wall as a long, coiled tube.
This great length of intestine suggests a typical
vegetarian animal, which the tadpole really is at
this stage. The hind limb buds appear as small
outgrowths on either side of the anal opening.
These buds will continue to grow by mitosis into
the real hind limbs.

The later stages of development follow. The front
limb buds appear and develop into typical front
legs. The tail gradually is resorbed and disappears,
the materials being taken to the liver and stored.
The internal gills are resorbed, and their place
taken with rapidly growing lungs. The adult frog
is not aquatic but has lungs similar to other land-
living (terrestrial) animals. The coiled intestine
gradually shortens, suggesting a typical carnivorous (flesh-eating) animal that the frog has
become.

EMBRYOLOGY OF MAN
Sperms produced by the testes are deposited at
copulation in the female vagina and swim by
means of their whiplike tails through the glandular secretions along the wall of the uterus and
finally to the paired fallopian tubes (oviducts).

The production of ova (eggs) by the ovary (Fig.

layer of the mesoderm, known as the splanchnic
layer (Gr. *splancnon*, entrail), lies next to the endoderm, whereas the outer layer, known as the
somatic layer (Gr. *soma*, body), lies next to the
ectoderm. The cells of both ectoderm and endoderm are now quite distinct.

Larval stages
The larval stage with external gills follows the
neural tube stage. A pair of oval, thick-lipped

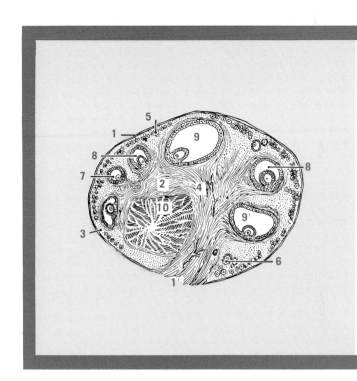

Fig. 28-6

Ovary of the cat (section). 1, Free border with its germinal epithelium, which will give rise to the graafian follicles; 2, stroma or framework of future follicles; 3, connective tissue stroma with 4, blood vessels; 5, graafian follicles of which 6, 7, 8, 9, and 9′ are successive stages of maturation; 10, corpus luteum (L. *corpus,* body; *luteus,* yellow), or "yellow mass." Cells of the ruptured graafian follicle give rise to the corpus luteum, which encloses a blood clot formed by blood escaping from the blood vessels of the ovary during liberation of the egg. If a human egg is fertilized, the corpus luteum enlarges and remains during the first 7 months of gestation; if the egg is not fertilized, the corpus luteum is reabsorbed in 2 to 3 weeks.

(After Schrön, from Tuttle, W. W., and Schottelius, B. A.: Textbook of physiology, ed. 15, St. Louis, 1965, The C. V. Mosby Co.)

28-6) is called ovulation. When the graafian follicle, which encloses the developing egg, collapses and the wall of the ovary breaks, the egg is passed into the abdominal cavity near the opening (ostium) of the fallopian tube. Through the action of the cilia that line the fallopian tube the egg is drawn into it, and fertilization usually takes place there, although on rare occasions, the egg may be fertilized while still in the abdominal cavity. In the latter case the developing embryo must be removed surgically. The egg will not develop unless fertilized within a week after it is produced. The egg extrudes its first polar body at the time of ovulation and a second polar body upon fertilization. The egg is now ready to divide (cleave).

Cleavage of the fertilized ovum (zygote) occurs in the fallopian tube. The first cleavage results in two equal cells (blastomeres), which adhere to each other and are surrounded by an albuminous layer called the zona pellucida. The second cleavage is accomplished by one of the blastomeres dividing longitudinally at right angles to the first cleavage, followed by cleavage of the other blastomere, thus

forming the 4-blastomere stage. Cleavage continues until a small sphere of blastomeres the size of a pinhead is formed; this is known as the morula stage. The morula is still in the fallopian tube but is approaching the fundus (base) of the uterus. Approximately 100 hours are required for the fertilized ovum to develop to the 16-cell stage of the morula. The morula consists of (1) an outer layer of cells called the trophectoderm, or trophoderm (Gr. *trophe,* nourishment; *ecto,* external; *derm,* covering), and (2) an inner cell mass. The cells of the trophectoderm increase so that the outer surface is much enlarged, thereby forming a fluid-filled cavity (blastocoel) within the morula, which is now known as the blastocyst (blastodermic vesicle). Within 10 days after fertilization the blastocyst has moved into the uterus. This has developed a thick, glandular layer, which is stimulated by a hormone, progestin, to produce a sticky fluid by which the blastocyst adheres to the uterine wall. After a few hours the blastocyst begins to sink beneath the mucous layer of the uterus (endometrium), which has been eroded by cytolytic

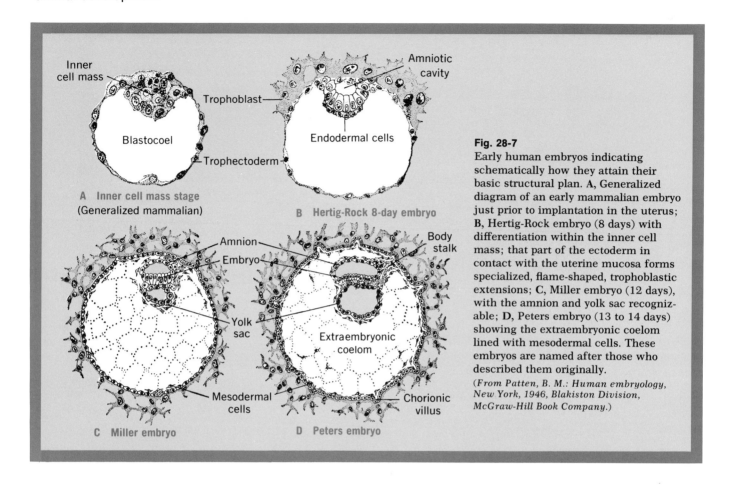

A **Inner cell mass stage**
(Generalized mammalian)

B **Hertig-Rock 8-day embryo**

C **Miller embryo**

D **Peters embryo**

Fig. 28-7
Early human embryos indicating schematically how they attain their basic structural plan. **A,** Generalized diagram of an early mammalian embryo just prior to implantation in the uterus; **B,** Hertig-Rock embryo (8 days) with differentiation within the inner cell mass; that part of the ectoderm in contact with the uterine mucosa forms specialized, flame-shaped, trophoblastic extensions; **C,** Miller embryo (12 days), with the amnion and yolk sac recognizable; **D,** Peters embryo (13 to 14 days) showing the extraembryonic coelom lined with mesodermal cells. These embryos are named after those who described them originally.

(From Patten, B. M.: Human embryology, New York, 1946, Blakiston Division, McGraw-Hill Book Company.)

action of the trophectoderm cells of the blastocyst. This phenomenon ensures nourishment for the developing embryo until it can secure its own food.

The inner cell mass, near the point where it contacts the trophectoderm, forms the hollow amniotic cavity (Fig. 28-7). The free, or unattached, region of the inner cell mass forms a yolk sac cavity along the inner surface of the trophectoderm. The space between the yolk sac cavity and the trophectoderm is quite large. The yolk sac cavity contains no yolk (food), but its upper region will later form the roof of the alimentary canal. The amniotic and yolk sac cavities are separated by a cellular embryonic disk (embryonic shield) that will form the embryo. The ventral layer of the embryonic disk is endodermal, its dorsal layer ectodermal, and the cells between the two are mesodermal. Mesoderm is also formed between the yolk sac cavity and the trophectoderm and between the ectodermal lining of the amniotic cavity and the trophectoderm. The trophectoderm is lined on the inside by a layer of mesoderm, known as somatic mesoderm. The trophectoderm and somatic mesoderm combined are known as the chorion, because the latter, through minute projections (villi), contacts the blood vessels of the uterus in order to supply nourishment until the future blood system of the embryo is developed. Hence, the endoderm of the yolk sac cavity is covered with splanchnic mesoderm, whereas the ectoderm of the amniotic cavity is covered with somatic mesoderm. The blastocyst also develops a third cavity, the extraembryonic

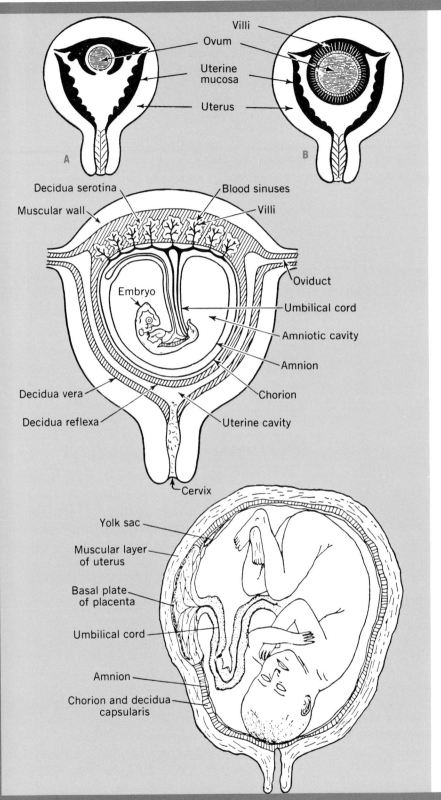

Villi

Ovum

Uterine
mucosa

Uterus

A

B

Fig. 28-8
Pregnant uterus (diagrammatic). **A,**
Embryo embedded in the mucous layer
of the uterus (uterine mucosa); **B,** later
stage with villi (for the absorption of
blood) of the chorion membrane covering
the embryo.
*(From Zoethout, W. D., and Tuttle, W. W.:
Textbook of physiology, ed. 11, St. Louis,
1952, The C. V. Mosby Co.)*

Decidua serotina

Blood sinuses

Muscular wall

Villi

Embryo

Oviduct

Umbilical cord

Amniotic cavity

Amnion

Decidua vera

Decidua reflexa

Chorion

Uterine cavity

Cervix

Fig. 28-9
Uterus at about the seventh week of
pregnancy (diagrammatic).
*(From Tuttle, W. W., and Schottelius, B. A.:
Textbook of physiology, ed. 15, St. Louis,
1965, The C. V. Mosby Co.)*

Yolk sac

Muscular layer
of uterus

Basal plate
of placenta

Umbilical cord

Amnion

Chorion and decidua
capsularis

Fig. 28-10
Human fetus shown in the normal
position in the uterus (section). The
chorion is the outer embryonic mem-
brane and the amnion is the inner.
*(Modified after Ahfeld; from Potter, G. E.:
Textbook of zoology, ed. 2, St. Louis, 1947,
The C. V. Mosby Co.)*

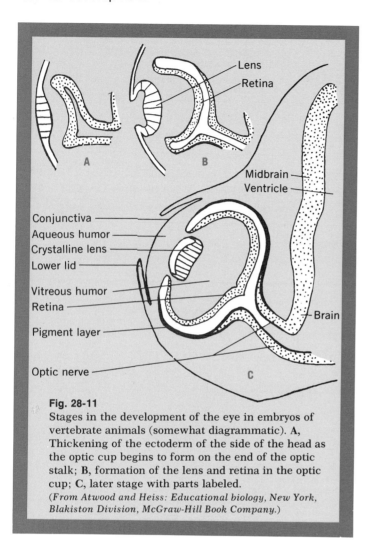

Fig. 28-11
Stages in the development of the eye in embryos of vertebrate animals (somewhat diagrammatic). **A,** Thickening of the ectoderm of the side of the head as the optic cup begins to form on the end of the optic stalk; **B,** formation of the lens and retina in the optic cup; **C,** later stage with parts labeled.
(*From Atwood and Heiss: Educational biology, New York, Blakiston Division, McGraw-Hill Book Company.*)

Labels in figure: Lens, Retina, Midbrain, Ventricle, Conjunctiva, Aqueous humor, Crystalline lens, Lower lid, Vitreous humor, Retina, Pigment layer, Optic nerve, Brain

ion. The yolk sac and allantois do not play as great a role in human embryologic development as they do in lower forms of organisms.

Between the third and fourth week of gestation the blastocyst has enlarged to form a bulge on the surface of the uterus. The embryo soon pushes into the cavity of the uterus, and is surrounded by the amnion membrane of the embryo (Gr. *amnion,* embryo covering). The body stalk functions as the umbilical cord, being continuous with the highly vascular, disk-shaped placenta, which in turn is in contact with the blood vessels of the walls of the uterus (Figs. 28-8 to 28-10).

About two thirds of the embryonic disk will form the future head, and the remainder will form the neck, trunk, and tail. The two cellular layers of the embryonic disk consist of (1) the lower endoderm layer (nearest the yolk sac cavity) and (2) the upper ectoderm layer, which gives rise to the brain, spinal cord, and outer skin. Between the ectoderm and endoderm, on either side of the median line and originating from both, two groups of cells are formed: (1) the membranous mesoderm (mesothelium) and (2) a loosely arranged meshwork of cells called the mesenchyme. The ectoderm, mesoderm, and endoderm are known as the three primary germ layers, because from them arise all the tissues and organs of the future organism.

From the ectoderm come (1) the epidermis and its derivatives, such as hair, nails, glands, and lenses of the eyes (Fig. 28-11); (2) nervous tissues, including the neuroglia; (3) the epithelium of the organs of special sense, of the mouth and its oral glands, of the hypophysis, of the anus, and the amnion; (3) the chorion; and (4) the smooth muscles of the iris (eye) and the sweat glands.

From the mesoderm come (1) the epithelial lining of the pericardium, pleura, peritoneum, and urogenital system; (2) striated muscles; (3) smooth muscles; (4) the notochord; (5) connective tissues, including cartilage and bone; (6) bone marrow; (7) blood; (8) the lining (endothelium) of the blood vessels and lymph system; (9) lymphoid organs; and (10) the cortex of the suprarenal gland.

From the endoderm come (1) the epithelium of the pharynx and its derivatives, the thyroid, parathyroids, thymus, tonsils, and auditory tube; (2) the digestive tract, including the liver and

coelom, located between the two separating layers of mesoderm. These two layers were originally one layer that was located between the endoderm of the yolk sac cavity and the trophectoderm.

In the next stages of development the embryonic mass (embryo) detaches itself partially from the inner surface of the chorion, grows rapidly, and forms a tubular outgrowth from the upper region of the yolk sac. This outgrowth, the allantois (Gr. *allanto,* sausage or tubular; *eidos,* form), grows toward the chorion and through the body stalk, by means of which the embryo is attached to the chor-

Table 28-1
Early embryology and germ layer derivatives

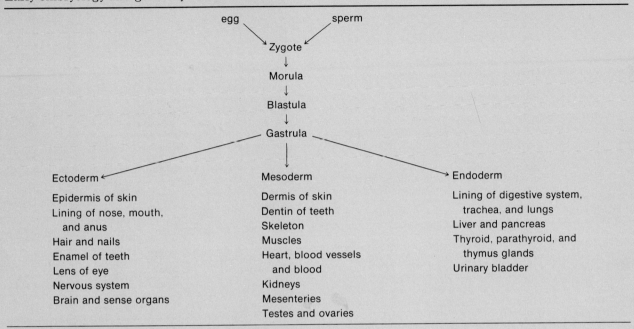

Ectoderm	Mesoderm	Endoderm
Epidermis of skin	Dermis of skin	Lining of digestive system, trachea, and lungs
Lining of nose, mouth, and anus	Dentin of teeth	Liver and pancreas
Hair and nails	Skeleton	Thyroid, parathyroid, and thymus glands
Enamel of teeth	Muscles	Urinary bladder
Lens of eye	Heart, blood vessels and blood	
Nervous system	Kidneys	
Brain and sense organs	Mesenteries	
	Testes and ovaries	

pancreas; (3) the respiratory tract, including the lungs, trachea, and larynx; (4) the bladder; (5) the prostate; (6) the urethra; (7) the yolk sac; and (8) the allantois.

The detailed description of the embryologic origin of each tissue and organ cannot be given here, but a few typical examples will be sufficient. As the embryo develops, the upper part of the yolk sac forms the tubular primitive mid- and hindgut, with its outgrowth, the liver, when the embryo is about 3 weeks old. The saclike "heart" begins to beat soon after this time. By the third week the yolk sac has numerous "blood islands" for developing the embryonic vitelline circulation. The embryonic disk infolds (invaginates) to form a troughlike groove (neural groove), the open upper side of which later closes to form a hollow tube (neural tube). From the anterior end of the tube will develop the various parts of the brain and cranial nerves, whereas the remainder forms the spinal cord with its spinal nerves.

Before 5 weeks the embryo externally shows a head, with rudimentary eyes, an external tail, and a neck, with four pairs of gill arches and four pairs of incompletely formed slits, which somewhat resemble the gill-bearing arches of a fish. There are numerous blood vessels here but no true gills. These gill arches give rise to such structures as the following: from the arches arise muscles used in chewing food, the middle ear bones, the hyoid bone (at the base of the tongue), certain facial nerves and muscles, and part of the cartilage of the larynx and its muscles; from the slits between the arches arise such structures as the eustachian tubes, the external ear passage, part of the tonsils, the thymus, and the parathyroids.

419

Early contributors to the study of animal reproduction

Marcello Malpighi (1628-1694)
An Italian scientist who studied the development of organs of chick embryos and erroneously concluded that the adult was preformed in a miniature form in the egg.

Jan Swammerdam (1637-1680)
Swammerdam, a Dutch naturalist, described the cleavage (division) in frog's eggs.

Charles Bonnet (1720-1793)
Bonnet, a Swiss naturalist, proposed the "Preformation Theory," which stated that an organism carried in its body many preformed individuals, in miniature form, for succeeding generations.

Kaspar F. Wolff (1733-1794)
Wolff, a German physiologist, disproved Bonnet's preformation theory by demonstrating that new embryonic structures develop where none had existed before. He stated that the particles that constitute all animal organs in their earliest inception are little microscopic globules.

Lazaro Spallanzani (1729-1799)
Spallanzani, an Italian biologist, attempted to induce frog's eggs to develop artificially by using vinegar and lemon juice (1785).

Karl Ernst von Baer (1792-1876)
A German embryologist who developed the science of comparative embryology. He studied eggs and tissue formation, which led to the formulation of the "Germ-layer Theory" (1828). He observed the resemblance between embryos of lower and higher animals and discovered mammalian eggs (1827). (*Historical Pictures Service, Chicago.*)

Francis M. Balfour (1851-1882)
An English embryologist who wrote a comprehensive book *Comparative Embryology*, thus ushering in this phase of the science (1880). (*Historical Pictures Service, Chicago.*)

Jacques Loeb (1859-1927)
A German-American biologist who caused frog eggs to develop into mature frogs (1899) by artificial parthenogenesis (development of an egg without a sperm).

Eduard van Beneden (1846-1910)
Van Beneden, a Belgian cytologist, studied the behavior of chromosomes during meiosis (reduction division) and helped to clarify their role in heredity and development (1883).

Thomas Hunt Morgan (1866-1945)
An American biologist who made many valuable contributions to experimental embryology.

Oskar Hertwig (1849-1922)
A German biologist who first observed (1875) all the steps in fertilization, including the union of egg and sperm chromosomes in sea urchins.

Hans Spemann (1869-1941)
A German zoologist who proposed the "Organizer Concept" in embryology (1921), which stated that certain parts of a developing embryo act as organizers and influence developmental patterns of the organism. (*Historical Pictures Service, Chicago.*)

Alexis Carrel (1873-1944)
A French surgeon and biologist who came to the United States in 1905, cultured tissues outside the body (in vitro) in 1912, and established a culture of chick embryo connective tissue that grew continuously for 25 years under laboratory conditions.

Ross V. Harrison (1870-1959)
An American biologist who devised a technique for culturing cells outside the body (1907) by mounting a fragment of tissue in clotting lymph fluid on a cover slip of a hanging drop (concave) slide.

von Baer

Balfour

Spemann

Review questions and topics

1 Discuss the stages of embryonic development (embryogenesis) in animals.
2 Explain the structures and functions of the embryonic development of a frog and man.
3 Discuss the tissues that arise from the ectoderm, mesoderm, and endoderm.
4 Review cleavage and gastrulation in the starfish.
5 How does cleavage in the starfish differ from cleavage in the frog? How is it similar?
6 Explain the relationship between the Latin or Greek origins of the name and the actual morphology of the organism in the following: morula; blastula; gastrula; archenteron; neurula.

Selected references

Alston, R. E.: Cellular continuity and development, Glenview, Ill., 1967, Scott, Foresman and Company.

Balinsky, B. I.: An introduction to embryology, ed. 2, Philadelphia, 1965, W. B. Saunders Company.

Barth, L. J.: Development; selected topics, Reading, Mass., 1964, Addison-Wesley Publishing Co., Inc.

Edwards, R. G.: Mammalian eggs in the laboratory, Sci. Amer. 215:72-81, 1966.

Etkin, W.: How a tadpole becomes a frog, Sci. Amer. 214:76-88, 1966.

Frieden, E.: The chemistry of amphibian metamorphosis, Sci. Amer. 209:110-118, 1963.

Gray, G. W.: Human growth, Sci. Amer. 189:65-76, 1953.

McElroy, W. D., and Glass, B., editors: The chemical basis of development, Baltimore, 1958, The Johns Hopkins Press.

Spemann, H.: Embryonic development and induction, New Haven, Conn., 1938, Yale University Press.

Sussman, M.: Growth and development, Englewood Cliffs, N. J., 1964, Prentice-Hall, Inc.

Tanner, J. M.: Earlier maturation in man, Sci. Amer. 218:21-27, 1968.

Waddington, C. H.: Principles of development and differentiation, New York, 1966, The Macmillan Company.

Waddington, C. H.: Principles of embryology, London, 1956, George Allen & Unwin, Ltd.

Whittaker, J. R.: Cellular differentiation, Belmont, Calif., 1968, Dickenson Pub. Co., Inc.

Willier, B. H., Weiss, P. A., and Hamburger, V., editors: Analysis of development, Philadelphia, 1955, W. B. Saunders Company.

chapter twenty-nine

Genetics

For thousands of years, mankind has been aware of the transmission of physical characteristics from parent to offspring. However it has been only in the last one hundred years that we have had any explanation of how this transmission works, and only in the last decade or so have we really begun to understand it.

Today the names of Mendel, Watson, and Crick are at least familiar to most people, and terms such as DNA and genetic code are not completely unknown. Conferences on the genetic effects of radiation or the possible control of heredity in man are not at all unusual.

The present chapter will describe the work of Mendel and the visible results of heredity, while the following chapter will deal with DNA, genes, and the genetic code.

GENERAL CONSIDERATIONS

Genetics (Gr. *genesis,* origin) is the study of how hereditary materials are transmitted through successive generations, and further, how the resultant produced traits have been influenced not only by their genetic inheritance, but also by internal and external environmental conditions.

Studies related to genetics have been conducted in many ways. Among the methods used are (1) the experimental crossing method, in which organisms of known genetic composition are mated

and the offspring observed; (2) by the pedigree method, whereby individuals in the same family line are studied and similarities and differences are analyzed; (3) by the cytogenetic method, involving the study of cells, nuclei, and chromosomes; and (4) by biochemical genetics, in which the relation of DNA and RNA to cell metabolism is considered. In some studies, the cell is subjected to various physical forces such as atomic radiation, x-rays, ultraviolet light, and various chemicals, and the results are then interpreted.

MENDEL'S LAWS

Gregor Mendel, an Austrian monk, gave the first scientific interpretation of the heredity mechanism in 1865 after 8 years of experimental work with garden peas. These interpretations led to the formulation of his famous laws and laid the foundation for future studies in genetics. Although the laws do not explain all types of inheritance, wherever they apply they are as valid today as when they were formulated.

There has been some speculation as to why Mendel was successful in his experiments. He produced hybrids among different types of peas and studied their heredity in successive generations. His method of study differed from that of previous workers, because he singled out and studied a particular trait of a plant instead of attempting to study the inheritance of the whole individual; for example, he studied seed color, plant height, flower location, and pod color. Fortunately, he also chose for study organisms that had rather simple methods of inheritance. He counted all of the thousands of individuals that were produced and kept complete pedigree records, making sure that he knew the ancestry of each individual plant and the traits displayed by each of its ancestors and offspring. He carefully made the particular artificial pollinations desired and prevented all undesired pollinations by variables such as winds or insects. In each generation where contrasting (alternative) traits appeared, for example, where both tall and short plants appeared among the offspring from a single cross, he counted the number of each type obtained. He then interpreted the data and formulated conclusions now known as Mendel's laws (mendelism).

Mendel experimentally crossed two pea plants with alternative traits. The resulting hybrids, each resembling one parent or the other, were then crossed with each other. In the hybrids he recognized the expressed trait and referred to it as the dominant trait; the other that was latent and did not express itself he called the recessive trait. When he crossed the hybrids, both traits reappeared in the next generation in a ratio of approximately three dominants to one recessive. The result based on outward appearance is called the phenotype ratio. This may be resolved into a genotype ratio, which is based on the actual genic compositions of the various individuals. When an individual contains two genes that are alike (as TT or tt), that individual is homozygous (Gr. homo-, same), whereas an individual with two different members of a pair (Tt) is heterozygous (Gr. heteros, different).

The laws that Mendel formulated may be summarized as the law of unit characters, the law of dominance, the law of segregation, and the law of independent assortment.

Law of unit characters

Genes* occur in pairs and control the inheritance of traits as a unit.

Law of dominance

One gene of a pair may mask or inhibit the expression of the opposite member of that pair. In an individual pea plant that has one gene for tallness (T) and one for shortness (t), the tall one will dominate the other and express itself. The one that expresses itself is called the dominant gene, and the other the recessive gene. The trait so developed is called the dominant trait. When two recessive genes (tt) are together, they will express themselves and exhibit the recessive trait.

Law of segregation

The genes that make up the different pairs are segregated (separated) from each other when gametes are formed in animals, or spores are formed in plants. Only one of each pair of genes goes into a sex cell or into a spore of plants.

*Mendel used the word "factor." The term "gene" was not used until 1909.

Law of independent assortment

The genes representing two or more contrasting pairs of traits are distributed independently of one another at the time of gamete formation in animals and at the time of spore formation in plants. The gametes unite at random.

HYBRID STUDIES

When two organisms, homozygous for opposite members of a trait, are crossed, the offspring are heterozygous for that character. These offspring are known as hybrids.

Monohybrid crosses are those in which only one pair of traits is studied, while dihybrid and trihybrid are those in which two or three pairs are considered simultaneously.

By using the pea plant and the following traits with their symbols the different crosses (monohybrid, dihybrid, and trihybrid) may be illustrated.

T, tall plant
t, short (dwarf) plant
R, round seed
r, wrinkled seed
Y, yellow seed color
y, green seed color

When the dominant (shown by the capital letter) is present, it expresses itself even though the opposite or recessive is present. Two recessive genes must be present together in order to express themselves.

In a monohybrid cross (Fig. 29-1), we assume the parents to be homozygous dominant and homozygous recessive, for example, tall and dwarf, TT × tt. Each homozygous parent produces only one gamete type, T or t. The offspring F (first filial generation) will all be heterozygous, Tt, having received one gene from each parent. Phenotypically these are all tall.

A cross between offspring results in an F_2 (second filial generation) composed of both tall and dwarf plants. This is because of each F_1 parent producing two gamete types, that is, both produce T and t. Since the gametes unite at random, possible F_2 genotypes are TT, Tt, Tt, and tt, hence the phenotype ratio of three dominant to one recessive.

The events of a dihybrid cross (Fig. 29-2) are

423

Parents (P)   Tall (TT) × Dwarf (tt)

Gametes produced   T   T   t   t

Offspring (first filial generation) (F_1)   Tall (Tt)

Gametes produced by both male and female   T   t
     individuals of the F_1

Offspring (second filial generation) (F_2) when various individuals are
     crossed and shown by so-called checkerboard (Punnett square)

	Male gametes	
	T	**t**
T	TT	Tt
t	Tt	tt

Female gametes {

Pheno 3 - 1 (handwritten)

Parents (P)   Round-yellow × Wrinkled-green
                 (RRYY)   (rryy)

Gametes produced   RY   ry

Offspring (F_1)   Round-yellow (RrYy)

Gametes for both male and female   RY Ry rY ry

Offspring (F_2) when (F_1) are intercrossed or self-pollinated (shown by the Punnett square)

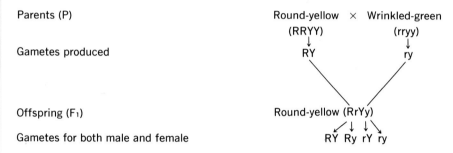

	Female gametes			
	RY	**Ry**	**rY**	**ry**
RY	RY RY	Ry RY	rY RY	ry RY
Ry	RY Ry	Ry Ry	rY Ry	ry Ry
rY	RY rY	Ry rY	rY rY	ry rY
ry	RY ry	Ry ry	rY ry	ry ry

Male gametes {

From these squares it is seen that the following offspring are obtained:

    9 round-yellow
    3 round-green
    3 wrinkled-yellow
    1 wrinkled-green

This is the **phenotype ratio** of 9-3-3-1 for a dihybrid cross in the F_2 generation.

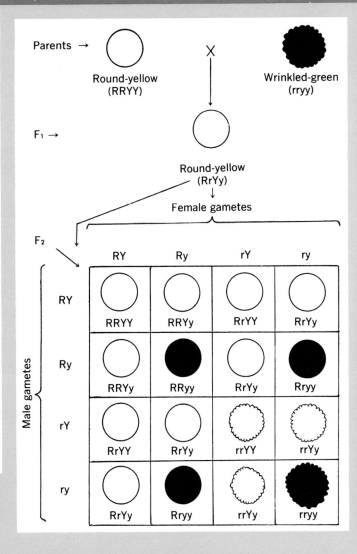

Fig. 29-3
Dihybrid cross in peas showing the gene content of each individual. Genes for various members of each generation are shown. When two similar F₁ individuals are crossed, results are shown in checkerboard (F₂).

Fig. 29-4
Trihybrid cross in peas.

Parents (P)	Tall-yellow-round (TTYYRR)	×	Dwarf-green-wrinkled (ttyyrr)
Gametes produced	TYR		tyr
Offspring (F₁)		Tall-yellow-round (TtYyRr)	
Gametes for both male and female	TYR TYr TyR Tyr tYR tYr tyR tyr		

Fig. 29-5
Bracket method for determining possible types of gametes.

RrYyTt
(Parent genes)

```
          ┌T ── RYT
        ┌Y┤
       │   └t ── RYt
     ┌R┤
    │  │   ┌T ── RyT
    │   └y┤
    │      └t ── Ryt
    ┤
    │      ┌T ── rYT
    │   ┌Y┤
    │  │   └t ── rYt
     └r┤
        │   ┌T ── ryT
         └y┤
            └t ── ryt
```

(Gametes)

Table 29-1

Genetic phenotypes of plants

Plant	Dominant	Recessive
Peas		
Plant height	Tall	Short (dwarf)
Pod color	Green	Yellow
Seed color	Yellow	Green
Seed shape	Round (smooth)	Wrinkled
Flower color	Colored	White
Four-o'clock plant		
Flower color	Colored (incomplete)	White
Snapdragon plant		
Flower color	Colored (incomplete)	White
Tomato		
Vine	Tall	Short (dwarf)
Fruit color	Red	Yellow
Summer squash		
Fruit color	White	Yellow
Fruit shape	Disk shaped	Sphere shaped
Corn		
Seed color	Colored	Colorless
Seed shape	Full	Shrunken (wrinkled)
Endosperm (stored food)	Starchy	Sugary (sweet)

essentially the same as those in a monohybrid cross. We assume that there are homozygous dominant and homozygous recessive parents, for example, round-yellow seeds × wrinkled-green seeds, that is, RRYY × rryy. Each parent produces one gamete type RY or ry. The F are therefore all heterozygous RrYy. If these are crossed, each parent produces four gamete types: RY, Ry, rY, and ry. Gametes again unite at random with sixteen possible F_2 phenotypes. These are: 9 round-yellow, 3 round-green, 3 wrinkled-yellow, and 1 wrinkled-green (Fig. 29-3). Thus the typical F_2 ratio of a dihybrid cross is 9-3-3-1.

A trihybrid cross involves the same stages as those in a dihybrid cross, in which each parent produces one gamete type; all F_1 offspring are heterozygous for all three traits. When crossed,

F_1 parents each produce 8 gamete types, which uniting at random, produce 64 possible F_2 phenotypes in a ratio of 27-9-9-9-3-3-3-1. It is important to understand that the phenotypes represent different genotypes. See Fig. 29-4 for details.

The same principles may be shown in any other organism in which dominant traits appear. The guinea pig is used as an example in Figs. 29-6 to 29-9. In this animal black hair is dominant over white, rough over smooth, and short over long. This illustrates the occurrence of several pairs of traits influencing the same structure.

■
Incomplete dominance

From studies made so far it has been noted that when opposite members of a pair of genes are present, one or the other is completely dominant. In

Table 29-2

Genetic phenotypes of animals

Animal	Dominant	Recessive
Fruit fly		
Eye color	Red	White
Body color	Gray	Ebony (black)
Wing length	Long	Vestigial (short)
Poultry		
Comb	Rose (as in Wyandottes)	Single (as in Leghorns)
Shank (thigh)	Feathered	Bare
Number of toes	Extra toes	Normal number
Guinea pigs		
Hair length	Short	Long
Coat	Rough	Smooth
Hair color	Colored	White
Cattle		
Leg length	Short	Long
Horns	Hornless (polled)	Horns present
Horses		
Hair color	Gray	Other colors
Running form	Trotting	Pacing

Parents (P)

Gametes produced

Offspring (first filial generation) (F₁)

Gametes produced by both male and female individuals of the F₁

Offspring (second filial generation) (F₂) when the various individuals crossed as shown by the so-called checkerboard, or Punnett square

Black (BB) × White (bb)

B B   b b

Black (Bb)

B b

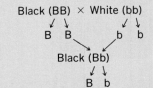

	Male gametes	
	B	b
B	BB	Bb
b	Bb	bb

Female gametes

Fig. 29-6
Monohybrid cross in guinea pigs.

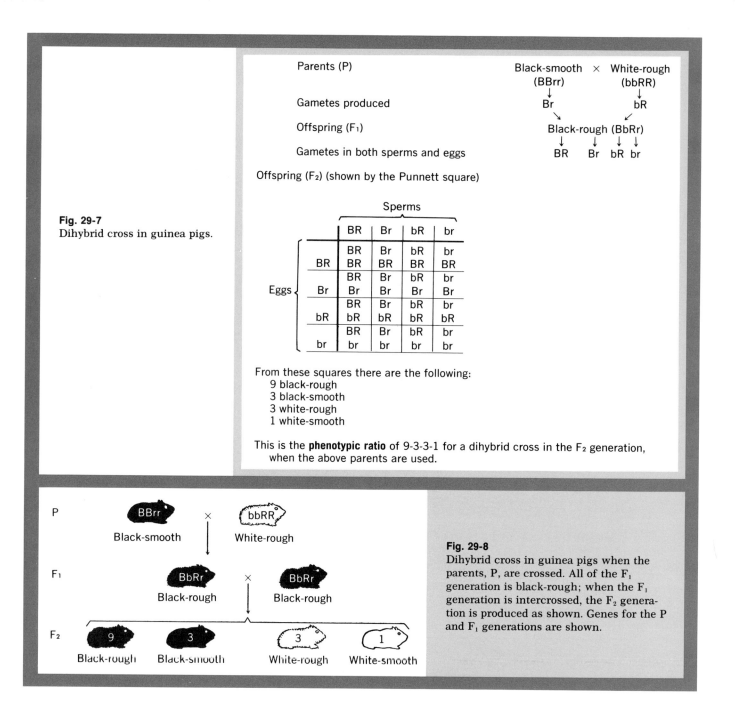

Fig. 29-7
Dihybrid cross in guinea pigs.

Parents (P)

Gametes produced

Offspring (F₁)

Gametes in both sperms and eggs

Offspring (F₂) (shown by the Punnett square)

Black-smooth × White-rough
(BBrr)   (bbRR)
↓   ↓
Br   bR
↘   ↙
Black-rough (BbRr)
↓   ↓   ↓   ↓
BR Br bR br

Sperms

	BR	Br	bR	br
BR	BR BR	Br BR	bR BR	br BR
Br	BR Br	Br Br	bR Br	br Br
bR	BR bR	Br bR	bR bR	br bR
br	BR br	Br br	bR br	br br

Eggs

From these squares there are the following:
  9 black-rough
  3 black-smooth
  3 white-rough
  1 white-smooth

This is the **phenotypic ratio** of 9-3-3-1 for a dihybrid cross in the F₂ generation, when the above parents are used.

P   BBrr × bbRR
     Black-smooth   White-rough

F₁   BbRr × BbRr
     Black-rough   Black-rough

F₂   9   3   3   1
     Black-rough   Black-smooth   White-rough   White-smooth

Fig. 29-8
Dihybrid cross in guinea pigs when the parents, P, are crossed. All of the F₁ generation is black-rough; when the F₁ generation is intercrossed, the F₂ generation is produced as shown. Genes for the P and F₁ generations are shown.

incomplete dominance (partial dominance, or absence of dominance) the F₁ does not resemble either parent exactly for the trait in question, neither gene of the pair being completely dominant. An example of this is the four-o'clock flower (Fig. 29-10), in which a homozygous white (rr) is crossed with a homozygous red (RR). The F₁ is neither red nor white, but is pink (Rr). When two F₁ individuals are crossed, there are produced in the F₂: one white (rr), two pinks (Rr); and one red (RR), with a ratio of 1-2-1. In this case the heterozygous (Rr) individuals can be detected visually, which is not possible in cases of complete dominance.

This phenomenon is also illustrated by the blue

428

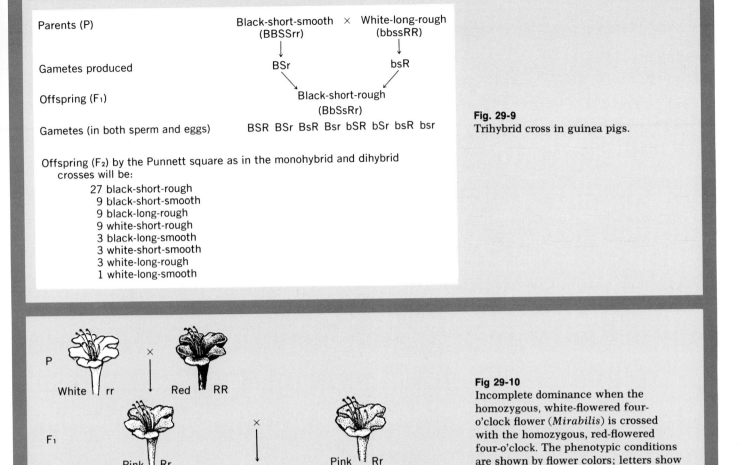

Parents (P)   Black-short-smooth × White-long-rough
           (BBSSrr)   (bbssRR)

Gametes produced   BSr   bsR

Offspring (F₁)   Black-short-rough
            (BbSsRr)

Gametes (in both sperm and eggs) BSR BSr BsR Bsr bSR bSr bsR bsr

Offspring (F₂) by the Punnett square as in the monohybrid and dihybrid
 crosses will be:
  27 black-short-rough
  9 black-short-smooth
  9 black-long-rough
  9 white-short-rough
  3 black-long-smooth
  3 white-short-smooth
  3 white-long-rough
  1 white-long-smooth

Fig. 29-9
Trihybrid cross in guinea pigs.

Fig 29-10
Incomplete dominance when the homozygous, white-flowered four-o'clock flower (*Mirabilis*) is crossed with the homozygous, red-flowered four-o'clock. The phenotypic conditions are shown by flower colors; letters show the genes involved. When two pinks of the F₁ generation are crossed, the results are shown in the F₂.

Andalusian fowl (Fig. 29-11). When a black and a white-splashed individual are crossed, the F₁ shows an intermediate shade of blue Andalusian that is heterozygous. When two blue Andalusian fowls are crossed, the ratio of offspring is: one fourth white-splashed; one half blue Andalusian; and one fourth black, that is, 1-2-1.

◼ Lethal genes

Lethal genes cause developmental or physiologic reactions that result in the death of an organism possessing them. They may affect the organism at any time during life, from the time of fertilization to maturity. The term lethal is usually applied to the killing in early life, whereas sublethal (semilethal) is reserved for describing those conditions that cause death before the reproductive age. Some lethals are dominant, others are recessive. Two kinds of recessive lethals are known. Those that are often called "dominant lethals" are actually recessive genes that have dominant, visible effects in heterozygous individuals of the phenotype.

429

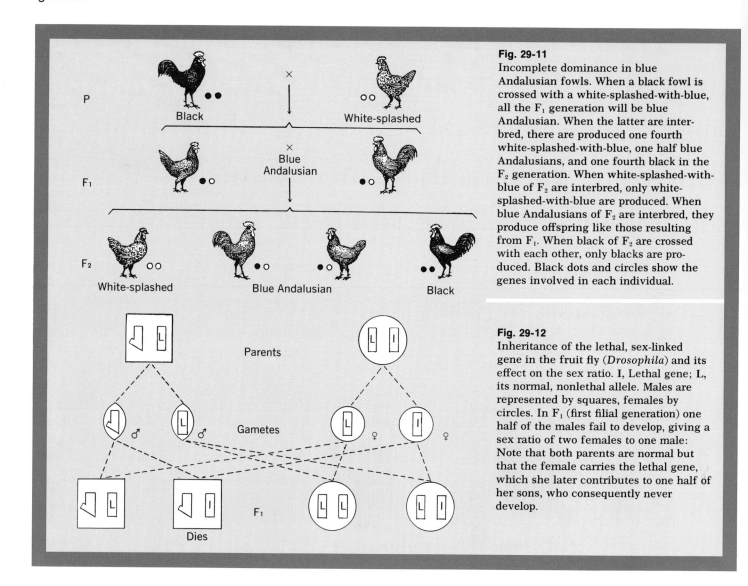

Fig. 29-11

Incomplete dominance in blue Andalusian fowls. When a black fowl is crossed with a white-splashed-with-blue, all the F₁ generation will be blue Andalusian. When the latter are inter-bred, there are produced one fourth white-splashed-with-blue, one half blue Andalusians, and one fourth black in the F₂ generation. When white-splashed-with-blue of F₂ are interbred, only white-splashed-with-blue are produced. When blue Andalusians of F₂ are interbred, they produce offspring like those resulting from F₁. When black of F₂ are crossed with each other, only blacks are produced. Black dots and circles show the genes involved in each individual.

Fig. 29-12

Inheritance of the lethal, sex-linked gene in the fruit fly (*Drosophila*) and its effect on the sex ratio. I, Lethal gene; L, its normal, nonlethal allele. Males are represented by squares, females by circles. In F₁ (first filial generation) one half of the males fail to develop, giving a sex ratio of two females to one male: Note that both parents are normal but that the female carries the lethal gene, which she later contributes to one half of her sons, who consequently never develop.

Other recessive lethals have no visible effect in the heterozygous condition. In the truly dominant lethals killing may occur even in the heterozygous condition.

In plants lethal genes may prevent the development of chlorophyll, so that photosynthesis cannot occur. Certain lethal genes in corn produce albinism (lack of chlorophyll), so that the young plant dies after the stored food of the seed is exhausted. Many mutations (change in a gene that is inheritable) that occur naturally, or are artificially induced by radiations, such as x-rays and radium, are lethal. Lethals may express themselves merely through the nonappearance of certain types of offspring.

A certain type of snapdragon has golden leaves rather than the normal green. When such plants are self-pollinated (self-fertilized), they may give a ratio of offspring of two golden to one green. This ratio is understood when the early seedlings are examined, for there may be three phenotypes with a ratio of one yellow, two golden, and one green.

The genes causing the abnormality apparently do not prevent sufficient nutrition when heterozygous (as in golden), but act as lethals (yellow) when homozygous, shown as follows:

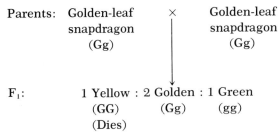

Parents: Golden-leaf   ×   Golden-leaf
          snapdragon   snapdragon
          (Gg)   (Gg)

F₁: 1 Yellow : 2 Golden : 1 Green
               (GG)   (Gg)   (gg)
               (Dies)

The inheritance of a lethal, sex-linked gene in the fruit fly (Drosophila) and its effects on the sex ratio are shown in Fig. 29-12. This lethal gene is a recessive and is present on the X chromosome (sex chromosome). Note that both parents appear normal, but the female carries the lethal gene (one), which she contributes to one half of her male offspring, which consequently never develop. In the F₁ one half of the males fail to develop, giving the adult sex ratio of two females to one male.

In mice yellow (Y) dominates gray (y) color. If crosses are made as follows, the lethal results are shown:

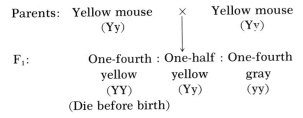

Parents: Yellow mouse   ×   Yellow mouse
          (Yy)   (Yy)

F₁: One-fourth : One-half : One-fourth
         yellow   yellow   gray
         (YY)   (Yy)   (yy)
         (Die before birth)

To explain the early deaths of one fourth of the mice it is surmised that when the lethal gene (Y), is present in duplicate (homozygous), it will cause death, which is not true in the heterozygous (Yy) condition. When matings as described are made, litters are about three fourths normal size, but the uteri of the pregnant mothers show normal numbers of embryos. About one fourth of these embryos die before birth.

Included in the human lethals and sublethals are: (1) Amaurotic idiocy (progressive mental deterioration and blindness with death usually within 5 years), which is caused by two recessive genes acting together and is not fatal in the heterozygous condition. Parents of an affected child may have good health and possess superior intelligence. There is no known treatment for heterozygous persons. (2) A rather rare condition of shortened fingers is called brachydactyly (brak i -dak′ ti ly) (Gr. *braxys*, short; *daktylos*, digit) and is caused by a dominant lethal gene. Homozygous offspring have grossly abnormal skeletal defects and die before birth. (3) Cancer of the eye retina in early childhood is called retinoblastoma (Gr. *blastos*, young; -*oma*, tumor) and is caused by a dominant lethal. Early surgical removal of the affected eye before cancer spreads may be successful. (4) The human sublethal hemophilia (Gr. *haima*, blood; *philos*, loving) is a sex-linked condition in which the blood of males clots so slowly that death may result from minor hemorrhages. (This condition is discussed under sex-linked traits.) There seem to be other types of hemophilic diseases caused by other genes, and the methods of inheritance are different. (5) The early death of a certain percentage of babies may be caused by the action of certain Rh blood types and is caused by multiple genes. This condition is called erythroblastosis fetalis and affects about 1 in 200 or 500 pregnancies and may be fatal. Some human abortions occur because of the expulsion of a dead fetus resulting from lethal homozygous recessive genes.

Multiple genes and interaction of genes

Many traits in plants and animals are determined by the interaction of multiple genes (more than one pair). The specific methods of inheritance vary, but the following examples may give a general idea of such genetic phenomena. When a so-called quantitative trait (one with various degrees of expression) is the result of the interaction of more than one pair of genes, the latter are known as multiple genes.

The Swedish geneticist, Nilsson-Ehle, found three pairs of interacting genes in certain types of red wheat. These possessed incomplete dominance, and various degrees of trait expression were produced because of the cumulative effects of the genes in question. Deepest red wheat is represented by R_1R_1 R_2R_2 R_3R_3, and white wheat by r_1r_1 r_2r_2 r_3r_3. The more genes represented by the capital letters in any individual plant, the darker the red. This is shown as follows:

431

Parents: Deepest   red   wheat × White   wheat
$(R_1R_1\ R_2R_2\ R_3R_3)$ $(r_1r_1\ r_2r_2\ r_3r_3)$

F_1:   Medium red wheat
$(R_1r_1\ R_2r_2\ R_3r_3)$

F_2: When two of the F_1 are crossed, the F_2 ratio
is:

   1 deepest red (6 red genes)
   6 very deep red (5 red genes)
   15 deep red (4 red genes)
   20 medium red (3 red genes)
   15 pale red (2 red genes)
   6 very pale red (1 red gene)
   1 white (no red genes, but 6 white
      genes)

Another, but somewhat different, example of
multiple gene inheritance is illustrated by the
flower color in sweet peas. When two dominant
genes located in different pairs of chromosomes
interact, they may complement each other (both
present and produce a visible effect). When two
strains of white-flowered sweet peas with the
genic composition shown here are crossed, the F_1
will be purple flowers. Purple flowers are represented by at least one C gene and at least one P
gene. White flowers are represented by at least one
C and pp; or by cc and at least one P; or by ccpp.
This may be shown as follows:

Parents: White-flowered × White-flowered
         sweet pea   sweet pea
         (CCpp)   (ccPP)

F_1:   Purple-flowered sweet pea
(CcPp)

F_2: When two of the F_1 are crossed, the F_2 ratio
is:

   9 purple (at least one C and at least
      one P)
   3 white (at least one C and pp)
   3 white (cc and at least one P)
   1 white (cc and pp)

This phenotype ratio is 9 purple-7 white.

Another example of multiple genes (two pairs)
is shown in the inheritance of the combs of chickens. In this case pea comb is represented by at
least one P and rr; rose comb by pp and at least one

R; walnut comb by at least one P and at least one
R, and single comb by pprr. The following crosses
show a homozygous rose comb crossed with a
homozygous pea comb with the genic contents
and ratios:

Parents: Rose comb × Pea comb
(ppRR)   (PPrr)

Gametes: pR   Pr

F_1:   Walnut comb
(PpRr)

F_2: When two walnut-combed individuals of
the F_1 are crossed, the F_2 shows:
   9 walnut comb (at least one P and at
      least one R)
   3 pea comb (at least one P and rr)
   3 rose comb (pp and at least one R)
   1 single comb (pprr)

This shows not only a new comb trait in the F_1
but also when intercrossed, a still different type,
namely single comb (pprr), is produced.

There are many human genetic traits, each of
which results from the interaction of more than
one pair of genes. Many of our so-called quantitative traits that have variable degrees of expression
are in this category. An example is the production
of skin color in which Negroes differ from whites
in two pairs of genes, which interact cumulatively
and show incomplete dominance. This is explained
by: Negro, AABB; dark mulatto, AABb or AaBB;
medium mulatto, AaBb, AAbb, or aaBB; light
mulatto, Aabb or aaBb; white, aabb. Using these
symbols, if a pure Negro (AABB) and a pure white
(aabb) are crossed, the F_1 offspring are medium mulatto (AaBb). Another crossing may be shown in
the accompanying diagram.

Parents: Medium mulatto × Medium mulatto
      (male)   (female)
      (AAbb)   (AaBb)
Gametes: Ab   AB Ab aB ab

F_1:   1 dark mulatto (AABb); 2 medium
      mulattoes (AAbb) (AaBb), and 1
      light mulatto (Aabb). This ratio
      is based on the one kind of male
      sperm and four different types of
      eggs in the female.

The actual process of color determination is much more complex and may involve more gene pairs.

Multiple alleles

Another multiple method of inheriting a human trait may be illustrated by the blood groups A, B, AB, and O, in which three alternative genes (triple alleles) are responsible. It is known that when blood from certain individuals is mixed frequently, this results in an agglutination (clumping) of the red blood corpuscles. Four types of human blood are known as groups A, B, AB, and O. Agglutination of human blood depends upon the presence of (1) the specific substance known as the agglutinogen (a type of antigen) in the red blood corpuscles and (2) the specific substance known as the agglutinin (a type of antibody) in the blood plasma. Both are necessary for blood to agglutinate, so naturally both cannot occur in the same person or his blood would agglutinate in his blood vessels. The two inheritable agglutinogens are known as A and B in man. Hence, the following human blood groups are possible: an individual of group A has agglutinogen A in his red blood corpuscles; group B has agglutinogen B; group AB has both agglutinogens A and B; group O has neither agglutinogen. Whichever agglutinogen an individual has in his red blood corpuscles, the corresponding agglutinin is absent in his blood plasma. When an agglutinogen is absent in his erythrocytes, the corresponding agglutinin is present in his plasma. A summary of the blood groups, their agglutinogens, and agglutinins, is given in Table 26-6.

Of the triple alleles, each person has only one pair of these genes, which consist of any one of the six possible combinations as follows:

Blood group	Genes and genotype
A	$I^A I^A$, or $I^A i$
B	$I^B I^B$, or $I^B i$
AB	$I^A I^B$
O	ii

In the above both I^A and I^B are dominant over gene i, and group O, being a recessive, must breed true. Group AB is a hybrid, whereas groups A and B may either breed true or be hybrids.

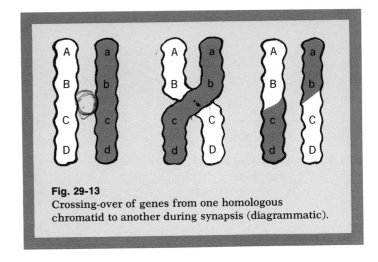

Fig. 29-13
Crossing-over of genes from one homologous chromatid to another during synapsis (diagrammatic).

In the inheritance of the preceding groups gene I^A produces agglutinogen A; gene I^B produces agglutinogen B; and gene i produces no agglutinogen. When both I^A and I^B are present, an individual of the AB group results. Blood types are inherited specifically and do not change from the embryo to old age, hence blood tests may be used in certain cases of disputed parentage. Such tests cannot prove that a certain man is the father of a particular child but whether he could or could not have been. In other instances a certain child with a specific blood group could not have been conceived by certain individuals with their specific blood group.

Additional agglutinogens, called M and N, are known in human erythrocytes, and these are inherited independently of other blood groups. They may be useful in identifying bloods but usually need not be considered in blood transfusions.

■ Linkage and crossing-over

Linked traits are developed from a linear series of genes that are associated (linked) together on the same chromosomes. Each gene occupies a particular position (loci) on a specific chromosome, and linkage tends to maintain this sequence. Linked genes have a tendency to be passed on to the next generation. However, linkage does not prevent segments of chromosomes, with their genes, from separating and crossing over (Fig. 29-13) during synapsis (temporary union of pairs of homologous chromatids at the time of meiosis). Crossing-over might be defined as the mutual exchange of blocks of homologous genes located on two members of a pair of chromatids.

433

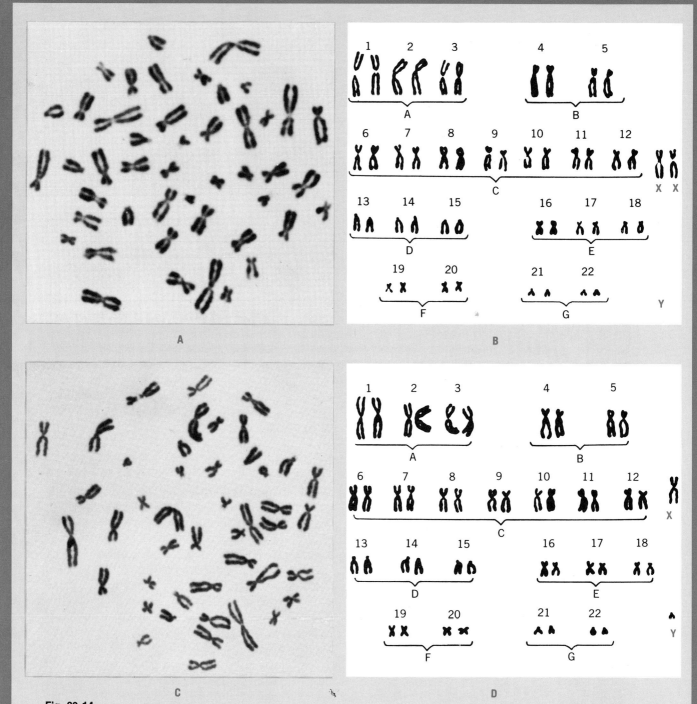

Fig. 29-14

Chromosomes from a culture of human white blood cells. **A,** Metaphase plate of a normal female; **B,** karyotype (paired chromosomes) made by photographing the plate at **A,** cutting out individual chromosomes, and arranging them in pairs according to the Denver system; **C,** metaphase plate of a normal male; **D,** karyotype made from **C.** *(Courtesy Willard Yarema and Cytogenetic Laboratory of St. Elizabeth Hospital, Dayton, Ohio.)*

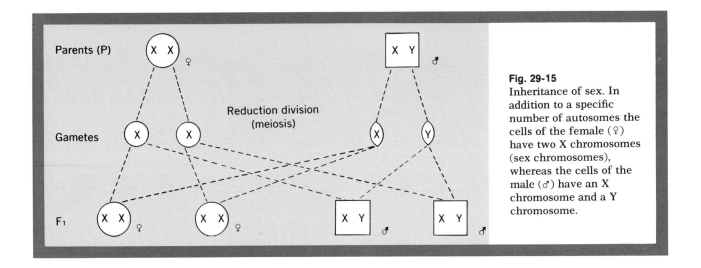

Fig. 29-15
Inheritance of sex. In addition to a specific number of autosomes the cells of the female (♀) have two X chromosomes (sex chromosomes), whereas the cells of the male (♂) have an X chromosome and a Y chromosome.

Parents (P)

Reduction division (meiosis)

Gametes

F₁

Under uniform environmental conditions the linked genes in a uniform stock of organisms cross over with a rather definite frequency. The exchanged segments consist of homologous pieces of chromatids belonging to homologous chromosomes, which result in two types of visible crossovers (recombinations) that occur in equal numbers, and also two types of noncrossovers that occur in equal numbers.

If all genes in all chromosomes were permanently linked, and no crossing-over were possible, variations in organisms would be limited. Crossing-over has the effect of increasing variations. Linkage tends to restrain a complete disorganization of the natural loci of genes, yet crossing-over permits recombinations of them, thereby permitting a greater variety of total traits with which the organism can attempt to adapt itself to its living conditions.

Linkage and crossing-over were first studied in England by Bateson and Punnett by experimenting with sweet peas. Thomas H. Morgan, in the United States, developed these two phenomena in 1910 by experimenting with fruit flies (*Drosophila*).

The frequency of crossing-over between two groups of genes depends for the most part upon the distance between the genes on the chromosome. In general, the farther apart two genes lie, the greater the percentage of crossing-over, whereas two genes that lie nearer each other have less opportunity for crossing-over. A crossover in a particular region tends to inhibit the occurrence of another crossover nearby through a process called interference.

Inheritance of sex and the sex ratio

Sex is an inheritable trait, as are many other characteristics of human beings. Each normal human body cell contains twenty-two pairs of autosomes and a pair of sex chromosomes (diploid number). In females the sex chromosomes are two similar X chromosomes, whereas in males the sex chromosomes are an X chromosome and a dissimilar Y chromosome (Figs. 29-14 to 29-16). When human sex cells are formed, through reduction division (meiosis) each sex cell (whether male or female) receives the single, haploid number of chromosomes. Thus by union of two sex cells the resulting zygote again has the diploid number. The sex of the offspring is determined by which of the two types of sperm (X or Y type) unites with the egg.

From cytologic studies of fruit flies we find that each somatic (body) cell contains three pairs of autosomes and one pair of sex chromosomes (Fig. 29-17). The female has a pair of similar sex chromosomes called X chromosomes, whereas the male has one X chromosome and one Y chromosome. When female sex cells are produced, each one contains three autosomes and one X chromosome. When male gametes are produced, one type contains three autosomes and one X chromosome, whereas the other type contains three autosomes and one Y chromosome. Since the two types of

435

male gametes are produced in equal numbers and if each type has an equal chance of fertilizing an egg, the ratio of male to female (sex ratio) should be approximately 50-50. *Drosophila melanogaster,* the common fruit fly, has the XX-XY type of sex chromosome–sex determining mechanism, the male being the heterogametic sex with XY. It is known that in this species sex determination is a result of a balance of female determiners on the X chromosome and of male determiners on the autosomes, the Y chromosome being inert, although necessary for male fertility.

Sex is determined at the time of fertilization. In general, the methods of gamete production, fertilization, and sex determination are somewhat alike in the fruit fly and man.

Mammals, with the exception of a few species, have an XX-XY type of sex mechanism, the male being the heterogametic sex with XY. In 1959 it was shown that mammalian sex determination is not quite like the *Drosophila* type in certain respects, but the Y chromosome, instead of being inert, is strongly male determining. The male-determining ability of the Y chromosome in the mouse was demonstrated by combined genetic and cytologic evidence. At about the same time there were parallel findings in the human species.

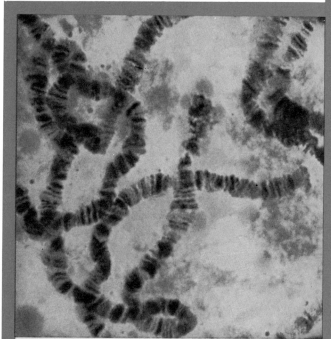

Fig. 29-16
Giant chromosomes from the salivary gland of the fruit fly (*Drosophila melanogaster*).
(*Courtesy General Biological Supply House, Inc., Chicago, Illinois.*)

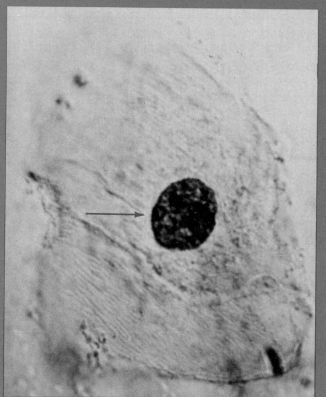

Fig. 29-17
Epithelial cell from the mouth of a normal human female. The dark spot, called a Barr body, at the left side of the nucleus is a chromatin mass of the second X chromosome.
(*Courtesy Willard Yarema and Cytogenetic Laboratory of St. Elizabeth Hospital, Dayton, Ohio.*)

The fact that the Y chromosome in mammals is strongly male determining suggests that it is not genetically empty, yet no other genes are definitely known to be located on the Y chromosome, although some are suspected.

Fig. 29-17 shows a cell taken from the lining of the mouth of a normal human female. The dark dot at the left side of the nucleus is called a Barr body, after its discoverer. It represents the second X chromosome in the female nucleus. The sex of developing embryos may be determined as early as twenty-one days after conception by culturing embryonic fluid and examining the cells for these Barr bodies.

Extra sex chromosomes

An error of meiosis, called nondisjunction, results in unequal numbers of chromosomes entering the gametes. With regard to sex chromosomes then, the following may result; a gamete with no sex chromosomes; a gamete with both an X and a Y chromosome; and a gamete with two X chromosomes.

If any of these abnormal gametes would unite with a normal egg or sperm, the resulting embryo could be: XO, YO, XXY, XYY, or XXX. All of these except YO have been found. (O means that the chromosome is missing.)

Two of these abnormal situations have been

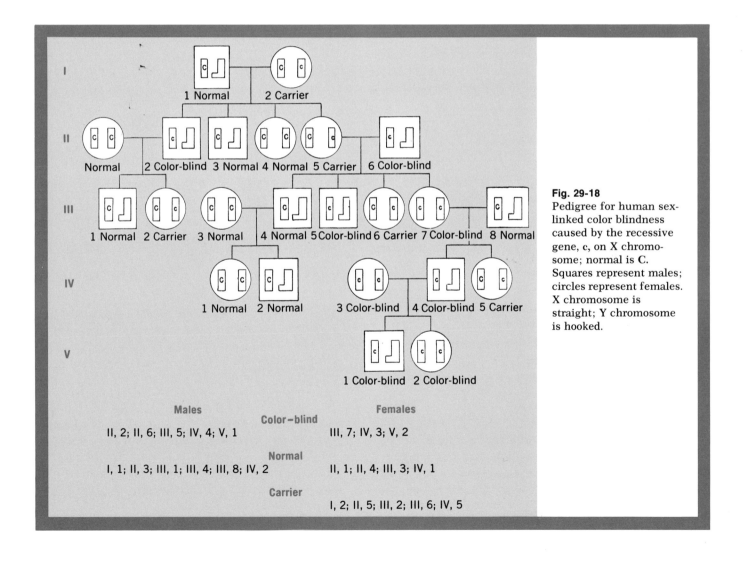

Fig. 29-18
Pedigree for human sex-linked color blindness caused by the recessive gene, c, on X chromosome; normal is C. Squares represent males; circles represent females. X chromosome is straight; Y chromosome is hooked.

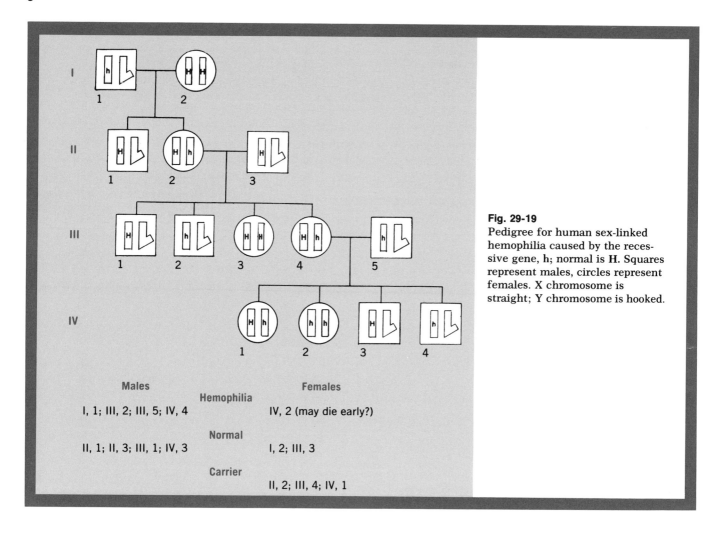

Fig. 29-19
Pedigree for human sex-linked hemophilia caused by the recessive gene, h; normal is H. Squares represent males, circles represent females. X chromosome is straight; Y chromosome is hooked.

	Males	Females
Hemophilia	I, 1; III, 2; III, 5; IV, 4	IV, 2 (may die early?)
Normal	II, 1; II, 3; III, 1; IV, 3	I, 2; III, 3
Carrier		II, 2; III, 4; IV, 1

studied extensively. The first of these, the XO condition, is called Turner's syndrome and produces females who are sterile and may show other abnormalities. It is interesting to note that the same syndrome exists in mice but that the female mouse remains fertile. The second case occurs in males with an extra X or Y chromosome. This is called Klinefelter's syndrome and the men are sterile with some degree of femaleness. Studies of this abnormality suggest that there is a connection between the condition and aggressive antisocial behavior. Recently (1968) in Australia, a man was acquitted of murder on the basis of "supermaleness" due to an XYY genetic makeup. The implica-

tions of this finding in the world's courts are interesting to ponder.

Sex-linked traits

Genes for some traits are located on the sex chromosomes. Naturally, the particular distribution of these sex-linked genes determines the appearance of such traits according to the sex of the individuals. Over fifty human sex-linked traits have been found.

One of these is a type of color blindness in which the individual is unable to distinguish red from green. The recessive gene, c, for this trait is carried on the X chromosome, while the Y chromosome lacks a corresponding dominant gene. This explains the different ratio of color blindness in the two sexes (Fig. 29-18).

A color-blind man (XcY) producing children with a normal woman who is not color-blind (XX), passes the gene (Xc) to his daughters. They carry the gene but have normal color vision because of their second, normal X chromosome. These females produce eggs with either (X) or (Xc) chromosomes. Assuming the father of their children has normal color vision, it is possible to produce normal girls (XX), carrier girls (XXc), normal boys (XY), and color-blind boys (XcY).

If a carrier girl married a color-blind male, the possibility of a color-blind girl is present (XcXc), along with a carrier female, normal boy, and color-blind boy.

Another human sex-linked trait is hemophilia (Fig. 29-19) caused by a recessive gene, h, linked on the X chromosome. Even minor injuries may result in excessive bleeding caused by very slow clotting of blood. The disease may be present in all social strata, although much publicity has been given to its presence in the royal Romanoff and Bourbon families in Europe. Many hemophiliacs die before maturity and many others choose not to produce children so that only a few family pedigrees are recorded. Few cases of hemophilic girls are known, and this is usually interpreted as resulting from the homozygous condition (hh), which usually causes death of the embryo before birth. Incomplete evidence for this comes from interrupted pregnancies in such marriages, and from the fact that hemophilic men have fewer than the expected percentage of daughters.

If a hemophilic man (h) lives to produce children and marries a normal woman (HH), all of his daughters will be carriers (Hh) and all of his sons will be normal (H), because the sons inherit their only X chromosome from their mother. If the same hemophilic man (h) marries a carrier woman (Hh), one half of his sons will be normal (H) and one half will be hemophilic (h); and theoretically, one half of his daughters will be carriers (Hh) and one half will be hemophilic (hh). It will be noted that the method of inheritance of this disease is the same as for the inheritance of color blindness, since both are sex linked on the X chromosome.

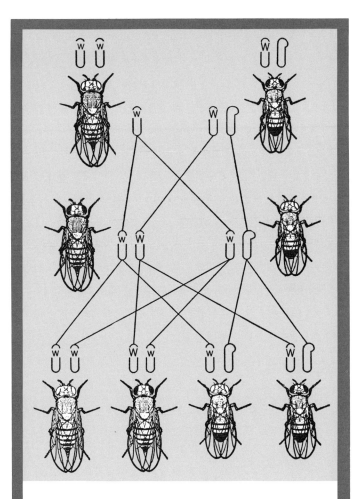

Fig. 29-20

Sex-linked inheritance in the fruit fly (*Drosophila* sp.). The red-eyed male (upper right) and white-eyed female (upper left) are crossed to produce a male and female of the F₁ generation. Gene W for red eyes is carried on the X chromosome of the male, whereas the curved male Y chromosome does not carry an eye-color gene. Gene w for white eyes is carried on the X chromosome of the female. When members of the F₁ generation are crossed, four kinds of the F₂ generation are produced (lower group): Note that a male has a larger tip of black on his abdomen than the female. Contrast these results with those in Fig. 29-21.

(From Morgan, T. H.: Evolution and genetics, Princeton, New Jersey, Princeton University Press.)

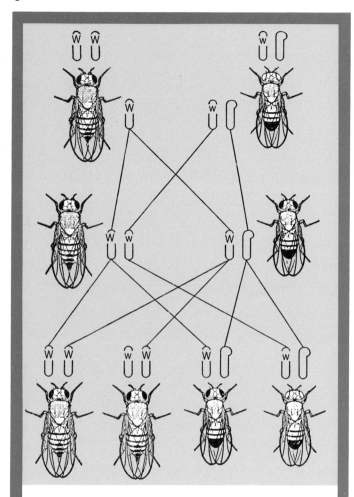

Fig. 29-21

Sex-linked inheritance in the fruit fly (*Drosophila* sp.).
A white-eyed male (upper right) and red-eyed female
(upper left) are crossed to produce a male and female of
the F₁ generation. Gene W for red eyes is carried on
the X chromosome of the female. Gene w for white
eyes is carried on the X chromosome of the male.
Curved male Y chromosome does not carry an eye-color
gene. When members of the F₁ generation are crossed,
members of the F₂ generation are produced, as shown in
the lower line: Note that the male has a larger tip of
black on his abdomen than the female. Contrast these
results with those in Fig. 29-20.

*(From Morgan, T. H.: Evolution and genetics, Princeton,
New Jersey, Princeton University Press.)*

Very few human traits show absolute Y linkage (do not appear in females and are not transmitted to them). A possible Y-linked trait is a
special type of webbed toe (webbed skin between
the second and third toes). A different type of
webbed toe exists in both males and females, with
a preponderance in males.

In the fruit fly the gametes are heterozygous in
males and homozygous in females. White eyes
are sex-linked traits that are recessive to normal
red eyes. When a white-eyed female (Fig. 29-20)
is crossed with a red-eyed male, the F₁ males are
white eyed and the females red eyed. In the F₂
one half of the individuals of each sex are white
eyed and one half are red eyed.

In the reciprocal cross in which a red-eyed female (Fig. 29-21) is crossed with a white-eyed
male, all the offspring (both male and female) of
the F₁ are red eyed; all the females of the F₂ are
also red eyed. One half of the males of the F₂ are
red eyed, and the other half are white eyed. Compare and contrast these two types of crosses (Figs.
29-20 and 29-21) to understand how sex-linked
traits develop.

Sex-influenced traits

Sex-influenced traits are inherited from genes
on the autosomes (not on the sex chromosomes),
but they are influenced or modified by the sex hormones of the male and female gonads and are responsible for differences in trait expressions in
the two sexes, even though the genes may be the
same in both.

An example is the inheritance of horns in sheep.
Neither sex of Suffolk sheep bears horns, but both
sexes of Dorset sheep always bear horns. When
Suffolks and Dorsets are crossed, the male offspring are horned and the females are hornless.
When offspring of the F₁ are interbred, the F₂ ratio
is obtained in which the males show three horned
to one hornless and the females show three hornless to one horned.

Parents: Suffolk (hornless) × Dorset (horned)

          (hh)   ↓   (HH)

                (Hh)

F₁: Hornless females and horned males

F₂: If 2 of the F₁ are crossed, the following ra-
     tios result:

Males: 1 horned : 2 horned : 1 hornless
                    (HH)   (Hh)   (hh)
Females: 1 horned : 2 hornless : 1 hornless
                    (HH)   (Hh)   (hh)

The gene (H) for horns expresses itself in heterozygous males, whereas its allele expresses itself in females.

Another example is the type of inheritable baldness in humans called pattern baldness, in which the hair recedes around the temples, then becomes thinner on the top of the head, finally leaving a fringe of hair low on the head. Analyses of pedigrees show that the genotype BB results in baldness in both sexes, whereas the genotype Bb results in baldness in men but normal hair in women. The genotype bb results in normal hair in both sexes. Many pedigrees show bald women, and this may be caused by a quantity of male sex hormone in their blood that is sufficient to produce baldness in women of genotype BB but insufficient to cause it in women of genotype Bb. Evidence for such phenomena has been obtained from adrenal cortex abnormalities, because this endocrine gland produces a small amount of male sex hormone. Secondary sex traits in general are examples of sex-influenced traits.

Cytoplasmic inheritance

After mendelian inheritance was formulated many biologists believed that the actions of chromosomes explained the phenomena of heredity adequately, but some believed that cytoplasm played more than a passive role. The classic experiments of Theodore Boveri (1862-1915) tested the role of cytoplasm in heredity. His first conclusions were that the specific traits were determined by chromosomes, but that the traits of higher organisms might be determined by cytoplasm. Jacques Loeb (1859–1924) even stated that cytoplasm determined "the embryo in the rough," but the chromosomes added the details. Later, Boveri's experiments seemed to support the chromosome theory of heredity, and for years biologists adhered to this theory for the most part.

In recent years much evidence has been obtained that nonchromosomal, or cytoplasmic, inheritance does occur and may play a more important role in heredity than has been suspected. Much work has been done with various organisms, including bacteria, algae, corn, protozoans, and mice. Both plastids and mitochondria have been shown to contain DNA and to be self-reproducing. The non-chromosomal entities are sometimes called cytogenes, or plasmagenes.

A well-known example in protozoans is the so-called killer trait of *Paramecium aurelia*, which was first studied by T. M. Sonneborn. He discovered that when individuals of one strain were mixed with those of another strain and no conjugation occurred, some of the animals were strangely affected and soon died. If conjugation occurred and the resulting exconjugants were immediately isolated into separate dishes, one animal gave rise, by fission, to a clone (Gr. *klon*, twig) of "sensitives" and the other to a clone of "killers." A clone consists of the offspring produced by asexual reproduction of a single organism. When the F_1 sensitives were crossed to produce the F_2, all the clones were sensitive. When F_1 killers were crossed, three "killer" clones were produced to one "sensitive" clone.

These results could be altered in the following manner. When conjugating pairs are about to separate, they do so normally at the anterior end first, and a minute or two later they separate at the oral region. Sometimes, the oral separation is delayed, either spontaneously or by the investigator, and when this happens, not only are nuclei exchanged between the members of the pair but there is also an exchange of cytoplasm. In the original cross, already mentioned, between the killer and sensitive strains in which separation has been delayed beyond 15 minutes, all resulting clones are killers.

How might such results be explained genetically? The 3-1 ratio in the F_2 killer-by-killer cross would suggest that the killer trait might be a dominant one, and therefore KK or Kk would constitute the killer and kk the sensitive. But this is only a partial answer since in the F_1 all individuals are Kk, yet only one member of a pair gives rise to a killer clone, the other being sensitive. Since there is a transfer of cytoplasm during delayed separation, killer clones result, even from the sensitive member of the pair. Coupled with this, when killer is crossed with killer, one out of three of the F_2

441

produces a sensitive clone, all of which suggests that some cytoplasmic factor may be involved.

When a killer *Paramecium* is stained by a special technique, small bodies containing DNA and called kappa particles can be found in the cytoplasm, but they are absent in the cytoplasm of sensitives. The results of these investigations support the following conclusions: (1) a dominant gene K and a ctyoplasmic kappa particle must be present if an animal is to be a killer; (2) kappa is dependent upon the presence of the K gene for its existence; and (3) the K gene does not produce kappa but merely creates a cellular environment that supports it and allows it to multiply.

In an attempt to understand the inheritance in the unicellular, green alga (*Chlamydomonas*) Ruth Sager and her co-workers have studied and analyzed both chromosomal and nonchromosomal systems of genetic determinants. Much of their research has been concerned with the nonchromosomal inheritance of two systems, namely chloroplast formation and streptomycin resistance. From studies of higher plants it is known that chloroplast formation is influenced by nonchromosomal and by chromosomal genes. They also studied a nonchromosomal genetic system exhibiting uniparental transmission and conferring streptomycin resistance. It was found that the two nonchromosomal systems for chloroplast formation and streptomycin resistance are different, because the chloroplast system is independent of the mating type in its transmission, whereas the streptomycin-resistance system is transmitted by one mating type.

Mutations

A mutation is an inheritable trait that appears suddenly because of some spontaneous change in the hereditary mechanism. Genes are able to mutate in most, if not all, types of cells, including sex cells. Mutations may affect any part of a plant or animal or any of its functions, and hundreds of different kinds have been discovered. The same kind of mutation may arise spontaneously in different individuals. A mutation may be insignificant, or it may be extensive. In the latter case the inheritable change may be sufficient to produce a new type, or possibly even a new species.

Although mutations occur naturally, they can be induced artificially by chemical or physical agents (known as mutagens), among them mustard gas, formaldehyde, x-rays, atomic radiations, radium, and ultraviolet light. Mutations have been induced in corn by using gamma rays. When mutations are induced artificially, there is no way to predict what type they may prove to be. In general, mutations are changes for the worse or are neutral, although a few are desirable. Mutations are often reversible, that is, a gene may mutate and then mutate back to the original (back mutation), as in certain bacteria and the mold *Neurospora*. If mutations occur in sex cells, the results can be inherited, but if genic changes occur in body cells, they cannot be inherited. Naturally, the time when the mutation occurs influences the extent of the results. The later in the process of development the mutation occurs the lesser will be the tissue that develops from the mutant cell. In man a brown segment in an otherwise blue iris of the eye may be caused by a rather late somatic mutation from blue to brown. Through irradiation a peanut plant can mutate to produce an erect stalk rather than a prostrate one.

Many plants can be propagated from cuttings or by other asexual methods, such as grafting or budding. A branch that contains a desirable mutant may be propagated asexually, and its results continued.

Thomas H. Morgan, in 1910, discovered the sudden appearance of a white-eyed mutant in a stock of true-breeding, red-eyed fruit flies (*Drosophila*). When the white-eyed mutant was crossed with a normal red-eyed fly, the white-eyed mutant trait was inherited as a recessive on the X chromosomes (sex chromosome). Other mutants in the eyes of fruit flies include purple and brown.

A desirable mutant arose abruptly in a sheep some years ago that resulted in the development of shorter legs, a trait which required that fences did not have to be built as high to retain the sheep.

Effects of radiations on heredity and the future of organisms

In the last two decades man has released and used atomic energies that increase the natural environmental level of radioactivity. Radiations affect the mutation rate of living organisms from viruses and bacteria to man. X-rays and related, high-energy, penetrating radiations are powerful mutagens that affect the heredity mechanism of living organisms. Many people today are exposed to high-energy radiations used both for constructive and destructive purposes. Harmful effects are far more likely than beneficial ones.

The effects of radiations are cumulative, so that continued exposure to even small amounts over several years may be as serious as receiving an equivalent dosage (same number of roentgens) in a short time. Mutations caused by irradiations may not be evident for several generations, because they may be masked by dominant alleles and may be seen only as homozygous recessives. The general level of radioactivity has increased as a consequence of the so-called atomic age. The biologic and social implications of the consequences are still being debated. What effects this will have on organisms, including man, cannot be foretold at present. However, if necessary precautions are taken, the grave and even irreparable harm to the genetic mechanism can be reduced.

Human populations and those of other organisms carry great numbers of harmful genes that may cause malformations, weaknesses, and tendencies to certain diseases. One way to diminish the harmful mutants would be to decrease the rates of mutations that produce them. However, man has learned how to increase mutations induced by high-energy radiations but not how to decrease mutation rates. The number of gene mutations induced by high-energy radiations is proportional to the amount of radiation received by the germ plasm. Acquired physiologic effects induced by radiations, such as burns and tumors, are lost when the organism dies. However, the genetic effects of induced mutants begin to appear in the progeny of exposed organisms, and they may persist in the gene pool of the population for many generations. Thus, as far as genetic effects are concerned, there is no such thing as a so-called safe dose.

Uncontrolled increase of mutation rates in human populations may produce a serious deterioration of public health and welfare for generations. All possible precautions must be taken to minimize the harmful effects.

Some contributors to the knowledge of genetics

J. Gottlieb Koelreuter (1733-1806)
A German who demonstrated the existence of sexes in plants and produced a plant hybrid by crossing two species of tobacco.

Gregor (Johan) Mendel (1822-1884)
An Austrian scientist and monk who described the inheritance of garden peas in a paper entitled "Experiments in Plant Hybridization" (1866). He formulated Mendel's laws by scientifically interpreting the results of experimental crossings. (*Historical Pictures Service, Chicago.*)

Mendel

Hugo de Vries (1848-1935)
A Dutch botanist who proposed the "Mutation Theory" based upon experiments with the evening primrose. However, most of the changes he observed were not gene mutations but a result of such other phenomena as chromosome changes and unusual combinations.

William Bateson (1861-1926) and R. C. Punnett
Bateson and Punnett, in England, performed experiments on the color of sweet peas and also on the comb shape of chickens. They discovered unusual ratios in these traits and attributed them to the action of two different genes. They also made observations that led to the discovery of linkage and crossing-over of genes (1905). Bateson postulated the production and transmission of discontinuous variations, and the results on the origin of species (1894).

C. E. McClung (1870-1946)
An American zoologist who established the role of chromosomes in sex determination in grasshoppers (1902), which later led to the discovery of similar phenomena in other animals.

Some contributors to the knowledge of genetics—cont'd

Friedrich Miescher (1844-1895)
A Swiss biochemist who discovered nucleic acids (1871), but whose work attracted little interest until the 1930's when several cell biologists, especially T. Caspersson, appreciated the importance of nucleic acids in cell physiology and applied new methods for their study.

Walter S. Sutton (1876-1916)
An American who explained fully (1903) the operation of Mendel's laws on the basis of chromosome behavior, derived largely from chromosome behaviors as explained by the German biologist Boveri.

Alfred H. Sturtevant (1891-   )
An American zoologist who made a chromosome map (1913) on which genes were located approximately on chromosomes by a scientific interpretation of the crossing-over percentages between chromosomes.

Herman J. Muller (1890-1967)
An American Nobel Prize winner (1946) who was the first to show (1927) that mutations could be artificially induced by radiations.

Thomas Hunt Morgan (1866-1945)
An American biologist and Nobel Prize winner (1933) who contributed to the knowledge of the mechanism of heredity. He reported the first gene mutation (white eye) in the fruit fly (*Drosophila*). (*Historical Pictures Service, Chicago.*)

Morgan

Albert F. Blakeslee (1874-1954)
An American botanist who applied the drug colchicine to dividing cells of plants, causing a doubling of the number of chromosomes with altered traits in the treated organisms (1927).

B. O. Dodge (1872-1960)
An American mycologist who suggested the importance of the pink bread mold (*Neurospora*), in genetic studies (1930).

G. W. Beadle (1903-   ) and E. L. Tatum (1909-   )
Beadle and Tatum, in California in 1941, studied the action of genes in the pink bread mold (*Neurospora*). It was concluded that the primary work of a gene is the production of a specific enzyme, which controls a living process.

Tracy M. Sonneborn (1905-   )
An American biologist who discovered plasmagenes (cytoplasmic genes) in *Paramecia* (1943) that could reproduce and initiate mutations.

Adolph Butenandt (1903-   )
A German biochemist and Nobel Prize winner (1939) who with the biologist Alfred Kühn experimented with the flour moth (*Ephestia*) and worked out a chain-like process of gene actions (1952) caused by the action of specific enzymes.

James D. Watson (1928-   ) and Francis H. Crick (1916-   )
Watson of Harvard University and Crick of Cambridge, England, found (1953) that the molecules which control heredity and growth in organisms are made of deoxyribonucleic acid (DNA) and consist of two long chains of atoms linked together and twisted spirally. They, with Maurice Wilkins, received Nobel Prizes in 1962.

Arthur Kornberg (1918-   )
An American biochemist who with his coworker, Severo Ochoa, received a Nobel Prize (1959) for work on the "Biologic Synthesis of Deoxyribonucleic Acid (DNA)." They found that an isolated enzyme catalyzes the synthesis of this nucleic acid in response to directions given from preexisting DNA.

Robert W. Holley (1922-   ), H. Gobind Khorana (1922-   ), and Marshall W. Nirenberg (1927-   )
These three Americans, working independently, explained how genes work. Their work led to formulation of the genetic code and its action in synthesis. They received the 1968 Nobel Prize in Physics.

Review questions and topics

1 Define genetics in your own words.
2 Describe the structure, composition, and functions of genes.
3 Define and give examples of: alleles, loci of genes, dominant, recessive, homozygous, heterozygous, synapsis, phenotype ratio, and genotype ratio.
4 State and give examples of Mendel's laws.
5 Explain and give examples of monohybrid crosses, dihybrid crosses, and trihybrid crosses.
6 What is the value of the checkerboard (Punnett square) in studying genetics?
7 Explain with examples complete dominance and incomplete dominance.
8 Discuss multiple genes and interaction of genes, describing each type with examples of each.
9 Discuss lethal genes and the results of their presence, including examples.
10 Explain the phenomena of linkage and crossing-over, with the results that follow in each case.
11 Explain how sex is determined genetically in plants and animals. What has this to do with heredity? Explain with examples sex-linked traits and sex-influenced traits.
12 Discuss the significance of cytoplasmic inheritance, including recent work that has been done in this field.
13 Problems in genetics:

Solve the following problems in peas, using the following symbols: T, tall plant; t, dwarf plant; Y, yellow seed; y, green seed.

a. Work out the entire monohybrid cross, using the proper symbols; carry through to the F_2 generation in each case: (1) homozygous tall × homozygous dwarf; (2) heterozygous tall × heterozygous tall.

b. Work the following dihybrid crosses as in a: (1) homozygous tall-yellow × dwarf-green; (2) heterozygous tall-yellow × heterozygous tall-yellow; (3) homozygous tall-yellow × heterozygous tall-yellow.

c. Work the following problems in guinea pig inheritance, using the following traits and symbols: R, rough coat of hair; r, smooth coat of hair; B, black hair; b, white hair.

(1) Work the entire monohybrid cross in the following by using the correct symbols; carry through to the F_2 generation in each: (a) homozygous rough × homozygous rough; (b) homozygous rough × smooth; (c) heterozygous rough × heterozygous rough; (d) heterozygous rough × smooth.

(2) Work the entire dihybrid cross, using the proper symbols; carry through to the F_2 generation in each: (a) homozygous black-rough × homozygous black-rough; (b) homozygous black-rough × white-smooth; (c) heterozygous black-rough × heterozygous black-rough; (d) white-smooth × white-smooth.

14 What are the phenotype ratios and genotype ratios in each of the preceding problems?

Selected references

Beadle, G. W.: Biochemical genetics, Chem. Rev. 37:15-96, 1945.

Boyer, S. H., IV, editor: Papers on human genetics, Englewood Cliffs, New Jersey, 1963, Prentice-Hall, Inc.

Gardner, E. J.: Principles of genetics, ed. 2, New York, 1964, John Wiley & Sons, Inc.

McElroy, W. D., and Glass, B.: The chemical basis of heredity, Baltimore, 1956, The Johns Hopkins Press.

Sinnott, E. W., Dunn, L. C., and Dobzhansky, T.: Principles of genetics, ed. 5, New York, 1958, McGraw-Hill Book Company.

Stern, C.: Principles of human genetics, ed. 2, San Francisco, 1960, W. H. Freeman & Co., Publishers.

Sutton, H. E.: An introduction to human genetics, New York, 1965, Holt, Rinehart & Winston, Inc.

Watson, J. D.: Involvement of RNA in the synthesis of proteins, Science 140:17-26, 1963.

Winchester, A. M.: Genetics, Boston, 1965, Houghton Mifflin Company.

chapter thirty

Genes and gene action

The term "gene" was originated by W. Johannsen of the University of Copenhagen in 1909. He also made the distinction between heritable and non-heritable traits, largely as a result of his work on the genetics of pure lines of beans. For most purposes we can consider a gene as (1) a unit of biochemical action, or a minimum section of a DNA molecule that controls metabolic activity in a cell; (2) a unit of mutation, or a minimum segment of a DNA molecule that has the inherent ability to mutate, or change, thus altering a heritable trait; (3) a unit of crossing-over, or a minimum segment of a chromosome that during certain stages in the life cycle of an organism, can cross over (transfer) to a neighboring chromosome.

DEVELOPMENT OF THE DNA CONCEPT

The best approach to understanding the function of the gene is through the history and study of deoxyribonucleic acid (DNA). Although DNA was first isolated from the cell nucleus in 1869 by a Swiss chemist, Friedrich Miescher, its recognition as the active principle of the gene is a fairly recent development. In 1914, another chemist, Robert Feulgen, developed a selective staining procedure that he used to identify DNA in a mass of ground-up cell components. Ten years later he used the same technique to show that the DNA of intact cells is concentrated in the chromosomes. Several investi-gators later demonstrated that the amount of DNA was fairly constant in all of the body cells of a particular organism. The egg and sperm of these organisms, however, contained only one half as much DNA as the somatic cells. This suggested that DNA might be a component of genes.

Still later, in 1928, an English bacteriologist, Frederick Griffith, made an intriguing discovery while working with the bacteria that cause pneumonia. He studied two different strains of pneumococcus, one virulent, or disease causing, and the other nonvirulent. The virulent strain was covered with a thick capsule, and the other was noncapsulated. When injected into mice, the virulent, capsulated strain caused death, while the other strain was harmless. Mice injected with heat-killed capsulated forms were not harmed.

However, when Griffith injected mice with a mixture of heat-killed virulent forms and live harmless forms, the mice died. When the mice were examined for the presence of bacteria, only live capsulated forms were present. Somehow the live, noncapsulated forms had been transformed into live, capsulated forms which were virulent. Griffith and others repeating his work believed that some heat resistant substance from the killed virulent strain had entered the nonvirulent noncapsulated forms and transformed them into the capsulated form.

It was not until 1944 that Avery, MacLeod, and McCarty showed that this transforming principle was actually DNA. They demonstrated that bacteria could be changed from harmless to virulent merely by supplying them with DNA extracted from a virulent strain.

This series of experiments, along with many others, showed that DNA is found in chromosomes, that it can be transmitted to other cells, that it acts in the same manner as the chromosomes during meiosis in that the quantity is halved, and that it is the hereditary substance.

DEVELOPMENT OF THE GENE THEORY

Once it was demonstrated that DNA was, indeed, the genetic principle the obvious question was, "How does it work?" For the answer to this we are indebted to George W. Beadle and E. L. Tatum and their work with the pink bread mold, *Neurospora*

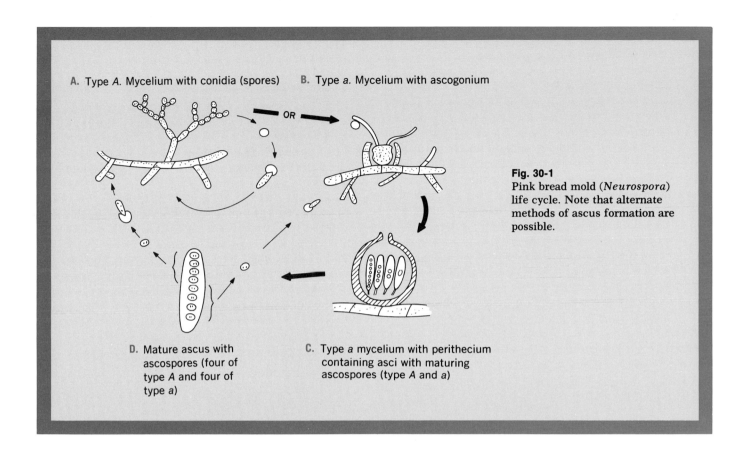

A. Type *A*. Mycelium with conidia (spores) B. Type *a*. Mycelium with ascogonium

OR

Fig. 30-1
Pink bread mold (*Neurospora*) life cycle. Note that alternate methods of ascus formation are possible.

D. Mature ascus with ascospores (four of type *A* and four of type *a*)

C. Type *a* mycelium with perithecium containing asci with maturing ascospores (type *A* and *a*)

crassa, in the early 1940's. This fungus, belonging to the class Ascomycetes, has a short life cycle, is propagated rather easily, and produces large populations of a given genotype.

Reproduction occurs asexually by spores called conidia, or simply through unspecialized fragments of the mycelium. The vegetative hyphae are segmented, with each segment containing many haploid (N) nuclei (Fig. 30-1). Sexual reproduction occurs only when there is a union of cells of opposite mating types, known as A and a. The resulting fusion nucleus is the only diploid (2N) stage. Nuclear fusion and meiosis occur within the saclike ascus. The ascus is sufficiently narrow so that the divisions of meiosis occur in sequence, and the resulting nuclei do not slip past each other but hold their original positions. The meiotic divisions give rise to four nuclei, with each in its particular position dividing once by mitosis, thus producing eight nuclei in a linear series in each ascus.

While still within the ascus, the eight linearly arranged ascospores provide an accurate record of what has happened at meiosis because (1) the position of a particular ascospore can be traced to the actual position of a nucleus in meiosis, as determined by the orientation of separating chromosomes on a spindle, and (2) all products of a single meiosis are retained within one structure and hence cannot be confused with products of other meiotic divisions.

With careful dissection each of the eight ascospores can be removed singly from the ascus yet retain its original position in the sequence. Each isolated ascospore can be cultured separately (in agar tubes), and the genetic characteristics of each such derived culture can be traced to segregation of genes at meiosis. If eight of these culture tubes are lined up in the same sequence as their corresponding ascospores in the ascus, it can be observed directly which alleles went to which nuclei at meiotic segregation.

Beadle and Tatum took advantage of the growth patterns of *Neurospora*, which forms a mycelium

447

with haploid (N) nuclei (Fig. 30-1). Any expressed trait was caused by one gene and not confused by possible dominants or recessives. *Neurospora* is very easy to grow, requiring only a sugar, some inorganic salts, and one vitamin (biotin) as nutrients. From this minimal medium of nutrients, it can then synthesize any other required substances.

Beadle and Tatum treated asexual spores with radiation and germinated them on an enriched medium containing additional vitamins, amino acids, and sugars. When all spores had produced growth, bits of each colony were transferred to minimal media. Some of the spores did not grow, indicating that they had lost the capacity to synthesize some essential substance presumably because of earlier radiation treatment.

Next Beadle and Tatum studied the original culture that had been used for the minimal media trials. Samples from the enriched medium were transferred to several tubes, each containing minimal medium plus either vitamins, amino acids, or sugars. By observing the growth characteristics on the new media, they were able to devise further tests to determine exactly which substance was required for growth by that particular strain of *Neurospora*. They reasoned that if a colony of *Neurospora* now required a particular amino acid that it had been able to synthesize prior to radiation treatment, some change had taken place in the genetic structure. In order to demonstrate that this was truly a genetic mutation, they crossed the new strain with the wild type (nonradiated) *Neurospora*. In the resulting ascus, four of the spores germinated to form wild type colonies while the other four formed colonies requiring supplemental amino acids.

Beadle and Tatum had therefore demonstrated not only that mutations could be induced and transmitted as genetic traits but also that each mutation resulted in loss of the ability to synthesize one type of molecule. Since the synthesis was controlled by an enzyme, they proposed that a gene works by controlling the synthesis of one enzyme.

This one gene—one enzyme hypothesis has been slightly modified by further studies but the work of Beadle and Tatum, for which they were awarded Nobel Prizes, is recognized as the basis of the gene theory.

SYNTHESIS OF PROTEINS

Roles of DNA and RNA

Once it was established that DNA was the genetic principle and that a gene controlled protein synthesis, the next task was to determine how this worked. Before going into this however, the chemistry of DNA should be reviewed.

Chemistry of DNA

This molecule, or rather macromolecule, is composed of large numbers of nucleotides, which in turn are composed of a five-carbon sugar, phosphoric acid, and a nitrogenous ring molecule (Figs. 6-10 to 6-12). The D in DNA refers to the sugar, deoxyribose, as the R in RNA refers to a similar sugar, ribose.

The nitrogenous compounds are double ring purines and single ring pyrimidines. Five kinds of these nitrogen bases are found in nucleotides. Two purines, adenine (A) and guanine (G) are found in both DNA and RNA. Three pyrimidines occur, cytosine (C) in both DNA and RNA, thymine (T) only in DNA, and uracil (U) only in RNA. Both DNA and RNA then are composed of four types of nucleotides that differ in the sugar and in the nitrogen base present. (See Table 6-9.)

The double helix

In 1950 it was shown by E. Chargaff that a consistent feature of the DNA molecule was that the amount of adenine always equaled the amount of thymine, and the guanine equaled the cytosine. By 1953 Maurice Wilkins was able to show by use of x-ray diffraction that the DNA molecule had a very regular and characteristic shape.

Later in 1953 James Watson and Francis Crick were able to put this information together to show that DNA was, in fact, two strands of DNA bonded together in the now famous "spiral staircase" or more accurately, double helix. (Details and personal recollections of the search for DNA are given in Watson's book, *The Double Helix.**)

In their model, Watson and Crick showed that

---

*Watson, J. D.: The double helix; a personal account of the discovery of the structure of DNA, New York, 1968, Atheneum Publishers.

the "steps of the staircase" were always composed of a purine bonded to a pyrimidine. This explained the earlier finding that the adenine content was always equal to the thymine and the guanine to the cytosine. The helical railings of the staircase were formed by the sugar and phosphate joined together. The significance of these findings earned the Nobel Prize for Watson, Crick, and Wilkins.

Once they had constructed their model, Watson and Crick suggested that the double helix structure had all of the criteria to account for mitosis and gene duplication. A double helix could unwind by breaking the hydrogen bonds between the base pairs of A-T or G-C, and thus provide two single strands. Each nucleotide on each strand would then attract its complementary nucleotide. Thus, an adenine nucleotide would attract a thymine nucleotide and a guanine would attract a cytosine nucleotide. In this manner each of the original strands would determine the structure of a new member of a double strand.

In 1957, this suggestion was proved true by the work of another recipient of the Nobel Prize, Arthur Kornberg. He showed that DNA, if provided with enzymes, adenosine triphosphate (ATP), and a quantity of nucleotides in a test tube, would synthesize these into new DNA molecules. This work was confirmed in living cells by growing them in media containing nitrogen bases with different isotopes of nitrogen and analyzing the distribution in later generations.

We are now ready to return to the problem of the control of protein synthesis by DNA.

Function of DNA

DNA has two functions: (1) replication, or formation of more DNA, and (2) transcription, or the formation of messenger RNA (mRNA).

The process of replication is thought to result from the unwinding of the double strand and the attracting of complementary nucleotides to the nucleotides of each strand. This process is most apparent just prior to cell division and results in each daughter cell containing an amount of DNA similar to that of the original parent cell.

Transcription results when one of the unwinding strands attracts ribose nucleotides instead of deoxyribonucleotides and uracil instead of thy-

mine as the complement to adenine. The resulting strand is detached as messenger RNA, which passes through the nuclear membrane into the cytoplasm where it is attracted to a ribosome, or group of ribosomes called a polyribosome. The ribosomes also contain a different type of RNA called ribosomal RNA. Its function is still in doubt.

Polypeptide synthesis

Proteins and smaller polypeptides are formed along the surface of a messenger RNA–ribosome complex. A third type of RNA, transfer RNA, is formed in the cytoplasm. This transfer RNA (tRNA) functions in bringing amino acids from the cytoplasm to the mRNA–ribosome complex.

The union of the various amino acids takes place as one tRNA attaches to its complementary site on the mRNA–ribosome complex, then is joined by the second tRNA attracted to the next complementary mRNA site. Transfer RNA has a point of attraction to mRNA, and at another site on its surface it has a point of attraction for a specific amino acid. This is determined by a particular sequence of nucleotides arranged in triplets and called the Genetic Code (Table 30-1).

The attachment of tRNA to the mRNA–ribosome complex results in the respective amino acids coming into proximity, with a peptide bond forming between them. Each tRNA adds one more amino acid to the chain. After the amino acid is delivered, the temporary attachment of tRNA to mRNA is broken and tRNA is returned to the cytoplasm. When the last amino acid has been delivered, the entire polypeptide is released into the cytoplasm. The total process of peptide formation is referred to as translation.

The entire process might be summarized as follows: DNA by transcription forms mRNA; messenger RNA carries the message spelled by the sequence of nucleotides, to ribosomes that translate the message in the sequence of attracting tRNA; the message is interpreted as a particular protein or polypeptide (Fig. 30-2).

Genetic Code

Much has been written about the way in which a protein or polypeptide is finally formed. We have seen how a nucleotide on the DNA strand attracts

449

Table 30-1

The Genetic Code*

Amino acid	Nucleotide triplets code
Alanine (Ala)	GCU, GCC, GCA, GCG
Arginine (Arg)	AGG, CGU, CGC, CGA, CGG
Asparginine (AspN)	AAU, AAC
Aspartic Acid (Asp)	GAU, GAC
Cysteine (Cys)	UGU, UGC
Glutamic Acid (Glu)	GAA, GAG
Glutamine (GluN)	CAA, CAG
Glycine (Gly)	GGU, GGC, GGA, GGG
Histidine (His)	CAU, CAC
Isoleucine (Ileu)	AUU, AUC, AUA
Leucine (Leu)	UUG, UUA, CUU, CUC, CUA, CUG
Lysine (Lys)	AAA, AAG
Methionine (Met)	AUG
Phenylalanine (Phe)	UUU, UUC
Proline (Pro)	CCC, CCU, CCA, CCG
Serine (Ser)	UCU, UCC, UCA, UCG, AGU, AGC
Threonine (Thr)	ACU, ACC, ACA, ACG
Tryptophan (TRYP)	UGG
Tyrosine (Tyr)	UAU, UAC
Valine (Val)	GUU, GUC, GUA, GUG
Stop ("punctuation")	UAA, UAG, UGA

*A = adenine, C = cytosine, G = guanine, U = uracil.

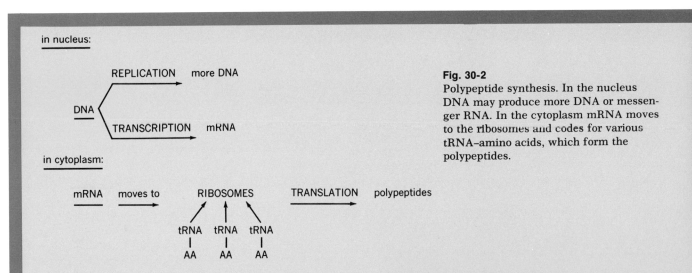

Fig. 30-2
Polypeptide synthesis. In the nucleus DNA may produce more DNA or messenger RNA. In the cytoplasm mRNA moves to the ribosomes and codes for various tRNA–amino acids, which form the polypeptides.

its complementary ribonucleotide in the formation of mRNA, as for example, adenine attracts uracil. By a series of ingenious experiments Robert W. Holley, H. Gobind Khorana, and Marshall W. Nirenberg, working separately, determined that a sequence of nucleotides on mRNA "codes" for a particular amino acid. The first of these discoveries, made by Nirenberg, showed that the sequence of three adenine nucleotides (AAA) on a DNA molecule resulted in three uracil (UUU) nucleotides on mRNA and that this sequence of three uracil nucleotides in turn was the part of the code, or the "codon," for the amino acid phenylalanine. Similarly, a sequence of CAA on the DNA molecule produced GUU on mRNA which in turn was the code for the amino acid valine, while a sequence of CTT on the DNA molecule produced GAA which was the code for the amino acid glutamic acid.

The sequence involved in the formation of a single polypeptide composed of these three amino acids would start with the DNA molecule in the nucleus. On this DNA molecule there would be a sequence of nine nucleotides as follows: A-A-A-C-A-A-C-T-T. Transcription would result in a mRNA molecule that contained the complementary sequence U-U-U-G-U-U-G-A-A. After moving to the ribosome site this sequence would be translated as the codon for the amino acids phenylalanine, valine, and glutamic acid.

For their work in this field, Holley, Khorana, and Nirenberg received the 1968 Nobel Prize in Physics.

Anticodons

The particular tRNA carrier of phenylalanine apparently has a complementary set of nucleotides for the codon UUU. This set is called an anticodon. Thus the codon UUU on a mRNA molecule, attracts an anticodon AAA on a tRNA molecule.

It is thought that tRNA molecules may be of several possible shapes formed by a strand of RNA bending back on itself and being held in shape by hydrogen bonds. The resulting molecule then has one or more loops and two ends. The most likely shape is probably that of a three-leaf clover. The anticodon recognition triplet is apparently on one of the loops and the amino acid attachment point is on a loose end.

Following the discovery that UUU was the codon for phenylalanine, many investigators began to look for more codons through the process of adding amino acids one at a time to a test tube solution containing a known type of synthetic mRNA. This led to the discovery that sixty-one of the possible sixty-four triplets (four possible nucleotides in three possible positions) do, in fact, code for particular amino acids. Some of the twenty amino acids are specified by more than one codon. It is believed that the three unknown triplets may be "punctuation marks" (stops) between polypeptide sequences.

DEFINITION OF GENES

We will conclude with a further definition of the gene as the minimum section of a DNA molecule that can be transcribed in forming a molecule of mRNA. The gene functions in the formation of polypeptides when the mRNA moves to ribosomes, attracts the coded tRNA–amino acids, and is translated as a particular amino acid sequence.

Review questions and topics

1 Distinguish between virulent and nonvirulent. Does there seem to be a connection between capsulated bacteria and virulence?
2 Trace the story of DNA from discovery to identification as the heredity substance.
3 What properties of Neurospora made it ideal for the work of Beadle and Tatum?
4 How did Beadle and Tatum state the way a gene works?
5 Review the chemistry of DNA and of RNA.
6 What are the component parts of a nucleic acid?
7 What are the genetic functions of DNA?
8 Starting in the nucleus trace the components involved in the synthesis of a polypeptide.
9 What is the significance of the Genetic Code?
10 Distinguish between the various types of RNA.

Selected references

Carlson, E. A., editor: Gene theory, Belmont, Calif., 1967, Dickenson Pub. Co., Inc.
Carpenter, B. H., editor: Molecular and cell biology, Belmont, Calif., 1967, Dickenson Pub. Co., Inc.
Clark, B. F. C., and Marcker, K. A.: How proteins start, Sci. Amer. 218:36-42, 1968.
Crick, F. H. C.: The genetic code: part III, Sci. Amer. 215:55-60, 1966.

451

Selected references —cont'd

Crick, F. H. C.: The structure of the hereditary material, Sci. Amer. Oct. 1954.

Haynes, R. H., and Hanawalt, P. C., editors: The molecular basis of life. Readings from Scientific American, San Francisco, 1968, W. H. Freeman and Co. Publishers.

Ingram, V. M.: The biosynthesis of macromolecules, New York, 1965, W. A. Benjamin, Inc.

Kornberg, A.: The synthesis of DNA, Sci. Amer. Oct. 1968.

Lerner, I. M.: Heredity, evolution and society, San Francisco, 1968, W. H. Freeman and Co. Publishers.

Mirsky, A. E.: The discovery of DNA, Sci. Amer. **218:**78-88, 1968.

Nirenberg, M. W.: The genetic code: part II, Sci. Amer. **208:**80-94, 1963.

Ravin, A. W.: The evolution of genetics, New York, 1965, Academic Press, Inc.

Watson, J. D.: The double helix; a personal account of the discovery of the structure of DNA, New York, 1968, Atheneum Publishers.

Watson, J. D.: Molecular biology of the gene, New York, 1965, W. A. Benjamin, Inc.

chapter thirty-one

Evolution and biogeography

THEORIES OF THE ORIGIN OF LIFE

Through previous studies of plants and animals you may have observed that living organisms are varying constantly, and many if not all, of their constituents are changing all the time. A study of evolution shows that it is the great organizing, centralizing, and unifying concept that underlies all life. This idea modifies our way of interpreting all biologic phenomena. Evolution cannot be appreciated fully until extensive knowledge of living organisms is obtained; however, it should not be considered in a single chapter but emphasized gradually and progressively throughout our studies. We should develop a sense of the universality of evolutionary processes, from the molecular level through the lower plants and animals to man, with his physical, mental, and sociocultural development. In this way the biologic and physical sciences, social sciences, and humanities can be integrated.

Abiogenesis

Until the seventeenth century people believed in abiogenesis (Gr. *a-*, not; *bios*, life; *genesis*, origin), which is the spontaneous generation of living organisms, especially lower types, from nonliving substances. For example, field mice were thought to arise spontaneously from mud of the Nile River, and flies were thought to originate from dirt, manure, and decaying meats. Empedocles (495-435? B.C.), a Greek philosopher, suggested that parts of animals had arisen spontaneously and separately from the earth and had assembled themselves at random into entire animals, thus explaining why certain organisms were so odd. Anaximander (611-547 B.C.) thought that air imparted life to all living things. Aristotle (384-322 B.C.) believed that eels (fishes) arose spontaneously from nonliving materials. He believed that complex forms of life evolved from simpler ones, although he had no clear idea of an explanatory, activating mechanism. Although Aristotle was a recognized authority in Roman and medieval times, his idea of evolution was neglected, and for hundreds of years many persons believed that each species of plant and animal was created separately and remained unchanged thereafter. There seemed to be little conflict between this belief and the ac-

ceptance of spontaneous generation of flies, fleas, and mice at that time. Kircher thought that animals arose through the action of water on the stems of plants. Von Helmont (1577-1644), a Flemish physician, thought that house mice were generated from pieces of cheese placed in bundles of rags.

One reason why abiogenesis was believed was because the complex life cycles of many animals and plants were not understood, and the microscopic stages of living organisms had never been seen. This lack of accurate information was a natural setting for such a theory. Experimental methods of attacking such problems were not yet developed, and proofs were attempted from hearsay, supposition, and frequently, illogical reasoning.

Joseph Needham, in 1749, believed that he had demonstrated the spontaneous origin of minute living organisms in infusions that he had boiled a few minutes in corked flasks. By boiling he thought that he had killed all living matter from which living organisms could arise later. However, the living organisms that later appeared in the flasks arose from living materials such as cysts and spores that had withstood the brief boiling temperature.

Biogenesis

Abbé Spallanzani (1729-1799), an Italian, stated that Needham's conclusions were incorrect because they were derived from incomplete sterilization of the infusions in the flasks. Spallanzani improved on Needham's experiments by sealing

453

the necks of the flasks with a blast lamp to prevent the entrance of anything, including air, and he boiled the infusions for hours instead of a few minutes. He found that under such conditions no living organisms could be found even after long periods of time (1776). Others stated that free oxygen, which is essential for life processes, was excluded in the experiments, thus preventing the possibility of spontaneous generation. To answer the latter point investigators during the first half of the last century showed that thoroughly sterilized infusions never developed living organisms even when sterile air was admitted. The air was sterilized by heating or passing through acids to remove the suspended "dust particles" on which living substances might be attached.

Francisco Redi (1621-1697), an Italian scientist, demonstrated by a simple experiment of protecting decaying meat (with a fine gauze) from contamination by flies that the maggots of these flies did not arise spontaneously from the dead meat, but that they developed from living eggs deposited by flies. Redi's results showed that "life must arise from living organisms."

Louis Pasteur (1822-1895), a French microbiologist, by means of a special type of flask showed that the source of life, which appeared in infusions exposed to the air, was the air itself, or rather the life existing in the air. Thus it was contended that much of the dust of the air contained microorganisms in a somewhat dormant condition that were ready to become active when suitable environmental conditions, such as proper food, temperature, and moisture, were encountered.

■
Biochemical theory

The rather intriguing notion that life has arisen from nonliving matter dates from the early Greeks. Suggestions concerning how this may have happened, and some data to support the theories, have been available only in recent years as a result of the work of many biochemists, among whom A. I. Oparin is outstanding. His work aided that of Dr. H. Urey, who in turn stimulated his graduate student Stanley Miller. Miller, as others after him, succeeded in establishing a primitive atmosphere in which some of the conditions present on the earth in its early history were duplicated.

The exact details of the formation of the earth are not known, and several conflicting theories exist, but there is some agreement on the course of development of complex molecules.

There were a number of elements in the atmosphere of early earth; among them we are most concerned with carbon (C), hydrogen (H), nitrogen (N), and oxygen (O) because of their prevalence in present-day organisms. The presence of large amounts of hydrogen resulted in the formation of compounds of water (H_2O), methane (CH_4), ammonia (NH_3), and later hydrogen cyanide (HCN).

These early molecules reacted under the influence of available energy sources, especially ultraviolet light, but also lightning, radiation, and volcanic activity. It is this early reaction that has attracted the attention of recent investigators. They have tried, and in most cases succeeded, to duplicate the various steps in molecular synthesis in present-day organisms (Fig. 31-1).

In general, it appears that the interaction of methane, ammonia, water, and hydrogen cyanide produced more complex molecules, including simple sugars, amino acids, purines, pyrimidines, and fatty acids.

The further synthesis of these compounds produced more complex molecules. If we consider this process as occurring in the early oceans, or at least in watery accumulations, that were being constantly fed by runoff, it is easy to picture the gradual buildup of the inorganic molecules involved in the process.

The laboratory synthesis involving the purine, adenine, and simple sugar resulted in the formation of a common nucleoside, which in a further reaction with phosphoric acid, formed adenosine triphosphate. It is significant that this nucleotide is the principal source of energy in metabolism. It has been suggested that the polymerization of adenosine triphosphate and related nucleotides led to the formation of nucleic acids.

The reaction involving amino acids led to the formation of proteins including primitive enzymes; lipids were formed from fatty acids and glycerol-like molecules.

These and similar reactions resulted in the formation of many molecules that are important to metabolism. Some of these molecules, or at least

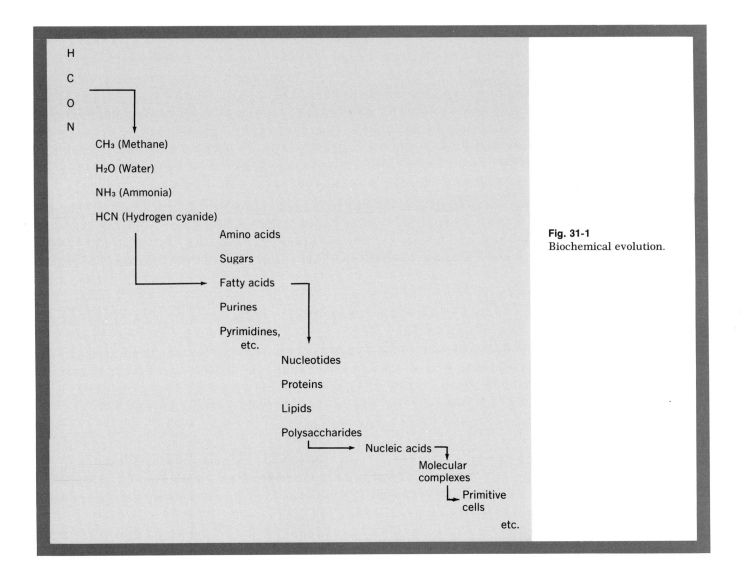

Fig. 31-1
Biochemical evolution.

H

C

O

N

CH₃ (Methane)

H₂O (Water)

NH₃ (Ammonia)

HCN (Hydrogen cyanide)

Amino acids

Sugars

Fatty acids

Purines

Pyrimidines,
etc.

Nucleotides

Proteins

Lipids

Polysaccharides

Nucleic acids

Molecular
complexes

Primitive
cells

etc.

the steps in their formation, have been prepared under laboratory conditions similar to the conditions of primitive earth.

The next step in this biochemical evolution appears to have been the gradual accumulation of rather simple groups of nucleic acids, surrounded by complexes of proteins and lipids—what we would recognize as a very primitive "cell."

The properties of the various molecules attracted to this complex would, in turn, determine its activities. There is reason to infer a process of change and development of the primitive complex into the forerunners of present-day bacteria and protozoa (Fig. 31-1).

The current interest in outer space has greatly stimulated the investigation of primitive environments and the possibility of extraterrestrial life.

EARLY THEORIES OF EVOLUTION

Jean Baptiste Lamarck (1744-1829)

Jean Baptiste Lamarck, a French zoologist, was one of the first to attempt to explain seriously the phenomenon of evolution. Lamarck observed the changes and adaptations of certain parts of animal bodies that were used continuously in certain ways. He noted that certain structures increased in size if used but grew smaller or atrophied if not used. In 1809, in his theory of acquired traits through use and disuse, he concluded that these gains or losses

455

through use or disuse were passed on to offspring in succeeding generations. Apparently, Lamarck never experimented to ascertain whether or not these acquired changes were hereditary. Use and disuse may change certain body traits, but the transmission of acquired traits, as such, to future generations has not been proved. Although the basic idea of his theory is no longer accepted, it did stimulate serious thought about evolution.

■
Erasmus Darwin (1731-1802)

Erasmus Darwin, the grandfather of Charles Darwin, recognized that the various types of organisms arose from each other, and he stressed the response of the animal to environmental changes as the basis for its modifications.

■
Sir Charles Lyell (1797-1875)

Sir Charles Lyell, a geologist, helped to foster sound thinking along evolutionary lines through his publication, *Principles of Geology* (1830). He formulated the theory of uniformitarianism, in which he stated that the forces that produced changes in the earth's surface in the past are the same as those that operate upon the earth's surface at present. Such forces over long periods of time could account for all the observed changes, including the formation of fossil-bearing rocks. This concept showed that the age of the earth was millions of years, rather than thousands of years. This great geologic work stimulated and influenced Charles Darwin's own thoughts and the formulation of his theory of organic evolution.

■
Charles Darwin (1809-1882)

While it is customary to refer to Darwin's presentations as the theory of evolution, he actually made two great contributions in his work, the first of which was so basic that it was easily overlooked. It was simply that evolution has occurred. The second theory was the mechanism by which it occurred, namely, natural selection.

The idea that evolution had occurred was not new with Darwin. His contribution was in stating the theory and then supporting it with massive evidence. He maintained that living species were descendants of similar, slightly different, forms that lived in the past.

Theory of natural selection (1859)

Charles Darwin's theory was based on the observation and study of many plants and animals in many parts of the world. He noted that all living plants and animals constantly display variations, great or small, structural or functional. He noted that the offspring of a pair of individuals, either plant or animal, showed they were not all alike. He also observed that, by selecting for breeding purposes those individuals that possessed certain desirable traits, there could be produced new kinds of domesticated animals and plants. He observed fossils, studied comparative anatomy and embryology, was familiar with geographic distribution of species and with geology, and he conducted breeding studies that produced variations in animals and plants.

His interests in nature and his lack of success in any other field led to an appointment as naturalist on a 5-year cruise of HMS Beagle around the world. This gave him the information that led to his belief in the truth of evolution.

Darwin was greatly influenced by Malthus' book *An Essay on the Principles of Population* (1798). Malthus stated that the human breeding potential far outstrips the limited supply of potential resources, leading to competition for the available goods and thus a struggle for existence. Darwin adopted Malthus' ideas of competition and struggle for existence in his theory of organic evolution.

Darwin was also greatly influenced by Alfred Russel Wallace (1823-1913), an English scientist, who had independently reached the same general conclusions concerning the theory of natural selection. However, when Darwin published his book *The Origin of Species* (1859), a year after Wallace had announced his conclusions in a short essay, most of the credit went to Darwin.

Evolution was explained as natural selection—a theory that was based on a few observations and the conclusions drawn from them.

First, it can be shown that the reproductive potential of any species is more than sufficient to replace the existing population. This is true whether in elephants, which require several years to produce offspring, or in flies, which require several days. A geometric progression for the increase is possible.

It can also be shown that this geometric increase

does not take place. For example, the offspring of a pair of flies, mating in early spring and producing many generations per season, have been calculated to be 1.91×10^{20}, which is more than sufficient to cover the earth with dead flies. Obviously they are not here. It may be concluded, therefore, that not all offspring survive. The next question is, then, which of them do survive?

Since variations among individuals, even close relatives, do exist, it becomes evident that some variations are "favored" over others in the survival race. Darwin believed that those variations that made the organism "better adjusted" were favorable ones and were transmitted to offspring, which then also had a greater survival advantage.

Some of the salient points of Darwin's theory of evolution might be summarized as follows: (1) Organisms of the same species are not all alike but vary slightly among themselves, and the offspring of a particular pair of parents tend to vary around the average of these parents. (2) More offspring are produced than can possibly survive. (3) In order to survive these offspring must compete with each other for existence. (4) As a result of this competition, those individuals best adapted to that environment tend to survive (survival of the fittest). (5) New species originate, over a period of many generations, by this natural selection of the best-fitted individuals. According to Darwin, whenever two groups of a plant or animal population are each faced with slightly different environmental conditions, they would tend to diverge from each other and in time would become sufficiently different from each other to form separate species. In a similar manner greater divergencies could arise at later times.

Although Darwin's theory of natural selection gives, in general, a satisfactory explanation of evolution, several parts of it must be altered in the light of more recent biologic knowledge. Darwin did not distinguish clearly between variations induced by environmental factors and those that involved chromosomal materials or alterations of germ plasm. Many of his types of variations are now known to be nonheritable, thus having little significance in evolution. Only variations that involve genes are inherited and furnish material on which natural selection can act. In fact, the mechanism of genetics was not appreciated until 1903, quite some time after Darwin's death.

NEO-DARWINISM

Modern studies of evolution are primarily the result of the rediscovery in 1903 of the work of Mendel, in which the mechanism of survival along the lines of differential reproduction and the genetic basis of variation was investigated.

Evolution takes place in populations of a particular species, not in its individual members. The breeding population is the potential unit of change. Recognition of this has led to the concept of gene pools, which consist of the sum of the different genes present in the entire population. In time any member of the pool may exhibit any combination of its genes.

A species of any organism would then consist of all the interconnected gene pools. Thus, not only does every member of population A potentially share in its genes but each member may contribute also to the gene pools of populations B, C, D, and so on, by migration.

Shortly after the rediscovery of Mendel's work, G. H. Hardy, an English mathematician, and W. Weinberg, a German physician, recognized the precision of the preservation of genes in a population. Stated quite simply the Hardy-Weinberg law is that, under usual conditions, the gene frequencies in a population remain constant from generation to generation.

Evolution occurs when gene frequencies change, thus upsetting the Hardy-Weinberg equilibrium. While many agencies enter the process, the major factors influencing gene frequencies are (1) mutation, (2) genetic drift, (3) selection or differential reproduction, and (4) migration.

Thus, the Hardy-Weinberg equilibrium is only an indication of gene frequencies and is applicable only when (1) mutations either do not occur or are balanced by back mutations; (2) genetic drift, or changes in gene frequency purely by chance in a small population, does not occur; (3) reproduction is random with no sexual selection; and (4) gene migration into or out of the pool does not take place.

Since it can readily be demonstrated that the preceding changes do occur, evolution can be defined simply as a series of changes in gene frequencies

457

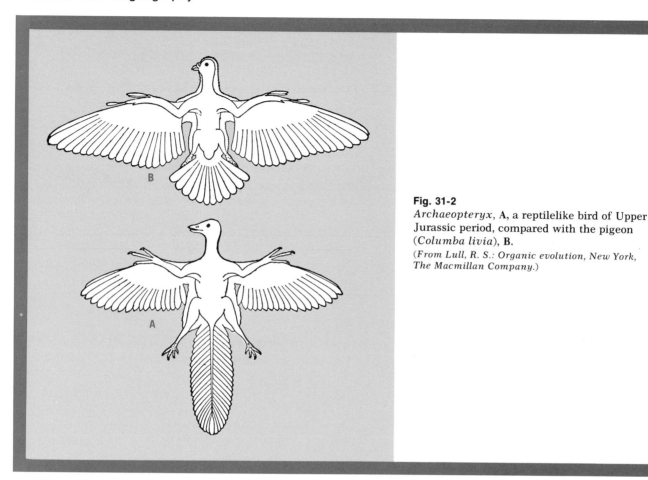

Fig. 31-2
Archaeopteryx, **A,** a reptilelike bird of Upper Jurassic period, compared with the pigeon (*Columba livia*), **B.**
(*From Lull, R. S.: Organic evolution, New York, The Macmillan Company.*)

in a population, or gene pool. When a mutation is selected or has a reproductive advantage, it becomes the norm of that particular population.

EVIDENCE OF EVOLUTION
Evidences of descent with change have been obtained from such sciences as (1) paleontology (science of fossil remains), (2) taxonomy (classification), (3) comparative embryology, (4) comparative anatomy, (5) comparative physiology, (6) biogeography (geographic distribution); and (7) genetics and variations.

Evidence from paleontology
The science that deals with fossil plants and animals is called paleontology (Gr. *palaios,* ancient; *onto-,* being; *logos,* study). Geologists can deter-

mine, in most instances with remarkable accuracy, the chronologic succession in time of the various strata composing the earth. The fossils of these strata testify to the order of appearance and disappearance of various types of animals and plants on the earth, the more recent appearing nearer the earth's surface. Three examples, birds, horses, and wheat, may be cited.

Evolution of birds
Birds seem to have evolved from a reptilelike ancestor, because despite superficial dissimilarities, such as the scaly-skinned, cold-blooded reptile and the feathered, warm-blooded bird, there are many fundamental structural and functional similarities not only between adult reptiles and birds but also between their embryonic stages (Fig.

31-14). Fossil remains of a reptilelike bird (*Archaeopteryx*) show a connecting link between reptiles and birds as they are known today (Fig. 31-2).

Evolution of horses

Modern horses have developed through a series of successive changes. Much of the evolution of the horse occurred in North America and extends back sixty million years ago to the Eocene epoch in geologic history (Table 31-1). Fossil records indicate that the descent proceeded in a definite direction, but there were undoubtedly many side forms that showed variations from the main line of descent. Many of these forms became extinct, probably through the process of natural selection. Important progressive changes in the horse included differences in the shape and size of the skull, in the bones of the feet, and in the structure and character of the teeth (Figs. 31-3 and 31-4). There also seemed to be a progressive increase in the size of most types in the direct line of descent.

The first known ancestor of the American horse is *Eohippus* (Eocene epoch), which was about 1 foot high at the shoulder and grazed on forest underbrush. Its forefeet had four toes (and a splint bone), whereas the hind foot had three well-developed toes (and two splints, which represented the first and fifth toes). Another type in the line of descent was *Mesohippus* (Oligocene epoch), which was taller. Each foot had three digits, with the middle one larger and better developed.

Merychippus (Miocene epoch) was probably the direct ancestor of later horses. It was three toed, with the lateral toes high above the ground. The skull was larger and the lower jaw was heavier than the ancestral forms. *Merychippus* was between 3 and 4 feet high at the shoulder and gave rise to several types, most of which became extinct. However, one that persisted was *Pliohippus* (Pliocene epoch), which was the first one-toed horse. The modern horse, *Equus*, arose from *Pliohippus* during the Pleistocene epoch in North America, and it spread to other continents. *Equus* finally became over 5 feet high at the shoulder, with only one toe on each foot (with two splint bones of former lateral toes). The horse became extinct in North America by the end of the Pleistocene epoch, but certain types persisted in Eurasia to become the ancestors of present-day horses. Wild types still

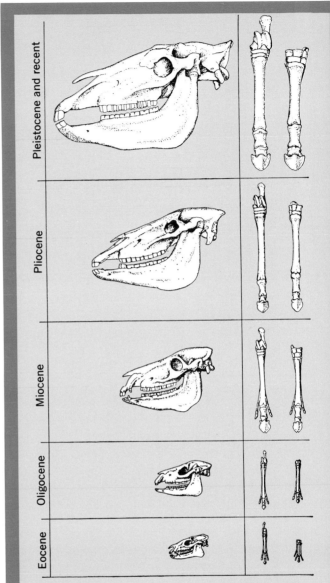

Fig. 31-3
Evolution of the horse. Comparisons of the shape and size of skulls, hindfeet, and forefeet of animals of successive geologic periods, indicating changes that led from a fox-sized ancestor in the Eocene period, sixty million years ago, to the modern horse (top): Note the changes from four toes and three toes of the early ancestors through successive stages to the one-toed type with splint bones that do not touch the ground.
(*From Crow, J. F.: Ionizing radiation and evolution, Sci. Amer., 201:138-145, 1959.*)

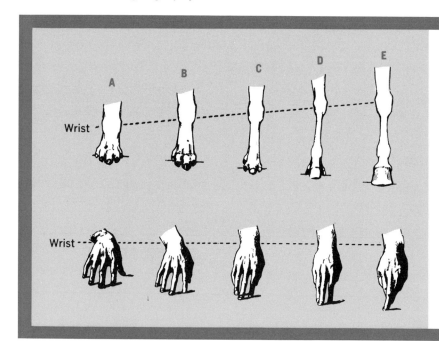

Fig. 31-4
Comparative stages in the elevation of the horse's foot to the tip of the middle toe, as shown by the human hand. (*Courtesy The American Museum of Natural History, New York, N. Y.*)

exist in Asia. Early Spanish colonists reintroduced horses to America after the time of Columbus, and some of them became the wild horses of our western plains and of South America.

Evolution of wheat

Wheat was one of the earliest plants cultivated by man. Carbonized kernels of wheat have been found by the University of Chicago archaeologist, Robert Braidwood, at Jarmo, Iraq, which may have been a birthplace of agriculture nearly 7,000 years ago. There is a remarkable resemblance between the ancient grains and modern grains, the latter having been carbonized to simulate the archeologic specimens. Two types of kernels were found in Jarmo, one almost identical with wild wheat still growing in the Near East, and the other almost exactly like present-day, cultivated wheat of the type called einkorn (Fig. 31-5).

Evidently, in certain types of wheat there has been little change in nearly 7,000 years. However, other types have evolved, so that we have today a number of distinct species of wheat, all belonging to the genus *Triticum* (L. *tres*, three). Some authorities recognize fourteen species, which fall into three distinct groups determined by the number of chromosomes in their reproductive cells. These numbers are, respectively, seven, fourteen, and twenty one, or double this number in body (somatic) cells. The chromosome numbers are closely associated with differences in such things as anatomy, resistance to disease, productivity, and baking qualities. The wheats with fourteen and twenty one chromosomes have arisen from the seven-chromosome wheat and certain related grasses through hybridization, followed by chromosome doubling. Hence, cultivated wheats are an excellent example of evolution, and the evidences collected involved studies of the genetics mechanisms.

Fossil record

A fossil may be defined as any trace, remains, or impression of a plant or animal of past geologic ages. Much of our present knowledge about ancient life has been gained by a study of these ancient organisms in the various strata of the earth. According to their origin, the rocks of the earth's surface are of two kinds: sedimentary and igneous. Sedimentary rocks, such as limestone, sandstone, and shale, may contain fossils and are formed by the transportation and deposition of small rock

460

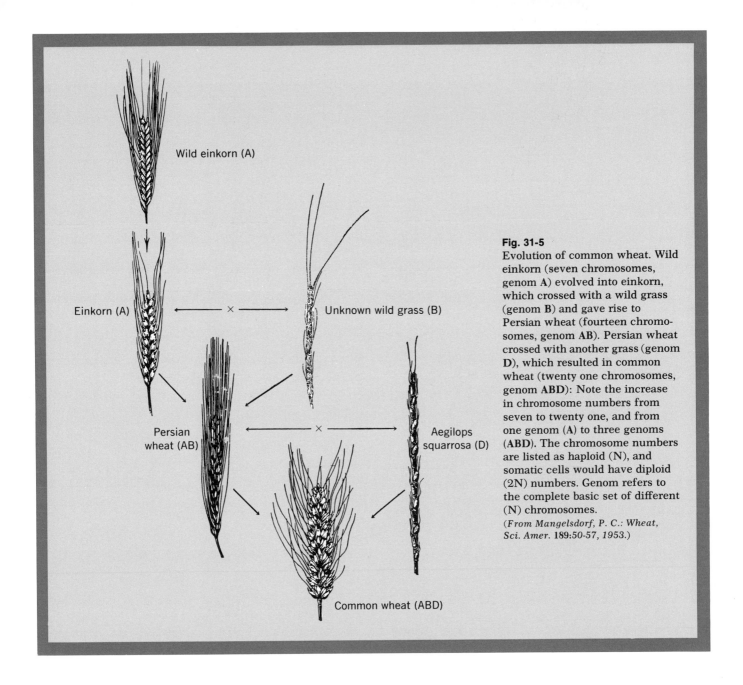

Fig. 31-5

Evolution of common wheat. Wild einkorn (seven chromosomes, genom A) evolved into einkorn, which crossed with a wild grass (genom B) and gave rise to Persian wheat (fourteen chromosomes, genom AB). Persian wheat crossed with another grass (genom D), which resulted in common wheat (twenty one chromosomes, genom ABD): Note the increase in chromosome numbers from seven to twenty one, and from one genom (A) to three genoms (ABD). The chromosome numbers are listed as haploid (N), and somatic cells would have diploid (2N) numbers. Genom refers to the complete basic set of different (N) chromosomes.

(From Mangelsdorf, P. C.: Wheat, Sci. Amer. 189:50-57, 1953.)

Figure labels: Wild einkorn (A); Einkorn (A) × Unknown wild grass (B); Persian wheat (AB) × Aegilops squarrosa (D); Common wheat (ABD)

particles, by the precipitation of materials from solutions, or by the secretions by certain organisms, as in the case of limestones. Igneous rocks (L. *igneus*, fire), such as volcanic rocks formed by the consolidation of molten lava, are produced as a result of heat and do not contain fossils. In the formation of sedimentary rocks the oldest occur at the bottom of a series of strata and the youngest near the surface. The most ancient fossils thus will be found in the oldest rocks, whereas the most recent fossils will occur in the youngest rocks.

Nature and kinds of fossils

The following are ways in which animals and plants of the past have left records: (1) by actual preservation of the organism, (2) by preservation of the skeletal structures, (3) by natural molds or

461

Fig. 31-6
Prehistoric mammoths, found frozen and well preserved in Siberia.

(Courtesy Chicago Natural History Museum; from Hickman, C. P.: Integrated principles of zoology, ed. 3, St. Louis, 1966, The C. V. Mosby Co.)

Fig. 31-7
Photographs of the remains of fronds of fernlike plants (*Pecopteris*). **A,** Two halves when opened; **B,** enclosed (upper right) and open (lower right).

incrustations, (4) by petrifaction, and (5) by leaving trails and impressions (imprints).

Actual preservation may occur by freezing and preserving in ice or soil. An example is the mammoth discovered frozen and perfectly intact in Siberia, with plants of the same period frozen with it. Many records of mammoths (Fig. 31-6) have been discovered. Remains of organs may be enclosed in rocks, as illustrated by a plant leaf (Fig. 31-7). The remains of animals and plants may be preserved, more or less intact, in tar, amber, or oil-impregnated soils. Amber is a yellowish plant resin that was originally soft and may have captured an animal or plant intact. The more volatile materials of the resin disappeared later, leaving the hard amber with its imprisoned organism. Certain organisms may remain intact by being buried in quicksands or swamps.

When the skeletal structure of an organism is preserved (Figs. 31-8 to 31-13), it remains almost in its original condition, except that it has lost most, if not all, of its organic material. Several skeletons of ancient mastodons have been found in different states of preservation. In some instances the skeletal remains may have added a chemical, such as carbonate of lime, which makes them heavier and more compact than the original. In other instances the shell-like skeletons of animals have become more porous and lighter than the originals.

In natural molds or incrustations neither the minute structures nor the materials of the original organism are preserved, but merely the general

Fig. 31-8
Photographs of remains of crustaceans called trilobites (Gr. *tri*, three; *lobos*, lobe), because the body was divided lengthwise into three lobes. Animals were also divided transversely into three body regions. Trilobites were marine animals with numerous jointed legs and breathing organs. They were probably able to walk or swim. **A,** Type of genus *Isotelus;* **B,** type of genus *Calymene;* **C,** six specimens of *Phacops,* shown in their rolled condition.

Fig. 31-9
Photographs of brachiopods (Gr. *brachion*, arm; *pous*, foot), so called because of two long "arms" enclosed between two shells or valves, by means of which food was obtained and respiration was carried on. They were also known as lamp shells because of their resemblance to Roman lamps.

Fig. 31-10
Specimens showing A, natural mold of the interior of an animal from which the shell has disappeared; B, original shell of a similar specimen.
(From Cleland, H. F.: Physical and historical geology, New York, American Book Company.)

outlines of form and shape are recorded. Animals and plants may be enclosed by incrustations of calcium carbonate or silica, which harden around the buried organism before it decays. The organic materials of animals are removed eventually by decay and percolation of dissolving waters. The cavity that remains eventually retains the general form and shape of the original organism. In some instances the skeleton has disappeared entirely, leaving only a mold of it as a record. The shells of mol-

463

Fig. 31-11
Prehistoric rhinoceros-like animal (*Brontotherium*) common during late Eocene and Oligocene epochs.
(Courtesy Chicago Natural History Museum; from Hickman, C. P.: Integrated principles of zoology, ed. 2, St. Louis, 1961, The C. V. Mosby Co.)

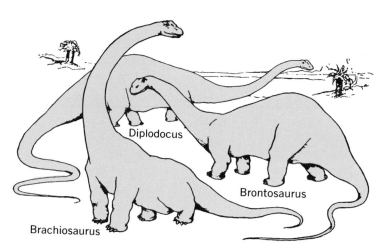

Diplodocus

Brontosaurus

Brachiosaurus

Fig. 31-12
Three enormous dinosaurs (Gr. *deinos*, terrible; *sauros*, lizard), extinct reptiles. *Diplodocus* was more than 80 feet long and weighed 40 tons; *Brontosaurus* was more than 65 feet long; *Brachiosaurus* was about 80 feet long.
(From Atwood, W. H.: Comparative anatomy, ed. 2, St. Louis, 1955, The C. V. Mosby Co.)

Fig. 31-13
La Brea tar pool near Los Angeles, California, with entrapped animals. An elephant and two wolves are caught, and a saber-toothed tiger is about to suffer the same fate.
(From Cleland, H. F.: Physical and historical geology, New York, American Book Company.)

lusks are covered with sediment, and the soft parts decay. The interior then is filled with sediment. Acidified waters then dissolve the limy shell, leaving only the molds of the exterior and interior. Sometimes the shell is removed, and the space left between the external and internal molds is filled with mineral matter, carried in by percolating waters. In this manner the form of the original skeleton is preserved but not its natural structure.

In petrifaction the original materials of the organism have undergone a certain amount of mineralization. In this case lime, silica, iron oxides, iron pyrites, and other substances have replaced the original materials of the organism, sometimes faithfully retaining its original shape, size, and even minute details. Usually, the older the fossil in time the greater the degree of mineralization. The harder parts of an organism, such as shells, teeth, tusks, bones, and the harder, woody parts of plants, are preserved most frequently. Petrified wood may show the minute structures just as they existed in the living trees but in which the cell walls are formed of the mineral silicon instead of the original cellulose. In this process as each particle of cellulose disappeared, it was accurately replaced by a particle of silicon, thus retaining the minute details.

Trails and impressions (imprints) might be called "fossils of living organisms," whereas other records are of dead organisms. Animals may leave their trails and imprints, such as footprints. The records must be left in soft materials that later become hardened and preserved. A leaf may leave an impression in soft mud that later hardens. If the material in which the plant or animal impression is made turns to rock, the result is a fossil. This type of fossil usually does not give much information concerning internal details but rather aids in determining the form and shape of the organism.

Conditions for fossil formation

In order for fossils to be formed there must be a rather rapid burial of the organism. This burial usually is accomplished by waterborne sediments. Organisms with hard parts are more likely to fossilize than those with softer ones. This may explain the absence of many fossils of the earliest, simpler organisms. The organism must also remain intact a sufficient length of time to permit fossilization

to occur. Air must be excluded in order to prevent oxidation of the organism and bacterial decay before the fossil is formed. During and after formation the fossil must withstand such natural conditions as the elevation or sinking of the earth's strata, pressure and heat, erosion processes, and the slow circulation of waters, especially acid waters, through the fossil.

Terrestrial plants and animals have less chance of becoming fossilized, unless they are placed in water and eventually covered. Most dead land plants and animals will be decomposed quickly on the earth's surface, thus leaving no extensive records. However, if covered with large quantities of volcanic ashes, sand or dust, earth through landslides or earthquakes, or materials from calcareous springs, even terrestrial plants and animals may leave some type of record. If the structures are thin, fragile, easily dissolved, and easily decayed, there is little opportunity to form a fossil.

Significance of fossils

Certain types of fossils indicate the boundaries and extent of former lands and waters. There is little land today that has not been below the level of the sea, sometimes repeatedly. This explains why we may find fossils of former marine organisms even on mountains. The Bering Strait between Asia and Alaska is about 35 miles wide and has a maximum depth of 200 feet. Studies of fossils on both sides of the strait show that these two lands were probably connected originally by land that sank beneath the water.

The character of fossils found in certain strata of the earth gives clues concerning their geologic ages and the time those particular sediments that formed these strata were laid down. Hence, certain animal and plant fossils are known as index fossils because through them it is possible to determine particular geologic ages and periods (Table 31-1).

A study of animal and plant fossils is important because it often includes the ancestors of modern species. The data obtained often explain relationships among various types of present-day animals and present-day plants. In some instances the ancient types serve to form connections between groups of organisms that at present seem to have no direct connections.

465

Table 31-1

Major divisions of geologic time (timetable)*

Era	Period	Epoch	Time when each epoch started (millions of years ago)	Life and general characteristics
Cenozoic (Gr. *kainos*, recent; *zoe*, life)	Quaternary	Recent	1/40	Modern man dominant; modern species of plants and animals; climate warmer; end of fourth ice age
		Pleistocene (Gr. *pleistos*, most; *kainos*, recent)	1	Extinction of giant mammals and many plants; development of man; cold and mild climates; four ice ages with glaciers covering much of North America, Europe, and Asia
	Tertiary	Pliocene (Gr. *pleion*, more)	12	Modern mammals; rise of herbaceous plants; invertebrate animals similar to modern types; dry, cool weather; volcanoes active; continents elevated
		Miocene (Gr. *meion*, less)	28	Grazing mammals; man apes; temperate types of plants; moderate climates; grasslands and plains developed; forests diminish
		Oligocene (Gr. *oligos*, little)	40	Primitive monkeys and apes; whales; forests widely distributed; temperate types of plants; mild climate; mountains formed
		Eocene (Gr. *eos*, dawn)	60	Modern mammals appear; first horses; subtropical forests; heavy rainfall; North America and Europe connected by land
		Paleocene (Gr. *palaios*, ancient)	75	Former mammals dominant; modern birds; dinosaurs extinct; subtropical plants; temperate to subtropical climates; mountains formed
Mesozoic (Gr. *mesos*, middle; *zoe*, life)	Cretaceous (L. *creta*, chalk)		130	Extinction of giant reptiles and toothed birds; origin of placental mammals; flowering plants arise; gymnosperms decline; mild to cool climates; mountains formed; inland seas and swamps forming

*This simplified presentation shows the earth's history briefly.

Table 31-1
Major divisions of geologic time (timetable)—cont'd

Era	Period	Epoch	Time when each epoch started (millions of years ago)	Life and general characteristics
	Jurassic (fine developments in Jura Mountains)		180	Giant dinosaurs and marine reptiles dominant; mammals appear; first toothed birds; angiospermous plants appear; conifers dominant; shallow inland seas
	Triassic (threefold development in Germany)		230	First small dinosaurs; marine reptiles; mammal-like reptiles; conifers dominant; seed ferns disappear; deserts formed
Paleozoic (Gr. *palaios*, ancient)	Permian (extensive in Perm, Russia)		260	Reptiles displace amphibians; many marine invertebrate animals extinct; modern insects; evergreens appear; cold, dry, and moist climate; mountains formed
	Pennsylvanian (well developed in Pennsylvania) (Upper Carboniferous) (carbon and coal-bearing rocks)		310	Reptiles originate; amphibians; giant insects; spore-bearing trees dominant; extensive coal-forming swamp forests; moist, warm climate; shallow inland seas
	Mississippian (well developed in Mississippi River valley) (Lower Carboniferous)		350	Amphibians; winged insects; sharks and bony fishes; horsetails and seed ferns abundant; early coal deposits; warm climate; hot, swamp lands; inland seas; mountains formed
	Devonian (common at Devon, England)		400	First amphibians; freshwater fishes; sharks; forest and land plants; wingless insects; corals; primitive horsetails, ferns, and seed ferns; arid land; heavy rainfall; small inland seas; mountains formed
	Silurian (Silures, ancient tribe of Wales)		430	Fishes with lower jaws; first land plants and arthropods; algae dominant; mild climate; European mountains formed

Continued

Table 31-1

Major divisions of geologic time (timetable)—cont'd

Era	Period	Epoch	Time when each epoch started (millions of years ago)	Life and general characteristics
	Ordovician (Ordovici, ancient tribe of Wales)		475	First vertebrates (ostracoderms); brachiopods; trilobites abundant; marine algae dominant; warm, mild climate; oceans increase in size; land submerging
	Cambrian (Latin for Wales)		550	Algae and marine invertebrate animals; first abundant fossils; trilobites and brachiopods dominant; mild climate
Proterozoic (Gr. *proteros,* early)	Upper Precambrian (L. *pre-,* before)		2,000	Bacteria; fossil algae; sponges; soft-bodied animals; start of autotrophic nutrition (making their own food); warm, moist to dry, cold climate; volcanoes active; glaciers; mountains forming
Archeozoic (Gr. *arch-,* beginning; *zoe,* life)	Lower Precambrian		4,500	Life presumed to originate; start of heterotrophic nutrition (obtaining nourishment from organic substances); no fossils found; lava flows; granite formed

Formerly, the classification of plants and animals was based primarily on living forms, but as the knowledge of fossils increased, the classification was made more accurate by incorporating the data contributed by paleontology. Many large groups of organisms of the past, although they may have disappeared completely, through their fossil records have thrown light on the relationships between present-day organisms. Other groups of organisms that were originally dominant have diminished in numbers and importance until they are represented by a limited number of types at the present time.

The earliest known organisms were simply constructed. Throughout geologic time there has been a continued evolutionary succession of plants and animals. With each era and period more complex and highly evolved organisms arose, only to be superseded later by newer and more complex groups.

A study of paleontology reveals certain geographic conditions and distributions of organisms of the past. Regions of the world that are united at the present time were once separated by barriers. Mountains have arisen or large land areas have been submerged beneath water. The presence of large numbers of plants at a certain period precludes a certain type of climate so that they might flourish. Studies reveal the former existence of luxuriant vegetation in regions that are at present more or less devoid of that type of plant. Such natural phenomena as floods, glaciers, volcanic eruptions, and earthquakes have affected plant and animal distributions in the past. A change in the atmosphere, water, foods, or soil in the past no doubt influenced the distribution of organisms. The greater and more extensive the changes the greater the effects on organisms.

Geologic time chart

By accurate studies of the strata of the earth scientists have divided the earth's history into eras (Table 31-1). Each era has been divided into periods, and the periods subdivided into epochs. Each division has specific characteristics, ages, durations, and types of life that were rather common during those times. The most recent fossils are found in the upper strata, and the more ancient are arranged in a series with the most ancient at the bottom.

The relative lengths of the eras and periods may be calculated in two ways. The age can be approximated by the thickness of the sedimentary rocks formed during each period. A definite time is required to form a certain thickness of sedimentary rock of a particular type. From these data it can be estimated how long it would take for a certain thickness to be formed.

The other method is by the radioactive disintegration method. Radioactive elements, such as uranium 238, uranium 235, and thorium, are slowly changed to lead. An analysis of the ratio between the lead and the uranium (or thorium) content of a rock gives its approximate age. Extimated by radioactive methods, the oldest rocks are thought to be over two billion years old. The rates at which uranium (and thorium) are converted to lead have been determined approximately as follows:

$$\frac{\text{Amount of lead}}{\text{Amount of uranium}} \times 7,600,000 \text{ years} = \text{Time}$$

Radioactive carbon (carbon 14), which loses one half of its radioactivity in about 5,760 years, has also been used in dating certain specimens. When bones are formed, small amounts of carbon 14 are incorporated, and upon death the radioactivity is gradually lost. The determination of the amount of radioactivity in bone makes it possible to approximate the time of death. By determining the age of trees killed by the last ice glacier that covered parts of North America, it is estimated that the time of its retreat was about 11,000 years ago. The study of carbon from the charcoal of fires shows that man entered North America shortly after the glacier

retreated and eventually migrated over the continent. By analyzing the wood of trees killed when a volcano exploded it has been estimated that Crater Lake, Oregon, was formed about 4,500 years ago.

Evidence from taxonomy (classification)

Comparative studies of various species of plants and animals reveal great similarities in many instances; in fact, similarities are often so great that it is difficult to decide where one species with its variations ends and another species with its variations begins. The intergrades (divergent individuals of a certain species) frequently are very similar functionally and structurally to those of closely related species. When we attempt to classify similar types of organisms, we observe the close anatomic and physiologic relationships between many of them.

Evidence from comparative embryology

A comparative study of the embryonic stages through which animals pass reveals a widespread general correspondence of the developmental stages in higher forms with the existing adult stages of lower forms. The history of the embryonic development of an individual frequently corresponds in a general and broad way to the history of the development of the race as a whole (Fig. 31-14).

A study of the embryonic development of a bird's or mammal's heart shows that the various stages through which it develops succeed each other in the same general way from the two-chambered to the four-chambered condition, as is shown when comparisons are made among the lower vertebrates, such as the fishes, up through the amphibia, reptiles, and birds, to mammals (Table 31-2 and Fig. 31-15).

A similar comparative study of the brains (Fig. 31-16), reproductive systems, skeletal systems, and digestive systems of these vertebrates shows a similar condition, in which the organs of higher animals during their development pass through stages that correspond in general with the larval or adult conditions of similar organs in the lower forms. Thus, knowledge of the anatomy of an animal gives a broad and general idea concerning its type of embryonic development.

Table 31-2
Chambers of the hearts of vertebrates

	Auricles	Ventricles
Fishes	One; receives blood returning through veins from entire body	One; receives blood from auricle and pumps it through gills on its way to all parts of body
Amphibia (frogs, toads)	Two separate; left receives blood from veins from lungs; right receives blood from veins from all parts of body	One; receives blood from both auricles and pumps mixture through arteries to all parts of body
Reptiles (lower types; lizards, snakes, turtles)	Two separate; left receives blood from veins from lungs; right receives blood from veins from all parts of body	Two partially separated; right receives blood from right auricle; left receives from left auricle; blood from two auricles mixed in partially separated ventricles and pumped to all parts of body
Reptiles (higher types; alligators, crocodiles)	Two separate; left receives blood from veins from lungs; right receives blood from veins from all parts of body	Two completely separated; right receives blood from right auricle; left receives blood from left auricle
Birds (adults)	Two separate; left receives blood from veins from lungs; right receives blood from veins from all parts of body	Two completely separated; right receives blood from right auricle; left from left auricle; left ventricle pumps blood to all parts of body; right ventricle, to lungs
Mammals (adults)	Two separate; left receives blood from veins from lungs; right receives blood from veins from all parts of body	Two completely separated; right and left have same functions as in birds

These similarities of lower and higher types of organisms found by embryonic studies suggest a similar inheritance as a basis and probably an actual relationship between them. The only other alternative is that the same "blueprint," with slight modifications and alterations, was used in the process of specially and individually creating each species.

■
Evidence from comparative anatomy

Evidences from comparative anatomy include those from gross anatomy and vestigial structures.

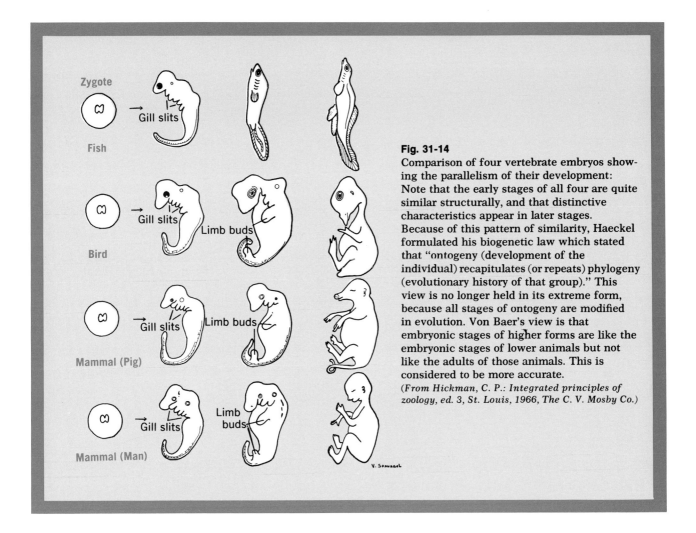

Fig. 31-14

Comparison of four vertebrate embryos showing the parallelism of their development: Note that the early stages of all four are quite similar structurally, and that distinctive characteristics appear in later stages. Because of this pattern of similarity, Haeckel formulated his biogenetic law which stated that "ontogeny (development of the individual) recapitulates (or repeats) phylogeny (evolutionary history of that group)." This view is no longer held in its extreme form, because all stages of ontogeny are modified in evolution. Von Baer's view is that embryonic stages of higher forms are like the embryonic stages of lower animals but not like the adults of those animals. This is considered to be more accurate.

(*From Hickman, C. P.: Integrated principles of zoology, ed. 3, St. Louis, 1966, The C. V. Mosby Co.*)

From gross comparative anatomy

A detailed comparative study of the anatomy of apparently different types of animals reveals a multitude of similarities that really overbalances their more obvious dissimilarities. For instance, the differences exhibited by the five classes of vertebrates are relatively slight when compared with the many fundamental resemblances that they all possess to some degree.

The forelimbs of the frog, bird, cat, horse, and man, for example, are constructed on the same general structural plan and arise in a similar way embryonically. They are thus homologous structures, and such differences or variations as exist are principally the result of the absence of some minor part or the transformation of a certain part, depending on the specific use to which that part has been put. In general, nearly all the bones, muscles, nerves, and blood vessels are constructed and arranged in a homologous manner in the forelimbs of the entire group, from the lower types, frogs, to the higher type, man. The same thing is true for the hind limbs, digestive systems, reproductive systems, and circulatory systems of this series.

From vestigial structures

Man has approximately one hundred vestigial structures that are also represented, and often useful, in lower types.

Illustrations of vestigial structures in man include the following: (1) The vermiform appendix is a remnant of an organ that is useful in certain herbivorous animals and may have had a specific function in man generations ago. (2) The third eyelid in the inner angle of the human eye corresponds

471

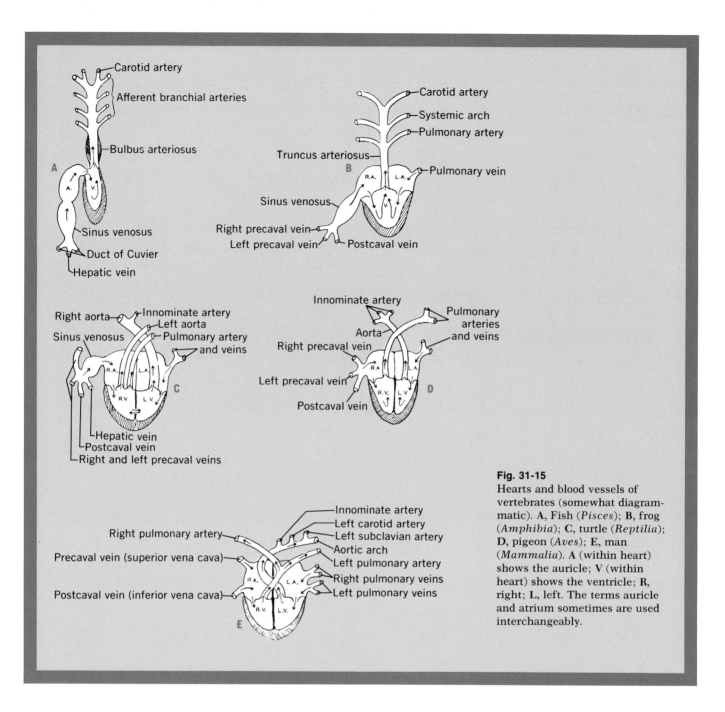

Fig. 31-15
Hearts and blood vessels of vertebrates (somewhat diagrammatic). **A**, Fish (*Pisces*); **B**, frog (*Amphibia*); **C**, turtle (*Reptilia*); **D**, pigeon (*Aves*); **E**, man (*Mammalia*). **A** (within heart) shows the auricle; **V** (within heart) shows the ventricle; **R**, right; **L**, left. The terms auricle and atrium sometimes are used interchangeably.

to the nictitating membrane, or lid, that moves laterally across the eye in such lower animals as the frog, bird, and dog. (3) Muscles of the external ear are useless for man but are used by lower animals to turn the ear in the proper direction to acquire the sound waves more accurately. (4) The terminal vertebrae (coccyx) are of no value to man, but they are the foundation for the external tail in lower animals. It is interesting to note that the early embryo of man (Fig. 31-14) possesses an external tail that is discarded before the advanced stages are reached. Only occasionally does the external tail persist in the adult man. (5) The lobe of the ear is of no practical benefit to man, although it may have had some function in the past. (6) The point of the ear, known as Darwin's point, on the

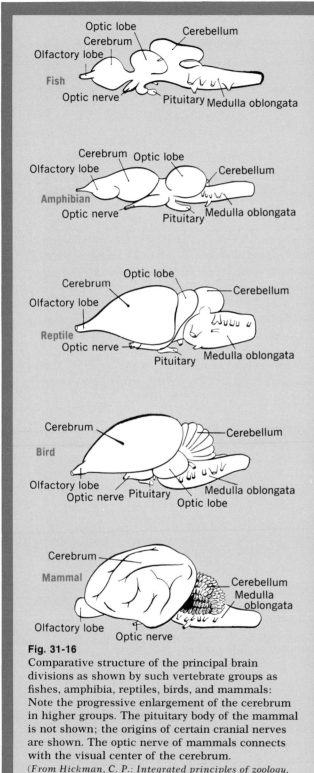

Fig. 31-16
Comparative structure of the principal brain divisions as shown by such vertebrate groups as fishes, amphibia, reptiles, birds, and mammals: Note the progressive enlargement of the cerebrum in higher groups. The pituitary body of the mammal is not shown; the origins of certain cranial nerves are shown. The optic nerve of mammals connects with the visual center of the cerebrum.

(From Hickman, C. P.: Integrated principles of zoology, ed. 3, St. Louis, 1966, The C. V. Mosby Co.)

edge of the upper roll or margin of the human ear, corresponds to the tip of the ear of animals that hold their ears upright.

Vestigial structures in other animals are illustrated by the following: (1) The splint bones of the legs of the horse are remnants of original toes. (2) The poison glands of certain snakes are modified, specialized salivary glands from which they have evidently developed through descent with change. (3) The gill slits of the embryos of higher vertebrates disappear except one pair, which develops as the eustachian tubes connecting the pharynx and the middle ear. (4) The milk glands of mammals are modified and specialized sweat glands of the skin. (5) Certain snakes bear small, useless hind limbs that structurally resemble those of other animals in which they are useful.

Evidence from comparative physiology

Since functions and structures are interdependent, one would expect to find fundamental physiologic similarities in organisms with structural similarities. The following examples illustrate this: (1) The blood of closely related organisms is more nearly alike chemically and physiologically than the blood of the more distantly related types. (2) The hormones of closely related organisms are quite similar and in some instances interchangeable. The insulin of the sheep pancreas may be used for an insulin deficiency in man. (3) The crystalline structures of bloods of similar organisms are more nearly alike than those of more dissimilar or unrelated organisms. In general, common properties persist in bloods of closely related types, and variations are greatest in distantly related forms.

Evidence from heredity and variations

Many evidences for evolution have come from the field of genetics. In the discussion of genetics in Chapter 29 the inheritance of variations in traits in living organisms was considered in connection with such phenomena as mutations, recombination of traits when parents with different traits were crossed, and crossing-over, and these might be reviewed advantageously.

A mutant gene that has been altered, thus changing some trait of its bearer, is an important founda-

473

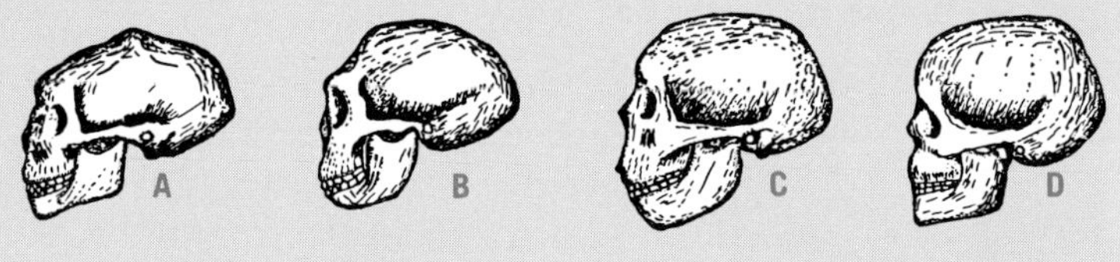

Fig. 31-17
Skulls of prehistoric man. **A,** Java man (*Pithecanthropus erectus*); **B,** Peking man (*Sinanthropus pekinensis*); **C,** Neanderthal man (*Homo neanderthalensis*); **D,** Cro-Magnon man (*Homo sapiens*). All skulls except **D** have been restored to a certain extent and are somewhat diagrammatic.

tion of evolution. Such mutants are known to occur naturally, but they have also been induced experimentally. In each case the resulting observations may have shed some light on the evolution of certain traits in such organisms. H. J. Muller discovered that radiation induces mutations in the fruit fly (*Drosophila*), and more than 30 years ago L. J. Stadler noted similar phenomena in plants. It is hypothesized that heredity information is encoded in various permutations in the arrangement of subunits of the molecule of DNA (deoxyribonucleic acid) in the chromosomes. A mutation occurs when this molecule fails to replicate itself exactly. The mutated gene, and its resulting trait, is reproduced thereafter until its bearer dies without reproducing or until, by chance, the altered gene mutates again.

Evolution of man

Early man has left many interesting and valuable records that enable us to get an idea of his physical and mental traits, achievements, and activities. Skeletons, implements, and tools form the basis for the knowledge of our remote ancestors. Early man did not always bury his dead, so that records do not exist before the Pleistocene epoch.

Pebble-culture man (Australopithecus prometheus)

One of the most interesting sites in the history of early man is in South Africa where the Makapan River has formed a series of limestone caves. In the lower strata are found small stones that have been chipped into crude tools. Geologic studies suggest that this primitive, so-called pebble culture may have flourished approximately 750,000 years ago. Raymond A. Dart found remains of *Australopithecus prometheus,* a smallish, erect-walking creature whose brain was large enough to equip him for making such tools. Since anthropologists define man as a "tool-making animal," the pebble-chipping *Australopithecus prometheus* may qualify as the most primitive creature that can be called definitely human.

The discoveries of *Zinjanthropus,* by Dr. L. S. B. Leakey, in eastern Africa, may extend the age of early man to 1,750,000 years ago.

Java man (Pithecanthropus erectus)

The skull cap, the left femur, and the lower jaw with three teeth of *Pithecanthropus erectus* were found in Java in 1891 by Eugene Dubois. It is thought that this type of man existed during the first glacial age of the early Pleistocene epoch. His cranial capacity was about 950 ml., which is approximately two-thirds that of an average, modern European, but half again as much as that of a large gorilla. His higher mental functions were limited because of the poorly developed frontal regions of his brain. The centers of taste, touch, and vision were probably well developed, and he may have used speech of some type. The skull cap was very thick, and his forehead was low, receding, and with massive supraorbital ridges. His skull was narrow, and his jaw projected in an almost snoutlike fashion (Figs. 31-17 and 31-18). His average height was about 5 feet 7 inches. He lived on land and seems to have been more similar to man than any ape. He used sharpened sticks and stones for implements. It is likely that this and the next type, Peking man, were closely related, perhaps subspecies of *Homo erectus.*

Fig. 31-18
Prehistoric men, photographs of restorations. From left: Java man, Neanderthal man, and Cro-Magnon man.
(*Courtesy Dr. J. H. McGregor; from Hickman, C. P.: Integrated principles of zoology, ed. 3, St. Louis, 1966, The C. V. Mosby Co.*)

Peking man (Sinanthropus pekinensis)

Skulls, teeth, and brain cases were found near Peking, China, from 1926 to 1928, by D. Black. Peking man is supposed to have lived during the first interglacial age of the early Pleistocene epoch. His cranial capacity was about 1,000 ml.; hence his head was larger than that of *Pithecanthropus*. The brain case shows the brain to be human but small and comparing rather favorably with normal human brains of primitive men of today. The walls of the skull were thick. The forehead was low, receding, and possessed heavy supraorbital ridges. This early man used fire, because charcoal and charred remains of various materials have been found buried with his remains. He used tools and implements of bone and stone, over 2,000 stone implements being present with the remains so far unearthed.

Neanderthal man (Homo neanderthalensis)

The skull cap and parts of the skeleton of Neanderthal man were found in the Neanderthal valley near Düsseldorf, Germany. This type of man is thought to have existed during the third interglacial and third glacial ages of the Pleistocene epoch. His cranial capacity was between 1,400 and 1,600 ml. His higher mental faculties were not well developed. The anterior region of his brain was not as highly developed nor as large as that of *Homo sapiens*. Neanderthal man had a low, broad forehead with massive supraorbital ridges. His eyes were large and round, his nose was broad, and he had a receding chin. His knees were bent, and his head was held forward when he stood or walked; his spinal column was slightly curved—character-istics that gave him a peculiar slouching attitude. His skeleton was not over 5 feet 4 inches tall, usually averaging less than 5 feet. His feet and hands were large, and his legs were longer than his arms. Many skeletons of this type of man have been found in England, Belgium, Germany, France, Spain, Italy, Palestine, Syria, Arabia, Iraq, Rhodesia, and China, suggesting a very wide distribution. From their remains it is thought that they lived at the entrance to caves rather than in them; that they used a language; that they used fire for warmth and cooking; that they were great hunters and ate the bone marrow of their captured animals; that they used implements of flint, bone, and unpolished stone; that they believed in a life after death because they buried flint implements and foods with their dead, probably clothing the hairy body in animal skins.

Additional skeletons of Neanderthal man were found in 1957 and 1960 by Columbia University anthropologists in a dry, well-protected cave at Shanidar, 250 miles north of Baghdad, Iraq. Probably these remains represent a "conservative" type of man that became extinct in Iraq about 45,000 years ago, as found by carbon 14 dating. Skeletons were found at different levels, the oldest at a 45-foot depth were estimated to be 70,000 years old. The skeletons were more complete than those usually found, almost entire ones were quite well preserved, and they give physical evidence of the stone-age (Mousterian) culture. Also discovered in a burial place were wild wheat, date palm pollen, and additional evidence of a change from cave to village life.

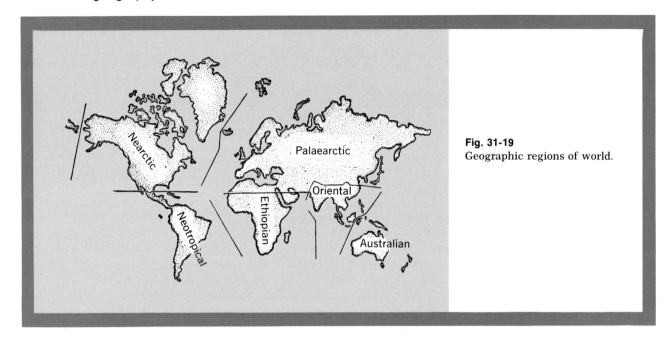

Fig. 31-19
Geographic regions of world.

Cro-Magnon man or modern man (Homo sapiens)

Five skeletons were found in the Cro-Magnon cave in Dordogne, France, in 1868. Cro-Magnon man is thought to have been present during the fourth glacial or ice age of the late Pleistocene epoch, even down to the recent epoch. His cranial capacity was from 1,400 to 1,500 ml., which is equal to if not greater than, that of the average European of today. The anterior part of the brain was large and well developed. The skull was large, long, and narrow. The forehead was high with moderate supraorbital ridges. The face was broad; the jaws, wide; the cheek bones, large; the eyes, large and far apart; the spinal column had four distinct curves. The male averaged 6 feet 2 inches in height, which suggests a strong, athletic race. The chin was well developed. In general, Cro-Magnon men were probably handsome people, comparing quite well with existing races. They lived in caves and rock shelters. They hunted and fished by means of skillfully made harpoons and spears. Many implements and ornaments of bone have been found. They developed an art in which they carved and made drawings in oil. They developed primitive industries in which they used bone more extensively than flint. The Cro-Magnon man is a good ancestor of modern man from a physical and mental standpoint. In the distant future, when the remains of some of us are unearthed, what type of record will we have left, and for what will our civilization be noted?

BIOGEOGRAPHY

Biogeography, which is the study of the distribution of organisms in space, not only examines the particular distribution of various organisms, but also attempts to explain the reasons for that distribution. The effects of environmental factors on the morphologic, physiologic, and developmental characters of organisms are also studied. In general, larger areas such as countries or continents are considered.

Because of its adaptation to a particular environment a species is restricted by its morphology and physiology to those areas in which that type of environment exists. However it may be found that an apparently similar environment in two different and widely separated places does not necessarily contain similar organisms. In the past the boundaries of sea and land changed, in some instances creating natural barriers, in others providing favor-

Table 31-3
Geographic regions of the world

Region	Location	Some typical animals
Nearctic (Gr. *neo-*, new or late)	North America (down to edge of Mexican plateau), Greenland	Blue jays, rattlesnakes, raccoons, opossums, skunks, prairie dogs, American water dogs
Palaearctic (Gr. *palae-*, ancient)	Europe, Africa (north of Tropic of Cancer), Asia (north of Himalayas), Japan	Nightingale, Japanese water dog, camel, dromedary (of central Asia and northern Africa)
Neotropical (Gr. *neo-*, new)	Central America, Mexico, South America, West Indies	Tapir, sloth, armadillo, llama, flat-nosed monkeys, tree anteaters, tree porcupines
Ethiopian (Gr. *aithiops*, black face)	Africa (south of Sahara Desert), southern Arabia, Madagascar	African elephant, hippopotamus, rhinoceros, zebra, giraffe, lion, leopard, gorilla, chimpanzee
Oriental (L. *orientalis*, eastern)	India (south of Himalayas), southern China, Phillippines, Borneo, Java, Sumatra	Indian elephant, rhinoceros, tiger, gibbons, cobra, jungle fowls (ancestors of domestic fowls)
Australian	Australia, New Zealand, New Guinea, Tasmania, Papua	Marsupial animals (with pouch for carrying young), such as kangaroo, duckbill platypus; certain wingless birds

able highways for dispersal and migration. It may be noted that many widely separated present-day types possibly had common ancestors in the past. The fossils of extinct North American camels are the remains of the ancestors of the present Old World camels and the llamas of South America.

Geographic regions of the world

The world has been divided into regions (Fig. 31-19), each with its different characteristics and typical flora and fauna. A brief consideration of these regions is given in Table 31-3. Many mammals of the Palaearctic region are similar to those of the Nearctic region and many trees and plants are common to both regions. These two regions are so similar that they are sometimes combined into what is called the Holarctic region (Gr. *holo-*, whole; Arctic). Animals common to both Nearctic and Palearctic regions include beavers, deer, hares, foxes, wildcats, and bears.

A region that is included here with the Australian region is sometimes called the Polynesian region (Gr. *poly-*, many; islands) and consists of the oceanic islands of the tropical Pacific, including the Hawaiian Islands, Samoa, and the Fiji Islands. They were formed by volcanic eruptions and are

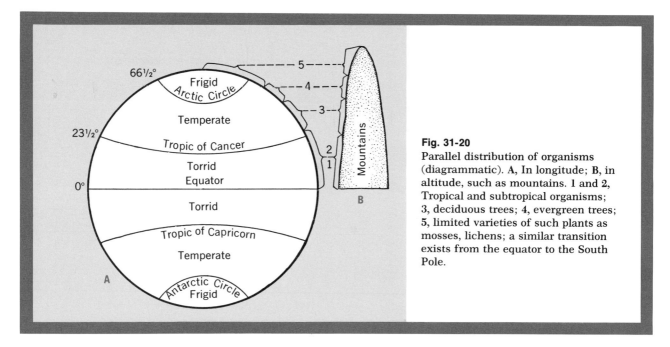

Fig. 31-20
Parallel distribution of organisms (diagrammatic). **A,** In longitude; **B,** in altitude, such as mountains. 1 and 2, Tropical and subtropical organisms; 3, deciduous trees; 4, evergreen trees; 5, limited varieties of such plants as mosses, lichens; a similar transition exists from the equator to the South Pole.

fringed with coral reefs. The vegetation is often large and herbaceous, such as palm and banana trees. There are no land mammals, except bats, and there are no amphibia. The types of living animals are quite limited.

Geographic distribution in space

There are two general types of geographic distribution of organisms in space: (1) lateral (longitudinal) distribution, in which the organisms are spread over the surface of the earth in the various geographic regions, and (2) vertical distribution, in which organisms are distributed throughout the various altitudes (Fig. 31-20). The latter type emphasizes the differences in distributions on mountains, in valleys, in caves, and in the depths of the sea. There are regions of distribution as we ascend from the lowest depths of the ocean to the top of the highest mountain.

Principles of geographic distribution

There are many principles of biogeography, but only a few will be considered here.

Principle of dispersion

The principle of dispersion illustrates the natural tendency of some animals to migrate (disperse) from their birthplace, because more offspring are produced than can be accommodated in that habitat normally. This reproductive pressure (population pressure) tends to overpopulation, and dispersal is an attempt to remedy it. Offspring and parents may be unable to occupy the same area and may destroy or at least compete with each other.

Principle of definite habitats

The principle of definite habitats shows that the habitat (home) of a particular species is determined by such factors as the following: (1) The quantity and quality of usable foods. Herbivorous animals, such as deer, must have suitable vegetation, whereas carnivorous animals, such as tigers and lions, must have suitable flesh foods. (2) The quantity and quality of water also affects the selection of a habitat. A certain amount of water is essential for all animals. Many species in dry climates prevent excessive evaporation by some type of covering. Some species found under rocks are not always there to shun the light but for moisture and protection. The depth, salinity, and hydrogen-ion concentration of water influences the selection of a definite habitat. (3) Oxygen content of the air is influential in the selection of a habitat by both terrestrial and aquatic types, particularly the former. (4) The presence of an optimum temperature also helps to determine the selection of a habitat. Animals tend to seek the temperature for which they are suited best. Many animals in the tropics pass the summer in a condition of aestivation, or semitorpid condition of semiactivity. Certain animals living in colder climates pass the winter in various ways: (a) hibernation, or a period of inactivity in some protected location; frogs, turtles, snakes, and the larvae and pupae of insects hibernate; (b) migration to warmer regions, as illustrated by species of birds that migrate from the Arctic regions to the tropics; and (c) continued activity in the cold habitats and increasing their fat layers as well as their coats of hair, or possibly changing their diets to include foods that will produce greater amounts of heat.

Principle of barriers and highways

The principle of barriers and highways is an important one. What is a barrier for one species may serve as a favorable highway for another. Some of the more common factors that interfere with dispersal are as follows: (1) There may be a lack of proper foods along the migration route. (2) Water may be a barrier for terrestrial forms but may be used successfully by aquatic types. The size, depth, temperature, acidity, and pressure in bodies of water influence dispersal of aquatic organisms. The aridity and humidity of terrestrial environments act as barriers or highways, depending on the type of animal. Salt water is a barrier for freshwater forms, and fresh water serves as a barrier for marine organisms. (3) Various kinds of land serve as barriers or highways, depending on the animal. Forests act as barriers for open-country or prairie-inhabiting species while deserts and open country act as such for forest-inhabiting types. (4) Interference by other animals, either through bodily struggles or competition for foods, also influences the dispersal of certain types of animals. (5) Winds, especially if strong, tend to carry certain types of animals in the direction of wind blow, which may result in their migration into more favorable or less favorable habitats. (6) Temperature prevents the dispersal of many animals, either by its direct effect on the migrant or by its effect on the vegetation upon which the migrant may depend for food and shelter. (7) The lack of inherent adaptive ability of an animal may prevent sufficiently quick adaptation to the new and changed conditions. The result may be extermination or an attempt to continue migration. This is frequently called a biologic barrier.

The following methods of dispersal are common in the animal world: (1) Driftwood may transport animals for great distances. William Beebe, on his Arcturus voyage, observed fifty-four species of marine fishes, worms, and crabs on one floating log. (2) Ships transport various types of organisms from port to port. Unknown numbers of rats have had free transportation in this manner. (3) Water and floods transport organisms mechanically, drive them from their original habitats, or change the food supplies sufficiently so as to require dispersal. (4) Aquatic animals transport other animals on or within their bodies; the larvae of clams may be carried on the gills and fins of fishes. (5) Terrestrial animals transport other animals on or within their bodies. Birds may carry eggs, larvae, pupae, or adults of smaller animals. (6) Winds direct the course of certain animals or blow objects to which certain types are attached. (7) Glaciers cause animal migrations by actually transporting them or by changing the temperature or food supply. (8) Man, either knowingly or unknowingly, aids in animal dispersal by means of automobiles, trains, and airplanes. A horned toad was transported from Texas to Springfield, Ohio, by a circus train, although after its arrival it found the rigors of city life too great. This was no fault of the method of dispersal.

Principle of discontinuous distribution

The principle of discontinuous distribution is illustrated by the presence of the same species of animal in two widely separated regions. It is thought that the distribution of that species was continuous between the regions originally. In the Pliocene epoch of geologic history tapirs were distributed over nearly all of North America, Europe, and

Table 31-4
Regions of vegetation of North America

Regions	Plants typically present	General characteristics
Tundras	Certain mosses, lichens, grasses, sedges, herbs, low shrubs, dwarf trees	North of latitude 55 to 60°; long, cold winters, with low soil temperatures and limited amounts of cold water; subsoil usually frozen; short growing season; strong winds
Deserts	Sagebrush, cacti, Yucca trees, bunch grasses, small herbs	Arizona, Nevada, parts of New Mexico, California, Texas, northern Mexico, and peninsula of southern California; small rainfall; intense heat and light; high loss of water from plants; fairly strong winds
Grasslands	Bunch grasses, cacti, various shrubs	Central Texas to Manitoba and along foothills of Rockies from New Mexico to Alberta; light rainfall; humus soil over sand and clay; few trees because of soil moisture and great evaporation because of heat
Forests (a) Northern evergreen	Cone-bearing trees, such as spruce, balsam fir, white pine, red or Norway pine, Jack pine, hemlock, arborvitae; deciduous trees, such as balsam poplar, aspen, white birch	From Atlantic to Pacific oceans between tundra on north and Great Lakes on south, and extending northwestward to Alaska
(b) Southern evergreen	Longleaf pine, shortleaf pine, bald cypress, magnolia trees, water oaks, gum trees	Southeastern United States from Texas to Florida and Virginia; many low, rolling sandy plains; also high coastal plains farther from the ocean; many swamps

Table 31-4
Regions of vegetation of North America—cont'd

Regions	Plants typically present	General characteristics
(c) Deciduous forests	White oaks, black oaks, hickories, chestnuts, walnuts, maples, ash, elm, birch, and certain cone-bearing trees	From central New York to Texas and Louisiana; from Wisconsin to Oklahoma
(d) Rocky Mountain forests	Western yellow pine; lodgepole pine, Douglas fir; western hemlock, western larch	Along Rockies from southern Mexico to Columbia; mountains of various elevations and climates, so variety of trees exist, but none above 10,000 feet where low, tundralike vegetation occurs
(e) Pacific Coast forests	California region—redwoods and sequoia trees (Fig. 13-1) Washington-Oregon region— with mild winters and good rainfall—Sitka spruce, Douglas fir, western hemlock, white pine; birch, maples, poplars; dense growths of ferns Canadian-Alaska region—Sitka spruce, Douglas fir, western hemlock	Extend along western slopes of mountains from California to Alaska
(f) Tropical forests	Various palms, tropical orchids, mangrove swamps, lianas (woody, climbing vines); dense jungles in certain places	West Indies, Central America, coasts of Mexico, southern tip of Florida

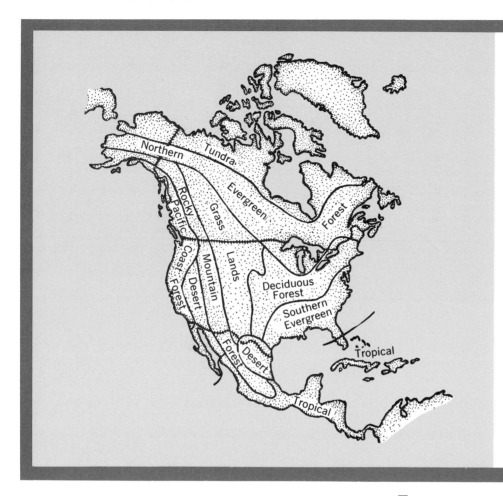

Fig. 31-21
Vegetation areas of North America (boundaries of various regions are given in a general way).

northern Asia, but today they are present only in Central and South America, southern Asia, and the Malay archipelago.

Principle of vertical distribution

According to the principle of vertical distribution the organisms of higher elevations of mountains tend to simulate those of the polar regions of the earth. As we proceed downward in altitude, the forms simulate those that might be found in traveling from the poles toward the equator. In general, the temperate zones not only extend laterally north and south of the equator but also vertically in parallel succession from the somewhat mild conditions at sea level to the somewhat frigid conditions at the tops of higher mountains.

Distribution of vegetation of North America

A study of the distribution of plants throughout the world shows that the world can be divided into geographic regions, each with its particular environmental characteristics and flora. For example (Fig. 31-21), the North American continent can be divided into various vegetation areas. These areas are summarized in Table 31-4, in which the general environmental characteristics and the plants typically present in each region are given. The environmental characteristics differ in tundras, deserts, grasslands, and forests. If the distribution of forests is studied in detail, it is apparent that different types of trees are present in various parts of the continent, depending upon the environment peculiar to each area.

Some early contributors to the knowledge of evolution and related topics

Thales (624-548 B.C.)
A Greek who proposed a theory that water (ocean) was the mother of all life.

Anaximander (611-547 B.C.)
A Greek who thought that life arose from a mixture of water and earth and that land forms arose from aquatic types, particularly under the influence of the sun's heat.

Heraclitus (510-450 B.C.)
A Greek natural philospher who stated that "struggle is life" and "all is flux"—thoughts that are basic to modern ideas in evolution.

Empedocles (495-435 B.C.)
A Greek who theorized that living organisms were generated spontaneously from scattered materials, being attracted by love and hate.

Aristotle (384-322 B.C.)
A Greek who theorized that in the living world there was a gradual change from the simple and imperfect to the more complex and perfect, thus suggesting the idea of evolution.

Saint Augustine (353-439 A.D.)
Saint Augustine interpreted the first chapter of Genesis as stating that in the beginning matter was created with the properties and potentialities to evolve into living and nonliving worlds as we know them today.

Francesco Redi (1621-1697)
An Italian who overthrew the theory of spontaneous generation, which stated that life arose from nonliving materials spontaneously, by discovering that eggs and larvae of insects originated from previous living insects, rather than from nonliving substances.

Georges de Buffon (1707-1788)
A French naturalist who excluded the possibility that species have the ability to evolve (change). He probably was influenced by the ancient ideas revived during the Renaissance that discredited such a possibility. (*Historical Pictures Service, Chicago.*)

Charles Bonnet (1720-1793)
A Swiss naturalist and philosopher who first used the term "evolution" but not quite as we do today, and who conceived that organisms could be arranged in a ladderlike, linear series.

Erasmus Darwin (1731-1802)
An English evolutionist who was the grandfather of Charles Darwin. He believed that acquired traits could be transmitted to future generations and he may have influenced Lamarck, who had similar views. (*The Bettmann Archive, Inc.*)

Jean Baptiste Lamarck (1744-1829)
A French biologist who was a student of organic evolution. He believed that environmental influences, and the effects of use and disuse of body parts, were causes of evolutionary changes—a theory that laid the foundation for the "Theory of the Inheritance of Acquired Traits."

Thomas Robert Malthus (1766-1834)
An English economist whose "Essay on Population", published in 1798, inspired both Darwin and Wallace in their theories of evolution.

Charles Lyell (1797-1875)
A Scottish geologist who laid the foundations for the science of earth structure in his *Principles of Geology* (1830-1832). He is considered an important contributor to the theory of organic evolution because of his influence on Charles Darwin and Alfred Russel Wallace.

Georges Cuvier (1769-1832)
A Frenchman who supported the "Cataclysmic Theory," in which he stated that there had been numerous creations, each of which was followed by a cataclysm that destroyed it, and its place was taken by new forms.

Buffon

Some early contributors to the knowledge of evolution and related topics—cont'd

Charles Darwin (1809-1882)

Darwin, in a voyage around the world in the sailing ship Beagle, indicated the descent of species by the development of varieties from common stocks. This process entailed a "struggle for existence," which resulted in a "natural selection of species" and a "survival of the fittest." He wrote *The Origin of Species by Means of Natural Selection* (1859).

Thomas Henry Huxley (1825-1895)

An English surgeon who actively supported the views of Charles Darwin and assisted in promoting them extensively.

Darwin

Alfred Russel Wallace (1823-1913)

An Englishman who studied animal geography in the East Indies and concluded that the life of wild animals is a struggle for existence. He worked on the problem of the origin of species and arrived at conclusions concerning evolution that resembled those of Charles Darwin. (*Historical Pictures Service, Chicago.*)

Louis Pasteur (1822-1895)

A French microbiologist and chemist who proved that only living organisms such as bacteria and yeasts could cause fermentations. He proposed a method of preventing this process by heating to a temperature high enough to kill the germs. This method is known as pasteurization. Pasteur's work ended the controversy regarding the possibility of spontaneous generation of living organisms from nonliving materials.

Wallace

August Weismann (1834-1914)

A German who distinguished between body cells and germ cells and proposed the "Theory of the Continuity of Germ Plasm From Generation to Generation" (1885). He opposed the idea that acquired traits might be transmitted.

Review questions and topics

1 Explain why abiogenesis was believed in the past, and when and how it was disproved.

2 Discuss biogenesis, including some early investigators who contributed to an explanation of this phenomenon.

3 Discuss some of the theories of the origin of life on earth. Which one, if any, seems most plausible? Why do you say so?

4 Discuss the various evidences that attempt to explain evolution, including examples of each type of evidence. Which science, if any, contributes the most logical evidence? Why do you say so?

5 Discuss the evolution of horses and common wheat in some detail.

6 What do comparative studies of the hearts and brains of vertebrates suggest?

7 Discuss the role of modern genetics in explaining evolution, including some examples to prove your statements.

8 Of what importance in everyday life is a knowledge of animals and plants of the past and their records?

9 List several reasons why certain softer types of animals and plants have left no fossil records.

10 How are we able to estimate the age of the earth by a scientific study of the fossils in the successive strata of the earth?

11 How have the various estimates of the age of the earth been made? How accurate are these estimates?

12 What is the estimated age of the earth from the Lower Precambrian period down to the present?

13 What percentage of the total age of the earth represents the time that human beings have inhabited the earth?

14 List the various types of fossils and records that ancient man has left and include the specific manner in which each has been preserved.

15 Make a table of the more representative types of ancient man, including the outstanding characteristics of each.

16 Explain how and where records of ancient man are discovered. What is the significance of where these records have been discovered? Are additional records being discovered at the present time? Where? (Read articles on present-day discoveries.)

17 What conclusions might you draw from studies of the sequence of records left by ancient man?

18 Give the general characteristics and boundaries of each of the regions into which the animal world may be divided. What factors might prevent migration from one region to another?

19 How may a certain condition act as a barrier to one type of organism and at the same time act as a method of transportation for another type? Give specific examples.

20 What is the effect of better methods of transportation by man on the distribution of certain types of organisms? Explain in detail giving specific examples.

Selected references

Boughey, A. S., editor: Population and environmental biology, Belmont, Calif., 1967, Dickenson Pub. Co., Inc.

Dobzhansky, T.: Mankind evolving, New Haven, 1962, Yale University Press.

Dobzhansky, T.: The present evolution of man, Sci. Amer. 203:206-217, 1960.

Eiseley, L. C.: Charles Darwin, Sci. Amer. Feb. 1956.

Howells, W. W.: Homo erectus, Sci. Amer. 215:46-53, 1966.

Huxley, P. M.: The confirmation of continental drift, Sci. Amer. April 1968.

Laughlin, W. S., and Osborne, R. H., editors: Human variation and origins. Readings from Scientific American, San Francisco, 1967, W. H. Freeman and Co. Publishers.

Lock, D.: Darwin's finches, Sci. Amer. April 1953.

Mayr, E.: Animal species and evolution, Cambridge, Mass., 1963, Harvard University Press.

Napier, J.: The evolution of the hand, Sci. Amer. 207:56-62, 1962.

Romer, A. S.: The vertebrate story, ed. 4, Chicago, 1959, University of Chicago Press.

Ross, H. H.: A synthesis of evolutionary theory, Englewood Cliffs, N. J., 1962, Prentice-Hall, Inc.

Ruibal, R., editor: The adaptations of organisms, Belmont, Calif., 1967, Dickenson Pub. Co., Inc.

Simpson, G. G.: The major features of evolution, New York, 1953, Columbia University Press.

Simpson, G. G.: This view of life, New York, 1963, Harcourt, Brace & World, Inc.

Solbrig, O. T.: Evolution and systematics, New York, 1966, The Macmillan Company.

Stebbins, G. L.: Process of organic evolution, Englewood Cliffs, N. J., 1966, Prentice-Hall, Inc.

Tax, S., editor: Evolution after Darwin, 3 volumes, Chicago, 1960, University of Chicago Press.

Wallace, B.: Chromosomes, giant molecules and evolution, New York, 1966, W. W. Norton & Company, Inc.

Wallace, B., and Srb, A.: Adaptations, ed. 2, Englewood Cliffs, N. J., 1964, Prentice-Hall, Inc.

485

part six

ORGANISM AND ENVIRONMENT

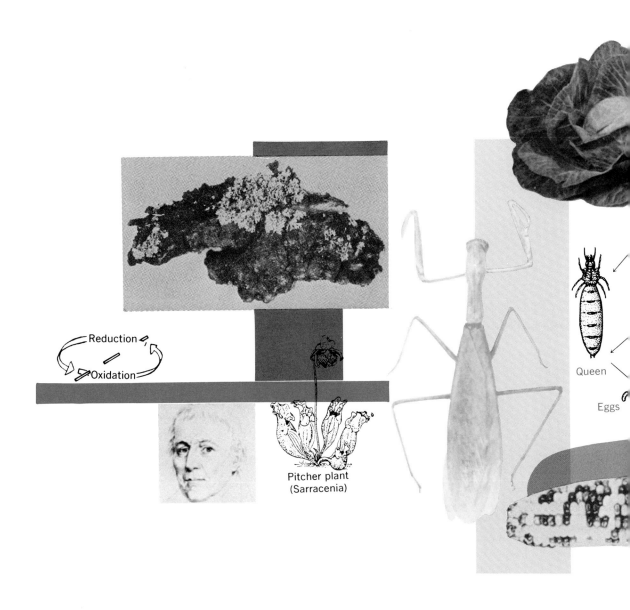

Reduction

Oxidation

Pitcher plant
(Sarracenia)

Queen

Eggs

Ecology
Interrelationships among organisms
Man and biology

chapter thirty-two

Ecology

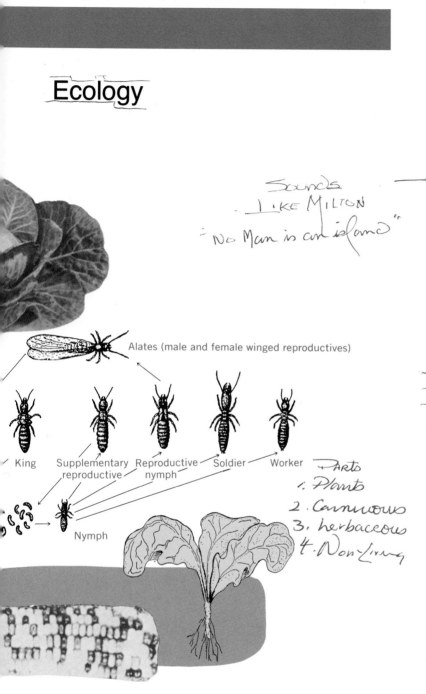

*Sounds
Like Milton
"No Man is an island"*

Alates (male and female winged reproductives)

King Supplementary Reproductive Soldier Worker
reproductive nymph

Nymph

*Parts
1. Plants
2. Carnivorous
3. herbaceous
4. Non Living*

All life in the world is so interdependent and closely related that it may be considered as a web (Fig. 32-1), composed of various individuals and groups of plants and animals that are more or less associated into a living unit. This web, or unity of life, is changing constantly as far as individuals are concerned, yet there seems to be a constancy in any given area. No living organism lives unto itself alone, but each affects other organisms and in turn is affected by them. The more we study biologic phenomena the more we appreciate the interdependence of living things. In any particular area each organism contributes something, large or small, to the total life of that area.

Ecology deals with the interrelationships between living organisms and their environments. The factors that influence these interrelationships are numerous and complex.

ECOSYSTEMS

The term "ecosystem" indicates a natural unit of living and nonliving parts interacting to form a somewhat stable system. Materials are exchanged between various living organisms, as well as between living organisms and nonliving matter, all in a cycle. Hence, organisms are not independent but are interrelated with other living plants and animals and with inanimate objects in the environment. Ecosystems vary in size from a small quantity of water, or a small cube of soil, to a large lake, or large tract of forest. An ecosystem, regardless of size, may be divided into (1) producers, which are those organisms that can manufacture organic compounds from simple inorganic substances by photosynthesis; (2) consumers, which are those organisms that consume, such as herbivorous (plant-eating) or carnivorous (flesh-eating) organisms; (3) decomposers, which are those organisms, such as bacteria and fungi, that decompose or break down organic compounds of dead protoplasm into organic substances that can be reused by green plants; and (4) nonliving components of the ecosystem that act as a reservoir from which the constituents may be drawn and to which they may be returned, to be used later. Different modes of nutrition of living organisms, various interactions between organisms, and cycles of chemical change are considered in some detail elsewhere.

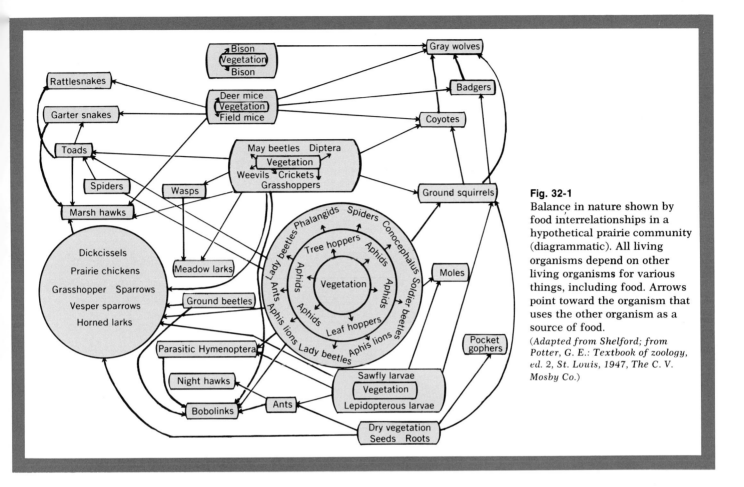

Fig. 32-1
Balance in nature shown by food interrelationships in a hypothetical prairie community (diagrammatic). All living organisms depend on other living organisms for various things, including food. Arrows point toward the organism that uses the other organism as a source of food.
(*Adapted from Shelford; from Potter, G. E.: Textbook of zoology, ed. 2, St. Louis, 1947, The C. V. Mosby Co.*)

HABITATS AND ECOLOGIC NICHES

The habitat of a plant or animal is the locality or external environment in which it lives. More than one plant or animal may live in a particular habitat, which may be large or small. The ecologic niche (F. *niche*, place or recess) is the status of a plant or animal within an ecosystem or community. The niche includes all the physical, chemical, physiologic, and biologic factors that an organism needs in order to live. This is influenced by the organism's structure, physiologic responses and abilities, and adaptations. When considering the niche of a particular organism, we must include such things as its food source, the organisms that use it as food, its effects on other organisms, its range of movement, and its effect on the nonliving surroundings. For example, in one habitat, such as a lake shore (Fig. 32-3 and Table 32-2), one finds many different kinds of organisms, each playing quite a different

role in the biologic economy of the lake and each thus occupying a different niche in that habitat. Organisms that may be present in typical environments are shown in Table 32-1.

FACTORS THAT AFFECT THE ECOLOGY OF ORGANISMS

In making an ecologic study of a particular species of plant or animal, or groups of them, the heredity of the organisms involved and all of the environmental factors must be taken into consideration. Some characteristics of representative environments and typical organisms present in them are given in Tables 32-1 and 32-2.

Unity and cooperation between various types of living organisms

When groups of living organisms are studied, a certain degree of cooperation and unity is apparent, not only among members of the same species

489

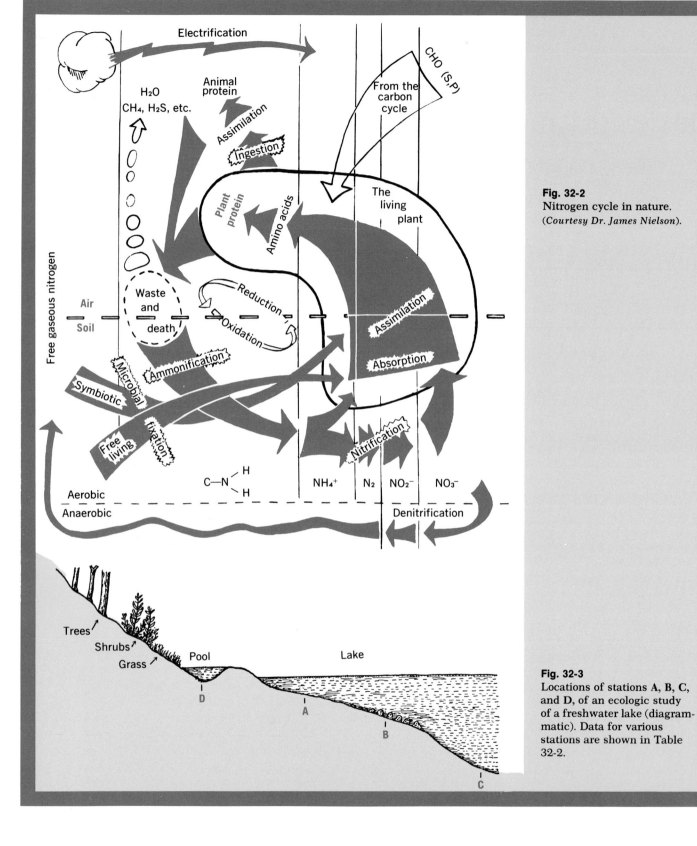

Fig. 32-2
Nitrogen cycle in nature.
(*Courtesy Dr. James Nielson*).

Fig. 32-3
Locations of stations **A**, **B**, **C**, and **D**, of an ecologic study of a freshwater lake (diagrammatic). Data for various stations are shown in Table 32-2.

Table 32-1

Typical environments and organisms found in them

Environments	Characteristics of environment	Typical organisms present
	Water or aquatic	
Rapid streams	Rapid flow of water; usually hard, firm bottom; usually loose rocks with crevices for protection; usually shallow, hence much light and oxygen	Larvae of mayflies, stone flies, caddis flies, midges (bloodworms); certain snails and fishes (such as darters); usually a minimum of vegetation
Pools (quiet parts of streams)	Slow flow of water; usually soft bottom of mud or sand for burrowing	Larvae of dragonflies, damselflies, midges; clams; various types of fishes, water snakes, and turtles; certain types of vegetation common
Ponds	Slow flow of water; usually soft bottom; many factors depend upon depth	Larvae of dragonflies, damselflies, caddis flies, midges; clams and water snails; crustaceans; fishes, water snakes, and turtles; certain types of vegetation common
Lakes	Wave action depends upon many conditions; bottom may be mud, sand, gravel, loose rocks, or solid, each influencing type of life found; sand and moving gravel undesirable for sessile organisms; sand may interfere with gill-breathing types; lakes larger and deeper than ponds	Types of animals present vary greatly, depending on specific qualities present in different areas of lake; vegetation may be limited because of sand movement, limited nourishment, wave action, and reduced light and oxygen at certain depths
	Land or terrestrial	
Open fields	Temperatures usually severe in summer and winter; moisture evaporation high; intense light; great wind action; little protection except in ground	Insects such as beetles, grasshoppers, leafhoppers; certain types of spiders, toads, snakes, birds; bees (if flowers present); vegetation usually short
Deserts	Temperatures severe in summer and winter; moisture evaporation **very** high; intense light; great wind action; limited protection	Insects such as beetles, grasshoppers, leafhoppers; certain types of spiders, toads, snakes, birds; vegetation usually sagebrush, cacti, yucca, bunchgrasses
Tundras	Winters long and cold; only upper limits of soil thaw; ground water cold and limited; plant growth season short; often strong winds	Few animals can tolerate the ravages of this polar or subpolar area; vegetation consists of lichens, mosses, certain grasses, herbs, and shrubs
Forests	Temperatures usually more moderate than surrounding areas; protection from light, heat, wind, and moisture evaporation; reduced wind action	Insects such as bees, beetles, crickets, certain grasshoppers and katydids; centipedes and millipedes; spiders; snakes; tree toads; numerous birds, mammals

Table 32-2

Summary of an ecologic study of a portion of a freshwater lake

Station number	Location of station	Depth of water	Temperature	General characteristics	Attachment and shelter	Food and oxygen	Number and types of animals per square foot	
A	Shoreline	6 in.	22° C.	Bottom of smooth, solid limestone Strong wave action; no crevices in bottom	Strongly attached No shelter No plants for protection	Very little sediment to interfere with animal respiration No aquatic plants for food or to supply oxygen Shallowness permitted light and oxygen from surface	Midge larvae Snails *(Goniobasis)* Roundworm Caddis fly larva	(50) (4) (1) (1)
B	6 ft. from shoreline	18 in.	22° C.	Numerous, irregular rocks of various sizes Medium wave action; back wave action pronounced Many large crevices	Fairly strongly attached to all surfaces of rocks Algae very abundant	Slight sediment Great masses of green algae to supply food, oxygen, and protection	Caddis fly larvae Midge larvae *Hydra* Snails *(Goniobasis)* Caddis fly pupae Mayfly larva	(36) (27) (15) (9) (6) (1)
C	12 ft. from shoreline	36 in.	21° C.	Smooth, solid limestone bottom with occasional, free, irregular rocks; surface wave action strong Few crevices	Fairly strong and uniform attachment Few plants	Slight sediment Few plants for food, oxygen, or protection	Caddis fly larvae Snails *(Goniobasis)* Midge larvae Mayfly larva	(35) (8) (4) (1)
D	Inland pool with connection with lake; 4 ft. inland from lake	6 in.	25° C.	Solid, smooth limestone bottom; no free rocks; no great disturbance by waves	Slight attachment Numerous plants (diatoms, desmids, algae)	No sediment Surface covered with vegetation that was actively emitting oxygen	Snails *(Lymnaea)* Midge larvae	(25) (4)

Table 32-3

Nitrogen transformations as produced and used by living organisms

	Bacterial organisms	Results
Nitrogen fixation Symbiotic	*Rhizobium*	Nitrates produced from free nitrogen of atmosphere in root nodules of leguminous plants (clovers, alfalfas, peas)
Nonsymbiotic	*Azotobacter, Clostridium*	Nitrates produced from free nitrogen of atmosphere within soil
Nitrification	Ammonifying bacteria	Change nitrogenous compounds into ammonia (NH_3) by process of ammonification
	Nitrosomonas, Nitrosococcus	Change ammonia to nitrites (NO_2)
	Nitrobacter	Change nitrites to nitrates (NO_3), which are usable by plants
Denitrification	Certain bacteria	Convert nitrates to nitrites, oxides of nitrogen, and free nitrogen

but also among members of different species. This is especially apparent in the cycles of those elements that make up most living protoplasm. There are many examples of these phenomena, but the following cycles are typical.

Nitrogen cycle

Nitrogen is an essential constituent of protoplasm, particularly for the construction of proteins. Some of the processes for obtaining and using nitrogen (Fig. 32-2) are shown in Table 32-3. Certain bacteria live symbiotically in the root nodules of leguminous plants where nitrates are formed for plant use.

When plants die, their nitrogenous compounds are reduced by bacterial actions to ammonia, which may be used in the process of nitrification again. When animals die, their nitrogenous compounds are reduced by bacterial actions to urea, which can be converted to ammonia to be used in the nitrification process. Hence, both plants and animals

are dependent on such processes for their sources of nitrogen.

Several blue-green algae, including *Nostoc* and *Anabena*, like certain bacteria, are able to use atmospheric nitrogen. Free-living, nonsymbiotic, nitrogen-fixing bacteria often inhabit the sheath of *Nostoc* and other blue-green algae of the soil, using the algae's sheath as food. Hence, it is necessary to use bacteria-free cultures in determining the nitrogen-fixing ability of these algae.

Carbon cycle

Carbon is also an essential constituent of protoplasm. The carbon dioxide of the atmosphere comes from animal respiration, from the burning of wood, coal, oil, gas, gasoline, from various manufacturing processes, and from volcanoes (Fig. 32-4). Chlorophyll-bearing plants combine carbon dioxide with water in the presence of energy-supplying light by the process of photosynthesis.

493

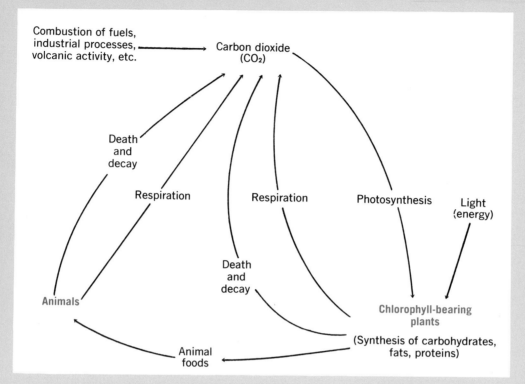

Fig. 32-4
Carbon cycle in nature (some parts of this cycle are not represented).

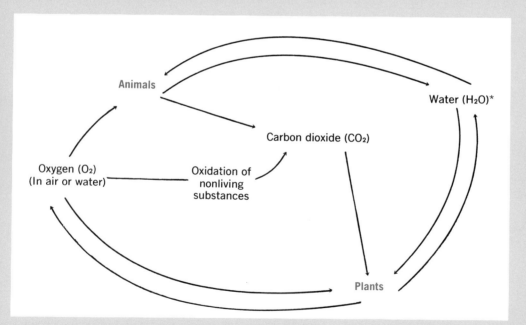

Fig. 32-5
Oxygen cycle in nature (some parts of this cycle are not represented). *See also Fig. 32-6.

Carbohydrates, such as sugars and starches, are formed, and oxygen is given off. Some of this oxygen is used by the plant. Plants and animals utilize such products to manufacture proteins and fats.

When plants and animals die, their carbon compounds are reduced to simpler carbon materials that may be used by living plants eventually. Bacteria, molds, earthworms, and certain insects aid in restoring these carbon materials to the soil for future use.

Oxygen cycle

Oxygen is also an essential constituent of protoplasm. It not only goes into the makeup of protoplasm but also combines with substances (oxidation), thus liberating the energy that held these substances together originally. Animals require oxygen and give off carbon dioxide; chlorophyll-bearing plants require carbon dioxide and give off oxygen (Fig. 32-5).

Environmental factors

Physical factors

The physical environmental factors include temperature, light, wind, gravity, soil conditions, pressure, and the presence or absence of natural barriers or of natural methods of dispersal.

Temperature

Most organisms have an optimum temperature at which their lives and their metabolic processes are maintained most successfully. They also have a minimum and a maximum temperature below and above which they cannot live. Hence, plants and animals will tend to select, as far as possible, those temperatures for which they are best fitted. The freezing of water in which animals live affects them in the following ways: (1) some become inactive during the frozen period; (2) some escape the freezing by burrowing deep in the earth; (3) some die under such conditions, but only after they have produced resistant stages in order to carry on the species. A covering of ice on a body of water not only affects the animals directly but also indirectly by altering the oxygen and food supplies. This may explain why aquatic animals come to holes that are cut in the ice.

Light

Certain animal protoplasms cannot tolerate excess light, whereas others require large quantities. Some species require the stimulation given by light in order to carry on many of their metabolic activities. Indirectly, animals are affected by the presence of plants that require light for their existence. Plants that depend on light supply animals with food, protection, and oxygen and thus influence animal distribution. Certain plants require a maximum of light, some require a medium amount, and still others require a minimum amount or possibly none at all, such as mushrooms and certain bacteria. Plants tend to locate in areas that contain the proper quality and quantity of light to meet their particular needs.

The distribution of certain animals is quite different in daylight from that at night. For example, certain insects can be observed only in the daytime (diurnal), and others are found more abundantly at night (nocturnal). In addition to the absence of light, a lower temperature and additional moisture at night influence the distribution of such animals. This is true particularly for those types that have no special abilities to prevent the rapid evaporation of moisture from their body surface.

Wind

The direction and velocity of the wind are factors in the dispersal of certain land animals. Strong wind also affects the distribution of aquatic animals. The wind affects animals in various ways: (1) it may cause injury to them directly; (2) it may stir up sediments in waters or place dust in the atmosphere, thus affecting their respiration; (3) it may influence the oxygen content of the water or atmosphere; (4) it may affect the temperature of the water or atmosphere; and (5) it may bring obnoxious gases that may affect their distribution.

Wind affects plants in various ways: (1) it may assist or prevent pollination; (2) it may help to disperse seeds or spores; and (3) it may distribute carbon dioxide, oxygen, or obnoxious gases that might influence plant activities.

Soil conditions

The various physical conditions of different types of soils are important factors in the distribution of living organisms in or on them. Among such factors are included moisture content, degree of aeration, exposure to the sun (heat and light), hardness or looseness, and presence or absence of specific nutrient materials in usable forms. The slope of the soil affects its drainage and erosion, thus influencing the distribution of organisms in or on it. Some animals require certain types of soils for protection or burrowing.

Pressure
Pressures in water, soil, and air vary with the depth. Air pressure is 15 pounds per square inch at sea level and decreases uniformly as one ascends from sea level to higher regions. In higher elevations air pressure may become too low to allow normal respiration in certain animals. Water pressure increases with the depth; in the ocean it is equal to the depth in feet multiplied by 0.434. Thus, at 200 feet the water pressure per square inch is approximately 87 pounds. This pressure influences the vertical distribution in deeper bodies of water because not all organisms are constructed to withstand such pressures.

Natural barriers and methods of dispersal
Different types of living organisms exist in certain kinds of environments that are conducive to their dispersal. Any natural hindrance to dispersal is known as a natural barrier. Something that is a barrier to dispersal for one species may be a natural method of dispersal (highway) for another species. Water is a natural method of dispersal for fishes, but it may prove to be a natural barrier for terrestrial forms if it is too deep or extensive. Mountains may be natural barriers even for terrestrial organisms because of altitude, snow, ice, and lack of proper vegetation. The absence of plants of certain types from a region may serve as a barrier to animal dispersal because some animals depend upon those plants for food, shelter, and homesites. Floods are barriers to certain forms but serve as methods of dispersal for others. Heavy seeds, which cannot be carried easily by wind or animals, may have difficulty in passing over a large body of water or a mountain, whereas lighter seeds may not be affected by such barriers.

Most animals and plants have particular methods

for their dispersal. Dandelion seeds are so constructed that they are carried easily by the wind. Some types of seeds, such as burs, that are distributed by animals have their dispersal affected if animals are absent. Seeds that depend upon birds for their dispersal may or may not be dispersed, depending on the bird population. If all usable types of dispersal are lacking, that particular plant or animal is limited in its distribution.

Chemical factors
Chemical environmental factors include quality of the soil and water (moisture), oxygen, carbon dioxide, and obnoxious gases, and quality and quantity of usable foods.

Quality of the soil and water (moisture)
The chemical composition of a soil affects not only animals and plants living in or on it but also the aquatic life living in any water that comes in contact with that soil. Highly acid waters or soils are not ideal for certain organisms, whereas alkaline or even neutral soils may be. The hydrogen-ion concentration (pH) of a soil or water is an important ecologic factor. The chemical quality of soil or water determines the kind of vegetation growing in or on it, and these plants affect animal distribution, because different vegetations supply a variety of foods, protection from the elements, and concealment from enemies. Earthworms may not abound in sandy soils because of the irritations induced by sand grains, a minimum of usable dead plant food, a lack of sufficient moisture, and interference of loose sand in making permanent tunnels.

Plants require an adequate amount of water (Fig. 32-6) and soil of the proper quality (chemical composition) to meet their requirements. In some instances the requirements are rather specific, and plants will not be found in areas that do not satisfy those needs. For example, cranberry plants grow primarily in acid soils. Certain plants may be eliminated by altering the acid-base reaction of the soil or water. A plant such as the dandelion is less specific in its requirements, growing in a variety of soils.

Living organisms require water for various purposes, although the quantity and quality satisfactory for one type may not be suitable for another.

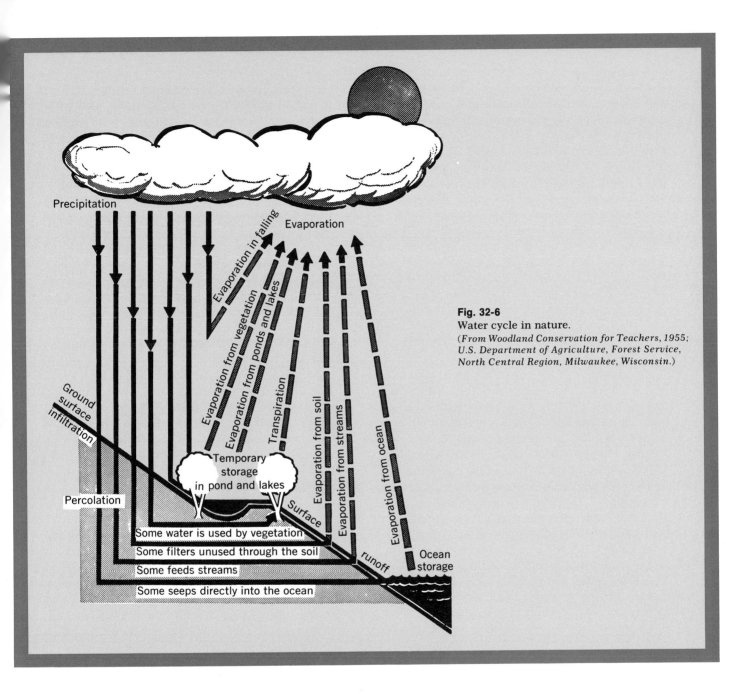

Fig. 32-6
Water cycle in nature.
(*From Woodland Conservation for Teachers, 1955; U.S. Department of Agriculture, Forest Service, North Central Region, Milwaukee, Wisconsin.*)

Protective substances and structures prevent excessive evaporation and thus permit such plants and animals to live in less than the normal requirement of moisture after they have once obtained their normal supply.

The pollution of waters with wastes and obnoxious materials affects the ecology of certain animals and plants, whereas the same conditions do not affect others. Certain industrial wastes are responsible for the elimination of fishes, aquatic snails, clams, crustaceans, and certain aquatic plants from polluted streams.

Oxygen, carbon dioxide, and obnoxious gases

All living organisms require a certain quantity of oxygen, and an insufficient supply will affect their distribution. Oxygen is required for the oxidation of foods to supply energy and other materials

497

necessary for their metabolism. Chlorophyll-bearing plants require oxygen, but they may obtain it from the process of photosynthesis. In this process the supply of carbon dioxide may be influential in their distribution. Excess carbon dioxide affects the distribution of terrestrial or aquatic animals; nonchlorophyll-bearing plants, such as bacteria and fungi, may not be affected as much by the carbon dioxide content of their environments.

Obnoxious gases, either naturally or artificially produced, affect the distribution of plants and animals. In fact, certain gases are utilized to combat many undesirable animals, such as insects, rats, moles, and gophers. Obnoxious gases may interfere with plant respiration, transpiration, and photosynthesis, thus influencing their distribution.

Quality and quantity of usable foods

All animals require foods of plant or animal origin. Some types require specific foods of definite qualities. If such foods are lacking, the animal may die, or it may attempt to move to a locality where they are present in sufficient quantity and quality. Other types of organisms are less specific in their food requirements and can exist on a greater variety, thus affecting their distribution accordingly. Organisms can be classified according to the types of foods utilized: those depending on animals for food being known as carnivorous (flesh eating), those depending on plants being herbivorous (plant eating), and those that use both being omnivorous (all eating). The distribution of organisms of these three types will be influenced accordingly by the kinds of foods available.

All plants require foods of one kind or another. Some types require rather specific kinds, whereas other types are more general in their food requirements. If an area has a limited amount of food of a specific quality and if plants require this kind of food in large quantities, the plant distribution is affected accordingly. In some instances the foods present may be in unusable forms, thus constituting an ecologic factor in plant or animal distribution.

Biologic factors

Biologic environmental factors include competition for food, light, moisture, and space; competition between the sexes; dependence of certain plants on insects for pollination; distribution affected by symbiosis, commensalism, parasitism, and predatism; dissemination and destruction of plants or their seeds by animals; and plants that are detrimental to certain animals.

Competition for food, light, moisture, space

If too many plants or animals with the same food requirements are present in an area with limited quantities of usable food, there will be a struggle for that food. The result will be either the migration of certain animals in order to obtain suitable foods or the death of some of the competitors. Because all organisms require foods, this struggle can have an important effect upon their ecology. Migrating animals in search of usable foods may upset the natural balance of the new community in which they locate.

Competition between plants of the same or different species resembles the struggle for existence as described for animals. Apparently nature sanctions this natural phenomenon in order to permit the fittest to survive and the unfit to be exterminated. Such competitions for food, light, moisture, space or position affect organisms in their ecologic distributions.

Competition between sexes

In order to propagate their species certain animals may travel long distances for the opposite sex, thus creating an ecologic factor during their journey. In other instances the competition of several members of one sex for a limited number of the opposite sex leads to dispersal or possibly extermination, either of which will influence their distribution in that area.

Dependence of certain plants on insects for pollination

Certain plants require insects to pollinate them. In some instances a specific type of insect is required. Bees are essential for pollinating clovers; if absent, clovers will bear a minimum of seeds, thus presenting a different ecologic phenomenon than if bees were present in sufficient numbers. Beehives are often placed in clover fields for pollination purposes, as well as for the honey supply. Plants that do not depend on insect pollination present an entirely different problem.

498

Distribution affected by symbiosis, commensalism, parasitism, and predatism

Sometimes, organisms are distributed in certain areas because of the help that they give or receive from organisms of a different species. If this help were unavailable, there would be an entirely different distribution of these species. Symbiosis, commensalism, parasitism, and predatism are all influential factors in ecologic distributions. These should be understood in order to explain this type of ecologic distribution. For example, termites (order Isoptera) are able to digest wood because they harbor in their digestive tracts certain flagellated protozoans that prepare the wood for absorption by the termites. In return for their labors the protozoans are given protection and distribution.

Dissemination and destruction of plants or their seeds by animals

Plants and their seeds may be distributed by insects and other animals by being carried in the digestive tract, on the external surfaces, or by mud on the feet. Many either useful or detrimental plants and their seeds are destroyed by animals that eat them, use them for making nests, parasitize them, or in some other way interfere with their normal habits and activities.

Human factors

Human environmental factors include animal and plant quarantine regulations, transportation of plants and animals by automobiles, trains, ships, and airplanes, and domestication of animals and plants.

Animal and plant quarantine regulations

Quarantine regulations enforced by a government prevent, to a great extent, the importing of many kinds of animals and plants from foreign countries. If imported, many of these might be antagonistic or even destructive to the native plants and animals and upset the natural balance or equilibrium of the present flora and fauna. This change in equilibrium might affect the ecologic relationships of many types of living organisms, either directly or indirectly. A disease that was first described in the Netherlands, the Dutch elm disease, was discovered in Ohio in 1930. If trees affected by the causal agent, a fungus of class Ascomycetes, had not been imported, many beautiful American elm trees might have been spared. The destruction of great numbers of these trees is not only a direct loss, but their absence also affects the ecologic relationships of other plants and animals in the affected areas.

Transportation of plants and animals

A little time spent on highways or wharves will show how plants and animals are transported long distances by any of a number of methods such as automobiles, trains, ships, and airplanes. After plants and animals have been imported suddenly into new regions their presence may influence the former populations to such extents that entirely new ecologic relationships will result. These methods of dispersal are man's inventions and enable the rapid transplantation of animals and plants.

Domestication of animals and plants

In many ways man's activities have created changes in areas of vegetation, and these in turn have influenced the distribution of animals dependent on or associated with such types of vegetation. The clearing of trees affects the animal population of that land in many ways. The introduction of domesticated or wild plants also directly or indirectly affects the animal distribution in an area. Man not only has taken domestic animals with him as he has traveled over the earth's surface, but these animals have taken their parasites with them. This has resulted in a redistribution of the populations into which the newcomers were taken. In general, what may seem to be a small, insignificant factor in the end may prove to be very influential as far as ecology is concerned. The destruction of a few apparently useless animals and plants may affect nature's balance, just as the introduction of harmful varieties may cause an ecologic readjustment.

Domestication of useful plants has resulted in their being protected and cultivated, and hence their wide distribution has been ensured. The cultivation of domestic plants tends to influence many wild types directly or indirectly. Many wild types are destroyed as weeds because they interfere with the normal production of domestic types.

An example is the destruction of corn plants by the imported European corn borer, which may spend part of its life cycle in a variety of weeds and other plants. Hence the number of plants harboring corn borers and surrounding a corn field may affect the domestic corn plants. The relationship between barberry bushes and the black stem rust of wheat is another example. Several stages in the life cycle of this wheat-damaging fungus are spent in the barberry plant. Thus the survival of both the fungus and the wheat depends on the availability of barberry plants.

Genetic factors

Because the genes in the cells of animals and plants largely determine what the organism is going to be, its structures, it abilities to use certain foods, and its necessity to develop in a certain type of environment, it is evident that genetics should be included in an ecologic study. Genes also determine the ability of a particular organism to develop variations that enable it to fit into its environment, especially if the environment should vary from time to time. The inheritance of an ability to move from place to place or to be stationary also influences the ecology of an organism. These and many other inherited factors largely determine the limits of distribution of a particular organism and hence its ecologic relationships. When environments change from their norm, the organisms living in them may attempt to vary sufficiently, enabling them to continue living there; if possible, they may move to more favorable environments, or they may die because they cannot accommodate to the changed conditions. Each living organism probably has an optimum environment, although in many instances life can continue in changed environments, provided the changes are not too great. Animals and plants that do not possess the inherited abilities to vary sufficiently to meet changed environmental conditions have less opportunity to survive under such conditions.

An organism, such as the common dandelion, is so constituted that it can live in a variety of environments, in many types of soils, in lowlands, and in rather high elevations. Cacti grow best in arid areas and poorly in wet, inadequately aerated soils. Because of the hereditary factors in a grain of corn, a corn plant will be a corn plant, but the specific way in which it develops will be determined also by many environmental factors. The common earthworm may be abundant in moist soils well supplied with humus and organic food, whereas it may be scarce or absent in dry, sandy, abrasive soils with little or no usable food.

Inherited abilities and reactions have influenced the ecologic relationships and distribution of the English sparrow. If it had not possessed in many successive generations a tendency to be unafraid, its present distribution might be quite different. The sparrow's fearlessness has ensured its protection and a generous supply of food because of its association with other animals whose foods were supplied by man. Sparrows are said to have entered the empty grain cars in the eastern United States, and after the car doors were closed, have rather contentedly "hitchhiked" their way to the West. Other birds with different inherited reactions might not have been transported across the continent so quickly and easily. The inherited ability to build nests anywhere and from all kinds of materials has influenced the sparrows' ecology; other birds require more specific types of nesting sites and nesting materials.

Such inherited structures as the gills of a fish naturally limit its distribution to water, whereas the lungs of men, birds, rabbits, and turtles necessitate their obtaining oxygen from the air. In order to develop their characteristic calcareous shells, aquatic snails cannot live in acid waters in which there is no lime. Some insects, such as the common walkingsticks, because of their resemblance to a twig, are usually found in bushes where they are protected by their inherited body form. These same insects, distributed artificially on smooth surfaces, are easily exposed and thus exterminated. Some animals, such as certain moths and butterflies, inherit definite color patterns that partially hide and protect them in one type of environment but may not do so in another.

Living plants have inherent tendencies to respond to certain stimuli in definite ways; for example, a certain species of plant responds in a rather definite way to gravity and specific quantities of moisture, light, and temperature. Certain plant seeds inherit definite structures (hooks, spines) by which they may be disseminated by animals or the wind. The inheritance of structures for controlling the process of transpiration will influence the distribution of plants in various environments. Cactus plants can exist in arid areas, but plants that do not possess their inherent abilities cannot. Hence, the distribution of these various types is somewhat predetermined. The root systems of some plants are such that they cannot supply materials and provide the necessary anchorage and support in certain types of soils, but that type of root system might be quite efficient in another type of soil.

Mutations and new types of organisms

Plants and animals that have been accustomed to a certain habitat sometimes mutate abruptly and spontaneously. The resulting mutant traits require an environment different from that of their parents, and the offspring have to develop in new habitats or be exterminated. Similarly, natural crossings may result in traits so different that the offspring have to develop in a different environment. In such cases an entirely new ecologic relationship is instituted, and consequently the distribution is affected.

BIOTIC COMMUNITIES AND POPULATIONS

Each area or region is inhabited by a number of plants or animals, and there are many interrlationships, including competition, parasitism, symbiosis, and predatism. Together with environmental factors these interrelationships assist in determining how many and what species may survive. All of these species of organisms living in a given area òr habitat constitute a biotic community, which is composed of intimately associated smaller groups called populations. A population is a group of plants or animals of the same or similar species that live in a given area.

Whenever organisms are found together in a community, the same environmental factors, such as temperature, light, oxygen and carbon dioxide, moisture, and foods, are shared by them, and all must attempt to adjust to these same factors. In general, communities may be widely separated from each other, but if their environmental factors are alike, similar kinds of living organisms will be found in them. Similar organisms will generally be found in somewhat similar environments, although there are exceptions. In the biotic community the idea of a natural association of organisms that are bound together by their requirement of the same environmental conditions is emphasized.

Communities do not always have definite boundaries but often overlap and grade into each other. Some animals shift from one community to another because of seasonal changes; others stay in one community during the day and spend the night in another.

Each community is organized and divided into definite horizontal and vertical strata. This principle of stratification is well illustrated by forest communities, terrestrial communities, and aquatic communities. In a forest community there may be vertical stratification, shown by certain organisms living on the forest floor, others only a few feet above the floor, and still others at greater heights above the floor. In aquatic communities some organisms also tend to live at certain levels (strata).

No community is balanced perfectly because many factors can upset and alter the interrelations of its various members. Changes in environmental conditions, the introduction of a new species, or a great decrease in the number of individuals of the species normally present, will influence the status of the other members of the community.

Most organisms bear some relation to the other organisms around them, and these interrelationships occur in many ways, but one of the most important is the so-called food chain (Fig. 32-7). In this there is a transfer of food energy from its source in plants through a series of organisms that eat plants or other animals that have eaten plants. At each step in the series there is usually a change in the amount of energy available. In the first step the capture of light energy by green plants is rather inefficient, with only about 0.2% of the available light energy being stored in the food produced. When an animal eats a plant or another animal, the percent of energy transfer is higher, varying between 5 and 20%. The percentage of food that is consumed and actually converted into protoplasm is the percent of efficiency of energy transfer.

Plants usually live in biotic communities or

501

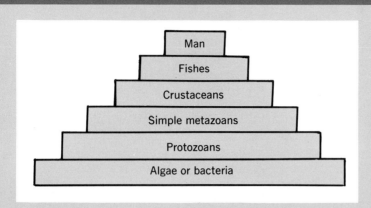

Fig. 32-7
Food chain (or pyramid), showing the dependence of various organisms. The relative sizes of the boxes show the approximate numbers, or quantity, of each form required.

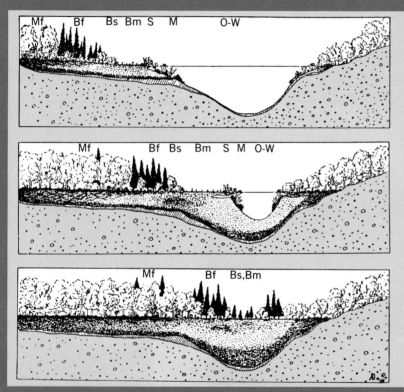

Fig. 32-8
Plant successions that replace one another (diagrammatic). In stages I, II, and III, a lake fills with deposits of peat and marl, and the original lake finally is covered by a forest. Successive stages are open-water plants, **O-W**; marginal plants, **M**; shore plants; **S**; bog-meadow plants, **Bm**; bog shrubs, **Bs**; bog (coniferous) forest, **Bf**; and mesophytic forest, **Mf**, consisting of deciduous trees whose water requirements are normal or medium. Deciduous trees are the climax vegetation.
(From Dachnowski, A.: Peat deposits of Ohio; courtesy Geological Survey of Ohio, Columbus, Ohio, Fourth Series, Bulletin 16.)

associations, which are composed of an assemblage of populations living in a defined habitat or area, where one or more species live under similar conditons and affect one another. Plant communities may contain only plants, but usually certain types of animals share that area with them. A community may consist of a mass of algae in water, some lichens on a rock, a swamp, the plants in a certain field, and a forest composed of certain types of trees.

In plant and animal communities, or in a combined community, many interacting factors that influence structures, functions, and behaviors of the various inhabitants are constantly at work. Certain plant species may adapt successfully to given environmental conditions, or they may become more or less interdependent upon one another. Most plant associations tend to expand their range to cover the entire environment to which they may be suited. Occasionally the presence of plants of certain types and in certain quantities so alters conditions that a different group of species supplants them. In addition, certain environmental conditions assist in markedly changing the distribution of plant associations. Plant successions (Fig. 32-8) result, in which certain types follow other types until the climax association is essentially permanent. Plant successions can be observed when a pond or lake is gradually filled, usually following in sequence such stages as aquatic plants, bog plants, shrubs, coniferous plants, and the climax vegetation—deciduous trees with their associated plants.

Population density and growth curve

A population has characteristics that result from the group as a whole and not from the individuals. Among such characteristics are population density (number of individuals per unit area or volume) as illustrated by the number of trees per acre in a forest or the number of human beings per square mile. Population density is an index of that population's successful living in that region under those conditions; it is often difficult to measure in terms of individuals but can be estimated by careful countings. A population growth curve is a graph in which the number of organisms is plotted against time. Such curves are characteristic of populations rather than of single species, and they are similar for the populations of most organisms, from bacteria and protozoans to man. In 1925 Raymond Pearl estimated that the human population of 2.2 billion would reach 2.6 billion in the year 2100, after which it would stabilize unless certain contributory factors should arise. We now know that this was a conservative estimate. Recent estimates by the United Nations show that the world population in 1970 is over 3.5 billion. These are not exact figures since the data from the world's two most populous countries, China and India, are very inexact.

Birthrate and mortality rate

The birthrate of a population is the number of new individuals produced per unit of time. The maximum birthrate is the largest number of organisms that could be produced per unit of time under ideal, or optimum, conditions when no limiting factors are involved. This rate is somewhat constant for a species and is influenced by such physiologic factors as the proportion of females and the number of eggs produced. It is difficult to ascertain the maximum birthrate because of the difficulty in eliminating all the limiting factors, and it is usually greater than the actual birthrate.

The mortality rate of a population is the number of individuals dying per unit of time. There is a minimum mortality, which is the number of deaths that would occur under ideal conditions, the deaths simply resulting from physiologic changes of old age. This is rather constant for a given population. The actual mortality rate varies with physical factors and size and composition of the population. In some countries man has so improved his average life expectancy by modern health measures that the curve for human survival approaches the curve for minimum mortality. Since the death rate is more variable and more affected by environmental factors than the birthrate, it plays a vital role in population control. Populations that differ in the relative numbers of young and old individuals, especially if past the reproductive age, will have quite different characteristics, birthrates, and death rates. Death rates usually vary with age, and birth rates are often proportional to the number of individuals of reproductive age. In general, rapidly increasing populations have a high proportion of young individuals.

Biotic potential and environmental resistance

The biotic potential (reproductive potential) expresses the inherent ability of a population to increase its numbers when the age ratio is stable and all environmental conditions are optimal. Under ordinary conditions when the environment is suboptimal (below best), the rate of population growth is less, and the difference between the potential ability of a population to increase and the actual observed performance is a measure of what is called environmental resistance. When a population is increasing rapidly, each individual organism of reproductive age reproduces at the same rate as before, the increased numbers resulting from increased survival. Environmental resistance consists of all the physical and biologic factors that prevent a species from reproducing at its maximum rate. When a species is first introduced into a new area, the environmental resistance is often low, resulting in a great increase in numbers, as illustrated by the introduction of the English sparrow into the United States or the rabbit into Australia. As a species increases in number, the environmental resistance may also increase in the form of other organisms that parasitize it or prey upon it and in the competition of members of the species for food, living space, and the opposite sex.

After a population has become established in a region and has reached an equilibrium level, the numbers may change from time to time, being affected by variations in environmental resistance, by factors inherent to the population, or by a combination of the two. The variations of some populations are very irregular, whereas others are regular and occur in cycles.

Balance in nature

When all organisms in an area are studied carefully, much competition and many struggles are evident, even though there is also much cooperation among the inhabitants. The various organisms and their activities somehow seem to counteract each other, resulting in a balance in nature. If one group of organisms is eliminated in an area, another group may take its place, or the remaining organisms may fill the vacancy, thus producing a new balance.

Some contributors to the knowledge of ecology

Anton Kerner (1831-1898)
An Austrian botanist who studied the vegetation of the Danube River Basin and was the first to observe one plant community being replaced by another on the same site. He reported an orderly, predictable succession of plants leading to a rather stable community. He is the "Father of the Climax Concept," in which there is a stable vegetation type toward which all successional types in a region lead.

Henry C. Cowles (1869-1939) and Frederic E. Clements (1874-1945)
Cowles and Clements, American botanists, accepted and established Kerner's concepts of plant succession. For years after 1900 Clements made great contributions to plant distribution knowledge by his emphasis of the climax community, as opposed to the serial communities leading up to it, and of the climatically controlled regional climax, as opposed to the special communities in special habitats.

Henry A. Gleason (1882-   )
An American who opposed Clements' views and proposed the "Individualistic Concept" of plant communities, in which he stated that each species has its own ecologic and geographic limits; that each community differs from all other communities, and that the transition between communities is in some cases gradual but in others quite sharp or abrupt. The recent trend in American studies seems to be a compromise between the views of Clements and Gleason.

Josias Braun-Blanquet (1884-   )
A great Swiss ecologist who dominated the work of ecology in Europe and suggested methods of sampling in making detailed analyses of the floristic composition of a community and the dominant species. In his method small areas of a community selected as samples for detailed study are called quadrats (typically, 1 meter square). The Braun-Blanquet system has not been accepted widely by American ecologists.

1 How can knowledge of ecology be beneficial in the successful cultivation of flowers and vegetables?
2 How can knowledge of ecology be helpful in the proper operation and care of an outdoor pool? Of an aquarium? Of a terrarium?
3 List some of the factors that might influence migrations of certain animals or dispersals of certain plants.
4 Explain how the distribution of certain types of plants might influence the distribution of certain types of animals and vice versa.
5 Discuss biotic communities, including advantages and disadvantages of such biotic relationships. What are the benefits of efficient communal life?
6 What is meant by the term population?
7 Explain what is meant by a so-called food chain, including its importance in a biotic community.
8 Explain what is meant by plant successions.
9 Explain the relationships of the nitrogen, oxygen, carbon, and sulfur cycles and soil fertility.

Selected references

Allee, W. C., and others: Principles of animal ecology, Philadelphia, 1949, W. B. Saunders Company.

Barnett, L., editor: The world we live in, New York, 1955, Simon & Schuster, Inc.

Carson, R.: The sea around us, New York, 1952, Oxford University Press, Inc.

Cooper, C. F.: The ecology of fire, Sci. Amer. April 1961.

Cole, L. C.: The ecosphere, Sci. Amer. April 1958.

Evans, F. C.: Ecosystem as the basic unit in ecology, Science 123:1127-1128, 1956.

Hutchinson, G. E.: Nitrogen in the biogeochemistry of the atmosphere, Sci. Amer. 32:178-195, 1944.

Kendeigh, S. C.: Animal ecology, Englewood Cliffs, N. J., 1961, Prentice-Hall, Inc.

Odum, E. P.: Fundamentals of ecology, ed. 2, Philadelphia, 1959, W. B. Saunders Company.

Polunin, N.: Introduction to plant geography, New York, 1960, McGraw-Hill Book Company.

Wecker, S. C.: Habitat selection, Sci. Amer. Oct. 1964.

Went, F. W.: The ecology of desert plants, Sci. Amer. April 1955.

Wessells, N. K., editor: Vertebrate adaptations. Readings from Scientific American, San Francisco, 1969, W. H. Freeman and Co. Publishers.

Woodwell, G. M.: The ecological effects of radiation, Sci. Amer. June 1963.

Woodwell, G. M.: Toxic substances and ecological cycles, Sci. Amer. 216:24-31, 1967.

chapter thirty-three

Interrelationships among organisms

Fig. 33-1
Herd of bison, Yellowstone National Park.
(*Courtesy General Biological Supply House, Inc.,
Chicago, Illinois.*)

There are many kinds of biotic relationships in the living world. They may exist within one species or among different species of plants, animals, or animals and plants.

GREGARIOUSNESS

In gregariousness (L. *grex,* flock) certain animals associate with each other for protection, for obtaining food, or for reproductive purposes. The herding of herbaceous mammals (Fig. 33-1), the flocking of birds, and the schooling of fishes are examples. Dogs and wolves often hunt in packs, enabling them to attack larger animals than if they hunted alone. The reasons for these aggregations are not always clear, and they are not always based on sex. In certain schools of fishes only one sex is present; in such a phenomenon food is possibly the controlling factor that brings about this association. Mass movements of the fishes result from imitation of a so-called leader. Herds of large mammals also show group or mass movements through the leadership of one individual, often a large or old male.

Social insects

Most insects live more or less solitary lives, coming together only for mating. Some types congregate in swarms, whereas others, such as bees, ants, and termites, have developed rather complex societies in which the labor is divided, and adults and young live in a cooperative community.

In the honeybee community more than 50,000 bees may live in one hive. This society consists of only one queen, a few hundred drones (males), and thousands of workers (infertile females). In a single season one queen may produce 200,000 eggs that are fertilized by the sperms from one drone. A drone fertilizes a queen by storing sufficient sperms in her spermatheca to last her lifetime, which may be five seasons. The drones usually live for one summer, then are killed or driven out by workers. The workers do the other work inside and outside the hive, and in their lifetime of a few weeks they gather nectar and pollen from flowers, produce beeswax, clean, guard, and ventilate the hive, and care for the young. During the winter the bees survive on stored honey. They maintain the temperature in the hive by forming clusters and rapidly beating their wings.

Ant colonies contain one or more fertile females (queen), fertile males (drones), and infertile females. The males supply sperms for fertilization and are not always present or very numerous. Some of the infertile females are food-gathering workers, some act as nursemaids, and others are soldiers to guard the colony.

The complex colonies of termites (Fig. 33-2) contain two castes: fertile females and males, and infertile females and males. Some reproductive

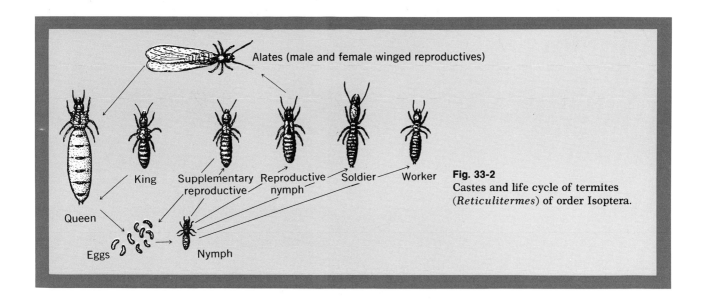

Fig. 33-2
Castes and life cycle of termites (*Reticulitermes*) of order Isoptera.

individuals have wings for flying and mating, and new colonies may be started when these winged reproductives swarm, mate, and crawl into holes in the ground. Other reproductive individuals have small, nonfunctional wings, and a few other reproductives are wingless. In emergencies the latter two types of reproductives may take the place of the queen and king. The sterile, wingless caste includes workers and soliders. The workers obtain food, build tunnels, may cultivate fungi for food (certain species), care for the eggs, the queen, and the young.

COMPETITION

The interaction of a number of organisms of the same or different species for the same limited resources is called competition. The limited resources may be such things as space, food, water, nesting sites, shelter, sunlight, or other similar factors. The competition may be caused by an increase in the number of organisms or by a decrease in the availability of the resources, or by a combination of both. Where members of two different species are involved, one of them usually survives and the other dies out. In cases where only one species is present, the population pressure leads to a limiting of the numbers of organisms that can survive.

The classic example of intraspecific (that is, one species) competition is that of the deer on the Kaibab Plateau in Arizona. At the beginning of the twentieth century this area was a thriving natural community of various types of plants and animals. The plentiful vegetation formed the chief food for a large herd of approximately 4,000 deer. The population of deer was controlled by predators such as pumas (mountain lion), wolves, and coyotes. There were naturally occurring fluctuations in the population, but the numbers were well below the potential limits determined by food supply and continued predation seemed to ensure a stable population.

However, during the years 1907 to 1923, a continuous campaign of extermination of predators was carried out. Records show that 3,000 coyotes, 600 pumas and 11 wolves were killed. As a result of this removal of the controlling pressure, the population of deer increased rapidly, reaching 100,000 by 1924. At this time an intense competition for food began. Since the available vegetation was not sufficient to support the entire population, over half of the herd starved during the next two years. As a further complication, much of the young vegetation never matured, because it was consumed by hungry deer. The herd size continued to decline, not because of predators but because of starvation. By 1940, the herd showed some signs of stabilizing its drastically reduced population; the added complication of the food supply was now the limiting factor.

The result of interspecific (two species) competition is frequently the extinction of one of them. G. F. Gause's work with two species of paramecia gives a good example of this. When grown in the

507

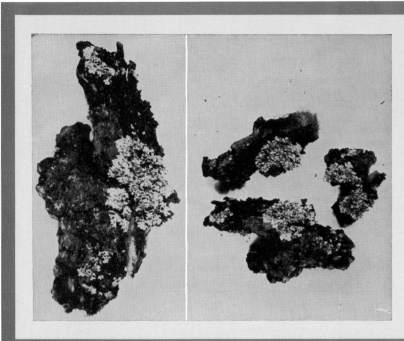

Fig. 33-3
Lichen, a type of plant composed of an alga and a fungus growing together symbiotically for mutual benefit. This lichen grows on the bark of a tree.

laboratory, *Paramecium aurelia* reaches a stable population size in about 8 to 10 days under certain conditions. Another species, *Paramecium caudatum* increases at a slower rate to reach a population peak of about one half that of *P. aurelia*. If the two are mixed in the same culture, competition for food and other resources begins. The result of this is that *P. aurelia* survives, and *P. caudatum* dies out. The population of *P. aurelia* eventually reaches the same size as when it is grown alone, but a longer time is required.

SYMBIOSIS

In true symbiosis (Gr. *symbioun*, to live together) there is an intimate relationship between two plants, two animals, or between a plant and an animal. This may involve a variety of relationships, and several types of symbiosis are therefore recognized.

MUTUALISM

When two species both gain from their association and are unable to live separately, the phenomenon is called mutualism (L. *mutus*, exchange).

In lichens (Fig. 33-3), in which an alga and a fungus are associated, the alga photosynthesizes food, part of which the fungus uses. The fungus furnishes water, essential elements, and probably some protection against drying, so that both benefit. Certain green algae live with several kinds of animals, including certain protozoans, freshwater sponges, and *Hydra*. For example, a green paramecium *(Paramecium bursaria)* may contain numerous algae cells.

Certain flagellated protozoans that live in the intestines of termites obtain food by assisting in the digestion of the cellulose of the wood eaten by the termites. In return, the protozoans are supplied with a stable environment and food and receive protection. Plant aphids secrete foods that are used by ants, and the ants give protection and shelter to the aphids.

The association between certain fungi and the root tips of higher plants, known as mycorrhiza (mi ko -ri' za) (Gr. *mykes*, fungus; *rhiza*, root), is another example. The fungi increase root absorption by the plant and receive some benefits in return. Another example is the symbiotic relation-

Fig. 33-4
Starfish attacking an oyster: Note the tube feet on the underside of the starfish arms.
(*Courtesy The American Museum of Natural History, New York, N. Y.*)

Fig. 33-5
Praying mantis, a predacious insect of order Orthoptera. This insect is so named because of the front legs folded as if in prayer, although it is also called the preying mantis.

ship between the nitrogen-fixing, root-tubercle bacteria and the leguminous plants in which they live.

COMMENSALISM

When organisms of different species live together habitually in such a way that one species benefits and the other is neither harmed nor benefitted, the phenomenon is called commensalism (L. *cum*, with; *mensa*, table), or literally "eating at a common table." Sea anemones may attach themselves to the shell of a hermit crab, thus obtaining food and giving some camouflage to the crab. The peculiar tropical fishes called remoras (*Echeneis remora*) have a modified dorsal fin that forms a sucker used to attach to sharks and other large aquatic animals. Their food consists of scraps of food left by the shark, who remains unharmed.

PARASITISM

In parasitism (Gr. *para*, beside; *sitos*, food) an organism known as the parasite lives on, or within, and at the expense of another living organism known as the host. When a parasite lives on the outside of the body of the host, it is an ectoparasite (Gr. *ekto-*, outside); when it lives within the body of the host, it is an endoparasite (Gr. *endon*, within). When the effects of parasitism on the host result in discernible, abnormal symptoms, the condition is known as disease production, or pathogenesis

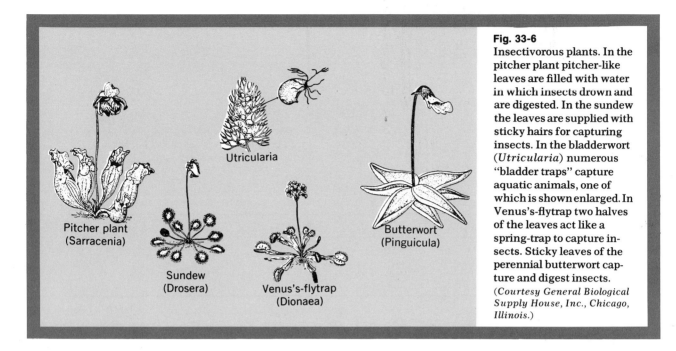

Fig. 33-6
Insectivorous plants. In the pitcher plant pitcher-like leaves are filled with water in which insects drown and are digested. In the sundew the leaves are supplied with sticky hairs for capturing insects. In the bladderwort (*Utricularia*) numerous "bladder traps" capture aquatic animals, one of which is shown enlarged. In Venus's-flytrap two halves of the leaves act like a spring-trap to capture insects. Sticky leaves of the perennial butterwort capture and digest insects.
(*Courtesy General Biological Supply House, Inc., Chicago, Illinois.*)

Pitcher plant
(Sarracenia)

Utricularia

Sundew
(Drosera)

Venus's-flytrap
(Dionaea)

Butterwort
(Pinguicula)

(Gr. *pathos,* disease or suffering; *genesis,* origin). The malaria parasite *Plasmodium vivax* is a good example of this.

In evolving their dependent mode of living, parasites at times lose some structures and abilities that are no longer used, or they develop modifications in certain systems. The loss of a digestive system and the development of a more complex reproductive system by the tapeworm (see Fig. 18-9) are illustrative. In general, parasites do not use complex equipments for obtaining foods, because they are usually supplied with ready-made foods.

Parasites are usually restricted to one or a few hosts, and they are thought to have evolved from free-living types rather than vice versa. The relationship between parasite and host may be quite harmful at first, but in time the forces of natural selection usually result in a decrease of the harmful effects, and in development of a give-and-take in the parasite-host relationship.

PREDATISM

Predatism or predaciousness (L. *praeda,* prey or booty) is a condition in which one organism, the predator, captures another living organism, using it for food (Fig. 33-4). These predatory habits are exhibited by a great number of animals, whose methods of capturing and devouring their prey vary greatly. *Amoeba, Paramecium,* and other protozoans capture a variety of living organisms for food. *Hydra* and other coelenterates devour aquatic organisms. The Portuguese man-of-war preys on fishes and crustaceans. Planarians feed on mollusks and arthropods; squids capture fishes; starfishes capture oysters and other animals; dragonflies destroy flies and mosquitoes; certain insects, such as praying mantises (Fig. 33-5), ground beetles, ladybird beetles, and aphis lions, destroy other insects, many of which are detrimental. Fishes devour worms, crustaceans, and insects; frogs capture worms and insects; snakes destroy frogs and birds; owls kill rabbits and mice. Predacious mammals include wolves and cougars, which kill sheep, cattle, and big game; dogs and cats are predacious when they kill animals such as rats and mice.

Insectivorous plants

The so-called insectivorous or carnivorous plants possess special organs, usually modified leaves or parts of leaves, by which they are able to trap insects or other small animals for part of their food

sorbed. In Venus's-flytrap (*Dionaea*) (Figs. 33-8 and 33-9) the specialized leaves possess a row of teeth on the outer margins of each half of the leaf blade. On the upper surface are sensitive hairs that when stimulated cause the two halves to spring together to entrap the insect. Digestion by enzymes somewhat resembles that in the pitcher plant. In the sundew (*Drosera*) (Fig. 33-10) the leaves are covered with long glandular hairs that secrete a sticky substance to capture and digest insects. There are numerous species of chlorophyll-bearing plants that are insectivorous or carnivorous. In the bladderwort plants, present in ponds and lakes, there are tiny "bladder traps" on the submerged stems. Each bladder has a one-way trapdoor through which aquatic animals enter and in which they are digested.

EPIPHYTISM

Epiphytes (Gr. *epi*, upon; *phyton*, plant) are plants that use other plants, poles, trees, or wires for support but do not derive nourishment from them; hence they are not parasitic. Epiphytes are primarily autotrophic (Gr. *autos*, self; *trophe*, nourishment), which means they photosynthesize their food. They obtain carbon dioxide and water from the atmosphere and moisture and nutrients from debris in crevices in which they may be anchored. Epiphytes sometimes injure the host plant by shading the leaves or by breaking limbs because of excessive weight.

(Fig. 33-6). The specialized structures secrete enzymes for the digestion of the food, which is then absorbed by the plant. The pitcher plants (*Sarracenia*) (Fig. 33-7), common in bogs, have a pitcherlike leaf that is filled with water in which insects drown. Escape is prevented by inwardly directed spines, and the digested insects are ab-

511

Fig. 33-9
"Trap" (modified leaves) of Venus's-flytrap (much enlarged).
(*Courtesy Carolina Biological Supply Co., Elon College, North Carolina.*)

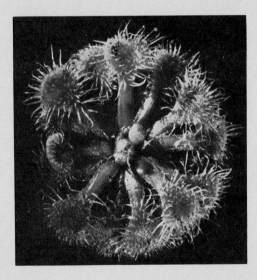

Fig. 33-10
Sundew (*Drosera*), an insectivorous plant.
(*Courtesy Carolina Biological Supply Co., Elon College, North Carolina.*)

Fig. 33-11
Spanish "moss" (not true moss), a flowering, epiphytic plant of the pineapple family that hangs in great masses from trees, poles, and wires in the southern United States.

The green alga *Protococcus* may grow epiphytically on the bark of trees. Certain species of algae grow among the hair of the three-toed sloth. Certain species of lichens, mosses, ferns, and tropical orchids may be epiphytes. Spanish "moss" (*Tillandsia*) which is a rootless, flowering epiphyte of the pineapple family found in the South, hangs in great masses from trees, poles, and wires (Fig. 33-11). Often the growth is so enormous that the tree is killed, even though the "moss" is not parasitic.

SAPROPHYTISM

A saprophyte (Gr. *sapros*, dead; *phyton*, plant) is an organism capable of utilizing dead, organic materials as foods. When the food change involves

Fig. 33-12
Indian pipe (*Monotropa*), a flowering plant
that contains no chlorophyll.

(bracket) fungi are common on dead trees, although parasitic species may kill living ones.

Saprophytism is less common in higher plants than in fungi. The common Indian pipe (*Monotropa*) is a flowering plant (Fig. 33-12) without chlorophyll, and for years it was considered to be a saprophyte. However, it seems that its stubby roots are completely surrounded by a compact layer of fungus mycelium from which it actually obtains its organic food. None of the roots have direct contact with the soil. If the Indian pipe derives its organic nourishment from the living fungus, it is a parasite. In turn, if the fungus obtains some benefit from the Indian pipe, there is a mutual relationship.

Review questions and topics

1 Discuss symbiosis, mutualism, and commensalism, including differences, similarities, and examples of each that you may have observed.
2 Discuss gregariousness, including the advantages and disadvantages of such a biologic relationship.
3 Discuss the effects of predatism, including examples from your own observations.
4 Discuss the unique biotic relationships displayed by insectivorous plants, including several examples with their specific activities.
5 Discuss epiphytism in plants, including examples.
6 Discuss the nature and significance of saprophytes in the living world.

Selected references

Ager, D. V.: Principles of paleoecology, New York, 1964, McGraw-Hill Book Company.

Baer, J. G.: Ecology of animal parasites, Urbana, Ill., 1951, University of Illinois Press.

Boughey, A. S., editor: Population and environmental biology, Belmont, Calif., 1967, Dickenson Pub. Co., Inc.

Burnett, A. L., and Eisner, T.: Animal adaptation, New York, 1964, Holt, Rinehart & Winston, Inc.

Caullery, M.: Parasitism and symbiosis, New York, 1956, John Wiley & Sons, Inc.

Chandler, A. C., and Read, C. P.: Introduction to parasitology, ed. 10, New York, 1961, John Wiley & Sons, Inc.

Cheng, T. C.: The biology of animal parasites, Philadelphia, 1964, W. B. Saunders Company.

Davis, D. E.: Integral animal behavior, New York, 1966, The Macmillan Company.

carbohydrates and is associated with the production of gas, it is called fermentation; when it involves proteins and is manifested by the production of foul odors, it is called putrefaction.

Plants without chlorophyll must absorb foods from the outside, which necessitates the presence of a certain amount of moisture. Many bacteria are saprophytic. Slime molds grow saprophytically on decaying plant materials, rotting woods, and leaf molds. Saprophytic true fungi usually live where they encounter suitable supplies of dead organic materials and moisture. *Rhizopus nigricans* and similar molds are common saprophytes on moist bread, ripe fruits, foodstuffs, and animal dung. The blue and green molds (*Pennicillium* and *Aspergillus*) grow on fruits, foodstuffs, leather, fabrics, and other organic materials in damp places. Mushrooms live saprophytically on dead leaves, dung, dead wood, and similar organic materials. Shelf

Selected
references
—cont'd

Fiennes, R.: Man, nature and disease, London, 1964, George Weidenfeld & Nicholson Ltd.

Linbaugh, C.: Cleaning symbiosis, Sci. Amer. Aug. 1961.

McGaugh, J. L., Weinberger, N. M., and Whalen, R. E., editors: Psychobiology; the biological bases of behavior. In Readings from Scientific American, San Francisco, 1967, W. H. Freeman and Co. Publishers.

Mykytowycz, R.: Territorial marking by rabbits, Sci. Amer. 218:116-126, 1968.

Petrunkevitch, A.: The spider and the wasp, Sci. Amer. Aug. 1952.

Platt, R. B., and Griffiths, J. F.: Environmental measurement and interpretation, New York, 1964, Reinhold Publishing Corp.

Rogers, W. P.: The nature of parasitism, New York, 1962, Academic Press, Inc.

Ruibal, R., editor: The adaptations of organisms, Belmont, Calif., 1967, Dickenson Pub. Co., Inc.

Shaw, E.: The schooling of fishes, Sci. Amer. June 1962.

Wynne-Edwards, V. C.: Animal dispersion, New York, 1962, Hafner Co., Inc.

514

chapter thirty-four

Man and biology

Consideration has already been given to many of the numerous applications of biology in various fields of endeavor; several ways in which applications of biology have been made in the world about us will now be examined. Additional applications will be left to the reader who, with some reflections and investigations, can easily add to the list.

AGRICULTURE AND HYDROPONICS

For many years man has utilized the animals and plants that he has found in nature, and in some instances he has improved and domesticated them. His early attempts to do this were somewhat crude and unscientific. With the advent of scientific biology he has made more progress, so that at the present time our food, clothing, shelter, furniture, and raw materials for many industries are in some way, directly or indirectly, influenced and made possible by the results of biologic studies.

Agriculture rests upon the knowledge of the structure, functions, inheritance, and development of domesticated plants and animals and of the great variety of environmental factors by which they are affected beneficially or detrimentally. Biologic studies have contributed much to each of these phases and have helped to make agriculture a more scientific and more successful operation.

Together with other sciences biology has given basic and practical methods to many other fields. The improved qualities of plants and animals by the proper selection of seeds and parents have produced better offspring. Knowledge of heredity has enabled the production of new types of plants and animals through hybridization.

Methods of controlling many plant and animal parasites and the prevention and treatment of many plant and animal diseases have evolved through biologic investigations. Plant and animal foods have been preserved more effectively and transported more efficiently. Biology and chemistry have enabled a more efficient use of animal and plant by-products that were formerly wasted, and biologic studies have contributed to many improvements in proper cultivation, fertilization, and care of soils.

Hydroponics (Gr. *hydor*, water; *ponos*, exertion) is a procedure whereby plants are grown in solution or sand cultures. The solutions must have the essential elements required for the specific plants, and these must be in proper proportions. Much information regarding the relative importance of various elements for plant growth has been obtained through experiments with solution and sand cultures. The elements that are essential to green plants for normal growth and for development of flowers and seeds include carbon, hydrogen, oxygen, nitrogen, phosphorus, potassium, sulfur, calcium, magnesium, and iron. Very small amounts of micronutrient (trace) elements include boron, copper, manganese, and zinc. The storage and probable functions of certain elements have been ascertained by introducing certain radioactive elements into different parts of plants and tracing their paths by means of Geiger counters.

FOODS, CLOTHING, FURNITURE, AND FUELS

Most of our food, clothing, shelter, furniture, and fuel are directly or indirectly influenced by the proper and effective application of biologic knowledge. The production, transportation, preparation, and efficient use of foods have been improved by increased biologic knowledge. Through animal experimentation the vitamin content and physiologic uses of foods have been determined and evaluated. The following plants, now common in America, had their origins in such countries as: banana (southeastern Asia), citrus fruits (China), coffee (Abyssinia), corn (Central and South America), Para rubber (South America), pineapple (South

515

America), sugarcane (southeastern Asia), wheat (southwestern Asia). The importation and improvement of such materials have required much scientific work.

Some Old World plants that have been cultivated for more than 4,000 years include apple, apricot, banana, barley, date palm, fig, flax, grape, hemp, olive, onion, peach, rice, sorghum, soybean, tea, watermelon, and wheat. Some Old World plants that have been cultivated for over 2,000 years include alfalfa, beet, carrot, celery, cherry, garden pea, lettuce, mustard, oat, plum, poppy, radish, rye, sugarcane, and yam. Some plants cultivated in America before the time of Columbus include avocado, cotton, peanut, pineapple, potato, pumpkin, and tomato.

Many articles of clothing originate directly or indirectly from plants or animals, for example, (1) silk from the silkworm, (2) Cotton from cotton plants, (3) wool from sheep and other animals, (4) linen from flax plants, (5) furs from rabbits, skunks, opossums, muskrats, beavers, raccoons, and minks, (6) leather from prepared skins of cows, horses, pigs, and alligators, (7) straw from the stalks of wheat, rye, oat, and barley, (8) felt from the wool, fur, and hair of animals, and (9) rubber from the juice of rubber trees.

Many of the materials for the construction of furniture are also of plant and animal origin and include (1) wood from plants, (2) willow furniture from willow plants, (3) glue from the hoofs and skins of animals, (4) leather from the prepared skins of animals, (5) excelsior from shredded wood, (6) paper from straw, bark, wood, and other fibers, (7) shellac from the resinous materials secreted by certain scale insects, and (8) stains, varnishes, and paints, at least in part, obtained from various plant materials.

Many fuels are closely related to plants that were formerly alive. Vegetation and stored solar energy, buries in swamps long ago, have undergone changes so that coal has been formed. Wood is plant tissue that was once alive. Natural gas is the product of biologic decomposition of plant and animal remains of the past. This has probably taken place under great pressures within the deeper strata of the earth. Peat is one of the intermediate stages in the formation of coal and is utilized in localities where coal is unavailable. Petroleum is also formed by the decomposition of materials that were largely of animal and plant origin; gasoline is obtained by refining petroleum. Paper is made from straw and other fibrous materials. The fuel, coke, is manufactured from coal by heating it in the absence of oxygen.

HUMAN WELFARE

Medicine and health

In few fields have the contributions of biology been greater than in those of health and medicine. Through animal experimentation many fundamental and basic truths of behavior, health, disease, and other phenomena have been ascertained and applied.

Some medicines are of plant or animal origin. In addition, the efficiency and proper use of medicines have been determined through animal experimentation. The parasites of plants, animals, and man have been studied scientifically, and the information acquired has been used successfully in making our environment a better place in which to live. Research in bacteriology and protozoology has contributed much information; the application of this information has resulted in lower morbidity and mortality rates among living organisms, including man. For example, biologic studies demonstrated the importance of the inspection and refrigeration of foods and the rigid inspection of oysters and other shellfish for possible contamination. The study of heredity has greatly contributed to the explanation of how we have come to be what we are, how certain abnormalities arise, and how some undesirable traits might be eliminated.

The science of endocrinology (study of the hormones) of certain animals has indicated the basic principles in the effective application of this field to human beings. Studies of pathogenic (disease-producing) bacteria, yeasts, molds, protozoans, worms, insects, and viruses have produced many methods of disease prevention, treatment, and cure. The production of antibiotics has greatly contributed to our conquest of diseases. What we, as individuals, owe in health and happiness to the thousands who have scientifically investigated diseases and their treatment is beyond our comprehension. Possibly some who are studying biology now will play an important role in the discovery of methods of disease prevention and treatment that are unknown at present.

■ Biology and wealth

People may not always associate biology and biologic products with a country's economy, but a few examples show their enormous contributions to the national wealth of the United States. Fourteen million bales of cotton (500 pounds each), valued at many millions of dollars, are produced annually. When 1,250 pounds of seed cotton are taken to the ginnery, one bale of cotton lint or fiber and about 750 pounds of cotton seed are produced, from which valuable products may be obtained.

The production of wheat in the United States averages about $1\frac{1}{2}$ billion bushels, and about $4\frac{1}{3}$ billion bushels of corn are produced annually.

Nearly 40 billion board feet of lumber are produced in the United States annually. About 26 million tons of wood are used for wood pulp annually, much of which is used in making paper and paper products. All of these products have values of many millions of dollars annually.

The value of all meat-producing animals in the United States is estimated to be several billion dollars per year. The value of woolen and worsted products is millions of dollars. The annual value of the products of our fisheries industry, especially the codfishes, salmon, and other widely used species, reaches many millions of dollars. The total value of all farm properties in this country is many billions of dollars. If we add to these such products as hay, vegetables, fruits, cereals, dairy products, clothing, fuels, and a multitude of others, the totals become enormous. In each, biology plays some role in production, preparation, and use.

Table 34-1 gives data concerning annual production of fruit in the United States. The figures in Table 34-2 show approximately the total numbers of certain livestock in the United States.

In addition to direct contributions to a country's wealth biology also makes appreciable contributions to methods of controlling unnecessary losses. Some parts of the United States are infested with rats that carry diseases and cause great damage to many valuable products. Annual losses of this

Table 34-1
Production of fruit in the United States

Fruits	Approximate production (1968)
Oranges	183,200,000 boxes
Grapes	3,578,850 tons
Grapefruit	57,300,000 boxes
Apples	5,403,000,000 pounds
Peaches	3,653,200,000 pounds
Pears	622,850 tons
Plums and prunes	312,400 tons
Lemons	19,300,000 boxes

Table 34-2
Livestock in the United States (1969)

Chickens	2,600,000,000
Cattle	109,000,000
Hogs	92,000,000
Sheep	28,000,000

kind are estimated at millions of dollars. The science of entomology (study of insects) has supplied us with information that, if correctly applied, would materially reduce losses produced by insects. Losses caused by termites are great, but proper control methods can reduce these losses also. Many millions of bushels of corn, oats, wheat, and barley are lost every year from plant diseases that affect both the yield and quality of the crops.

■ Water supplies and sewage disposal

The value of water has been recognized by peoples of all times, because they have tended to settle where the supply has been of the proper quantity and quality. In the past the quantity and quality were not always the problem that they are now. Primitive peoples used water for drinking, for preparing food, and a small amount for washing and for their primitive handicrafts. At the present time, because of a greater use of water and greater centers of population, together with increased industrialization, in many areas the water problem has become acute. In some communities we find water sanitation so inadequate that people actually consume their own or their neighbor's "purified" sewage in order to obtain a sufficient supply. The per capita daily consumption of water in cities in the United States is more than 100 gallons, although it is much higher in certain cities. The following factors influence the amount of water consumed: the cost of water; the number and types of industries; the chemical and physical properties of the water; the amount used in cleaning streets; the number of fires to be extinguished; whether the water is sold by meter or not; the number of leaky fixtures and pipes; the temperature and humidity of the climate; and the amount used in watering lawns and gardens.

Water comes to the earth in the form of rain or snow; some is used by animals and plants and some evaporates. The remainder collects on the surface of the earth (rivers, lakes, reservoirs) or penetrates into the subsurface or ground (wells, underground rivers and lakes). Waters may be classified as (1) potable water, which is safe from the health standpoint and is desirable from an odor, taste, or appearance standpoint; (2) polluted water, which contains substances not necessarily harmful but of such a character as to offend sight, taste, or smell (pollution usually refers to such physical characteristics as unpleasant tastes and odors, undesirable color, or excessive turbidity); (3) contaminated waters, which contain substances harmful to health (pathogenic microorganisms, inorganic or organic poisons); and (4) pure waters, which are chemically and physically pure; such waters do not exist naturally but can be secured by distillation.

Waters may become polluted and contaminated by picking up all types of materials in suspension and solution; they may acquire silt by passing through fertile lands; they may be hard from in-corporating chemicals as they flow through limestone; they may have undesirable tastes and odors by contacting decaying plant and animal matters; they may be rendered undesirable by industrial wastes, seepage from mines, domestic sewage, and wastes from oil wells. From a sanitary standpoint human excrements play a most important role in the contamination of water.

Waters may be treated by (1) filtration through sand filters, with or without previous coagulation induced by the use of chemicals that precipitate undesirable materials; (2) disinfection by the use of certain chemicals, usually chlorine; (3) some kind of water-softening process; or (4) a combination of these processes. The specific method used is determined by the quality of the raw water and the quality of the water expected after treatment. If water contains little dissolved or suspended matter, chemical disinfection may be sufficient; if it is soft but contains suspended matter and microorganisms, filtration and chlorination may be necessary; if it contains dissolved salts, such as those of calcium or magnesium, it may require softening by some means.

The two types of sand filtrations used for removing pollution from waters are: (1) slow sand filtrations (Fig. 34-1) and (2) rapid sand filtrations. The former have been used extensively in Europe since 1830 but have not been used extensively in the United States. Slow sand filters were made of concrete, covered about 1 acre each, and were filled with sand to a depth of 1 to 4 feet. Bacteria and other microorganisms were removed by mechanical filtration and also by their destruction in the gelatinous, zoogleal mass that covers the filter surface after it has operated for a few days. Criticisms of the slow sand filter include the following: (1) even if operated at full capacity, the rate of purification is only 2 to 3 million gallons per acre; (2) it is inefficient if water contains large amounts of suspended materials, because the surface soon becomes coated, thus interfering with the delivery of sufficient, desirable water; and (3) it is expensive, especially if large quantities of water are required, because large areas of expensive land are necessary to build sufficient filters.

Rapid sand filters, introduced into the United States at about 1890, are extensively used in mod-

518

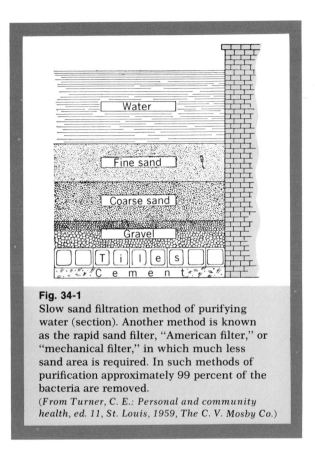

Fig. 34-1

Slow sand filtration method of purifying water (section). Another method is known as the rapid sand filter, "American filter," or "mechanical filter," in which much less sand area is required. In such methods of purification approximately 99 percent of the bacteria are removed.

(From Turner, C. E.: Personal and community health, ed. 11, St. Louis, 1959, The C. V. Mosby Co.)

ern water purification plants. The process, in brief, is as follows: (1) screen the raw water to keep out sticks, leaves, and animals; (2) mix the water with flocculating chemicals, such as aluminum sulfate or iron sulfate, and allow the suspended floc to settle out in settling tanks; (3) pass the clarified supernatant water through rapid sand filter beds; (4) disinfect the water by means of chemicals, usually liquefied chlorine gas.

In recent times the contamination of water supplies by sewage has become a major problem because of the increased population and industrialization in cities, the increased demand for more water for homes and industries, and the greater difficulty in efficiently disposing of large quantities of sewage (see box, p. 2). Sewage may be considered as the used water supply to which have been added (1) human excrements (urine and feces) and water used for bathing and washing; (2) industrial wastes from laundries, creameries, breweries, chemical plants, slaughterhouses, tanneries, and many other similar industries; and (3) water from streets and sidewalks.

Sewage may be disposed of by (1) dilution, (2) irrigation, or (3) stabilization of the sewage through bacterial actions. In the process of disposition by dilution the sewage is placed in a body of water sufficiently large to render the sewage more or less harmless. This old method is inexpensive, and if the body of water is large enough to dilute the sewage properly and is not to be used for other purposes, it is reasonably satisfactory. Sewage disposal by irrigation or by running raw sewage over land is not commonly used in the United States. The stabilization of sewage by the actions of various bacteria is based upon the fact that organic and inorganic substances in sewage are excellent foods for bacteria. When bacteria use these substances, they oxidize them more or less completely, forming new substances with less energy and lower molecular weights. When most of the energy of sewage is consumed, bacteria no longer grow rapidly, and the sewage is stabilized. The specific method for the bacterial treatment of sewage depends upon many factors, such as the quality of the sewage, the quantity to be disposed, and the nature of the body of water or soil into which the treated sewage is allowed to run.

IMPROVEMENTS OF PLANTS AND ANIMALS

The improvements of plants and animals for human needs probably started soon after man began to cultivate crops and domesticate animals. Although the crossing of plants and the breeding of animals have been practiced for a long time, only since our knowledge of genetics has developed have they been practiced on a scientific basis.

Careful observations reveal that all kinds of living organisms show more or less diversity among themselves. This universal variability enables man to improve plants and animals. It must be remembered that not all variations are capable of transmission from one generation to another. Most variations directly influenced by the environment are not heritable although environmental factors may induce changes. On the other hand,

519

inherent variations that are determined by the genic constitution of an organism are transmittable. Only heritable variations are of great importance to the plant and animal breeder, because they make racial improvements possible. Among the many traits in cultivated plants that are influenced by heredity are yield, vigor and rate of growth, size and shape of parts, hardiness, quality of fruits, flower color, disease resistance, and drought resistance.

■ Plants

The many types of cultivated plants that serve human needs have been developed by selection, hybridization, or a combination of both. The two important methods of selection are mass culture and pedigree culture. Mass culture, the oldest method of plant breeding, involves breeding from a selected group of individuals that vary in some desirable direction. Seeds collected from superior plants that possess the desired traits are sowed en masse. The selective process is continued for many generations until an improved race is obtained. Although mass culture has been largely replaced by newer methods, it has been used widely in the breeding of corn, cotton, and other crops. Some limitations of mass culture include the following: (1) it does not produce new traits but merely improves existing ones; (2) unless selection is continued for a long time, the improved race usually deteriorates; (3) it is slow, only slight improvement being made in each generation; (4) it does not isolate the best individuals but merely raises the average quality of a large group; (5) selection made on outward appearance is often erroneous, especially if heterozygous individuals are involved; and (6) the selected individuals often owe their improved qualities to favorable environmental influences, and such variations have little permanent value in racial improvement.

In pedigree culture individual plants with desirable traits are selected and form the basis of the improved race. Seeds from each selected plant are usually planted in isolated areas to prevent undesirable intercrossing. An accurate record (pedigree) is kept of the offspring of each selected individual, and after several generations, the best strain is preserved. Hence, selection is based on hereditary behavior rather than outward appearance, as in mass culture. Some attributes of the pedigree culture method include the following: (1) There is a greater uniformity among the members of the improved race than in the mass culture method. However, the improved plants must be prevented from crossing with plants possessing inferior, undesirable traits. If the original plant were homozygous for its selected trait, the new type of plant will breed true. (2) Pedigree culture is well adapted for plants that normally self-pollinate, and it has been employed successfully with wheat, oats, beans, peas, tobacco, and potatoes. (3) Pedigree culture is useful in preserving mutants, individuals that appear spontaneously and show some striking variation. If isolated, nearly all mutants breed true. Many new horticultural varieties have arisen as mutants. Examples include plants bearing seedless fruits, double flowers, variegated leaves, drooping branches.

Hybridization is a method that is used extensively in bringing together the traits of two different parents to form a new combination. For example, a tender variety possessing superior fruit may be crossed with a hardy variety possessing inferior fruit, and possibly some offspring will be formed that are hardy and have superior fruit. Because such a desirable combination depends on chance, usually a great many crosses of the same kind have to be made, with the hope that in some of the offspring the desired combination of traits will occur. Hybridization is valuable for plants where vegetative propagation is possible, because with no sexual reproduction there is no segregation of genes, so the traits of the hybrid are preserved for a long time. Where vegetative propagation cannot be used, the hybrid offspring may be diverse and exhibit many new trait combinations. If some of these are desirable, they may be isolated and raised under the pedigree culture method until a pure-breeding race is secured.

Hybridization often produces increased vigor. Many hybrids reach a larger size than their parents, grow more rapidly, have larger parts, or have greater resistance to adverse conditions. Hybrid corn (Figs. 34-2 and 34-3) is now grown extensively in the United States. Seeds for planting are produced by crossing two or more inbred strains that

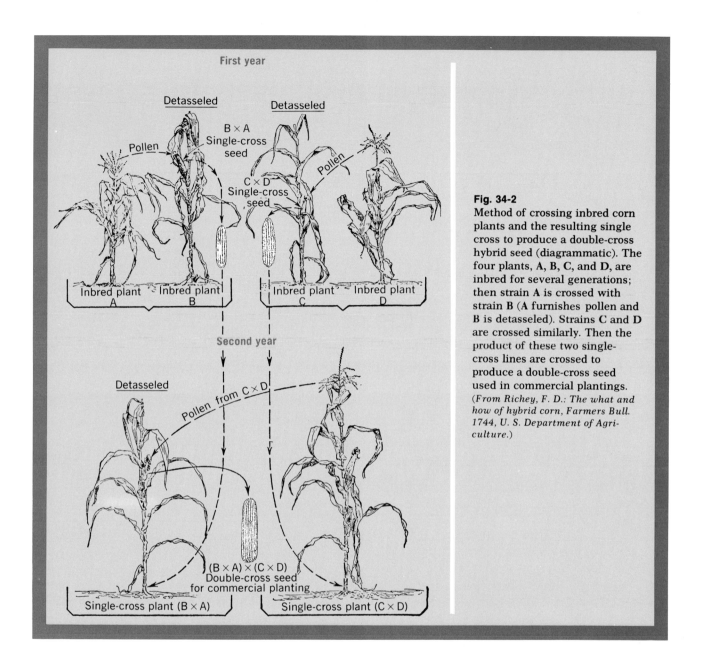

First year

Detasseled Detasseled

B × A
Single-cross
seed

Pollen

C × D
Single-cross
seed

Pollen

Inbred plant A Inbred plant B Inbred plant C Inbred plant D

Second year

Detasseled

Pollen from C × D

(B × A) × (C × D)
Double-cross seed
for commercial planting

Single-cross plant (B × A) Single-cross plant (C × D)

Fig. 34-2
Method of crossing inbred corn plants and the resulting single cross to produce a double-cross hybrid seed (diagrammatic). The four plants, **A, B, C,** and **D,** are inbred for several generations; then strain **A** is crossed with strain **B** (A furnishes pollen and **B** is detasseled). Strains **C** and **D** are crossed similarly. Then the product of these two single-cross lines are crossed to produce a double-cross seed used in commercial plantings. (*From Richey, F. D.: The what and how of hybrid corn, Farmers Bull. 1744, U. S. Department of Agriculture.*)

have been developed by self-pollinating selected plants for about seven generations. These inbred strains decrease in vigor, but through constant selection they become uniform for their desired traits. When finally crossed, hybrids are produced that excel the parents in yield, size, and vigor. The yield of grain is often one third greater than that of the original corn plants from which the inbred strains were developed. If seeds from the hybrids are planted, the quality is markedly deteriorated.

Because corn hybridization is one of the most recent and successful examples, it might be considered briefly. Corn is normally cross-pollinated. When self-pollinated for at least seven successive generations, the corn plants tend to become progressively smaller and less productive. Eventually, when two such self-pollinated corn plants are crossed, the resulting hybrid is more productive

521

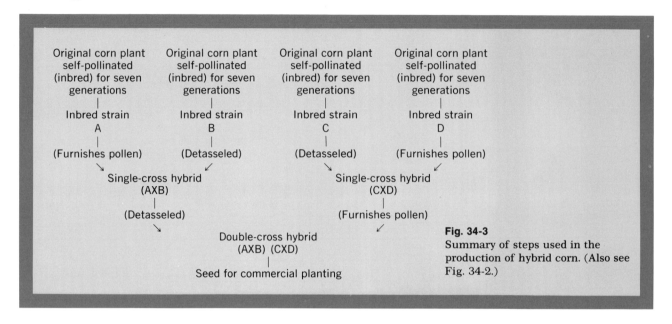

Fig. 34-3
Summary of steps used in the production of hybrid corn. (Also see Fig. 34-2.)

Fig. 34-4
Ears of corn, showing methods of inheritance between one pair and two pairs of traits. **A,** Ratio of three purple to one white (aleurone color); **B,** nine white-starchy, three white-sweet, three purple-starchy, one purple-sweet (aleurone color and endosperm texture); **C,** nine purple-starchy, three purple-sweet, three white-starchy, one white-sweet (aleurone color and endosperm texture); **D,** nine purple to seven white (aleurone color). Starchy grains are smooth, whereas sweet grains are wrinkled.

and larger than the ancestors. The procedure, in brief, is as follows: (1) inbreeding (self-pollination) for about seven successive generations tends to produce homozygous strains; (2) two such homozygous strains that possess the desired traits are then cross-pollinated to produce the F_1 hybrids known as single-cross hybrids. However, such seeds are not sold, primarily because the yield is usually low and grains are of variable size; (3) two single-cross hybrids, produced from different

Head cabbage

Cauliflower

Kohlrabi

Brussels sprouts

Fig. 34-5
Some domesticated plants derived from the cliff cabbage found in Europe. The wild ancestor is sketched diagrammatically at lower left. (Not shown to true scale.)
(Courtesy W. Atlee Burpee Co., Philadelphia, Pennsylvania.)

Cliff cabbage

Broccoli

homozygous strains, are now crossed, producing double-cross hybrids, which produce higher yields of uniformly larger seeds (Fig. 34-2).

Many genetic improvements in plants have been made, including such common examples as: fiber length in cotton, sugar content of melons, yellow color of peaches, seedless grapes, improved tobacco plants, and resistance of plants to diseases (wheat stem rust, corn blight, oats smut, and tomato wilt),

and to insect pests. Some of the results of experimental crossings of corn are shown in Fig. 34-4.

Most of our important species of presentday cultivated plants were domesticated by early man. He recognized certain valuable properties, or parts, in the wild ancestors. For some time he collected these plants in the wild state; later he cultivated particular species and noted variations in their heredity that had also occurred in the wild state.

Under conditions of cultivation he selected seeds or vegetative parts for propagation, and these were more widely used. Continued variations accompanied by selections for hundreds of years resulted in cultivated varieties that differed from their wild ancestors in many ways. For example, our present cultivated varieties of head cabbage, kohlrabi, cauliflower, broccoli, and Brussels sprouts were all derived from a mustardlike wild ancestor (cliff cabbage) in Europe (Fig. 34-5).

The evolution of cultivated plants has occurred over such a long period of time and has resulted in such marked varieties that it is difficult to discover the wild ancestor of some of them. The wild ancestors of the domesticated plants of Eurasia are better known than those of America. Corn is the only important cereal that originated in America, and its wild ancestor was a grass that is still unknown. When Columbus arrived in America, the Indians were cultivating several varieties of corn in local areas between what is now Canada and Argentina. They also cultivated several varieties of potatoes, sweet potatoes, beans, peanuts, pumpkins, tomatoes, tobacco, and cotton. These plants were domesticated by early man in the highlands of Mexico, Central America, and in northwestern South America, the regions in which the Incas, Mayas, and Aztecs later developed their remarkable civilizations.

In the eastern hemisphere the regions in which the earliest civilizations became most highly developed also were centers in which basic crop plants originated. Among the cultivated plants on which these early civilizations of Eurasia depended were the cereals (wheat, rice, rye, barley, oats, and sorghum), forage crops (clovers and grasses), soybeans, and several common edible fruits and vegetables.

Civilization in the past depended upon the plant supply as the source of foods, just as it does today. Migrations were often determined by the abundance or scarcity of wild plants. The Indians who lived in what is now the United States and Canada used over 1,000 species of plants as foods, only a few of which were ever cultivated. Many others were used as sources of materials for weaving, medicines, shelters, and canoes. One of the greatest accomplishments of prehistoric man was his discovery of the methods of plant propagation. With this knowledge he no longer needed to live as a nomad, raiding other tribes when supplies of wild plants and animals became depleted. He could establish permanent abodes in the fertile agricultural areas. With the discovery of plant propagation by transporting and planting seeds and vegetative plant structures primitive man could occupy new areas permanently. Without this knowledge man could not have occupied certain areas of the world that he does now. With few exceptions inhabitants of the United States and many other countries at the present time are dependent largely upon plants that were first domesticated in other parts of the world.

■
Animals

It is not known in detail just how or when early man utilized wild animals or when he first attempted to domesticate some of them. As in the case of wild plants, he used them for food and for making clothing and shelters. The abundance or scarcity of certain types undoubtedly influenced his migrations and settlements. Man's early attempts in the racial improvements of animals were doubtless unscientific, but since the development of genetics, he has made rapid progress.

Most of the problems of the animal breeder are rather complex, because many traits that have commercial application are dependent upon the interaction of multiple genes. Hence, most techniques for the improvements of commercial types involve principles of selection based on quantitative variations. Selection is the most widely used method of improving and maintaining domestic animals. Long before man understood the basic principles of genetics he learned that the use of desirable animals as breeding stock tended to produce offspring with more desirable traits. However, successful application of this type of artificial selection is difficult. For example, by proper selection we might obtain offspring in which the size was increased but the other traits were less desirable. It is not always possible for all of the most desirable traits to be combined in one individual because of the segregation of genes. Selection on the basis of one trait alone often defeats its purpose by producing organisms that are deficient in other desirable qualities. If selection is to be valuable, all desirable qualities of the organism must be considered. The successful product must contain a maximum number of desirable traits and a minimum of undesirable ones.

Some of the more important traits to be considered in animal improvements include the following: (1) Productivity (the quantity produced) is of great importance to the breeder, and selection for this trait often has priority over many other traits. For example, the quantity of eggs, milk, or wool per animal is an important trait in any program of selection for animal improvement. (2) The quality of the product is also highly important. If a sheep is developed that produces a greater quantity but inferior quality of wool, the results are of little value. (3) The hardiness of the organism is an important factor and includes such points as the ability to resist diseases, or to withstand adverse environmental factors. If a chicken is developed with increased egg-laying ability but also with much susceptibility to disease, the former will be of questionable value. (4) The body form is an important factor in selecting for racial improvement in some cases. A certain body form in beef cattle is associated with the greatest percentage of dressed beef and the finest quality of meat, hence there is great market value in such animals. (5) Early maturity is a valuable trait, because the sooner an animal matures to the productive age the less the cost. If a hen matures early and begins egg production, she is more valuable than one that matures later. (6) Economy in the use of food is also an important factor in selection. If the amount of food required to produce a certain quantity and quality of animal product is great, the commercial value of such an animal may not be as high as that for an animal in which this efficiency is greater. Hogs may be selected for maximum ham and bacon production in relation to food intake. These are only a few of the factors that are to be considered in racial improvement, but they show that the principles of artificial selection for improvements of domestic animals are quite complex.

Inbreeding

Inbreeding must almost always accompany artificial selection for racial improvements of animals.

It is used to retain desirable genetic traits in animals that possess them. When outstanding individuals are produced, inbreeding is required in order to retain as many of the desirable traits as possible by keeping the combination of genes of the progenitors as intact as possible. Ignorance of genetics has prompted the belief that inbreeding actually causes harmful traits. Most genes that produce harmful effects are recessive, and any individual is likely to be heterozygous for many of them. However, inbreeding may result in homozygous recessive genes combining together to express the harmful phenotypic trait.

If a race is relatively free of such harmful recessive genes, the detrimental effects of inbreeding will be reduced. The likelihood of the inbreeding being harmful depends on the genic content of the individuals. In most prize-winning, pedigreed animals the pedigree record frequently shows that individual animals are used repeatedly in the line of ancestry, suggesting the frequent desirability of inbreeding.

Outbreeding

The benefits that are to be gained by selection and inbreeding within any particular strain of a population are naturally limited to those specific genes within that strain which determine the desirable traits. The results of inbreeding can be no greater than those permitted by the genes of the inbreeds. One of the problems of the animal breeder is to introduce new genes into a population in order to increase the possibilities of selection for certain traits. One method is through outbreeding. In this method it is often possible to outbreed a desirable type to another less desirable type and then, through proper selection, to increase the degree of desirable traits. Thus new, valuable genes may be added, even though they may come from a variety of the stock that in many other ways is somewhat inferior.

Outbreeding in animals is often used to produce some specific valuable trait. In swine, sows of bacon-producing breeds may be bred to boars of lard-producing breeds, thereby producing a cross that yields a larger number of vigorous, early-maturing pigs possessing a higher market value than offspring of pure breeds. Beef cattle may be crossed with dairy cattle to produce calves for superior veal production.

Outbreeding may also be used to create new breeds. It is possible to establish a breed with desirable traits that will appear in the hybrid by the selection from succeeding generations, and thus eventually obtain a new breed that incorporates the desired traits from the two original breeds. This is expensive and requires time, and despite careful selection, there is usually a loss of vigor and fertility because of the necessary inbreeding. However, through hybridization and selection from other breeds many present-day breeds have been developed.

In some cases different species are crossed to yield a hybrid of superior vigor and value, as illustrated by the mule, produced by crossing a mare and a jack. Mules usually are superior to horses in strength, endurance, resistance to disease, and ability to work under unfavorable conditions. The mule is a hybrid and, like many interspecific hybrids is sterile, the sperms in the seminal fluid of the male being nonfunctional. In very rare cases, when a female mule is crossed with a jack, a colt is produced, because all of the necessary chromosomes of the horse or donkey may segregate into a viable, usable sex cell, which is used in the production of this colt.

Another method of introducing new traits is by the production of mutations that have been induced by artificial means. However, because many mutations are harmful, this expensive method is impractical. Mutations may also occur in nature, and any desirable natural mutations can be used in racial improvement because of their heritability. A sheep in New England some years ago is said to have mutated in the direction of desirable shorter legs and thus to have formed the basis for racial improvement. Silver and black foxes represent color phases of inherited variations of the red fox. New color mutations of the red fox that are being used commercially are platinum and pearl, which sell for high prices because of their scarcity. Mutations in mink fur colors include sapphire, steel blue, silver blue, platinum blond, and Aleutian blue.

The improvements in animals are extensive, as reference readings will reveal. Representative

examples include poultry that is resistant to white diarrhea caused by the bacterium *Salmonella pullorum*, resistance to abortion in rabbits, increased egg production in fowls, hornless (polled) cattle, increased butterfat in milk, improved qualities in race horses, better meat qualities in turkeys, and improvements in various breeds of cats and dogs.

CONSERVATION

Any constructive program for the conservation of natural resources will be largely biologic. Many instances of shortage occur because of man's extravagance and shortsightedness. No matter how adept man becomes in producing artificial substitutes, there are instances where natural resources cannot be supplanted. They must be conserved and restored because of their beauty and aesthetic value. With greater leisure, man will have more time to enjoy nature; proper conservation and provisions must be made to meet the ever-increasing demands in this direction. The biologic reclamation of nonproductive lands also will be an important factor in supplying sufficient products for various daily needs. One of the greatest of natural resources of a nation is the health of its individuals. Biology in its various medical and health phases can contribute greatly toward our future physical and mental well-being.

Natural resources may be classified as (1) irreplaceable, or those that, once used or destroyed, cannot be replaced, such as coal, oil, natural gas, and minerals, and (2) replaceable, or those that are used or destroyed but may be replaced if the proper conservation measures are employed, such as forests, soils, water, fishes, and wildlife. Increasing demands by homes, industries, and wars have used great quantities of irreplaceable natural resources so that the supply will be exhausted eventually. In some instances substitutes may be found, but in most cases they will be insufficient. Mineral resources, in addition to those essential for industries, are necessary for plant growth and health maintenance of all living things. Irreplaceable resources must be used carefully and conserved properly to ensure future needs. In most instances the replaceable resources will not replace themselves on their own but only when man institutes and follows certain conservation measures. In order to better understand the problems of conservation and the necessary remedial measures, the following resources are considered briefly.

■
Destruction and conservation of forests

Unless forests are conserved, they will soon be unable to supply the necessities of life. In years past, when a smaller population required less forest and the per capita supply was much greater, the problem was not so important. In addition to the extensive cutting of forests, they have been destroyed by such things as fires, improper cutting, and animals. Fires are often caused by lightning, careless campers, hunters, and vacationists. Ninety percent of the 200,000 forest fires in the United States each year are caused by man, and the loss totals millions of dollars. Forests require more than 50 years to be replaced, and in the meantime soil erosion may have started. Fires kill or injure trees, destroy plants, leaves, and soil, deprive birds of nesting sites, and deprive other animals of desirable protection. Fires injure trees or their young seedlings, exposing them to destructive bacterial and fungal diseases. The improper cutting of trees includes the cutting of small, immature trees, the destruction or injury of young trees during cutting operations, and the nonreplacement of cut trees by new young trees. Animals destroy young trees, seedlings, and seeds by grazing in forests, particularly if the forest has been burned off for grazing purposes.

Forest conservation measures include the following: (1) replanting burned-off areas; (2) replanting forests that have been cut; (3) removal of undesirable trees and vegetation to permit better growth of desirable varieties; (4) prevention of forest tree diseases; (5) providing basic protection for all forests qualifying for cooperative federal and state protection; and (6) education of hunters and campers. The following are rules the public should follow faithfully, to prevent forest fires: (1) hold and pinch all matches until they are cold; (2) crush out cigarette, cigar, and pipe ashes (use a rock or an ashtray); (3) drown all campfires, stir, and drown again; (4) learn and obey laws about burning grass, brush, and trash. In addition, building more fire lookout posts and patrol stations, discontinuing the practice of burning off forests for grazing purposes, wisely and more economically using timber, and selecting better species of trees for particular areas so that better qualities can be raised in a shorter time will help in preserving and improving forests.

■
Loss and conservation of soils

Soils are lost by wind erosion and water erosion. Wind erosion is the result of insufficient surface vegetation to hold the soil particles while water erosion is caused by currents washing away great quantities of soil. It is estimated that 40 tons of soil per acre are washed away from land with a 2% slope during one rainy season. In the United States dust storms have removed large amounts of soil for hundreds of miles, causing not only the loss of soil where it is needed but also the placement of it where it is not needed. The removal of trees, grasses, and other vegetation produces water and wind erosion, with the loss of the water-retaining humus and minerals essential for plant growth. Soil washed into streams interferes with plant and animal life and fills up rivers, lakes, and ponds, rendering them useless. This silt will quickly fill a body of water behind a dam, defeating the original purposes of the dam. Soil erosion also interferes with the water supply of cities, because much effort is required to remove these soil particles from it. Many streams are polluted because of soil particles that interfere with the normal growth and development of plants and animals inhabiting them.

Soil conservation measures include the following: (1) reestablishment of proper types of vegetation (trees, grasses, and crops) to hold the soil particles; (2) building level spaces, known as terraces, on lands where the slope is great enough to permit erosion; (3) using the correct type of contour plowing and the practice of strip cropping to reduce erosion to a minimum; (4) establishment of permanent grasslands by planting the proper types of grasses in a soil supplied with the proper fertilizing ingredients to ensure growth; (5) establishment of permanent woodlands where other crop growth is not feasible; (6) building dams across streams and gullies; and (7) increase in fertility of the soil, either by natural or artificial methods, to promote greater plant growth. In brief, there should be no barren soils, and each soil should promote the type of vegetation for which it is best fitted.

Loss and conservation of water

As we look at the ocean or a large lake we may wonder if it is necessary to conserve water. There may be about as much water now as there has ever been, but it is not located in the right places. Soils must contain a certain amount of water to ensure proper plant growth. Any factors that permit the rapid loss of water from the soil must be corrected if we are to have sufficient supplies. Larger quantities of water are now being used in homes and industries than formerly, and this has aggravated the problem still more. The greater use of water has reduced the natural water level of the soil. This, in turn, diminishes the amount of plant growth, which results in greater loss of water, so that the cycle is complete.

Another important influence upon the quantity and quality of available, usable water is the pollution of water supplies with wastes from oil wells, coal mines, various industries, and sewage. Sometimes an apparently sufficient supply may be unsatisfactory for the purposes desired. Many streams have been altered, so that their waters run off too quickly to permit their retention by the soils. Vegetation on their banks has been removed, thus permitting more water to enter the streams quickly. The presence of wastes and silt, the lowering of the natural water level, the consequent changes in animal and plant foods, and the destruction of natural feeding and breeding areas are among the influences responsible for the diminished supply of fishes and other aquatic life in our waters.

Water conservation measures include the following: (1) restoration of streams and other bodies of water to their natural conditions as far as possible; (2) prevention of unnecessary pollution of water by wastes; (3) wiser and more economical use of water; (4) replacement of the "vegetation blanket" (trees, grasses, and crops) in order to retain a maximum of soil moisture; (5) building dams and dikes to conserve the supply until needed; (6) employment of the correct types of plowing and cultivating to retain a maximum of water in the soil; (7)

reduction of evaporation by a covering of vegetation; and (8) institution of a system of flood preventions, thus alleviating flood damage and conserving water for future uses.

Loss and conservation of animal and plant wildlife

Many wild animals and plants have disappeared because of loss of natural resources. Many species of wild flowers no longer exist because their natural habitats are no longer present. The removal of a forest results in a destruction of its wild plant and animal life. Fishes, seals, deer, buffalo, birds, beavers, and wild flowers may gradually diminish, and probably disappear eventually, unless conservation measures are promptly instituted. It must be remembered that all living things require more or less specific environments for their optimum growth and development. When these are interfered with or destroyed, the living organisms must perish if they cannot adjust to another type of environment. The loss of each type of wildlife constitutes a unique problem in conservation, but the following will illustrate general measures: (1) regulation and control of fishing and hunting, and of collection of wild flowers; (2) restoration of streams and other bodies of water to their natural conditions as far as possible; (3) restoration of forests, fields, and swamps to invite the growth of inhabitants normally found there; (4) prevention of pollution of bodies of water by industrial wastes; (5) prevention of destruction of plant and animal life by industrial fumes; (6) building bird sanctuaries and providing nesting sites, proper foods, and protection; (7) protection of such animals as fishes, seals, deer, pheasants, and buffaloes by proper hunting and fishing regulations; (8) increase of state and national parks and preserves in which animals and plants have a natural environment protected by laws; (9) prevention of unnecessary destruction of wildlife by educating man concerning causes, results, and remedial measures; (10) educating man that picking wild flowers, especially varieties that are scarce, will soon lead to their extinction, because each flower picked is the prospective parent for future flowers; (11) increased support for state and national governmental agencies for the conservation of natural resources; (12)

prevention of devastating forest fires; and (13) placing big game in large forests where they are protected by law.

Conservation of human resources

The greatest of all resources are healthy and happy human beings. No nation can become, or remain, great if its inhabitants are physically unfit or socially and psychologically maladjusted. Other resources are unimportant if man cannot properly enjoy and utilize them. To maintain itself a population must show more births than deaths over a period of time. However, excessive growth in population may prove to be a liability instead of an asset. Of the total number of deaths in the United States each year, few result from natural senility; a large proportion are preventable if the proper conservation measures are followed. Often as many persons suffer from various diseases as die from them; hence, the efficiency and happiness of men could be greatly increased if the ravages of diseases and accidents were controlled.

Human resources conservation measures include the following: (1) reduction of the rate of infant mortality; (2) control and prevention of communicable diseases; (3) guarantee of pure foods and water in sufficient quantities for the individual needs of each person; (4) proper elimination of sewage and industrial wastes; (5) proper growth and inspection of foods to prevent transmission of diseases to man; (6) greater support of city, county, state, and national governmental health agencies, which are doing much to educate the general public concerning physical and mental health problems; (7) proper enforcement and acceptance of quarantine regulations; (8) institution of a program of physical activity to develop and maintain a maximum of physical health for each person commensurate with his inherited abilities; (9) elimination of infectious organisms, especially from crowded places; (10) proper control of carriers of disease germs to prevent transmission of germs to others; (11) better public education regarding the causes, transmission, prevention, treatment, and effects of human diseases; (12) decrease in the occupational diseases through better working conditions and a reduction of the number of accidental deaths and injuries; (13)

better understanding of the deficiency diseases; (14) more rigid enforcement of properly formulated pure food and drug laws; (15) more research in the fields of bacteriology, protozoology, immunology, and public health; and (16) maintenance of our individual lives so that we can develop and sustain the maximum of physical and mental health commensurate with our inherited abilities.

National parks

Some of the most outstanding and widely used natural resources in the United States and other countries are the numerous nature reservations. In the United States they are operated by the National Park Service, of the U.S. Department of the Interior. This includes thirty-three national parks and more than 230 national monuments, historic sites, national memorials, and recreational areas. Over 27 million acres of the most beautiful land in the United States have been set aside for use by the public, and visitors total many millions annually. In addition, states have state parks, state forests, and similar areas for use by the general public.

Review questions and topics

1 Investigate the early beginnings of agriculture. What were the first attempts at plant cultivation and animal breeding? What were the food sources for man in early times? Did he always cultivate crops and raise domestic animals?

2 Discuss the dependence of man upon the soil, both in the past and at the present time.

3 List the ways in which scientific biology contributes to the advancement of agriculture.

4 Discuss the values of hydroponics from a food production standpoint.

5 Discuss the relationship between biology and each of the following: (1) foods, (2) clothing, (3) furniture, (4) fuels, (5) medicine, (6) health, (7) wealth, (8) water supplies, and (9) sewage disposal.

6 Discuss the relationships between biology and each of the following: (1) dentistry, (2) nursing, (3) pharmacy, (4) medical technology, and (5) psychology and psychiatry.

7 Discuss the relationships between biology and each of the following: (1) industries, (2) forestry, (3) horticulture, (4) floriculture, (5) landscape gardening, and (6) out-of-door pools.

8 Discuss the relationship between biology and (1) a greater appreciation of the out doors, and (2) the formulation of a philosophy of life.

Selected references

Amerine, M. A.: Wine, Sci. Amer. Aug. 1964.

Brewbaker, J. L.: Agricultural genetics, Englewood Cliffs, N. J., 1964, Prentice-Hall, Inc.

Braidwood, R. J.: The agricultural revolution, Sci. Amer. 203:131-148, 1960.

Champagnat, A.: Protein from petroleum, Sci. Amer. 213:13-17, 1965.

Dasmann, R. F.: Wildlife biology, New York, 1964, John Wiley & Sons, Inc.

Herber, L.: Crisis in our cities, Englewood Cliffs, N. J., 1964, Prentice-Hall, Inc.

Higbee, E. C.: American agriculture, New York, 1958, John Wiley & Sons, Inc.

Hutt, F. B.: Animal genetics, New York, 1964, The Ronald Press Company.

Mangelsdorf, P. C.: The mystery of corn, Sci. Amer. July 1950.

Mangelsdorf, P. C.: Wheat, Sci. Amer. July 1953.

Pirie, N. W.: Orthodox and unorthodox methods of meeting world food needs, Sci. Amer. 216:27-35, 1967.

Williams, W.: Genetical principles and plant breeding, Philadelphia, 1964, F. A. Davis Co.

The following list is by no means complete, but it will serve as a basis for additional types. The following abbreviations are used: Gr., from the Greek; L., from the Latin.

Prefixes, suffixes, and combining forms used in biology

A

a- or **an-** (Gr., without or absent), *asexual*, without sex; *anaerobe*, organism that lives without free oxygen.

ab- (L., away from or without), *aboral*, away from the mouth.

ad- (L., toward, upon, or equal), *adrenal*, relating to the kidney; *adduct*, to draw one part toward another.

-ae (L., plural ending of singular Latin nouns ending in a) *alga* and *algae* (pl.)

aer- (Gr., air), *aerobe*, organism that requires free air.

alb- (L., white), *albino*, organism exhibiting no pigment.

-algia (Gr., pain), *neuralgia*, pain in a nerve.

ambi- (L., both), *ambidextrous*, being able to use either hand.

amphi- (Gr., on both sides), *Amphibia*, class of vertebrate animals living in water and on land.

amyl- (L., starch), *amylase*, enzyme that changes starch to sugar.

ana- (Gr., back or again), *anabolism*, building-up process of metabolism.

angio- (Gr., enclosed), *angiosperm*, plant with enclosed or protected seeds.

ante- (L., before in time or space), *antedorsal*, placed before dorsal.

anti- (Gr., opposed or opposite), *antitoxin*, antibody opposed to or neutralizing a toxin.

antr- (L., cavity), *antrum*, cavity of a bone.

apo- (Gr., away or separate), *apodeme*, ingrowth from the exoskeleton of most arthropods.

aqua- (L., water), *aquatic*, living in water.

arch- (Gr., early or chief), *archenteron*, early digestive tract or enteron; *Archeozoic*, earliest era of geologic history.

areol- (L., space), *areolar*, containing minute spaces.

arthr- (Gr., joint), *Arthropoda*, phylum of invertebrate animals with jointed appendages or feet.

asco- (Gr., sac or bag), *Ascomycetes*, class of sac-bearing fungi.

-ase (suffix designating an enzyme), *protease* enzyme that acts on proteins.

aster- (Gr., star), *Asteroidea*, class of echinoderms resembling stars.

auto- (Gr., self), *autosynthesis*, self building up.

B

bacter- (Gr., *baktron*, a stick), *bacteria*, rod-shaped organisms.

basi- (Gr., base), *basidiospore*, spore formed at the base of a basidium.

bi- (L., double), *bilateral*, similar on both sides.

bio- (Gr., life), *biology*, science of life.

blast- (Gr., bud or young), *blastoderm*, primitive germ layer.

brachy- (Gr., short), *brachydactyly*, abnormal shortness of the digits.

brevis (L., short), *adductor brevis*, short adductor muscle.

bryo- (Gr., moss), *bryophyte*, plant of the phylum comprising the mosses.

C

caec- (L., blind), *cecum (caecum)*, blind pouch.

calci- (L., lime), *calcareous*, containing lime.

-carp (Gr., fruit), *pericarp*, wall around the plant ovary.

cata- (Gr., down), *catabolism*, breaking-down process of metabolism.

cauda- (L., tail), *caudal*, relating to a tail.

cav- (L., hollow), *vena cava*, hollow vein.

ceno- (Gr., recent), *Cenozoic*, recent era of geologic history.

centr- (L., center), *centrosome*, center of activity during mitosis.

cephalo- (Gr., head), *cephalic*, relating to, or toward the head.

chlor- (Gr., green), *chlorophyll*, green coloring matter of plants.

-chondro (Gr., granular), *mitochondria*, small, granular parts of protoplasm.

chondro- (Gr., cartilage), *chondrocranium*, part of the cranium developing from cartilage.

chrom- (Gr., color), *chromatophore*, color-bearing cell.

-cide (Gr., kill), *insecticide*, agent that kills insects.

cili- (L., eyelash), *cilia*, minute, hairlike processes.

circum- (L., around, *circumesophageal*, around the esophagus.

cloaca (L., sewer), *cloaca*, outlet for excretions.

cnido- (Gr., nettle), *cnidoblast*, nettle cell of certain animals.

coel- (Gr., hollow), *coelom (celom)*, hollow body cavity.

coeno- (Gr., common), *coenosarc*, common tissue in certain animals.

coleo- (Gr., sheathed), *Coleoptera*, order of sheathed insects, such as beetles.

com- (L., together), *commensalism*, living together.

con- (L., cone), *conifer*, cone-bearing tree; or (L., with), *concretion*, something that has grown together.

cotyl- (Gr., cup shaped), *cotyledon*, cup-shaped seed leaf.

creta- (L., chalk), *Cretaceous*, chalk period of geologic times.

cyan- (Gr., blue), *Cyanophyta*, phylum of blue-green algae.

cyst (Gr., sac), *cyst*, pouch or sac.

cyt- (Gr., cell), *cytology*, branch of biology studying cell structure and function.

531

D

de- (L., off), *degenerate,* to lose generative ability.

dendr- (Gr., brush or tree), *dendrite,* treelike structure of a nerve cell.

derm- (Gr., skin), *dermis,* part of the skin.

di- (Gr., twice), *diploblastic,* possessing two germ layers; *dicotyledon,* plant possessing two cotyledons.

dis- (L., away), *distal,* away from the point of origin.

dors- (L., back), *dorsal,* pertaining to the back.

dura- (L., tough), *dura mater,* tough, outer covering of the brain and spinal cord.

E

e- (L., out of, without), *egest,* to pass outside.

ec- (Gr., house or environment), *ecology,* study of the habitats of an organism.

ecto- (Gr., outside), *ectoderm,* outer layer of cells.

-ectomy (Gr., cut), *appendectomy,* removal of the appendix.

-emia (Gr., blood), *anemia,* blood deficiency.

en- (Gr., in or within), *encyst,* to cover with a membranous cyst.

endo- or **ento-** (Gr., within), *endoderm,* inner layer of cells.

eo- (Gr., dawn, or early), *Eocene,* early geologic period.

epi- (Gr., upon), *epidermis,* epithelial layer upon the dermis.

equi- (L., horse), *Equisetineae,* class to which the horsetails belong.

eu- (Gr., good or well), *eugenic,* being fitted for the production of good offspring.

ex- (Gr., external), *exoskeleton,* external skeleton.

extra- (L., beyond), *extracellular,* beyond or outside the cell.

F

-fer (L., to bear), *Porifera,* phylum comprising pore-bearing sponges.

fil- (L., thread), *filiform,* threadlike.

flex- (L., bend), *flexor,* muscle that bends joints.

-form (L., shape), *uniform,* all one shape.

G

gam- (Gr., marriage), *gamete,* reproductive cell.

gastr- (Gr., stomach), *gastric,* pertaining to the stomach.

-gen (Gr., to produce), *pathogenic,* capable of causing disease.

geo- (Gr., earth), *geology,* science of the earth.

-gest (Gr., to bear or hold), *ingest,* to take in.

-glea (Gr., jelly), *mesoglea,* middle, jellylike layer in certain animals.

glyc- (Gr., sweet or carbohydrate), *glycogen,* animal starch.

gono- (Gr., seed or reproduction), *gonad,* organ of reproduction.

gymn- (Gr., naked), *gymnosperm,* class of seed plants whose seeds are not enclosed in an ovary.

H

haem- (Gr., blood), *hemoglobin (haemoglobin),* substance in the blood.

hemi- (Gr., half), *hemisphere,* one half of a sphere.

hepat- (Gr. liver), *hepatic,* pertaining to the liver.

hetero- (Gr., other or different), *heterogeneous,* consisting of different constituents.

hex- (Gr., six), *hexagonal,* six sided.

homo- (Gr., same); *homogeneous,* of a similar kind.

hyal- (Gr., glass), *hyalin,* something that is transparent or glasslike.

hydr- (Gr., water), *dehydrate,* to remove water.

hymen- (Gr., membrane), *Hymenoptera,* order of insects with membranous wings.

hyper- (Gr., above), *hypersensitive,* especially sensitive.

hypo- (Gr., under), *hypoglossal,* situated under the tongue.

I

in- (L., in, into, not, without), *invaginate,* to infold one part within another.

infra- (L., below), *infraorbital,* below the orbit.

inter- (L., between), *intercellular,* between cells.

intra- (L., inside), *intracellular,* within a cell.

is- (Gr., equal), *isothermic,* having equal temperatures.

-itis (Gr. inflammation), *appendicitis,* inflammation of the appendix.

J

-juga (L., join), *conjugation,* a process of reproduction in which two animals are joined.

K

kata- or **cata-** (Gr., down or destroy), *catabolism,* breaking-down process of metabolism.

kine- (Gr., move), *kinetic,* as in kinetic energy, which is energy of movement.

L

labi- (L., lip), *labium,* lip.

lac- (L., milk), *lactose,* milk sugar.

later- (L., side), *lateral,* relating to the side.

-lemma (Gr., covering), *neurilemma,* covering of a nerve.

lepi- (Gr., scale), *Lepidoptera,* order of insects with scale wings.

leuko- (Gr., white), *leukocyte,* white blood cell.

lip- (Gr., fatty), *lipoid,* fatty substance.

-log (Gr., study), *zoology,* study of animals.

luci- (L., light), *luciferin,* light-producing material.

lysis (Gr., destroy), *bacteriolysis,* destruction of bacteria.

M

macro- (Gr., large), *macronucleus,* large nucleus.

mal- (Gr., bad), *malnutrition,* bad nutrition.

mega- (Gr., large), *megaspore,* large spore.

mens- (L., table), *commensalism,* eating at a common source of food.

-mere (Gr., part), *micromere,* small part.

meso- (Gr., middle), *mesoderm,* middle cellular layer.

meta- (Gr., after), *metaphase,* later phase of mitosis.

micro- (Gr., small), *micronucleus,* small nucleus.

milli- (Gr., thousand), *millipede,* animal with a "thousand" legs.

mio- (Gr., less), *Miocene,* less recent period in geologic history.

mito- (Gr., thread), *mitosis,* cell division with the formation of threadlike structures.

mono- (Gr., one), *monograph,* something written about one subject.

morph- (Gr., form), *morphology,* study of form.

multi- (L., many), *multicolored,* of many colors.

muta- (L., to change), *mutation,* abrupt hereditary change.

myco- (Gr., fungus), *mycology,* study of fungi.

myxo- (Gr., slime), *Myxomycophyta,* phylum comprising the slime molds.

N

nema- (Gr., thread), *nematocyst,* threadlike structure of coelenterates.

neo- (Gr., young or recent), *Neotropical,* constituting a recent biogeographic region in the tropics.

nephro- (Gr., kidney), *nephridium*, tubular excretory organ.
non- (L., not), *nonirritant*, not irritating.
nuc- (L., kernel or center), *nucleus*, central portion of a cell.

O

octo- (L., eight), *octopus*, animal with eight appendages.
oedo- (Gr., swollen), *edema (oedema)*, swollen condition.
-oid (Gr., like), *ameboid (amoeboid)*, like an *Amoeba*.
oligo- (Gr., few or little), *oligotrichous*, having few cilia.
-oma (Gr., swelling or tumor), *carcinoma*, malignant growth (cancer).
oo- (Gr., egg), *oogenesis*, formation and development of an egg.
or- (L., mouth), *oral*, pertaining to the mouth.
ortho- (Gr., straight), *Orthoptera*, order of insects with straight wings.
os- (Gr., bone), *osseous*, pertaining to bone.
ovi- (L., egg), *ovum*, egg.

P

palaio- (Gr., ancient), *paleontology*, study of ancient life.
para- (Gr., beside), *parapodia*, appendages beside others.
path- (Gr., disease), *pathogenic*, capable of causing disease.
ped- (L., feet), *pedal*, pertaining to the foot.
peri- (Gr., around), *peristome*, region around an opening or mouth.
phaeo- (Gr., dark or brown), *Phaeophyta*, phylum of brown algae.
phago- (Gr., to eat), *phagocyte*, cell that eats or destroys.
phor- (Gr., to bear), *sporophore*, part of a sporophyte that bears spores.
photo- (Gr., light), *photosynthesis*, formation of carbohydrates in the presence of light.
-phil (Gr., loving), *thermophile*, heat-loving organism.
phyco- (Gr., alga, or seaweed), *Phycomycetes*, algalike fungus.
-phyll (Gr., leaf), *mesophyll*, middle part of a leaf.
physio- (Gr., nature), *physiology*, study of the nature or function of living matter.
-phyte (Gr., plant), *sporophyte*, spore-bearing plant.
-plasm (Gr., formed), *ectoplasm*, outer region of the cell cytoplasm.
-plast (Gr., living), *chloroplast*, green body in certain living plants.
platy- (Gr., flat), *Platyhelminthes*, phylum of flatworms.
plio- (Gr., more), *Pliocene*, more recent geologic period.
poly- (Gr., many), *polymorphous*, having many forms.
post- (L., after), *postnatal*, after birth.
-pous (Gr., foot), *octopus*, animal with eight feet or appendages.
pre- (L., before), *prenatal*, before birth.
pro- (Gr., before), *prostomium*, portion of the head situated before the mouth of certain worms and mollusks.
proto- (Gr., first or essential), *protoplasm*, essential material of all plant and animal cells.
prox- (L., nearest), *proximal*, nearest.
pseudo- (Gr., false), *pseudopodia*, false feet.
-ptero (Gr., wing), *Diptera*, order of insects with two wings.

R

re- (L., again or back), *regenerate*, to form again.
ren- (L., kidney), *renal*, pertaining to the kidney.
rept- (L., creeping), *reptile*, creeping animal.
retro- (L., backward), *retrolingual*, backward from the tongue.
rhizo- (Gr., root), *Rhizopoda*, subclass of animals with rootlike appendages.
rhodo- (Gr., red), *Rhodophyta*, phylum of red algae.
roti- (L., wheel), *rotifer*, animal with a wheel-like structure on its head.

S

-sarc (Gr., flesh), *ectosarc*, outer flesh or layer of protoplasm.
schizo- (Gr., to divide), *Schizomycophyta*, phylum of fission fungi (bacteria).
scler- (Gr., hard), *sclerotic*, hard.
-scope (Gr., see) *microscope*, instrument enabling one to see minute objects.
-sect (L., to cut), *dissect*, to cut.
semi- (L., half), *semicircle*, half of a circle.
sept- (L., wall), *septum*, partition.
set- (L., bristle), *seta*, bristlelike structure.
sinu- (L., hollow), *sinus*, hollow cavity.
soma- (Gr., body), *somatoplasm*, protoplasm of the body.
spor- (Gr., seed), *spore*, reproductive structure.
stoma- (Gr., opening), *stoma*, opening, such as is found in leaves.
sub- (Gr., under), *submaxillary*, under the maxilla.
super- (L., over or above), *superior*, higher, upper, or above.
supra- (L., above), *suprarenal*, above the kidney.
sym- (Gr., together), *symbiosis*, living together.
syn- (Gr., together), *synapsis*, association or union.

T

telo- (Gr., complete or end), *telophase*, end stage of cell division.
terato- (Gr., marvel, or monster), *teratology*, study of monstrosities or deviations from the normal.
tetra- (Gr., four), *tetrapod*, something that has four feet.
-thec (Gr., case), *spermatheca*, sperm case.
thermo- (Gr., heat), *thermotropism*, reaction to heat.
thigmo- (Gr., contact), *thigmotropism*, reaction to contact.
-tom (Gr., to cut), *microtome*, instrument to cut small sections.
toxi- (Gr., poison), *toxin*, poison.
trans- (Gr., across), *transfer*, to carry across.
tri- (Gr., three), *trilobed*, having three lobes.
tricho- (Gr., hair), *trichocyst*, hairlike structure.
trop- (Gr., reaction), *tropism*, reaction to stimuli.

U

ultra- (L., beyond), *ultramicroscopic*, so small that it is beyond the microscope.
uni- (L., one), *unilateral*, on one side.
-ur (Gr., tail), *Anura*, animals without tails.

V

vas- (L., vessel), *vas deferens*, vessel to transmit male sex cells.
ventr- (Gr., belly), *ventral*, pertaining to the lower or belly side.
vit- (L., life), *vital*, essential to life.
vorti- (L., to turn), *Vorticella*, animal that turns as it moves.

Z

zoo- (Gr., life or animal), *zoology*, study of animals.
zyg- (Gr., union), *zygote*, cell produced by the uniting of male and female sex cells.
zym- (Gr., a ferment), *zymase*, enzyme that acts on a certain carbohydrate to produce carbon dioxide and water, or alcohol and carbon dioxide.

Index

534

537

539

543

545